THE EARTH
A Physical and Human
Geography

Harm J. de Blij
National Geographic Society
University of Miami

THE EARTH

A Physical and Human Geography 3/E

JOHN WILEY & SONS
New York • Chichester • Brisbane • Toronto • Singapore

To my father
Hendrik de Blij
on the occasion of
his 80th birthday
October 3, 1987

Cover and Interior designed by Carolyn Joseph
Cover photo of countryside in Yugoslavia
by Charles Weckler/Image Bank.

Library of Congress Cataloging-in-Publication Data:

De Blij, Harm J.
   The earth: a physical and human geography.

   Rev. ed. of: The earth, a topical geography.
2nd ed. c1980.
   Bibliography: p.
   Includes index.
   1. Geography.   I. De Blij, Harm J.   Earth, a
topical geography.   II. Title.
G116   1987   910   87-14727
ISBN 0-471-85495-6

Printed in the United States of America

10 9 8 7 6 5 4 3 2 1

This is the third edition of a book that has undergone changes in title as well as in contents. First published as *Man Shapes the Earth* (1977), it has grown in scope as well as dimension. The original objective still stands: to provide an overview of basic substance and theory in geography, physical as well as human, systematic as well as regional.

The present edition differs from its predecessors in several ways, and many of the changes result from suggestions made by readers. One of these concerns the book's title; that of the second edition (*The Earth: A Topical Geography*) was interpreted by some to mean a physical geography. The changed title better reflects the book's broad contents.

The major structural changes involve the addition of nine color plates, the deletion of the final chapter dealing with world realms, and the addition of a concluding essay that encourages students to consider a scholastic career in geography, with some practical information on how to proceed. The crucial material of the deleted regional chapter, dealing with regional concepts and constructs, has been incorporated into an expanded Chapter 7 (Culture Worlds). Readers who are familiar with the earlier editions will note that additional heads and subheads have been used to enhance clarity.

Individual chapters have received careful and detailed attention, and once again I have benefited enormously from the constructive criticisms made by colleagues and students. The introductory chapter has been expanded and brought up to date, although it retains its emphasis on several of the historic figures who propelled the discipline in earlier years and who ensured its place among the sciences. In our enthusiasm for "new" geographies, we sometimes lose track of our sources, and this is perilous. Let us engender

again an appreciation for the work of those who, in another era, moved our field forward.

Many readers who made suggestions for the improvement of the physical chapters will see their recommendations implemented. New opening segments now link the physical and human themes more effectively, and historic events (such as Santorini's eruption) and modern phenomena (such as the El Niño problem) strengthen the connection. Spatial and environmental topics anchor these introductory chapters. New material on geographic implications of crustal plate movement, ocean floor topography, and karst landforms augments Chapter 2. The chapter on the geography of soils and vegetation (the biospheres) has a new introduction and a conclusion that addresses biogeography more comprehensively. Chapter 5 focuses on resources and has been much modified. The soil, studied in the previous chapter as a part of the biosphere, now is treated as a renewable, but not inexhaustible resource. Natural vegetation is accorded similar treatment. Energy resources are considered with a new and stronger focus on North America, and with attention to the attractions and risks of nuclear power. A rewritten conclusion draws attention to the interconnectedness of our world.

The second half of the book retains its basic structure (but without the concluding regional chapter); topics range from demographic to political and from cultural to economic. Significant changes include (1) the addition of a segment on realms, regions, and the regional concept; (2) the addition of new material on urban centers; (3) the restructuring of major parts of the chapter on political geography; (4) the condensation of material on international boundaries on land; (5) the revision of the segments on colonialism; and

v

(6) the updating and recasting of material on international boundaries at sea.

All the chapter-ending bibliographical sections were rewritten and updated, and attention has been given to several of the major journals in the discipline that deal with topics covered in individual chapters of this book. The glossary has been considerably expanded, but it would have been impractical to include in the glossary every term that is printed in boldface in the text. Terms defined and discussed in considerable detail in the text may not be defined again in the glossary.

vi

A C K N O W L E D G M E N T S

My editor, Katie Vignery, urged me to revise *The Earth* and kept the project on schedule; without her assistance in many areas, the book would not have been produced. I am deeply grateful to her. My friend, colleague, and co-author Peter Muller discussed aspects of this project with me over a period of about three years, much to my benefit; he also permitted me to use material from our joint publications, for which I am indebted to him. I could not have written Chapter 1 without having read virtually everything written by our leading biographer and commentator on the geographic scene, Professor Geoffrey J. Martin of Southern Connecticut State University. I also have had the opportunity to incorporate suggestions made by Professor Thomas Giambelluca of the University of Hawaii (Manoa). I further acknowledge with gratitude the detailed comments made by Professor Anthony F. Grande of Hunter College, New York; Professor W. C. Jameson of the University of Central Arkansas; Professor D. Knudsen of Indiana University, Bloomington; Professor R. Marionneaux of Eastern Kentucky University; and Professor Hubert B. Stroud of Arkansas State University.

At John Wiley & Sons, I benefited from the work of Ellen Brown, who directed this project; Joshua Spieler, who supervised the editing; Carolyn Joseph, who designed the book; Gigi Ghriskey and Lisa Heard, who solved many difficult problems of cartography; and Sara Lampert, who was in charge of photographs. I deeply appreciate their work and I am grateful for their commitment to this project.

H. J. de Blij

Coconut Grove, Florida
February 1987

vii

CONTENTS

viii

ix

x

Geography!

For years I have been asking people about their favorite place in the world. Where is it? What is there about it that makes this so special? To a geographer, the answers are always interesting. A student talks longingly of that comfortable, spacious house in the suburb where she grew up. "I couldn't wait to leave for college, and now I can't wait for vacation so I can go back to spend some time in my old room." A businessman loves his condominium on a high floor: "You should see the view over the city," he says. "Even after all these years, I never tire of watching all that activity, day and night." To one of my friends, her favorite place is a beach house in Florida. One time a young African villager took me on a long walk through the bush. Suddenly, we came to the top of a scarp, and the view was magnificent. He pointed to some rocks worn smooth by many hours of sitting. There was no need to explain.

All of us carry such images. Some will fade over time (that favorite room may be soon forgotten), and others will take their place. These new ones may be more significant. Imagine that you were asked to decide in which state (of the United States) you would prefer to live. That question has been asked so many times that the responses, say, of your geography class this term is quite predictable. Many will rank the state where they grew up at the top of the list (there, again, is that comfortable familiarity). California and Florida also rank high. People feel that they *know* such places. Isn't the weather good? The scenery nice? Aren't the cities exciting? But what if a new employer, an international corporation, gave you a choice of countries for your first appointment: Senegal, Ecuador, Sri Lanka, or Somalia. What would you decide? On what basis?

As you will soon discover, the study of geography is an excellent way to learn to make informed decisions of this kind. Ours is an infinitely complex world, but there are ways to decide what information is important, and there are means of organizing such information. Once we have these global frameworks in mind, we can begin to fill in some of the details that interest us especially. That is, we begin to *specialize*. You will find that geographers specialize in all kinds of topics: climate and weather, soils and slopes, the layout of cities, the roles of political boundaries.

Geographers have even concentrated their attention on such subjects as tourism, sports, and . . . wine!

Thus the scope of geography ranges far and wide. No body of facts or data belongs exclusively to the field of geography, but geographers often combine information from various other fields for their studies and research. When you read a professional or popular journal in geography, you will note that geographers share an interest in spatial relationships on and near the surface of the earth. The term **spatial** (also sometimes spelled *spacial*) comes, of course, from the noun *space*—not the space of the solar system and beyond, but the space on the surface of our earth. Where are places located, and why? How far is that capital city from the country's international border? Such questions have to do with the location of people, places, and things *relative to each other*, and this is a central theme of geographic study and research. Geographers have devised methods to analyze and solve many problems involving some aspect of *relative location*.

Places, small and large, also are studied for their own sake, and this can be a very rewarding aspect of geography. What is a West African village like? How do people there make a living from the surrounding countryside? Why is Paris so much larger than any other French city? When we try to answer questions of this kind, it soon becomes clear that geographers are interested in the relationships between human societies and natural environments. Those African villagers have learned to use the soils of their farmlands to the best advantage given their farming equipment (or technology), and they know that the seasonal rains are the key to good crops and harvests. Paris, like other places in France, started as a small village, but this village had geographic advantage: a favored spot along the Seine River. Natural and human factors combined to make Paris what it is—a beneficiary of geography. So, when we answer the question "What is it like?" the geographic answer is a blend of natural (physical) and human (social) characteristics.

This brings us to another dimension we geographers must consider when doing our study and research: *scale*. When we describe that West African village, it is possible to go into much more detail than when we generalize geographically about the advantages enjoyed by the growing city of Paris. On a map, detail translates to scale. The map of an entire country cannot show much detail; whole cities are mere dots on such a map. Therefore, the scale is small and is represented by a fraction (say, 1:1,000,000) that confirms this arithmetically. A map of Paris alone shows much more detail; it is therefore on a larger scale and represented by a larger fraction (1:100,000). A map of a village may have a scale of 1:10,000, a scale that is large enough to show individual houses and yards. We will look into the development of mapmaking (cartography) later, but we should realize that scale is not just a matter of maps: it also enters into our descriptions of places. A mental scale enters into the picture when we generalize about countries and places. Let us return for a moment to that question the prospective employer asked about four possible countries of employment. To live in any one of those countries can mean a variety of choices. In Senegal, living in the capital, Dakar, would be an experience quite different from an assignment in a town in the interior. In Ecuador, life in the coastal city of Guayaquil contrasts considerably vis-à-vis Quito in the Andes Mountains. So, we can generalize to some degree (e.g., French would be needed in Senegal, Spanish in Ecuador), but we would need a larger-scale mental map for other decisions.

Maps are among geographers' favorite means of communication. Sometimes they pose questions: Why do these crops yield so much better in this area than in another? Sometimes they provide answers, for example, by showing patterns that appear to be related in some way, thus pointing the need for further research. It has been said that a picture is worth a thousand words; if that is true, then a map is worth a million. Maps can convey complicated information with great efficiency. Imagine a weather reporter on television's nightly news trying to explain the approaching weather systems *without* that background map and enhanced satellite image!

In this book (as in other geography books and journals), there are many references to maps. This is not just because maps convey information so efficiently; there are other reasons. Maps can demonstrate that vital rule of geography referred to earlier: places on the earth's surface have their own distinc-

tive properties that, taken together, give each of these places its own character. No place is exactly identical to any other, and the map is the best way to prove this.

Maps constitute one aspect of communication in geography. Although there is no body of facts that is exclusively geographic, there certainly is a vocabulary of geography. You find certain terms used frequently. These will become familiar as our study progresses because they reflect geography's focus on matters spatial. For example, geographers focus on such questions as the effect of *distance* on migration patterns, the influence of wind *direction* on the expansion of a desert, the relationship between the rate of *diffusion* (**spreading**) of a new idea and the *density* of the population through which it spreads, the impact of low *accessibility* on the development of a frontier farming area. In one way or another, aspects of relative location dominate the language of geography: region, pattern, proximity, isolation, distribution, remoteness, concentration, core, periphery, shape, clustering, and so on.

Such terminology might suggest that geography is exclusively a social science, but this is not the case. Many geographers apply their locational interest to physical phenomena, seeking to learn more about the earth's surface itself or about the climate and weather that affect it. Physical geographers, too, often are attracted by especially interesting patterns and clusters, forms, and shapes. Why does a regional river system fail to conform to the structures of the rocks below? Why does a corner of West Africa experience a monsoon-type deluge of rain every year when other stretches of the coast do not?

Virtually all geographic research and study contribute to our understanding of the earth as humanity's habitat. There were several million years of gradual evolution, then, during the past ten thousand years, human communities experienced revolutionary changes that witnessed the domestication of plants and animals; the development of villages, towns, and cities; the growth of transport networks; and the unprecedented exploitation of the earth's resources. During the last two hundred years, these changes have accelerated almost unimaginably, fueled by the Industrial Revolution and its aftermath and accompanied by explosive population increases. In the pro-

5

cess, the earth has been—and continues to be—transformed, reshaped by its human occupants. Plains are plowed and planted, rivers are diked and dammed, hills are topped and terraced. The expanding asphalt jungle claims ever more of the countryside. By studying the earth's surface, its natural environments, and its populations in geographic perspective, geographers (physical as well as human) contribute to the comprehension of the interactions between societies and the lands on which they depend.

Geography, then, is a physical (natural) science as well as a social science. Some would argue that it is also a very lively art. Indeed, geography's bibliography includes some memorable literature, ranging from vibrant regional descriptions and exciting accounts of early and modern explorations to impressionistic discussions of geographic dimensions in art and music. Geography's many lineages have produced a field with few limits and countless opportunities.

Cartography is more than five thousand years old: this map of rivers and fields in northern Mesopotamia was drawn in clay about 3800 B.C.

•   Origins   •

6

Geography is a modern science, but its traditions extend to the dawn of scholarly inquiry and research. More than five thousand years ago, Mesopotamian cartographers drew maps of rivers, farm fields, and towns in soft clay; when the clay hardened in the sun, the heat-baked tablet could be lifted up and moved indoors. Some of those ancient maps have survived, and we can pause at a museum display to wonder at the observational skills of the pioneer cartographers of the Middle East. They were the first to use symbols on maps: one symbol for farm fields, another for streams, still another for paths and streets. They were, evidently, the first also to cope with the problem of scale: How much of the real world can be represented on a limited drawing surface? Sitting there, on a hillslope overlooking the valley being mapped, those early cartographers began an alliance with geography that continues to this day.

Actually, nobody was calling this science geography in those days. Certainly there were practitioners of the field in Mesopotamia, in ancient Egypt, and later in Crete and Greece. Homer wrote beautifully

and incisively in his *Odyssey* about the character of distant places and areas. Plato and Aristotle wrote treatises, which can only be described as geographic, that raised questions about the relationships between human communities and physical environments.

The works of Aristotle are of special interest because they reveal the nature of early geographic thought and debate. Aristotle (born in 384 B.C.) studied at the school Plato had founded near the city of Athens. But Aristotle was less mathematically inclined than Plato; he was more interested in purposes and ideals. Geographers who study the development of their field often refer to Aristotle's essay *The Ideal State* as one of the earliest geographic treatises ever written. In this essay, Aristotle outlines the geography of what he viewed as a well-functioning country. He uses some terminology that was to become standard in geography: the definition of a national territory by boundaries and its protection by inaccessibility from beyond, for example. He also touches on a matter that still causes debate (e.g., in the modern United Nations): the appropriate size of the population of a state.

Such a population, he argued, should be neither too small—and thus weak—nor too large—and thus ungovernable. Aristotle, in addition, correctly characterizes cities as market centers with external trade connections (all that is missing is the term *hinterland*, but modern German had not yet been invented!). He mentions a factor that we will discuss in detail later, something that has much to do with the prosperity or the decline of cities, their *situation* (a good definition of relative location). As noted earlier, Aristotle also speculated about the effect of natural environment on the behavior of people, an issue that still interests social scientists today.

Still, geography was not known as such. The ancient philosophers, Aristotle among them, established lines of scientific and scholarly enquiry that created foundations for many sciences that would emerge centuries later. One of the first of these sciences to acquire a name was geography, and it happened not too long after Aristotle's lifetime.

The scholar who is credited with the naming of *geo* (earth) *graphy* (description) was a Greek philosopher, Eratosthenes. He was born about 276 B.C. in what is today Libya. Eratosthenes had many scholarly interests: he was a scientist, poet, librarian, explorer, mathematician, and cartographer. A follower of Aristotle, he developed many of Aristotle's notions further in a book about the geography of the known earth. But his most important work, and the calculations for which he is most remembered, had to do with the circumference of the earth. This was quite an innovation; first, not everyone believed that the earth was a sphere. But Eratosthenes drew the correct conclusions from what he could observe of the relationship between sun and earth, the behavior of the sun through the seasons, and the changing sun angle with latitude. A round earth, Eratosthenes reasoned, would have an extremely hot equatorial zone, two very cold polar zones, and between these opposites, two temperate zones of which the known inhabited world was one. He even proposed latitudinal limits for these environmental zones, which was a major step forward.

The measurement for which Eratosthenes would become permanently known was made in Egypt. He observed that at noon on about June 21 of a given year the bottom of a deep well in the city of Syene was completely lit by the sun. Thus the sun's ray was exactly perpendicular to the surface of the earth at that place on that date. The next year, on the same date, Eratosthenes measured the noon sun angle at Alexandria, where he worked in the library. The sun was not quite perpendicular, but 7.2° off the vertical. Thus the distance between Syene and Alexandria made a 7.2° difference, or one-fiftieth of the 360° of a complete circle. All Eratosthenes needed now was the surface distance between Syene and Alexandria; by multiplying this figure by 50, he would have the circumference of his round earth. But now he faced a problem: that distance was not easily established. In those days, distances were estimated from the time it took for travelers to go from one place to another and were given in *stadia* (the plural of stadium, the measured length of a sports arena). It is not certain what stadium Eratosthenes used, but he estimated the Syene–Alexandria distance to be 5000 stadia. Some prominent stadiums in Greece had a length of slightly more than 600 ft (185 m). If he used that figure, the result of his calculation would be 26,700 miles (43,000 km) for the circumference of the earth. The actual circumference is now known to be about 25,000 miles (40,000 km). Eratosthenes may have been even closer, a remarkable achievement and one for which he is justly remembered by geographers.

Eratosthenes did more than measure and calculate, however. He was a skilled cartographer, and one of his maps of the known world (about 220 B.C.) is especially interesting. Note that this map shows the outlines of the Mediterranean Sea quite well; Greece, Italy, the Iberian Peninsula, and North Africa ("Libya") are in their approximate positions. Perhaps more remarkable is the shape of Britain and Ireland and the way they are shown. It looks as though Eratosthenes knows their actual outline, but the projection he uses (in his mind, not on the map by latitude and longitude lines) compels him to distort it. Also on the map are "Aethiopes" for Africa south of Egypt, a good representation of the actual course of the Nile River, and "India" toward the eastern edge of the known world. Even the Euxine (Black) and Caspian seas are on the map. Eratosthenes had a remarkable capacity for synthesis as well as mathematics.

The exact circumference of the earth may be known much better today than it was in Eratosthenes' time, but there are still other calculations and mea-

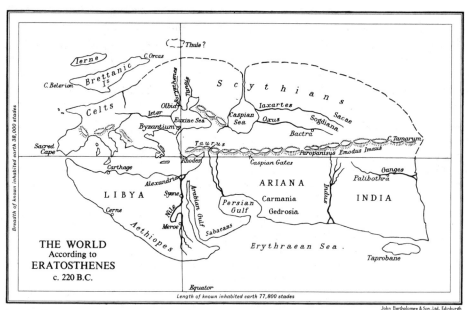

Reproduced by kind permission of John Bartholomew & Son Limited, Edinburgh.

John Bartholomew & Son, Ltd., Edinburgh

Eratosthenes' map of the known world, constructed about 220 B.C.

surements to be made. The exact heights of mountains (above mean sea level) and the exact distances between places require ever more precise measurement. Today we are able to measure these so exactly that a remarkable fact has come to light: those distances actually *change*, as do the heights of some mountains! Of course, we know much more now about the fragile crust of our earth than Eratosthenes could have known. But the branch of geography that involves such measurements as he made—*geodesy*—still is an active and important science.

Many other contributions made to geography by ancient Greek philosophers have unfortunately been lost. The works of one, Strabo, are an exception: they survive in almost complete form. To Strabo, who was born about 63 B.C. in present-day Turkey, geography was no mere sideline. He must have been one of the world's first professional geographers, determined as he was to write a compendium of the field that would set down everything that had been achieved as well as establish new directions. He traveled to Rome, studied in the library where Eratosthenes had worked centuries earlier in Alexandria, and read everything geographic that had been written. Based on this re-

search, he wrote a 17-volume *Geography*, a monumental work that still constitutes a guide to the field in ancient times. Ironically, Strabo's great effort attracted rather little attention when it first appeared; it was to become a classic, but not until centuries later. Today it remains a window to the geography of ancient Greece.

ROMAN CONTRIBUTIONS

Compared to the ancient Greeks, Roman scholars did not prove nearly as diligent or successful as geographers. There were exceptions, of course, and one Roman scholar with geographic inclinations was Pliny the Elder (born A.D. 23), who single-handedly assembled and wrote the world's first encyclopedia. Pliny studied in Rome, then entered the army, and followed his service with a period of law study. But he was attracted to research in science and history and eventually produced a giant, 37-book *Natural History* (numerous other writings were lost). Part of this encyclopedic work consists of a compendium of current geographic knowledge (Books III through VI), including valuable descriptions of ancient cities that have since withered or been destroyed.

Pliny the Elder had a lasting influence on science for centuries to come through his *Natural History*, a work that, in effect, summarized what Greek and Roman scholars had learned. He remains known as Pliny the Elder because his adopted son, Pliny the Younger (born about A.D. 61), also was an author. The lives of the two men are joined by an event that killed the father and was survived—and chronicled—by the son: the eruption of Mount Vesuvius, the volcano that destroyed Pompeii, Herculaneum, and other towns in the Naples area in A.D. 79. Pliny the Elder, from a distance, saw the mushroom cloud rising over the area and rushed to the scene to render assistance. He was overcome by the toxic fumes and died on the volcano's slopes. Pliny the Younger, on a boat in Naples Bay, saw the same cloud "rising like a pine [tree] of Rome." His vivid description of the historic event remains an important account for researchers to this day.

Another Roman geographer to be remembered is Ptolemy, who in the second century A.D. produced an eight-volume *Guide to Geography*. But Ptolemy shows both the strengths and the weaknesses of Roman geographers. Although he worked for many years in the library at Alexandria, his *Guide* does not even mention Strabo's important work. Also, Ptolemy's arithmetic was not as good as Eratosthenes', although he tried to "correct" previous calculations. Ptolemy wanted to establish the latitude and longitude of places in the Roman Empire. Rather than do this on the basis of Eratosthenes' measurements, he did his own—and got it wrong by about 30 percent. That is, he estimated the earth to be one-third smaller than it really is, thus his determinations of latitude and longitude are far off. In other respects, too (e.g., in the analyses of climate, vegetation, relief, and the river systems of the countries it describes), Ptolemy's *Guide* is not as good as much of what was written previously. Why, then, is Ptolemy remembered as an important scholar? Because Roman influence became paramount and dominant in Europe for more than a thousand years after his lifetime, and what the Romans had written and said had great weight—even the rule of law. To challenge Roman beliefs (as transmitted to

Ptolemy's map of the world constituted a summary of Greek and Roman geography. This map has a specific (conical) projection, a major innovation; note also the climatic-latitudinal zones on the left side.

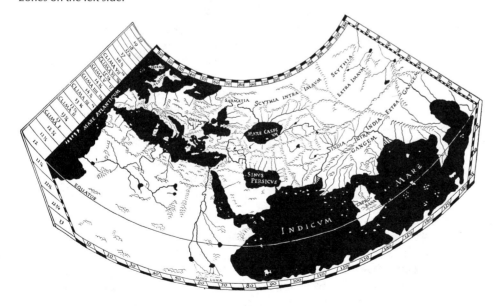

Western Europe) scientists often took great personal risks. Thus thirteen centuries later when Columbus sailed for the New World, he undertook his transatlantic voyage in the belief that the earth was much smaller than it really was, trusting Ptolemy rather than Eratosthenes. (Who knows . . . Columbus might not have tried but for Ptolemy's miscalculation!)

Stagnation: A.D. 500–1500

In Europe, the systematic study of geography progressed little during the thousand years following the breakdown of the Roman Empire. While research, exploration, and the development of geographic traditions continued in China and while Arab scholars were translating Greek and Roman works for the libraries of the rising Moslem world, geography during Europe's Middle Ages was a dormant field. Cartographic practices deteriorated, scholarship stagnated. Strands of geographic interest were kept alive by the reports and descriptions from European travelers to distant lands, but there was no one to synthesize such new information into a new worldview. Still, speculations about undiscovered lands and oceans and increasing maritime traffic sustained the attention of some scholars and mapmakers, so that by the fifteenth century, there were signs that the long period of dormancy was coming to an end.

ARAB SCHOLARSHIP

In large measure, this reawakening was due to the progress made by geographers and other scholars in the Arab world and in China. Europe still held to ideas developed by the ancient Greeks, but the Arabs had translated the Greek works and built their scholarship on those foundations, correcting many misconceptions in the process. For example, the ancient Greeks believed that the earth's equatorial zone was so hot that human life there would be impossible (after all, the sun had scorched black the skins of peoples living even as far from the equator as North Africa). But the Arabs sailed southward along the East African coast,

reaching Zanzibar and Moçambique before the close of the tenth century, thus they crossed the equator and proved equatorial regions to be habitable. Arab travelers (ibn-Batuta prominent among them) brought back to Moslem universities information from as far away as West Africa and eastern Europe. Arab scholars could therefore improve on Greek concepts and correct Roman misinterpretations. Eventually, physical and human geography emerged as discrete fields of Arab scholarship, and Arab geographers made giant strides in the interpretation of the evolution of mountain ranges and the deposition of sediments as well as in the analysis of the atmospheric processes that produce particular weather patterns.

CHINA

In China, too, geography made progress while it lagged in Europe. Record keeping was a hallmark of early Chinese scholarship, and during the Han dynasty (206 B.C.–A.D. 220) China embarked on the world's first population census. Centuries of weather records as well as data on river regimes, soil productivity, and other long-term information provided Chinese scholars opportunities to analyze patterns and processes; ancient Chinese geography was marked by more precision and less speculation than the geography of ancient Greece. Moreover, while the Greek and Roman worlds remained comparatively confined, Chinese explorers discovered the Mediterranean civilizations in the second century B.C., so that Chinese scholars had firsthand accounts of a far wider world than did the Greeks or the Romans. Marco Polo did not reach China until fourteen centuries after the Chinese geographer Chang Ch'ien visited Europe.

Revival

By the early fifteenth century, scholarship in Europe was beginning to ferment: even Roman ideas were questioned by some scientists. Greek works, translated into Arabic and then revised by Arab scholars, were translated once again (into Latin) and hence found their way into Christian Europe. This was one of several stimuli that revived scholarship in general and geography in particular in Europe; another was the increased pace of exploration during the fifteenth century and the need for more accurate information and better maps. Beginning with the Portuguese voyages along African coasts, European navigators from Columbus to Cook traversed oceans, charted coastlines, and penetrated lands hitherto unknown to Europe. An avalanche of new information confronted European scientists, and at the same time, old scientific assumptions were challenged. From the beginning of the sixteenth century a state of turmoil stirred European scholarship as Copernicus challenged Ptolemy's concept of an earth-centered universe, and in the seventeenth century, Kepler, Galileo, and Newton ushered in a scientific revolution that is still continuing today.

But even in those days of scholarly ferment and scientific revolution, it was possible for one scholar to make substantial contributions to several different fields. Leonardo da Vinci—in the ancient Greek tradition but with unprecedented genius—was a painter, sculptor, architect, and engineer. Galileo studied medicine but was a mathematician, astronomer, and physicist. Geographers who use mathematical methods to solve their problems remember Galileo as the father of the important probability theory.

In England, William Petty was trained in medicine, taught anatomy at Oxford, and produced (in 1685) a detailed geography of Ireland, including an atlas that remained the standard of reference for that country for more than a century. In Germany and the Netherlands, Bernhard Varen (also known as Varenius) turned from a career in medicine to geography, and in his brief lifetime (1622–c. 1651), produced two impressive works. The first, *Description of the Kingdom of Japan* (1649), was a regional geography of Japan and Siam (Thailand), with chapters on other topics. The second, *General Geography*, was published in the following year and placed the field of geography on a scientific foundation; it was the first truly systematic geography and the standard work until well after its last English translation in 1765. Varenius had realized that regional descriptions such as his work on Japan and Siam must fit into a systematic framework containing universal and general laws and principles. Geography's new identity was beginning to take shape.

ALEXANDER VON HUMBOLDT

Still, the practitioners of geography during the eighteenth and early nineteenth centuries were more than solely geographers. Alexander von Humboldt (1769–1859) was an accomplished botanist, geologist, historian, astronomer, cartographer—and geographer. He was, in addition, an inveterate traveler and keen observer of an almost endless range of phenomena. His life was a scientific treasure hunt, and his travels took him to Ecuador and Peru, to the Amazonian interior, to Middle and North America, and to Russia and Siberia. Born into the comforts of an aristocratic landowning family in Prussia, von Humboldt overcame the most severe environmental hardships in pursuit of his numerous scholarly interests. After extensive travels in Europe, he inherited enough money to make overseas explorations possible. He took a canoe some 1800 miles (3000 km) along the uncharted Orinoco

Alexander von Humbolt, explorer–geographer, last of the great generalists.

11

River in South America, collecting thousands of rock samples and plant specimens and fighting off danger and disease. In the Andes of Colombia, Ecuador, and Peru, he recognized that altitude, temperature, vegetation, and crop cultivation are interrelated. He climbed mountains, entered the craters of active volcanoes, measured the temperature and velocity of the ocean current named after him, and carried out hundreds of other experiments. After visiting Cuba and the United States (in 1804), he returned to Europe, taking up residence in Paris to write 30 volumes on his American explorations. The topics and titles range from botany and zoology to historical and physical geography, from astronomy to politics. He worked on this monumental series from 1805 to 1827.

Alexander von Humboldt had not exhausted his appetite for travel and writing, however. In 1829, he visited Russia and Siberia and began writing his famous *Cosmos*, a six-volume series published between 1845 and 1862 (the last volume appeared after his death). This giant scholarly and literary achievement constitutes a summation of the state of scientific knowledge in von Humboldt's lifetime and his numerous contributions to its various branches. Von Humboldt was the last of the great generalists.

Johann Heinrich von Thünen, first of the theorists.

FIRST SPECIALISTS

While von Humboldt was working in Paris, the man who could lay claim to the title of geography's first specialist was busy elsewhere. Johann Heinrich von Thünen (1783–1850) was owner of a large farming estate not far from the German town of Rostock in the state of Mecklenburg. As he farmed his fields and sent his harvests to market, von Thünen became interested in a subject that still excites economic geographers today—the effects of distance from the market and transportation costs on the location of productive farming. Would a farmer located near a market center cultivate different crops than a farmer located farther away on the same soils? If so, how far from the market center would farmers have to be to make the decision to grow something else?

For four decades, von Thünen kept meticulous records of all the transactions of his estate; while doing so, he began to publish books on the spatial structure of agriculture. This series, which he called

The Isolated State, in some ways constitutes the foundation of spatial theory. By making certain assumptions (see Chapter 8), von Thünen was able to develop a model for the spatial arrangement of farming zones around a market center. He concluded that high-priced, perishable goods would be produced nearest the market; farther away the farm products would be more durable. Still farther away, transport costs would become so great that production would become unprofitable. There, land would lie unused. Von Thünen's rings still merit discussion in geography's literature. He was an innovator well ahead of his time.

Another geographer, Carl Ritter, also was a contemporary of von Humboldt, but his career took a very different course. Born 10 years after von Humboldt (1779), Ritter died in the same year as the last of the great generalists (1859). While von Humboldt was traveling and recording geographic data in various areas of the world, Ritter was holding a position as professor of geography at the University of Berlin (from 1820 until his death). His was the first appointment of its kind in Germany, so that Ritter's role in the development of the discipline has special importance.

Ritter made it a point not only to practice geography, but also to define and prescribe the nature and scope of the field. Von Humboldt never worried whether his field work was or was not geography; much of it was geology, or botany, or even archeology. But Ritter tried to define a new, more scientific geography in which spatial relationships were to be understood and the earth viewed as the home of humanity. He always gave strong emphasis to history (some of his critics contended that he made geography a subsidiary of history), and he wanted to avoid a geography that consisted of a mere summation of unrelated facts.

Ritter's own fieldwork remained confined to Europe, but his great written work *Die Erdkunde* (literally, earth knowledge or earth science) was intended to cover the entire globe. The first volume, on Africa, appeared in 1817 and resulted in his appointment at Berlin. Eventually Ritter completed 19 books in the series, but he never got beyond Africa and Asia; he had set himself too enormous a task. By all accounts, Ritter was a superb lecturer who had a large following and a wide reputation, but his writings were often turgid and obscure. Unlike the adventurous von Humboldt, Ritter depended in his *Erdkunde* on the work of others, not his own observations and records. Yet he was the one to set definitions for the field of geography. It was a sign of things to come.

EUROPE TO AMERICA

A critical aspect of the development of an academic field of study is its transmittal from teacher to student and its perpetuation through generations of strongly committed workers. Von Humboldt had a legion of admirers, but he stimulated no lasting school of geography. Ritter, on the other hand, had devoted disciples who were both sure that he was right about his definitions of geography and who wanted to carry on his work on the unfinished *Erdkunde* series after his death. Two of his leading students, Élisée Reclus from France and Arnold Guyot from Switzerland, carried on the traditions Ritter had developed. As it hap-

pened, both Reclus and Guyot found their way to the United States. Reclus made a brief visit, but Guyot lectured on Ritter's "new geography" at Harvard University in 1848 and was appointed professor of physical geography at the College of New Jersey (later Princeton University) in 1854, a position he retained until he retired in 1880. European geographic ideas began to take hold in America.

Identity

The second half of the nineteenth century proved a critical time in the development of a geographic discipline. The center stage remained Germany, where several major universities appointed geography professors and where geography faculties thus began to develop. Two leaders in this movement were Ferdinand von Richthofen, who did much to define the field of physical geography, and Friedrich Ratzel, who concentrated on the development of human geography.

Ratzel (1844–1904) was trained in zoology, chemistry, geology, anatomy, and journalism. Like von Humboldt, he always wanted to engage in fieldwork, but at first these ambitions were frustrated. Ratzel was deeply impressed with the work of Darwin and the concept of evolution and wrote a thesis about Darwin's new ideas. But it was his journalistic skill that provided the resources to travel. He took a job as a reporter and did well enough to be able to visit the United States and Mexico in 1874 and 1875. Adding his American experiences to his impressions of Europe, Ratzel took a teaching position in geography that led to his appointment as professor at the University of Leipzig in 1886.

Von Richthofen had focused his attention on the physical world, whereas Ratzel was especially interested in the earth's human population, its societies, and cultures. In 1882, he published the first volume in the series he called *Anthropogeography*, a title that underscored his primary concern. Later, he turned to the properties of national states, still mindful of Darwin's revolutionary ideas about evolution. Applying

13

Darwin's biological concepts to the life cycles of states, Ratzel in 1897 published a book, *Political Geography*, that introduced "the state as a land-based organism." Just the preceding year he had produced a rather Aristotelian article, "The Laws of the Spatial Growth of States," but in his new book he set forth his theory about the struggle for survival among the world's national states. Like biological organisms, Ratzel argued, states need sustenance to grow and thrive. Just as an organism requires food, states need territory, space into which to grow, and cultures to annex and absorb.

Ratzel remains most strongly identified with this single (very debatable) idea, which is unfortunate because he made many other substantial and lasting contributions. Not the least of these was his impact on colleagues and students. His lectures were famed and packed; original ideas flowed freely and debate was encouraged. Among Ratzel's students were many who later took academic positions elsewhere in Germany and in other European countries. An American geographer, Ellen Churchill Semple, attended Ratzel's classes during the 1890s and carried his geographic viewpoints to the United States. In 1911, Semple published a book entitled *Influences of Geographic Environment, on the Basis of Ratzel's System of Anthropo-Geography*. In this way, still another set of geographic ideas was transplanted from Europe to the United States.

NATIONAL SCHOOLS

The German school of geography produced the first real theoretical notions as well as the earliest specializations in the field. Actually, much of this happened even before there was a united German state. It is well to remember that when von Thünen was doing his calculating and when Ratzel studied his first geography, Germany still was a loose confederation of states; and there was much conflict among them. In the 1860s, a powerful German leader, Otto von Bismarck, led the Prussians to victory over northern Germany. Next, he defeated France in the Franco-Prussian War of 1870–71, and only then was a united German Empire (*Reich*) established. Ratzel, the political geographer, could observe the birth and death of states firsthand.

German geographic ideas radiated out to other European countries, but those countries had their own geographic traditions, too. In France, predictably, there was a reaction against Ratzel's opinions about the natural environment controlling and determining human activity and behavior. Paul Vidal de la Blache, a professor of geography at the famous university in Paris, the Sorbonne, took the opposite position. He suggested that human will and decision making can overcome obstacles posed by the natural environment and that people, not the environment, determine the course of events. Certainly nature could set limits, but within these, people adjust and make use of the opportunities. In this way, de la Blache answered Ratzel's **determinism** with his concept of **possibilism**. French geography differed from German geography in another way. Whereas physical and human geography developed rather separately in Germany, they were much more closely intertwined in France. This was due mainly to the work of two of de la Blache's students, Jean Brunhes (a human geographer) and Emmanuel de Martonne (a physical geographer). Each knew the other's field well, and they both saw the uniting perspective of geography: its *spatial* viewpoints.

Another important aspect of geographic scholarship in France was the existence of several major societies and associations. These geographic organizations (notably the Paris-based Geographical Society) were very active in public as well as academic affairs. They promoted geography and geographic ideas vigorously, and their meetings and lecture series were highly publicized. The Geographical Society, for example, advised French governments to expand France's colonial empire not only to strengthen the state, but also to confer the advantage of French civilization on colonized people. Such advice was taken seriously, because geography had a central position in scholarly circles.

British geography, like geography in France, was vigorously promoted by geographic societies and associations. The Geographical Society in Paris had much influence on French intellectual life; in Britain, the Royal Geographical Society kept geography in the public eye. Largely through the Royal Society's efforts on behalf of the field, geography faculty were appointed at Oxford and Cambridge in the 1880s, and soon strong geography departments developed at

14

other major British universities. The leading figure in British geography in the late nineteenth and early twentieth centuries was Halford J. Mackinder, whose *Britain and the British Seas*, published in 1902, was still being assigned to geography students at Commonwealth universities a half century later. In 1904, Mackinder presented a paper before the Royal Geographical Society, "The Geographical Pivot of History," that thrust him into national and international prominence. In this paper, Mackinder took a global geographic view of history and concluded that ultimate world power would concentrate in a region protected by distance and terrain from the fleets of competitors. This region lay in interior Eurasia, and at its center was the so-called **heartland**. If a national state could unite East Europe and neighboring Russia, it would dominate the world.

In the United Kingdom in 1904, Mackinder's paper made headlines but generated little practical concern: Britain ruled the oceans and much of the world. But in Germany, there were those who listened attentively, especially students of Ratzel. Mackinder seemed to confirm several of Ratzel's opinions, and his paper appeared to have relevance to German national policy. Geography as global strategy (**geopolitics**) had a ready market in Germany.

Given the nature of the times—political rivalries and conflicts, revolutionary changes in government, colonial expansion—it is not surprising that political geography was an important driving force in scholarly circles. Ratzel's ideas gained wide acceptance not only in Germany, but also beyond (e.g., in Sweden); Mackinder's prophecies made headlines in the papers. But not all geographers pursued this area of specialization. Some of Ritter's followers continued to take a broader view and tried to find a definition of just what geographic research was—a definition that would satisfy all or almost all geographers. In France, the balance between physical and human geography continued, and a student in those days (the late nineteenth and early twentieth centuries) had a well-rounded education. In Britain, physical geography became the mainstay of geographic training at the universities, and there was much stress also on field work. The British, perhaps more than any other geographers, believed that field observation and recording were critical skills that all must learn.

REGIONAL GEOGRAPHY

This also was a time when one of geography's most important organizational concepts gained ground. If you read books and articles published by geographers in various Western European countries, a unifying theme soon becomes apparent: the *regional* concept. It was obvious to geographers in Germany, France, Britain, and elsewhere that certain areas of the world (at several levels of scale) had a certain uniformity, a sameness. This could be seen in the field; it could be recorded on maps. But establishing the boundaries of regions turned out to be quite a challenge: Which contents of regions have relevance in geography? Which do not? Such questions still interest regional geographers today, nearly a century later. We will have occasion to refer to theoretical and practical aspects of regional geography throughout this book.

• Transition •

15

A set of geographic traditions also was emerging in the United States during the late nineteenth and early twentieth centuries, and it would be incorrect to assume that American geography developed only as a transplant of European geography. Explorer–scientists, supported by the government or by universities and private interests, traversed the western United States and wrote reports about the complex surface features of the region and its indigenous cultures. Prominent among these scholars were John Wesley Powell, who explored the Rocky Mountains and traveled the Colorado River through the Grand Canyon, and Grove Karl Gilbert, who studied river erosion in the West and whose report on the Henry Mountains (dated 1878) remains a classic.

WILLIAM MORRIS DAVIS

American geography really came into its own, however, through the pioneering work of William Morris Davis (1850–1934). Davis who was trained in geology, became interested in the processes that

William Morris Davis, pioneer of American geography.

raphy in the United States and, indeed, around the world. Davis stimulated discussion and debate in Germany, France, England, and wherever else his geographical cycle was published and read. He participated actively in these exchanges, publishing numerous papers in support of his concepts. He was a persuasive teacher, especially in the field, and he was not afraid to criticize the field he loved. "[N]othing seems to me clearer than that geography has already suffered too long from the disuse of imagination, invention, deduction, and the various other mental faculties that contribute towards the attainment of a well-tested explanation."

Davis was a physical geographer, but he took a wider view of the subject. He was less successful when he attempted to introduce ideas in human geography into the discipline, because he gave credence to the same social Darwinist views that had influenced Ratzel. Nevertheless, Davis was one of those unusual scientists who also concern themselves with teaching and education, and he influenced many a college administrator to consider modern geography for inclusion in the curriculum. Moreover, he took a leading role in the founding of the Association of American Geographers, serving as president in its inaugural year (1904) and twice thereafter.

Davis also taught and influenced many contemporary geographers. Among his students were Mark Jefferson and Ellsworth Huntington, two geographers who had great impact on their field of study, though in different ways. Jefferson, an indefatigable teacher and prolific author, gave human geography much stronger emphasis than Davis had, but he retained Davis's rigor and discipline. From 1901 to 1939 Jefferson taught geography at Michigan State Normal College (now Eastern Michigan University) in Ypsilanti. Several of his numerous articles are still quoted today, including "The Civilizing Rails" from *Economic Geography* (1928) and "The Law of the Primate City" from the *Geographical Review* (1939). Jefferson did important work on the geography of cities over a lengthy productive period.

ELLSWORTH HUNTINGTON

Ellsworth Huntington, who did fieldwork in Asia from 1897 to 1901 and again from 1903 to 1906, may be

shape landscapes. He observed relationships between the geologic *structures* beneath existing landscapes, the *processes* of erosion (by stream water especially) that acted upon these rocks and structures, and the *stage* these processes had reached. This combination of structure, process, and stage was called by Davis the **cycle of erosion**. As the cycle progresses (so reasoned Davis) mountains will be worn down to plains. We can observe this happening in the field—where we note that not a single plain is absolutely flat, in the geometric sense. So, Davis called his plains *peneplains*, or near-plains.

Davis's innovations were not speculations about global geostrategy and the future of nations. He was a meticulous field researcher who sketched and measured, observed, and recorded; his model of the erosional cycle was firmly based on field data. It was an enormous forward step, and Davis, who taught geography at Harvard University from 1878 to 1912 (he was appointed assistant professor of physical geography in 1885) had an unprecendented impact on geog-

the best-known scholar American geography has ever produced, even though many of his ideas could not withstand the test of time. Huntington focused his research on the influences of large-scale climatic change on human societies. He asserted that climate was the critical factor in determining the level of civilization of the world's peoples: the stimulating, variable climatic regimes of the middle latitudes would generate the most advanced societies, but the hot, enervating tropics would never support such cultures. Huntington's most comprehensive statement of this hypothesis of environmental determinism is contained in *Civilization and Climate* (1915). In those days, information about climatic changes over time was far less complete than it is today (it is still inadequate), and data concerning the culture and civilization of peoples in many parts of the world were sketchy. As a result, Huntington's generalizations were often shaky. Furthermore, speculations of this

Ellsworth Huntington, controversial geographer, prolific author.

kind tend to lead to premature conclusions about superior and inferior peoples and races. Many scholars, thus, were highly critical of Huntington's work, but few could ignore his huge output. He produced about 300 publications, ranging from nearly 30 scholarly books such as *The Pulse of Asia* (1907) to articles in newspapers and magazines. Huntington raised some questions that still have not been satisfactorily answered today, but many of his modern critics appear not to have read the mass of his work.

During Huntington's rise to prominence, his work was effectively supplemented by that of Ellen Churchill Semple, who had studied with Ratzel in Germany in the 1890s. Although she did not accept the strict determinism inherent in Ratzel's organic theory, she did take an environmental determinist's view of human society's evolution. Semple's first book was *American History and Its Geographic Conditions* (1903); her more famous work involved the transmission of many of Ratzel's concepts to the English-speaking audience in *Influences of Geographic Environment* (1911). Later, she did extensive fieldwork in the Mediterranean region and published numerous articles, eventually combining two decades of writing in her last book, which appeared in 1931.

17

Expansion

By the 1930s geography in the United States was no longer the pursuit of a few dedicated enthusiasts stimulated by a glimpse of the new geography of Germany or sustained by an explorer's instinct. Geography in the 1910s and 1920s attracted a growing cadre of scholars who held positions in an ever-larger number of geography departments at colleges and universities. From Harvard (where Wallace W. Atwood carried on the traditions Davis had established) to Chicago and from Yale to Berkeley, geography prospered. Geographers could now afford the luxury of specializing, of developing the field's academic and professional directions, of discussing the merits of various teaching curricula, and of debating geography's scope and limits. It was an exciting time of development and expansion, lively exchange, and the beginning of the formation of schools of geography in several parts of the country at individual universities or groups of universities.

The development of schools of geography was an important development. Typically, in such a school, there would be several professors who shared a set of viewpoints and approaches. They might not totally agree on scholarly matters, but their publications, teachings, and debates would vigorously push forward their area of specialization. Such active professors attracted an unusually strong and productive group of graduate students, and these students, in turn, invigorated the department. Sometimes the school (really a school *of thought*) was dominated by an especially powerful figure whose pronouncements could alter the course of the whole discipline.

Such schools of geography still exist today, and then as now, viewpoints and positions vary widely. This can be a blessing and a curse. A blessing because a variety of viewpoints leads to productive interaction in which one group learns from the findings of another. A curse because there are always some who want to set the discipline on one particular course, saying that all else is not worth pursuing. Unfortunately, such people are sometimes believed by too many followers.

A good example of such an effort to redefine geography is an article written by a University of Chicago geographer, Harlan H. Barrows, that was published in 1923. Entitled "Geography as Human Ecology," this article proposed that geography should focus on humanity's efforts to make a living on the earth, so that economic geography would be the core of the discipline. Notwithstanding Davis's contributions, Barrows suggested that physical geography in all its forms (including climatology) should be abandoned by geographers.

CARL O. SAUER

As it turned out, this extreme view persuaded few geographers to reorient their work. A more consequential article appeared two years later, when Carl O. Sauer published "*The Morphology of Landscape.*" Sauer was professor of geography at the University of California, Berkeley, from 1923 to 1957, and this early statement constituted the beginning of one of geogra-

phy's most respected and productive schools. The article was a warning against the sometimes blind acceptance of environmental determinism as the central theorem of geography without adequate proof or verification. It also put forward a new framework for geographic inquiry as the study of "the forms superimposed on the physical landscape by the activities of man, the *cultural landscape.*" This alternative viewpoint held that deeply entrenched traditions and customs, including religious beliefs and practices, systems of government and law, and attitudes toward the natural environment would constitute the dominant influence in the development of patterns of land use, cities and towns, trade, communications, and industry—the total cultural landscape. In a sense, Sauer's school proposed the replacement of environmental determinism by a kind of cultural determinism.

Thus the cultural landscape school of geography focused attention on the transformation of landscape by human action. This might seem so comprehensive and far-reaching a concept that it is not really useful; in fact, it had a major impact on the discipline, an impact that continues to be felt today more than a half century later. It made a thorough grounding in anthropology as necessary for human geographers as an acquaintance with geology was for physical geographers. It led to new ways of looking at elements of the cultural landscape, at the cultural imprints left by earlier societies, at the relationships between human society and natural environment. As the researchers worked and published their findings, the arguments over theory went on. Should all things in the cultural landscape be measurable? If they can be measured and mapped, can predictions be made about future changes? How does one deal with those intangible qualities of the cultural landscape so easily perceived but so difficult to define?

PHYSIOGRAPHY

While the human geographers were engaged in this debate, physical geographers (or physiographers, as they were also called) had their own theoretical arguments. One prominent issue had to do with Davis's concept of landscape reduction and just how this process took place. Did slopes wear *down*, as Davis had suggested, or would erosion drive them *back*, as

18

the German physiographer Walther Penck proposed? (This dispute has deeper implications than might at first appear and is discussed in some detail in Chapter 2.) The debate did not, however, deter physical geographers from continuing their work on a major project: the unraveling of the complex physical geography of the United States and the delimitation of the country's physiographic regions. For example, everyone uses Rocky Mountains, Great Plains, and many other such names to identify regional segments of the United States. But where, exactly, are the boundaries of these regions? What are the components of each? And what are the relationships between them? Nevin M. Fenneman, who taught geography at the University of Cincinnati from 1907 to 1937, produced a two-volume series to answer such questions. The first, *Physiography of the Western United States,* appeared in 1931. The second book in this monumental effort, *Physiography of the Eastern United States*, was published in 1938. Another physical geographer, Armin K. Lobeck, contributed a meticulously detailed physiographic diagram of the United States, a map that became a standard reference for physiographers and is used by thousands of students to this day.

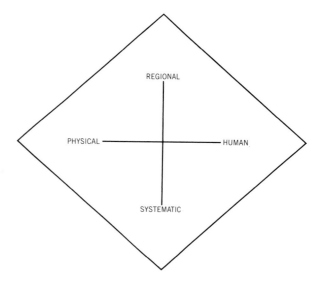

FIGURE 1.1

Two dualisms of geography.

•   Dualisms   •

By midcentury, geography was thriving in universities and colleges, government and business. This strength was displayed in many ways: a well-established critical mass of scholars, several developing schools of geography, and a growing literature with a number of rigorous journals, including the *Annals* of the Association of American Geographers and the *Geographical Review* published by the American Geographical Society. Geographers did not limit themselves to teaching in colleges and universities, but were also in government and business where they proved the practical usefulness of their field as urban and regional planners, cartographers, transportation and marketing analysts, field surveyors, and in other capacities. Geographers worked on boundary commissions in the aftermath of the Second World War, in the Office of

the Geographer at the U.S. Department of State, and as geographic attachés in embassies abroad.

Notwithstanding all this progress, geography still continued to suffer from a problem nearly a century old: its deep division and fragmentation. This division can be conceptualized as a kind of double dualism that made a complete education in geography very difficult (Fig. 1.1). All too often, a geographer trained thoroughly in one area of geography would be inadequately prepared in another; this, in turn, made it difficult for other scholars and scientists to understand just what geographers do best.

The first of geography's two dualisms involved the old division between physical and human geography. There were geographers who focused their attention almost exclusively on the earth's physical surface or on the workings of climate and weather, paying little attention to the human modification of these natural features. At the other end of the spectrum, human geographers studied cultural imprints and activities, many of them knowing little about the work of their colleagues in physical geography. There were always geographers whose work put them somewhere between these physical-human poles, but they were in a minority. The human-physical dualism was an inescapable reality of geography at midcentury.

The second dualism that marked geography divided the discipline into regional and systematic camps. Regional geographers were interested in the geography of a particular area or region of the world. Their interest might focus on some specific aspect of that region, perhaps its farming systems, or its transport networks, or its physical landscapes, but regional geographers were identified *first* by their locational pursuits. This might take the form of a world realm (for an African specialist, or Africanist) or some other regional—possibly environmental—association (for someone who studies tropical rain forests, or deserts, or glacial landscapes wherever they might occur on the earth). At the opposite end of this axis lay the emerging cluster of systematic or topical geographies. Here the focus of study was on some aspect of human behavior without necessary reference to any region. Thus the development of cities and towns is a global phenomenon, and urban geography studies the urbanization process and its spatial manifestations wherever it occurs—in America, Africa, Asia, and elsewhere. Regions are incidental, if not irrelevant, concepts to historical geographers, political geographers, cultural geographers, economic geographers, and others interested in systematics.

CORE AND CIRCUMFERENCE

Does geography have a core in which the various and different geographic pursuits converge? Some (not all) geographers continue to view regional geography as the essence of the field. That is nothing new: regional themes were unifying elements in Western European geography a century ago. As long ago as 1919, Nevin Fenneman (the physical geographer who wrote the two monumental volumes on the physiography of the Eastern and Western United States) published an article entitled "The Circumference of Geography" in the *Annals*, the principal publication of the Association of American Geographers. In that article, Fenneman presented a schematic diagram of what he viewed as the core of geography—regional geography—and the systematic fields that contribute to regional studies (Fig. 1.2). On the "circumference" lie economics (which in spatial, geographic terms becomes economic geography, anthropology [cultural geography], meteorology [climatology], etc.). Since this article was pub-

lished, geography has, of course, undergone many changes. In some respects, however, the diagram retains a certain validity. Take that old topic of Ratzel and Mackinder, political geography. Today there is a full-fledged social science called political science, the study of political institutions, political behavior, and related topics. In spatial context, this becomes political geography, the study of the imprints of political behavior on the ground. This can involve political boundaries and their effects on nearby communities; it also can take the form of studies of voter preferences, say, in urban, suburban, and rural settings. In this book, several of Fenneman's topics on the "circumference" are the subjects of individual chapters.

If these debates over the dualisms, core, and "circumference" of geography posed problems of identity and cohesion, they also had their advantages. By its very breadth, geography attracted a body of scholars with a wide range of interests but with a common curiosity about things spatial. When geographers assembled at their professional meetings, physical, human, regional, and systematic specialists discussed and debated a variety of topics unmatched by any other scholarly gathering; at times, these conferences were as exciting as any ever held. This quality continues to mark geography today: anyone attending a meeting of geographers may come away unsure of the scope and limits of the field, but no one will have doubts about its variety and vitality.

•   New Directions   •

During the second half of the twentieth century, geography has developed in several new directions—but some old traditions also have proved to be very durable. Among the most significant changes are: (1) new techniques in the gathering, storage, and retrieval of information; (2) new methods for the analysis of this mass of data; (3) a stronger orientation toward problem solving and a growing interest in systems and processes; and (4) unprecedented expansion, proliferation, and diversification of geography; the emergence of new branches of the field; and closer cooperation with scholars in related disciplines.

FIGURE 1.2
A modern version of Nevin Fenneman's diagram of 1919.

In several of these spheres, geography has shared experiences with other disciplines such as anthropology, sociology, political science, economics, geology, botany, and others. The technological revolution in the acquisition and recording of information has transformed the state of knowledge. At midcentury, air photography remained the most advanced aid in the recording of terrain, land use, crop patterns and yields, surface communications, and similar data. Then the use of infrared photography made it possible to survey extensive areas for specific purposes—for example, forest density, soil type, or water availability. But the major technological breakthrough came when satellites carrying cameras were put into earth orbit and began to send back masses of earth data with ever-increasing detail and accuracy. Geographers who had depended on field sources alone to gauge de-

velopments in distant regions (urban sprawl, road quality, farm fields) now had access to an entire new bank of information.

REMOTE SENSING

The acquisition and recording of information from high above the earth is known as remote sensing. In his book *The Surveillant Science: Remote Sensing of the Environment*, Professor Robert Holz of the University of Texas describes advantages of such systems. "[They] can provide *regional views* not easily gained from ground observation or observation from lower altitudes. Such views might show an entire mountain range, a massive geologic formation, the extent of a forested region, or all of a large urbanized area. Remote sensing can expand our perceptions of the energy environment by making use of radiation that is otherwise invisible to us. For example, instrumentation allows sensing in the ultraviolet, infrared, and microwave fields. Remote sensing can give us repetitive looks—daily, weekly, monthly, or yearly—at the same area or target [so that] we can gain valuable information about events that change over time such as the development, biomass and vigor of agricultural crops; the 'green-up' in spring; the extent of snow cover; and changes in water currents, sediment loads, or pollution levels. . . . Because much of the remote sensing imagery generated today is in an electronic format, it can be processed electronically. This makes possible the use of modern, high-speed computers that can store, quantitatively analyze, and display remote sensing data."

Undoubtedly the most important remote sensing system as far as geographers are concerned is Landsat. Beginning in 1972, a series of satellites was launched under the designation ERTS, for Earth Resources Technology Satellite; in 1975, ERTS was renamed *Landsat*. Its cameras and other recording devices have sent back to earth magnificent spatial displays of environmental conditions, topography and geology, vegetative patterns, and more. Many underdeveloped countries were able to purchase Landsat imagery that was far superior to any available maps. Superb evidence of the value of Landsat is the *Atlas of North America* published in 1985 by the National Geographic Society, forerunner of a new generation of atlases using remotely sensed data. Remote sensing has, indeed, brought an information explosion.

The almost incredible explosion in the volume of information has produced its own problems: how to assemble, code, and store the data and how to make reasonably fast retrieval possible. There is no point in amassing data without preparing some way to have access to them; so the electronic computer became an indispensable aid to many geographers. Not only can the computer serve as a data bank of practically unlimited capacity, it also performs almost instant calculations on the available data when instructed to do so. Thus geography in the 1950s experienced a so-called quantitative revolution when many geographers turned to mathematical and statistical techniques to help solve spatial problems. Actually,

geographers had been using quantitative methods to attack certain numerical problems much earlier, but in the 1950s there was a major resurgence of interest in statistical analysis. Today the training of an undergraduate geographer is likely to include a course in quantitative methodology; at midcentury, it would have been difficult to find a geography department offering such a course addressed specifically to spatial problems.

And the quantitative revolution had its impact on other, more established courses in the curriculum as well. Cartography, for example, was affected in several ways. New demands arose for the representation of the results of quantitative analysis; thus computer mapping made its appearance.

At the same time, certain major areas of study and research became modern geography's focal points. The new methodologies resurrected some old problems and produced valuable new insights. Central place theory was one of these. As we see in more detail in Chapter 9, a German geographer, Walter Christaller, published a book in 1933 entitled *The Central Places of Southern Germany*. In this volume, Christaller laid the groundwork for central place theory as he sought to discover the laws that govern the distribution, or spacing, of urban places in various size categories. With the new methodologies, Christaller's work could now be subjected to mathematical analysis, and assumptions he had had to make in words could be translated into numbers and indexes. Many geographers in the United States in the 1950s and 1960s used the new analytical techniques to study some aspect of the subfield called location theory.

Another area of increased interest was the study of environmental perception. This would seem to be an eminently geographic pursuit, involving human behavior, the physical world, and patterns of interaction. Geographers borrowed from psychologists some of the techniques involved in measuring perception and found ways to explain how different peoples assess and evaluate their natural environment. Thus geographers were able to do important research on the perception of hazards as part of the natural world: Why do people continue to live in the shadow of frequently active volcanoes? On flood-prone plains below vulnerable levees? Along coasts that have a history of hurricane devastation?

Still another group of geographers focused on diffusion in its several forms and processes. Carl O. Sauer had studied cultural origins and dispersals long before, but now a new, more precise view could be taken through the use of mathematical models. Diffusion studies contribute to theory in geography, but they have practical significance as well. Knowledge of diffusion processes helps a government predict how well some innovation (a better plow or improved seed) is likely to be accepted and adopted by its farming peoples or it may assist doctors in directing their attack on a spreading disease.

Location theory, perception analysis, and diffusion study are a few of the several areas of concentration in contemporary geography. In one way or another, the new techniques and methodologies affected almost every branch of the discipline, but the fundamental traditions of the field were little changed. The school of cultural geography Sauer had founded still thrived in the 1980s. Historical geographers continued their work. Regional science, the application of new ideas in a regional context and the mathematical study of regional systems, supplemented (but did not supplant) traditional regional geography. Courses in the geography of Latin America or Europe still remained part of the university curriculum. The new directions have enriched geography, but they have not transformed it.

In one respect geography did change during the nearly 40 years following 1950: in its expansion and proliferation and its inevitable transition from a closed society of scholars to an open community of practitioners of many kinds (although teaching still dominates). Specializations unheard of a few years ago now diversify the field: medical geography, the geography of sport, the geography of crime, marine geography, and so on. This, in turn, signals a closer cooperation with workers in related fields and less concern about exceeding the "limits" of geography. Geography always was an interdisciplinary field, and this advantage now facilitates cooperation with physicians, criminologists, sociologists, economists, political scientists, botanists, zoologists, and geologists (by no means an exhaustive list) as scholars join to solve common problems rather than confine themselves behind disciplinary walls. Geography, after all, is the discipline of synthesis.

• Reading about Geography •

Books about academic disciplines tend to be heavy going, but there are exceptions. One of these is *The Geographer at Work* (London: Routledge & Kegan Paul, 1985) by P. R. Gould, a professor of geography at Pennsylvania State University. This book is not only about geography as a field, but also about the work, adventures, and accomplishments of geographers in the United States as well as in other countries. Another very readable work is *All Possible Worlds: A History of Geographical Ideas* by G. J. Martin and P. E. James (New York: Wiley, 1981). This volume traces the historical development of the discipline from the ancient Chinese and Greeks to the 1980s and contains useful bibliographies and sketches of leading geographers. For a more technical treatment of geography, see R. Hartshorne's monumental *The Nature of Geography* (Lancaster, Pa.: AAG, 1939) and his *Perspective on the Nature of Geography* (Chicago: Rand McNally, 1959). Also see R. Murphey, *The Scope of Geography* (Chicago: Rand McNally, 1973).

It does not take long to discern the geographer in Aristotle, see "The Ideal State" (in *The Works of Aristotle* [London/New York: Oxford Univ. Press: (Clarendon), 1921]). Your library's card file will lead to translations of the prominent writings of Strabo, Pliny, Ptolemy, and Varenius. A. von Humboldt's *Kosmos* was translated into English by E. C. Otté and published in a series (London: Bohn, 1849–61). No comprehensive translation of Ritter's *Erdkunde* has been made. A translation by R. L. Bolin of Ratzel's "Laws of the Spatial Growth of States" appears in R. E. Kasperson and J. V. Minghi (eds.), *The Structure of Political Geography* (Chicago: Aldine, 1969). For a fascinating account of von Humboldt's life, see L. McIntyre, "Humboldt's Way," *National Geographic*, Vol. 168, No. 3, September 1985, pp. 318–51.

23

Three influential works appeared early in the century: E. C. Semple's *Influences of Geographic Environment* (New York: Holt, 1911); H. J. Mackinder's *Britain and the British Seas* (New York: Appleton, 1902); Mackinder's famous paper "The Geographical Pivot of History" appeared in the *Geographical Journal* (Vol. 23, 1904). A worthwhile volume is a collection of Mackinder's thoughts and writings, *Democratic Ideals and Reality* (New York: Holt, 1919; reissued in paperback by Norton, 1962).

W. M. Davis's bibliography is lengthy; consult instead D. W. Johnson (ed.), *Geographical Essays by William Morris Davis* (New York: Dover, 1954). This collection includes many of Davis's memorable publications, including some of his excellent field sketches. (The quotation on p. 16 is from "The Geographical Cycle," p. 252.) The best picture of Mark Jefferson emerges from an excellent biography: G. J. Martin, *Mark Jefferson, Geographer* (Ypsilanti: Eastern Michigan Univ. Press, 1968).

E. Huntington's *Pulse of Asia* appeared in 1907 (Boston: Houghton Mifflin) and his *Civilization and Climate* in 1915 (New Haven: Yale Univ. Press). G. J. Martin's *Ellsworth Huntington: His Life and Thought* (Hamden, Conn.: Shoe String Press [Archon Books], 1973) is a superb biography.

H. H. Barrows's article "Geography as Human Ecology" appeared in the *Annals* of the Association of American Geographers (Vol. 13, 1923). C. O. Sauer's seminal paper "The Morphology of Landscape" was Volume 2 of the *University of California Publications in Geography* (Berkeley, 1925).

In the 1930s, N. M. Fenneman's volumes became available: *Physiography of the Western United States* (New York: McGraw-Hill, 1931) and its companion *Physiography of the Eastern United States* (New York: McGraw-Hill, 1938). It is of interest to compare with these W. W. Atwood's *The Physiographic Provinces of North America* (Boston: Ginn, 1940) and W. D. Thornbury's later *Regional Geomorphology of the United States* (New York: Wiley, 1965).

As modern fields within geography have expanded, their literature has grown. On remote sensing, see R. K. Holz, *The Surveillant Science: Remote Sensing of the Environment* (New York: Wiley, 1985). The quotation on p. 21 comes from "Remote Sensing: Coming of Age in the '80s," p. 30. Current and contemporary references to other works will be given at the end of each topical chapter to follow.

JOURNALS

A large number of journals in geography can be found in the library. In the United States and Canada, the most widely seen is the *National Geographic*, a magazine that presents topics in geography (and related to geography) in popular form, often accompanied by excellent maps and photographs. It reports on research, deals with current issues (e.g., urban congestion in Mexico City, energy problems, acid rain, South Africa) and gives geographic perspective to the news. This magazine does not concern itself with professional debates about the nature of geography as a discipline; rather, it publishes articles that, in one way or another, inform about places and people, from the familiar (to us) to the little known.

Geographical Magazine, published in England, is sometimes seen on newsstands in North America. This journal keeps a bit closer to "real" geography. More professional geographers write articles in this journal than in the *National Geographic*.

Among the professional journals, the leading North American publication is the *Annals* of the Association of American Geographers. The association also publishes the *Professional Geographer* and a *Newsletter* with news about geography departments at universities and colleges, the kind of work geographers are doing, and jobs available in the field.

The Association of American Geographers has a number of regional divisions, and these divisions also publish their own journals such as the *Southeastern Geographer* and the *Yearbook of the Association of Pacific Coast Geographers*.

The American Geographical Society in New York published a professional journal, *Geographical Review*, and the popular *Focus*.

Another professional organization, the National Council for Geographic Education, publishes the *Journal of Geography*. Addresses of these and other organizations are given at the end of this book. You may wish to write for information and perhaps some sample materials.

Literally hundreds of other geographic journals

are published around the world, many in English. You may enjoy exploring the library to get a glimpse of what can be found in, say, the *Scottish Geographical Magazine*, the *New Zealand Geographer*, and similar journals from India, the Philippines, Australia, and South Africa. Although many of the articles in these journals are technical, you will be pleasantly surprised at the number that can be understood fairly easily. And, of course, it is always fascinating to read in back issues the actual writings of geographers who were then beginners and who are now legendary figures.

25

CHAPTER TWO

The Geography of Natural Landscapes

⊷

To societies the world over, ancient and present, the physical earth has been mysterious, even threatening. Gods were seen in clouds, their anger vented in the lightning of thunderstorms. Ancestral spirits were (and still are) held to occupy hills, trees, even animals. Sacrifices were made to raging rivers and rumbling volcanoes. Understanding the earth's surface and its landscapes, its restless oceans, and its changeable climates has been a journey of thousands of years. It is far from finished.

Looking back, the association between physical earth and menacing (but sometimes benevolent) deities is not surprising if we consider where Western civilization has its roots. The eastern Mediterranean is an area where earthquakes, volcanic eruptions, violent storms, and other natural hazards were part of life. Surely the timing and impact of such disasters appeared ordained by some supernatural force!

Perhaps the most stupendous natural disaster ever seen by any civilization happened in the present-day Greek islands about 3500 years ago. Imagine the scene: the island of Crete with its thriving culture, the Minoan culture. Magnificent palaces had been built. Farmlands produced copiously. Markets bustled. Arts flourished. Feasts and festivals enlivened the annual calendar. From the north coast of Crete, about 70 miles (110 km) away, one could see a tall volcanic mountain rising from the blue Mediterranean Sea. Smoke puffed from the summit. Occasionally the mountain rumbled, but its slopes were fertile and life was good. In the main port lay a fleet of boats, some Minoan and others Phoenician, with brightly colored sails. Trade with Crete and the Mediterranean's east coast was profitable.

And then, on a certain day in what may have been the year 1635 B.C., that island-mountain exploded with a force beyond all description. The entire center of it (now known as Santorini or Thera) was hurled, catapulted into the air. The largest pieces of pulverized rock fell back on the outer slopes and into the sea; ash and dust rose many miles into the atmosphere. The skies darkened and stayed dark for days. A succession of great waves turned the Mediterranean into a cauldron. Ash rained on Crete and reached Egypt, hundreds of miles away. In the Bible's Old Testament, the event is described as an act of God

in retribution against the pharaoh . . . "thick darkness in all the land of Egypt three days." When calm returned, only a shell of Santorini remained, the outer slopes of a great volcano whose heart had been torn out. The effect of Santorini's dust, in orbit around the earth for years, was to change the climate; whatever the cause, Minoan civilization died. And perhaps the legend of Atlantis, the lost land, was born.

Today the forces that produced the catastrophe of Santorini are better understood. A little more than a century ago, in 1883, a similar event occurred in Indonesia when Krakatoa exploded. Located between Djawa (Java) and Sumatra, Krakatoa, too, was an island volcano. Its explosion, on an August day, buried thousands of people under rock and ash, and the sea waves it caused killed an estimated forty thousand along nearby coasts. Yet Krakatoa's blast (which was heard more than 2000 miles away in Australia) has been estimated as only about one-quarter the magnitude of Santorini's. Krakatoa's behavior prior to its eruption was well recorded, as the mountain had been spewing lava and ash, rumbling, and shaking. The same may have happened at Santorini, and some residents probably left the island before the major explosion occured.

Volcanic mountains that erupt as Santorini and Krakatoa did generally provide warnings that interior pressures and temperatures are reaching an explosive phase. Sometimes those signals are misread. In 1902 on the Caribbean island of Martinique, Mount Pelée, towering over Saint-Pierre (then the island's capital), gave ample warning: a huge solid spire protruded from its summit and was pushed daily higher by the forces inside the mountain's central vent. Smoke, gases, and ash squeezed from the crater, blocked by that rock-hard spire. Still the local geologist calmed local citizens' fears by saying that the mountain was only "moody." But on the morning of May 8, Mount Pelée's top blew off, the spire and crater rim falling in millions of pieces on the city, followed by waves of burning gases. About thirty thousand people perished—one sixth of the population of all of Martinique—and Saint Pierre, the "Paris of the Caribbean," was no more. Water in the harbor became so hot that boats disintegrated. Ever since, the name of the mountain has been given to explosive eruptions: *Peléan*.

Wherever volcanoes stand in tropical or moderate climates, their lava soon converts into fertile soils, people cluster in their very shadow—and in danger. Looming over the town of Soufriere on the Caribbean island of St. Lucia are the twin plugs of a volcano in a scene chillingly reminiscent of Mt. Pelée and St. Pierre (Martinique).

Heeding warning signs is one thing; predicting the timing of earthquakes and volcanic activity is another. Geologists knew that Mount Saint Helens would soon erupt with explosive force as that event approached in May 1980. Geographers know that people often ignore the most serious of hazards, continuing to live and work in high-risk locales—where earthquakes strike, floods threaten, hurricanes prowl, and volcanoes menace. Even after evacuation orders were given, more than sixty persons died near Mount Saint Helens in an eruption most knew would come.

These events illustrate the relationships between what we may designate as surface, structure, and society. To learn about the *surface* of the earth—its landscapes—would be impossible without studying what lies below, the rocks and the forms into which they have been fashioned. The construct of the entire planet earth even has relevance, but more directly significant are the *structures* that support the visible landscape. These structures—expressed as mountain ranges, plateaus, plains, and other landscapes—in turn, affect the distribution of climates, soils, and vegetation and hence *society*, the human world. Thus the study of landscape can involve the surface itself, what lies below, or the relationship between this part of the natural environment and human communities . . . all, of course, in spatial context.

• Planet Earth •

Ours is not a large planet. Its familiar dimensions—a diameter of nearly 8,000 miles (12,750 km) and a circumference of almost 25,000 miles (40,000 km)—seem tiny against the vastness of a universe in which distances are measured in millions of light years. And even in our particular corner of the universe, the solar system, the earth is no giant. The planet Jupiter has a diameter more than 11 times as large as Earth's. Saturn is much larger. True, Earth is slightly larger than Venus and Mars, but more than 70 percent of our planet is covered by water, making our living space smaller still.

What Earth lacks in size, however, it makes up in the wealth and variety of its contents. More than 6 billion years ago, when it was born from a gaseous mass of material that slowly collapsed and congealed, the earth was endowed with an almost infinitely complex chemistry. Throughout the planet's life, that chemistry was modified continuously. The molten earth mass, turning and cooling, differentiated into several shell-like layers, of which the crust is but one. Gases escaped and began to collect in low-lying areas as liquids, starting the formation of the oceans. Other volatiles created the earth's initial atmosphere. Eventually that atmosphere supported the beginnings of life, and late in the planet's history, Earth attained levels of oxygen content that made possible the evolution of complex life forms. Humanity is the culmination of this process, and human communities began to make the first organized use of the earth's accumulated natural wealth.

As was noted in Chapter 1, there are geographers who concern themselves exclusively with the physical earth, the surface and the forces that shape and sculpt it. These physical geographers study such subjects as the role of ice sheets in the creation of landscape in the upper Midwest, the functions of waves in the formation of coastal landscapes, the relationships between wind direction and the shape of sand dunes, and the effect of stream volume and velocity on the properties of river valleys. Sometimes such research is grouped together in a special field of physical geography called *geomorphology* (*geo* = earth + *morphology* = form or shape). We may call this field the study of the evolution of the natural landscape. If you have the opportunity to go out into the field with a geomorphologist, do not miss it. The landscape has much to reveal, and you will be amazed at the conclusions a physical geographer can draw from the appearance of a low ridge, a simple-looking hill, a creek, or even a road cut. To a student of geomorphology, every trip into the countryside can be an adventure.

One question to concern us is this: to be able to study the surface, the landscapes of the earth, how

much do we need to know about the underlying geology and the geo*physics* of the earth? The answer is that physical geographers often turn to geologists during their research. The surface of the earth, after all, is the outer skin of the **crust**, the uppermost layer of the planet. The crust consists of solid rocks, both on the continents and below the waters of the oceans. But liquid rock (lava) can penetrate through the vents of volcanoes and through fissures to flow out onto the surface. And solid as the crust's rocks may be, they are violently shaken and even fractured by earthquakes. To know the surface we should study the crust of our planet as well as the water, wind, glaciers, and waves that attack it from above.

THE INTERIOR EARTH

To understand the crust we need to know what, in turn, supports it from below. In fact, the interior structure of the earth is quite relevant to geomorphology. Beneath the crust, temperatures and pressures rise to such levels that the rock material becomes viscous (sticky, rather like hot tar). Chemical change generates the heat that keeps rocks in such a viscous or even molten state, and much of the earth's interior is continuously in motion. Resting on this unstable interior, the crust averages from about 6 to about 25 miles (10 to 40 km) in thickness and is thus vulnerable to failure and fracturing—as people living in earthquake zones or near volcanoes can affirm.

It is something of an irony that scientists have carried rocks back from the moon, but that no one has seen the inside of the earth itself beyond a mere 2 to 3 miles (3 to 5 km) from the surface in the deepest mine shafts. Boreholes for oil wells go deeper, but not enough to penetrate the crust, much less the layers below. Scientists must use delicate instruments, careful extrapolation, and considerable ingenuity to construct their models of the earth's interior structure. The key to their success comes from an unexpected source: earthquakes. Major earthquakes send shock waves not only through the crust, but also through the core of the earth itself. By monitoring the type and velocity of these waves, geophysicists have been able to deduce the kind and condition of the materials through which they travel as they penetrate the globe.

30

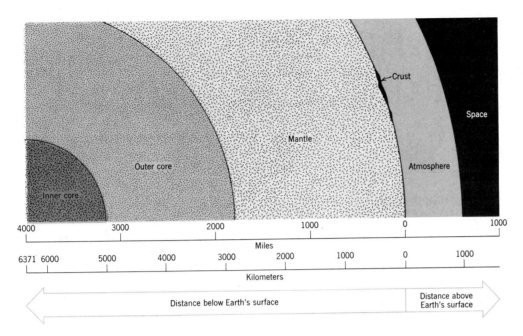

Miles

Kilometers

Distance below Earth's surface

Distance above
Earth's surface

FIGURE 2.1

Cross-sectional view illustrating the layered structure of the earth. The inner core, outer core, and mantle are drawn to scale; the crust and its features have a vertical exaggeration of three times.

In this way, the zones of the earth's interior (Fig. 2.1) have been identified.

As Figure 2.1 shows, the earth consists of a series of layers, with an extremely dense, heavy ball known as the **inner core** at its center. This solid inner core has a radius of about 780 miles (1250 km) and is surrounded by an almost equally heavy liquid layer forming the **outer core**. Incredible heat and pressure keep the heavy metallic material of this outer core in a molten state to a thickness of nearly 1400 miles (over 2200 km). Outside the core lies the **mantle** of the earth, where the rock material is lighter and less dense than in the core; the complex mantle contains zones of viscous and liquid matter as well as solid rock. Some earthquakes are known to originate as much as 400 miles (650 km) into the mantle, proving the existence of rigid material in this earth layer.

Overall, the mantle is about 1800 miles (2900 km) thick. It is a complex layer, and many questions about it have remained unanswered. In the late 1980s, scientific research based on more refined recording and interpretation of earthquake (seismic) waves began to unlock some of the mantle's secrets. It is known, for example, that the mantle's material is in continuous motion in giant convection cells. The moving material in these cells drags along the bottom of the solid crust and pushes and pulls pieces of the crust along. Seismic waves also indicate that in the mantle there is a significant change (a *discontinuity*) at a depth of some 420 miles (670 km) from the earth's surface. Do the convection cells that keep the mantle in motion stop at that level or do they cross it? The answer will soon be known—and perhaps that answer will involve the recognition of not one, but two mantles beneath the crust.

Whatever the outcome of this research on the mantle will be, it will contribute much to our understanding of the behavior and appearance of the crust, both geologically and geomorphologically. It is already clear that interior forces change the upper surface of the crust: volcanic eruptions and earthquakes prove that dramatically. But the slow, continuous movement of material in the mantle causes more subtle changes in the crust, slowly pushing, pulling, warping, and even bending it.

The Earth's Crust

The ancient Greeks called the crust of the earth the **lithosphere**, a remarkably good name: *lithos* means rock, and *sphere* comes from *sphaira* or ball. Thus the earth, to the Greeks, was a "rocky ball," and of course the scholars of old were unaware that the rocky crust is (compared to the whole planet) only eggshell thin.

That discovery was many centuries away, and not until earthquake records began to unveil the secrets of the interior was the thinness of the crust realized. Even now, there are questions about just how deep down the rocks of the crust remain solid and when do they yield to the properties of the mantle. The crust does not simply merge gradually into the mantle: a clear break in the physical state and chemical composition of the rocks occurs, and on this basis, the crust may be recognized as a distinct earth layer. This break was first identified by the Yugoslav geophysicist, Andrija Mohorovicic, who deduced its existence from earthquake data. The break between crust and

mantle became known as the Mohorovicić discontinuity, conveniently shortened to **Moho discontinuity**.

Compared to the heavy, dense material of the core, the rocks that make up the crust are very light. The upper layer of the crust contains the lightest rocks, the granitic rocks, and rocks derived from these granitics. Deeper into the crust, the rocks tend to become somewhat heavier. The upper layer's rocks contain a high percentage of *si*licates, many of them rich in *al*uminum. Using the first two pairs of letters, this becomes the **sial**, the crust's uppermost zone. The continental landmasses—North America, Africa, Australia—are made of sial. Below the sial, *si*licates continue to prevail in the rocks, but *ma*gnesium takes over from aluminum as the second most common substance, and (again, using the first two pairs of letters) here lies the **sima**. The sial that makes up the continents is discontinuous, for it does not extend into the ocean basins. The floors of the oceans are made of sima, and the sima extends below the sial under the continental landmasses (Fig. 2.2). Therefore, the crust under the continental surface is much thicker than it is below the oceans. The continental crust may be 25-miles (40-km) thick, whereas a sima-only oceanic crust may be a mere 6-miles (10-km) thick. At the base of the sima is the Moho discontinuity, marking the transition into the mantle.

Together the sial and the sima constitute the

FIGURE 2.2

Detail of the earth's crust, showing relationships between sial and sima, the position of the Moho discontinuity, and the postulated relationship between crust and mantle beneath continents and oceans.

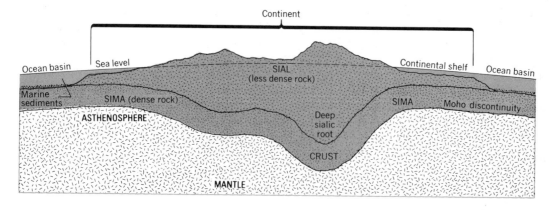

lithosphere, the layer of solid rock. This does not mean that no viscous or liquid rock material can exist in this layer (volcanic eruptions at the surface do occur, proving the existence of rock that is not in the solid state), but it implies that, as a whole, this outer shell behaves as a rigid substance. Below it, the upper mantle has the character of material in a softer, plastic form (not hard plastic as in a toy or model airplane, but in the hot, sticky condition it is before being cooled into those shapes). In the upper mantle, this hot plastic condition is beyond doubt; what happens deeper down in less certain. Geophysicists have called the upper layer of the mantle the **asthenosphere**, the weak and flexible layer of the mantle.

If you could descend in a deep shaft through the upper crust (say, to a depth of 40 miles [65 km] into the earth) and started your journey on a continental landmass, you would first see layers of lighter rock. These would change to heavier rocks (and they would be darker in color, too). All the time as you descend, temperature and pressure rise until eventually you arrive at the Moho discontinuity, a sign that you are leaving the crust. Penetrating the Moho, you enter the asthenosphere, a signal that the upper mantle has been reached. No longer are the rocks solid and distinguishable by their texture and color; it is now so hot that the earth's material is tarlike, with bubbles of equally hot (or even hotter) gas. But do not expect to see the flow in the asthenosphere. It proceeds, but very slowly.

ROCKS IN THE CRUST

Ultimately, the study of the natural landscape is the study of rock surfaces—how earth forces create and then deform rocks and how processes of weathering and erosion fashion them into the scenery we observe. Thus, physical geographers want to know the regional geology under the landscapes they study and, more particularly, the properties of the rocks they encounter there. Some rocks are known to withstand forces of weathering and erosion, whereas other rocks yield more quickly to water, ice, and wind.

A rock is a natural mass that forms part of the earth's solid crust. When we think of rocks, names such as granite and basalt and marble come to mind; but technically, even sand (on a beach) and gravel (in a river bed) are also rocks. Like all rocks, sand and gravel are also made up of **minerals**, the building units of rocks.

When you look at a rock under a microscope, these mineral components can be clearly seen. It is almost as though the minerals (such as quartz, feldspar, and mica) can be seen to struggle against each other for every last bit of space. And sometimes that is exactly what has happened when the rock formed. Certain rocks have formed from the cooling and solidification of hot, mineral-rich *magma* from the inner earth. As the temperature dropped in that hot mass, groups of minerals formed and occupied available space. Those that solidified first had ample space to form large, well-developed crystals. Those coming next had less space. Eventually there were only little corners left for those minerals that solidify at the lowest temperatures. By the time the magma had cooled into solid rock, every space was taken. The minerals are tightly packed, for example, as they are in granite. Such tight packing makes the rock strong and resistant to erosion; it can even withstand crushing from earthquakes. Old granites are among the earth's longest-lasting rocks.

Rocks that solidify from the molten state (e.g., from liquid lava around volcanoes) are called *igneous* rocks (igneous means origin by fire). Rocks also can form from the accumulation of sand, silt, clay, and even the skeletons of organisms (e.g., shells in the sea). These are the *sedimentary* rocks, and by and large they are softer and more quickly worn down than the igneous rocks. A third kind of rock develops when preexisting igneous or sedimentary rocks are changed by high temperature (e.g., when a lava flow covers the sand on a beach and bakes it into a hard layer) or high pressure (e.g., when a major movement in the crust, perhaps caused by a pushing force in the asthenosphere, crushes a rock mass). Such action creates *metamorphic* rocks (a good name for them since *meta* means change and *morphe*, again in the old Greek, means form or shape).

Igneous Rocks

Igneous rocks (as we have noted) result from the hardening of molten magma and consist of assemblages of tightly welded minerals that crystallized as the magma cooled. In granite, for example, it is known that the

33

feldspar crystals form first, and then, as the magma cools further, mica crystals develop. Finally, quartz crystals occupy the niches still remaining. When the magma cools very slowly, there is much time for crystallization to take place, and the crystals are large and well shaped. But when cooling is rapid, crystals tend to be smaller and more haphazard. Obsidian—one type of igneous rock—cools so fast that no crystals form at all, and a kind of natural glass results. From this glassy obsidian, Stone Age peoples fashioned sharp hand axes to skin animals, cut meat, and even to trim hair.

The formation of igneous rock from silicate-rich magmas can take place in various ways. Magma can flow out onto the surface as lava, and the resulting igneous rocks are described as **extrusives**. On the other hand, upwelling magma may never reach the surface, so that it hardens into solid rock below the bedrock. When this happens the rocks are called **intrusives**. Since the intrusives are surrounded by other rocks in a confined chamber, cooling there tends to be slower than in the free-flowing extrusives; intrusives are coarse-grained compared to most extrusives. Eventually intrusive rocks may appear at the crust's surface—not by force but through the erosion of the bedrock above. The great granite domes of Yosemite, for example, are thought to have formed as intrusive, bulbous, massive **batholiths** of solidified magma, later to be exhumed by erosion. The Palisades, overlooking the Hudson River between New York and New Jersey,

FIGURE 2.3

New York's Palisades, overlooking the Hudson River, are formed from the edge of an intrusive sill, exposed by erosion of the overlying rock. Redrawn from a sketch in Donn, *The Earth*, 1972.

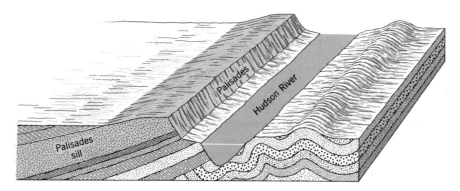

34

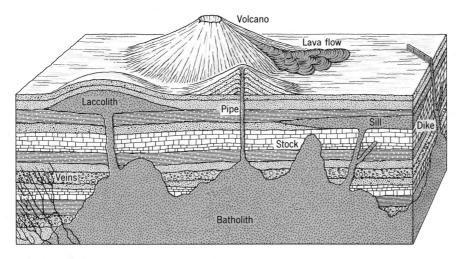

FIGURE 2.4

Intrusive and extrusive igneous rocks. Note the ridge formed by the dike at right.
Redrawn from a sketch in Donn, *The Earth*, 1972.

are part of an exposed sill or layer of basaltic rock that squeezed between existing strata but did not penetrate to the surface (Fig. 2.3). Sometimes the magma does not squeeze between layers or along existing cracks, but simply cuts through the rock, melting away obstructions and breaking sections of the bedrock apart. In this way, the magma creates a **dike**. When erosion reaches it and removes the softer rock around it, the dike stands out in the landscape as a persistent ridge (Fig. 2.4).

Igneous rocks such as granite and basalt are usually quite resistant to erosion, but they do afford the forces of weathering and erosion opportunities to attack. When the magma cools, it creates planes of weakness and separation called **joint planes**; jointing affects all igneous rocks (and other rocks as well). Jointing can be observed in many rock quarries where

Joints occur as natural planes of weakness in bedrock. The face of this crystalline mass reveals the blocks into which it will eventually break down. Near the upper surface, trees and smaller plants already have established themselves in the joint cracks, speeding up the weathering and erosion of this slope. The photograph was taken in the Rocky Mountains west of Boulder, Colorado.

even the freshest rock face, just exposed by the quarrying equipment, displays a pattern of lines that represent natural breaks. In granite, this pattern tends to be square, rectangular, and blocky; in basalt, it is hexagonal and columnar (hence the columns of the Palisades). These joint systems are the places where weathering attacks the igneous rocks. Moisture can penetrate here, and roots of trees wedge the blocks apart. Igneous rocks are hard, but they have their built-in weaknesses.

Sedimentary Rocks

When igneous rocks are weathered and their crystals loosened, erosion carries the loose particles away to deposit them elsewhere. Long-term deposition of loose material produces **sedimentary rocks**. Sedimentation occurs over wide areas and in many places—on the continental shelf, in mountain valleys, in river deltas, in desert basins. Eventually the accumulated sediment is compacted, or *lithified*, into harder rocks.

Most but not all sedimentary rocks are made from the particles of other rocks. The exceptions are chemical deposits such as salt beds and biological accumulations, among which the most important is limestone—formed from the deposition on the ocean floor of the tiny shells of marine organisms. Sedimentary rocks made from particles of other rocks are called **clastic sediments** (from the Greek *klastos*, broken) and range from extremely fine-grained shale to cobbly conglomerate. The lithification process involves compaction by weight and cementing by seeping solutions carrying binding agents such as silica or calcium carbonate.

Clastic sedimentary rocks are most conveniently classified according to the size of their grains, ranging from silt to boulders. The coarsest-grained of all the sedimentary rocks is conglomerate, a rock containing pebbles, cobbles, even boulders, which are all cemented together. Whatever their size, the fragments in a conglomerate are well rounded, indicating that they were moved quite a distance down a river valley or, perhaps, washed back and forth on a beach. Another common sedimentary rock is sandstone, in which the grains are sand-sized, usually made of silica (quartz) and cemented by a silica solution or by calcium carbonate. Often there are some open spaces left between the grains, and these spaces may be filled with water. Sandstone layers sometimes act as water reservoirs beneath the earth's surface; oil reserves also occur in porous sandstone. Neither sandstone nor an even softer and finer-grained sedimentary rock, shale, is normally very hard or resistant to erosion under humid climatic conditions. Shale is compacted mud, and the rock has a tendency to split into thin layers. In the Appalachian Mountains, the valleys often are underlain by shale or soft sandstone, and the ridges are supported by other rocks.

A fault plane can become a canyon. The fault that displaced these sedimentary strata in the Cape Ranges at the tip of the African continent pulverized the rocks in the contact zone. The loose fragments were soon removed by running water, and next the fissure thus created began to be widened. Low resistance, jointing, and high relief in the area contribute to the effectiveness of the denudational processes.

This butte in the Utah landscape, eroded from a previously larger mesa, illustrates the effect of strong, erosion-resistant caprock. In the surrounding area, erosion already has reduced the surface to a much lower level, but the process is slowed by the hard, protective caprock on top of this landform.

When sedimentary strata form scarps or stand as relatively steep slopes, it is likely that some stronger supporting layer (e.g., a sill of igneous rock) protects them. In many mesas and buttes in the dry country of the American West, the top layer (the **caprock**) is such a sill. In dry regions, limestone can act as a strong inhibitor of erosion as well and stands out as a **positive force** in the landscape, although limestone is susceptible to solution and disintegration in humid zones.

Sedimentary rocks provide vital information for the effort to reconstruct landscapes of the past. Unlike other rocks, sedimentary rocks contain fossils, and much of what is known about earth history is based on the fossil record. The color, texture, and structure of sedimentary rocks suggest the kind of environments that prevailed when they were deposited, permitting the reconstruction of climatic conditions that existed in parts of the world millions of years ago. By observing the deposition of sediments today and comparing what we see to the rocks already in the landscape, we can deduce the changes that have affected the older rocks. For example, sedimentary rocks are laid down horizontally, like a stack of magazines. When sedimentary strata are tilted, warped, bent (folded), or in some other way deformed, this must result from postdeposition disturbance. It was not part of their original condition.

Metamorphic Rocks

Metamorphic rocks are the rocks of change, of modification. They result from the alteration of other rocks through increased pressure and greater heat. Metamorphism can remelt granite and transform it into gneiss; it can bake sandstone, cement and all, into quartzite; it can squeeze shale into slate. Even metamorphic rocks themselves have been modified.

As a process, **metamorphism** is not yet completely understood. One cannot see it taking place as we can observe lava cooling and hardening into solid igneous rock. But the results are evident everywhere because metamorphism leaves telltale signs. One of these is a certain banding of minerals, so that crystalline rock that might otherwise be mistaken for granite is recognized for what it is: remelted and banded, or **foliated**, gneiss. And the effects of metamorphism can be seen adjacent to chambers of hot intrusive magma. Next to batholiths, for example, the bedrock is likely to be metamorphosed into harder rock; this is **contact** metamorphism. Metamorphism can affect vast areas, too. The ancient continental shields, which began as granitic cores, now have been metamorphosed almost everywhere. This is the process of **regional** metamorphism, in which time may be a crucial factor.

In the landscape, metamorphic rocks normally have strong resistance to erosion. Certainly the products of metamorphism of sedimentary rocks are much harder than the sedimentary rocks themselves. Quartzite often supports prominent ridges; sandstone does so under unusual circumstances or when protected by a caprock. Marble is much harder than limestone. And gneiss is certainly not weaker than granite. The overall effect of metamorphism is to strengthen existing rocks, although original structures may be destroyed beyond recognition.

When we study regional landscapes, the underlying rocks are key evidence. The size, shape, and orientation of fragments in sedimentary rocks can reveal whether water, wind, or ice laid them down. The direction of glacial (ice) movement tens of thousands of years ago can be disclosed by striations (scratches) made in hard igneous rocks (e.g., granite) as the ice dragged slowly across it. To interpret landscape, geographers want to know all they can about what rocks can reveal.

The Earth and Time

Rocks sometimes form almost overnight, for example, when a lava flow cools and solidifies. But other rocks are deposited over thousands, even millions of years. When geologists began to realize that there is a degree of order in the massive accumulations of rocks that make up the earth's crust, they realized that it was necessary to establish a timetable according to which newly identified rocks could be permanently labeled. The geologic time scale was first developed more than a century and a half ago by British geologists. They were able to see how a geologic cross section of England involved several discrete phases of accumulation of rocks and deformation by bending and breaking of the crust. They called the oldest sequence of rocks they were able to recognize the *Primary* and the next two the *Secondary* and *Tertiary*. Later the very youngest rocks such as sediments now accumulating in river deltas were separately identified as *Quaternary*. If we used these original terms, our geologic time scale would logically look like the following:

4	Quaternary	(Recent)
3	Tertiary	(Young)
2	Secondary	(Intermediate)
1	Primary	(Oldest)

Field research in the decades after the establishment of the first geologic time scale produced such a vast quantity of detail that it soon became necessary to subdivide these original major units. Bit by bit, the geologic time scale became more exact, and more complicated. More was learned about fossils and correlations were made with discoveries in distant countries. How important the discoveries pertaining to the evolution of life were can be seen from the names in the geologic time scale now in use: Primary has become *Paleozoic* (era of ancient life); Secondary is now *Mesozoic* (era of medieval life). We can still find Tertiary and Quarternary in the modern time scale, but they are now subdivisions of a third great era, the *Cenozoic* (era of recent life). It has also been realized that there are rocks older than the Paleozoic (older than about 600 million years); surprisingly, this era is not called the Prepaleozoic, as would seem reason-

TABLE 2.1   The Geologic Time Scale

ERA	PERIOD	EPOCH	MIL YRS AGO	EVENTS	FEATURES	AGE OF
CENOZOIC	QUATERNARY	Recent (Holocene)			Civilizations	Humanity
		Pleistocene		Glaciers overspread large areas in several stages	Rise of humanity	
	TERTIARY	Pliocene	2–3 6	Cascades and Sierra Nevada formed	Large mammals	Mammals
		Miocene	25	Large deposition of sediments on coastal plain of Gulf of Mexico and Atlantic Ocean; also in Pacific coastal areas	Monkeys Horses	
		Oligocene	40		Grasses spread; grazing animals	
		Eocene	60			
		Paleocene			Mammals develop	
MESOZOIC	CRETACEOUS		70 130	Rocky Mountains formed Mountain-building movements and great granitic intrusions in Sierra Nevada region	Dinosaurs extinct Flowering plants	Reptiles
	JURASSIC		190		Dinosaurs Birds	
	TRIASSIC		220	Appalachian Mountains formed		
PALEOZOIC	PERMIAN		270		First dinosaurs Primitive mammals	Amphibians
	PENNSYLVANIAN		310	Coal-bearing layers deposited	Conifer forests Reptiles spread	
	MISSISSIPPIAN		350	Much crustal activity and granitic intrusion in New England Much deposition of marine sediments; crustal folding in eastern and western North America	Sharks	
	DEVONIAN		400		Amphibians appear First forests Fishes develop	Fishes
	SILURIAN		440			
	ORDOVICIAN		500	Great sequences of deposition and crustal deformation; not well known	Marine life First vertebrates	Marine invertebrates
	CAMBRIAN		625			
	PRECAMBRIAN				First life forms	

39

able, but *Precambrian*, which means that it is older than the oldest period of the Paleozoic.

A simplified geologic time scale is given in Table 2.1. It is useful to become well acquainted with at least the eras and periods (and the epochs, too, for the Tertiary and Quaternary) because geologists and physical geographers use these terms about as commonly as we refer to the months of the calendar and the days of the week. The table shows that more is known about relatively recent geologic times than about older periods; subdivisions at the top of the table are more numerous, and they cover shorter time intervals. This is not surprising, for the youngest rocks tend to be closest to the surface. Some of them still are being deposited; others such as those layers left in the northern United States and Canada during a recent glaciation are fresh and have hardly changed, geologically speaking.

The time scale shows the three major *eras* divided into *periods*, which, in the case of the Tertiary and

Quaternary, are shown further subdivided into *epochs*. Even so, the time scale can be seen to cover only the last 600 million years of the earth's life in any detail. This may be less than 10 percent of the total lifetime of this planet. But in those 600 million years occurred momentous developments that shaped the landscapes of the world as we know it and witnessed the rise of mammalian life and humanity.

Surface and Space

We have now viewed the building materials of the earth's crust and are also mindful of the time factor in the evolution of landscape. Let us turn to some of those favorite spatial perspectives geographers like to take; we will do this, first, at the smallest of scales—the whole planet. How often have you looked at a globe, perhaps to find some prominent place or country? Look at the globe anew—this time with a more general approach. The globe has some striking distributional and morphological (shape) features that quickly lead us to ask *why*. Why do three of the continents have roughly triangular shapes, with the apexes pointing in the same southerly direction? Is it really an accident or does this reveal something basic about the structure of the crust? And why, when the earth is a sphere and everything on it so mobile, are the continents clustered on one half of it, the land

hemisphere, with a sea hemisphere that is virtually all ocean? And looking somewhat more carefully, why can we see long, linear mountain ranges on all the major continents except one, Africa? There is an Andes in South America, a Rocky Mountains in North America, an Alps in Europe, a Himalayas in Asia, a Great Dividing Range in Australia. Why is Africa's equivalent missing? And there is a question of location. Those elongated mountain chains are on the Pacific side of the continents. South and North America's major mountains lie in the west, on or near Pacific shores; Australia's ranges lie near the eastern margin of that continent, again on the Pacific side. Do the continents share a relationship with the Pacific Basin that generates this situation?

It may be, of course, that the answers to such questions must be found by scientists in disciplines other than geography. That does not matter. The important thing is to be alert enough to recognize the need for a solution in the first place. The questions we just posed have been attacked by geologists, geophysicists, paleontologists, physical geographers, and others. The focus of our discussion is on the geographic dimension.

CONTINENTS AND OCEANS

The continental landmasses that protrude above the ocean waters at present constitute just 30 percent of the total surface of the globe. It is appropriate to say "at present," because sea level varies through time,

FIGURE 2.5
Schematic (nonscale) drawing of the general features of continental margins and ocean basins.

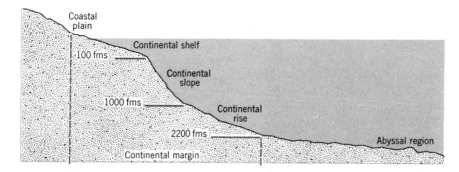

A landscape of high relief. In this area of Canada's Rocky Mountains, glaciers and rivers
have cut deeply into the original surface. High-relief landscapes can develop at low as well as
high elevations.

rising and falling as earth environments change. During the ice ages of the Pleistocene, vast quantities of ocean water were locked up in enlarged ice caps and sea level stood much lower—possibly more than 600 ft (as much as 200 m) below its present level. This lowered sea level exposed large plains adjacent to the continents, but today these areas are submerged again. These are the **continental shelves** (Fig. 2.5). The continental shelf is especially extensive around Eurasia and northern and eastern North America. If the shelf area were added to that of the continents, the continental landmasses would amount to 35 percent of the total surface of the globe. Since the sialic material extends to the edge of the continental self (called the **continental slope**), it is reasonable, in geologic terms, to view the continents this way. From other points of view, it is appropriate also: not only are the continental shelves among the world's richest fishing grounds, but also more and more of the resources below the shelf, in the **subsoil**, are being brought within human reach by advancing technology. There has even been talk of permanent underwater settlements. The continental shelves are part of the continents in more than one way.

The landscapes beneath the oceans will be discussed separately later. First we examine the topography and relief of the exposed continents. The term **topography** refers to the overall form or terrain of the surface, its general configuration. By **relief** is meant the vertical difference between the highest and lowest elevations in a given area; thus a mountainous area has high relief, whereas a plain has low relief. The term **elevation** denotes the altitude of a point or area above sea level, but it says nothing about relief. A plain, for example, may lie as high as 6500 ft (2000 m) in elevation but still have low relief; a mountain range's highest crests may never reach 6500 ft (2000 m), but its relief may be very high. This is the context in which we should see Figure 2.6, a generalization of the world's landscapes.

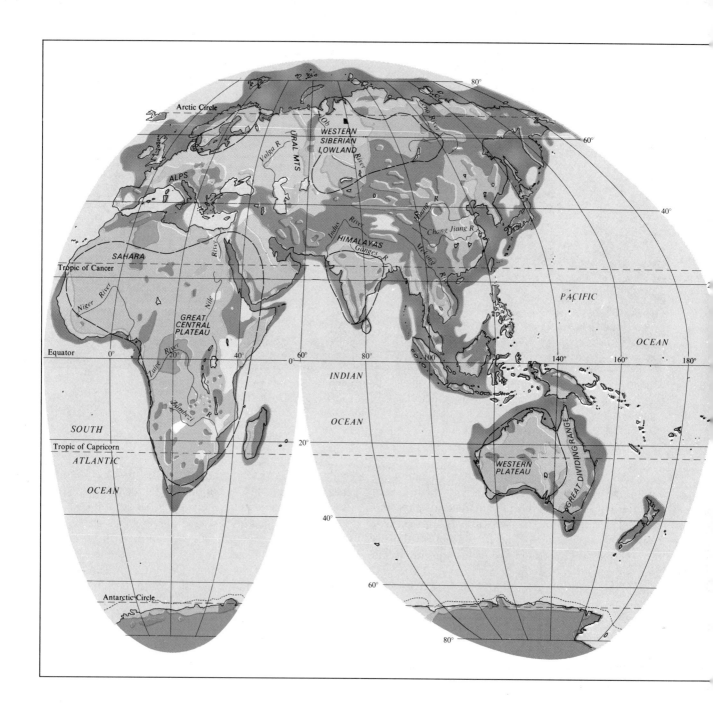

Arctic Circle

WESTERN
SIBERIAN
LOWLAND

80°

60°

Volga R

URAL MTS

Ob River

Yenisei River

Huang R

40°

ALPS

Indus River

HIMALAYAS

Ganges R

Chang Jiang R

Mekong R

SAHARA

Tropic of Cancer

Niger River

Nile

GREAT
CENTRAL
PLATEAU

PACIFIC

Zaire River

20°

40°

Equator   0°

0°   60°   80°

100°

140°

160°

180°

INDIAN

Zambezi

OCEAN

SOUTH

Tropic of Capricorn

ATLANTIC

20°

OCEAN

WESTERN
PLATEAU

GREAT DIVIDING RANGE

40°

60°

Antarctic Circle

80°

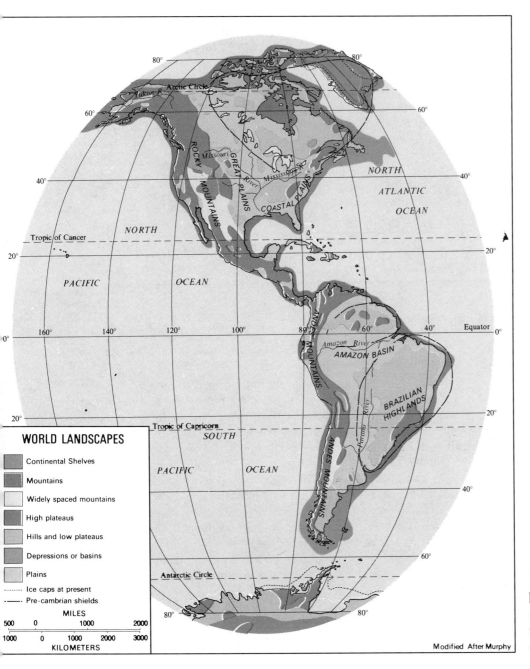

43

FIGURE 2.6
Landscape regions of
the world (modified
after Murphy).

WORLD LANDSCAPES

- Continental Shelves
- Mountains
- Widely spaced mountains
- High plateaus
- Hills and low plateaus
- Depressions or basins
- Plains
- .... Ice caps at present
- —·—· Pre-cambrian shields

MILES
500   0   1000   2000
1000   0   1000   2000   3000
KILOMETERS

Modified After Murphy

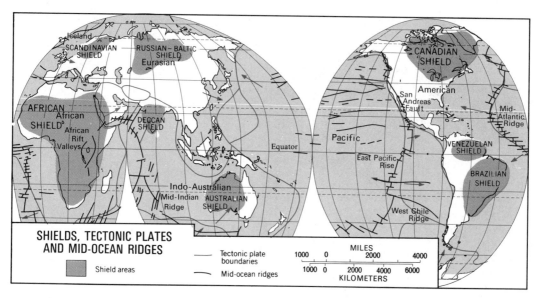

FIGURE 2.7

World distribution of shield zones, tectonic plates, and midocean ridges.

Shields and Orogenic Belts

The earth's landmasses consist of two basic geological components: shields and orogenic belts. These, in turn, are represented by two kinds of landscapes: plains and mountain belts. The **shields** form the cores of the continents. These vast shields are made of very old rocks. On the time scale (Table 2.1), they would lie in the Precambrian and thus be older than 600 million years, but most are well over 1 billion years old, and some shield rocks are thought to be between 3 and 4 billion years old. These rocks are mostly of the granitic type, though some have been altered by susequent volcanic activity. Apparently, the shield areas represent the original heartlands of the continents. Today they are typically low, undulating plains that bear witness to the lengthy erosion and degradation they have undergone. North America's core area is the *Canadian* Shield; Asia's is the *Russian-Baltic* Shield. In the Southern Hemisphere, all three of the continents have large shield areas. South America has two: the *Venezuelan* and the *Brazilian* shields. Africa's great central plateau is a vast shield that today supports an enormous upland, and Western Australia also is underlain by a major shield (Fig. 2.7).

Adjacent to the shields lie mountain ranges, also known as **orogenic belts**. Whereas the shield areas are the stable zones of the earth, the orogenic belts are places where the crust is deformed, where earthquakes occur frequently, and where volcanic mountains exist.

The rocks that underlie shield areas differ from those underlying mountain belts. Shields are crystalline (as we noted), but mountain ranges are in large part made of sedimentary rocks that, along with other rocks, have been crushed and folded into gigantic ranges. This happens during certain periods of earth history, and the term **orogeny** is used to identify these. For example, on the geologic time scale, where the formation of the Rocky Mountains is indicated the **Laramide orogeny** occurred. Just why orogenies have taken place throughout earth history during certain periods of strong activity and were followed by times of comparative quiescence is not clear. But it is now known that such mountain building takes place when great, rigid pieces of the earth's crust are moved by convection currents in the asthenosphere and collide. These rigid segments, called *plates*, move slowly—but not harmoniously. Also known as *tectonic* (structural)

44

plates, they sometimes carry whole continents or, when in an ocean basin, are the size of continents. When they collide, the "stronger" plate overrides the "weaker" one, folding, crushing, and pushing the rocks in the contact zone into huge, linear flexures. It is happening today along the east coast of Asia, where the plate that makes up part of the Pacific Ocean floor is pushing against the plate that constitutes much of Asia. In time, a great mountain range will rise along Asia's eastern margin; already, that process is going on along a line from north of Japan to south of the Philippines. Most of the topographic evidence still lies below sea level.

The Ocean Floor

A major reason that interest in the oceans and ocean floors has increased so much in recent decades lies in the existence under the seas of resources also found on land. Because the continental shelves are extensions of the sialic continents, they contain mineral deposits like those of the continents—including large oil reserves. Undersea explorations have greatly increased our knowledge of the relief of the ocean floors.

The continental shelves vary greatly in width. Off the shores of western South America, for example, the seafloor drops off within a few miles of the land, and there is almost no continental shelf at all. Off eastern North America, Western Europe, and Southeast Asia the shelf is hundreds of miles wide. The edge of the shelf comes at a depth of 400 to 600 ft (120 to 180 m), sometimes slightly deeper, where the seafloor drops off quite rapidly in a scarplike **continental slope**. This continental slope is thus the true limit of the continental landmasses, and it carries ocean depths down to an average of 6000 ft (1800 m). At the foot of the continental slope is a more gently inclined **continental rise**, signaling the beginning of the real deep-ocean floor. This vast ocean floor is marked by extensive plains, some of them at depths of as much as 15,000 to 20,000 ft (4500 to 6000 m) and others rising, like great submerged plateaus, to higher levels. These are the marine plains known as the **abyssal plains**; research progress notwithstanding, little detail is yet known about these regions. It is pitch-dark and totally quiet; there may be entire fish fauna about

which we know as little as we do about life on other planets. Even less known are the huge **foredeeps**, or **trenches**, that cut across the marine plains off East Asia, adjacent to western South America, and in isolated places elsewhere. Hundreds of miles long and tens of miles wide, these deep-ocean troughs reach more than 30,000 ft (9000 m) below sea level. Structurally they are the deeply folded rock strata along the collision zone between lithospheric plates. What the marine environment is like at those depths is less certain.

One of the most important discoveries relating to the ocean floor, made during recent decades, concerns the **midocean ridges**. It had long been known that linear ridges existed below the ocean surface, and scientists were best acquainted with the Mid-Atlantic Ridge. It was assumed that these ridges represented mountain ranges, as on land, but submerged beneath the sea. Now, however, the Mid-Atlantic Ridge is known to be only one segment of a worldwide system of midocean ridges that extends across the Indian and Pacific oceans as well. It is more significant that these ridges are quite unlike mountain ranges on the continents. They are, in fact, fracture zones in the earth's crust where subcrustal material wells up to the surface—below the ocean's water, that is.

This discovery is one of the most far-reaching scientific finds of the twentieth century, because it revealed where new rock is created. True, rock can be seen to form from volcanic lava, but not in the quantities needed to explain the map of plates and mountain ranges. We can see how rocks are folded, crumpled, and crushed along orogenic zones. When that happens, the folded rock takes up much less space than it did before it was pushed into folds (in the same way a towel takes up more space lying flat than piled in a heap). The question always was: What fills the vacated space? The midocean ridges gave the answer. There, new rock is made as magma wells up along giant fractures. As soon as it solidifies, it becomes part of the sima of the ocean floor, and it begins to move away from the midocean ridge, because newly upwelling, still-liquid magma forces it away. The earth functions like a giant conveyor belt along these midocean ridges, bringing "new" rock to the surface and pushing it away (Fig. 2.8).

That outward movement, in turn, has much to do

45

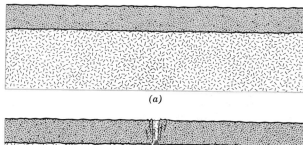

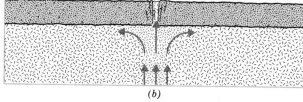

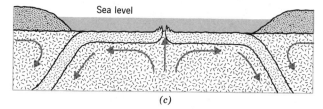

FIGURE 2.8

Crustal spreading and the separation of sialic landmasses. In (a) the predrift stage is represented. In (b) the sialic crust has begun to rift and separate; basaltic lava penetrates through the fissures. In (c) the fragments of the continent have drifted far apart; an ocean has begun to form between them. On the ocean's floor is the ridge marking the original separation. Redrawn from a sketch in Donn, *The Earth*, 1972.

with the constant mobility of the tectonic plates. Along the midocean ridges the movement is outward, but this means, inevitably, that plates will collide somewhere else. The map of tectonic plates (Fig. 2.9) shows where this happens. That map is a rather good guide to the earthquake-prone and volcanism-affected zones of the earth.

• Earth's Plates •

The construction of the map of the earth's **tectonic** (or lithospheric) **plates** has been a major scientific development, and it is of great significance not only to

geology and geophysics, but to geomorphology as well. We now know that the earth's crust is fragmented into a number of these plates, some very large, others smaller (Fig. 2.9). The plates are in motion, in some areas converging and colliding, in other places diverging and separating, and in still other locales sliding past each other. Major recent orogenic zones, earthquake belts, and volcanic activity are related in various ways to these movements.

If you have read popular scientific magazines over the past several years or so, you will have noted that opinions about the plates—their outlines, boundaries, thickness, and so on—continue to change. For example, the African Plate, once assumed to encompass not only the whole continent, but oceanic zones adjacent as well, now is known to consist of two plates (at least): the East African Plate and the larger African Plate. Between the North American Plate and the South American Plate lie two smaller plates, the Caribbean and Cocos Plates. Largest of all is the Pacific Plate, consisting almost entirely of rocks characteristic of the ocean floors. It is moving northwestward (hence the major subduction zone along East Asia's margin). The plate that carries Australia also supports India, but it has more oceanic than continental lithosphere. The Eurasian Plate rivals the Pacific Plate in size, but it is mainly continental in character. It spreads eastward from the Mid-Atlantic Ridge and moves generally eastward, so that its collision with the northwestward-moving Pacific Plate is especially intense. The Antarctic Plate, a piece of continental lithosphere encircled by oceanic crust, is almost completely surrounded by spreading midocean-ridge boundaries.

Thus the earth appears to have seven major plates: the North American, South American, Eurasian, African, Pacific, Australian (also called Indo-Australian), and Antarctic. The lesser plates are the Nazca (largest of these), Cocos, and Caribbean (all adjacent to the South American Plate); the Arabian and East African Plates; and the Philippine Plate. But do not be surprised if future maps show additional plates. There is some evidence that the African Plate may have other neighbors, not yet identified in 1988. For example, the very small Juan de Fuca Plate was recognized recently, squeezed between the Pacific and American Plates. A plate map representing a final consensus is some years away.

46

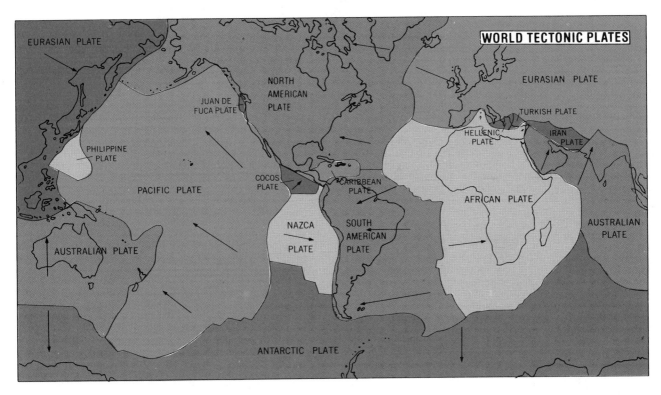

FIGURE 2.9
Tectonic plates.

Nor is the question of the thickness of the plates finally resolved. New interpretations of earthquake data (seismic records) indicate that continental "roots" of lighter material extend as far as 120 miles (200 km) down into the asthenosphere, perhaps much deeper. These roots may be the "sails" before the asthenospheric convection currents that push the continental plates along and create the great orogenic/subduction collisions. If this is so, then the average thickness of the crust may be greater than we have long believed.

EARTHQUAKES AND VOLCANOES

The earth's crust is in constant motion, imperceptibly in some areas, palpably in others. As Figure 2–10 reveals, the earth's most quake-prone belts surround the Pacific Ocean and cross Eurasia along the Alps and the Himalayas. The midocean ridges can also be discerned as belts of frequent earthquakes. The shield areas of the continents, on the other hand, are much less affected.

Earthquakes originate within the crust as well as the upper mantle, but most begin within 3 miles (5 km) of the surface. The point of origin is the earthquake's **focus**, and the location directly above this focus, at the surface of the crust, is the **epicenter**. The map of earthquake distribution (Fig. 2.10) shows the recorded epicenters for a series of years. An earthquake results from the sudden movement of rock that has been subjected to prolonged stress. When two lithospheric plates collide, stresses are set up that cause certain rocks to fracture. Such fractures in the crust are called **faults**, and some faults such as the San Andreas Fault in California are well known as the source of repeated severe earthquakes. It was a movement along this fault that caused the 1906 earthquake that destroyed much of San Francisco.

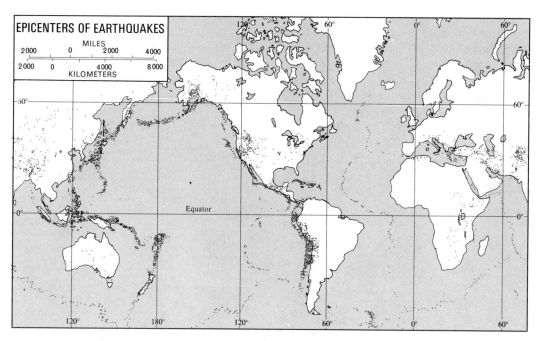

EPICENTERS OF EARTHQUAKES

MILES
2000   0   2000   4000
2000   0   4000   8000
KILOMETERS

Equator

FIGURE 2.10

World distribution of earthquake epicenters.

48

Repeated earthquakes along a fault zone can produce cliffs called **scarps**, and if the movement along the fracture continues, a major **escarpment** may develop. Earthquakes also generate landslides that block streams and change the character of river valleys. Occasionally a powerful earthquake with a submarine focus creates a mighty ocean wave, or **tsunami**, capable of doing severe damage to coastal settlements.

The field of seismology (a branch of geophysics) has contributed much to the unraveling of the mysteries of the earth's interior. In 1935, Charles F. Richter, seismologist at the California Institute of Technology, devised a scale of earthquake magnitudes that is still in use. It ranges from 0 (the smallest recordable tremor) to 9, and the numbers represent the calculated energy released at the earthquake focus. Earthquakes measuring from 0 to 4 are minor, from 4 to 7 moderate, and over 7 severe and destructive. Quakes with a magnitude over 7 are recorded all over the world, and these severe shocks generate the waves that penetrate the globe and permit analysis of the interior. The 1906 San Francisco earthquake is esti-

mated to have had a magnitude of 7.8, and the 1964 Alaska earthquake about 8.5. Even though Anchorage was severely damaged, this earthquake's epicenter was 75 miles (120 km) from the city. In 1976, an earthquake with a magnitude of 8.4 struck east of Beijing, the capital city of China. This was one of the century's most destructive earthquakes.

When such a major earthquake occurs, several different kinds of seismic waves are radiated in all directions. First come the "push-pull," or primary, waves that travel through the crust, through the mantle, and even through the liquid outer core (although they are refracted in the process). Next the "shake," or secondary, waves, very damaging and destructive, reach the seismographs—but not those in a certain zone around the opposite side of the earth. The "shake" (also called shock) waves are not propagated through fluids, and because they fail to reach that distant zone, a liquid core can be assumed to exist inside our planet.

A comparison between the map of earthquake incidence (Fig. 2.10) and one of world population distribution (see Plate 5) indicates why so many peo-

The San Andreas Fault zone in the San Francisco area is clearly visible on this Landsat image. Downtown San Francisco lies on the peninsula in the upper left of the picture; the fault passes south of the Bay and enters the Pacific Ocean to the west.

Devastated San Francisco shortly after the 1906 earthquake. The earthquake of 1906 is estimated to have had a magnitude on the Richter scale of 7.8. Although buildings are stronger and precautions better, a similar quake today would nevertheless cause enormous damage and loss of life in the densely populated Bay Area.

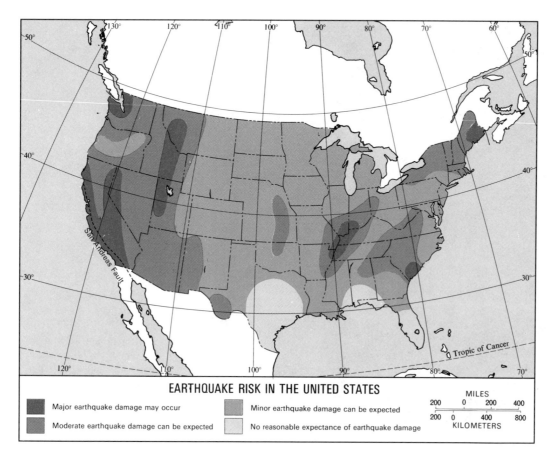

FIGURE 2.11

Regions of potential earthquake damage in the United States.

ple are killed by earthquakes. The high-incidence zone lies through regions of dense population in underdeveloped countries where buildings are often weak and susceptible to collapse. But loss of life occurs even in areas where buildings are modern and strong. This was demonstrated by the devastating earthquake that struck Mexico on September 19, 1985, with a magnitude of 8.1 on the Richter scale. This earthquake's epicenter was in the zone off the west coast of Mexico where the Cocos Plate is sliding under the North American Plate. The "shake" or shock waves did comparatively little damage in the regions of igneous and metamorphic rock of the west-

ern mountains. But when they reached Mexico City, the capital city built in large part on alluvial sedimentary strata, the ground shook so violently that many tall buildings collapsed. More than twenty thousand people died and twice that number were injured. If an earthquake of magnitude 8 were to occur near San Francisco or Los Angeles, there also would be much destruction and loss of life. Figure 2.11 indicates the earthquake risk in various areas of the United States. Where the risk is greatest, especially in the populous West, earthquake prediction is a matter of urgency, and two promising research directions are being followed. The first involves the detection of recurrent

such strain can be measured and breaking points predicted, an oncoming earthquake may be forecast and the evacuation of threatened areas ordered. But the science of earthquake prediction is still in its infancy and hundreds of millions of people continue to live under a continuous threat.

Volcanic Landscapes

The world distribution of volcanic activity (Fig. 2.12) shows a considerable similarity to world earthquake patterns—so much so that it is appropriate to speak of a **Pacific Ring of Fire**, a Pacific Ocean-girdling zone of crustal instability where volcanic eruptions as well as earthquakes are common. **Volcanism** occurs when rock in the molten state flows from the earth's interior onto the crust's surface through vents or fissures.

The mountains created by volcanic activity are classified according to their shape and appearance that, in turn, are related to the material from which they are built. This material comes to the surface through a narrow conduit called a **pipe**, which leads to the crater. The molten rock, called **magma** while it is still below the surface and **lava** when it flows out, may be fluid and flow easily or, depending on its composition, it may be pasty and pile up near the vent. When the lava is fluid, the volcanic cone is broad, with very gently curved slopes; such volcanoes are called **shield volcanoes**. The Hawaiian Islands are the tops of such shield volcanoes, Mauna Loa rising from 15,000 ft (4500 m) below sea level to 14,000 ft (4200 m) above.

In other volcanoes, the lava is more viscous, containing pent-up gases that often explode, sending large amounts of debris hurling into the air. These **pyroclastics** (from ancient Greek *pyr*, fire, and *klastos*, broken) fall on the mountain's slopes and are covered by layers of the slow-flowing, gluey lava. The resulting volcanic cone tends to be steeper-sided than the shield volcano. Made of layers of pyroclastics alternating with lavas, such a mountain is known as a **composite volcano**. Famous volcanoes such as Vesuvius in Europe, Kilimanjaro in East Africa, Fuji in Japan, and Hood in North America are composite volcanoes.

51

Central America's series of volcanoes are part of the Pacific Ring of Fire. Here in Nicaragua, a small town, laid out by Spanish planners in rough squares with a central *plaza* and cathedral, lies at the foot of an active volcano. Lethal gases from the central vent could reach the inhabitants in minutes. But the soil is productive, and farm fields crawl up the mountain's slopes.

tremor patterns that may precede severe shocks. More sensitive recording equipment and computer calculations may lead to a recognition of such warnings of nature. The second concentrates on measuring the growing strain on rocks in earthquake-prone areas. If

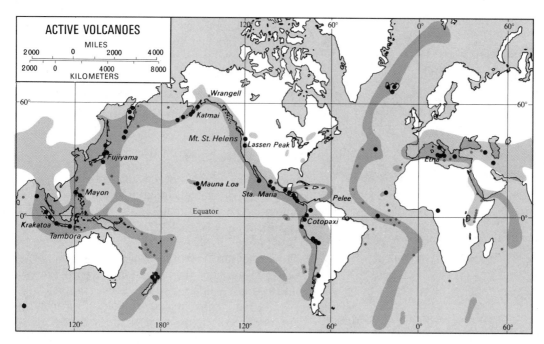

FIGURE 2.12

World distribution of active volcanoes.

Fujiyama, Japan, is one of the world's great volcanoes, a composite cone.

A Mobile Lithosphere

It is becoming more evident that the study of natural landscapes is a complicated matter. Already, it is clear that the surface expression of the crust is affected by various kinds of movement: the horizontal motion of the tectonic plates, the collision and subduction process (where one plate slides under the other, being reabsorbed in the mantle), the continuous trembling and shaking caused by tremors and earthquakes, and the eruption of lava from vents and fissures.

Nor is this all. Other forms of movement also affect the crust, less dramatic and not as visible as, say, a mountain range building along a zone of plate contact, but of major consequence where the evolution of landscapes is concerned. These movements may be combined under the rubric of *crustal equilibrium*. There is evidence that the crust and mantle are in dynamic adjustment—when there is nothing to disturb that adjustment. But because the lithospheric plates are constantly in motion and because conditions at the surface also change, that equilibrium is frequently disturbed.

How does it work? Take, as an example, an old, long-established delta. For millions of years, a major river (say, the Mississippi) has been depositing sediment at its mouth, enough to build mountains and to clog the river's mouth. But nothing of the sort happens. The delta may grow larger, but its landscape stays the same: low-lying, swampy, crossed by many channels. The crust below adjusts downward, sagging slowly as the weight of the sediments increases. At the surface, things stay pretty much the same. But below, adjustment takes place to restore the equilibrium between crust and mantle.

Can a reverse process be imagined? Does the crust ever become lighter rather than heavier? Here are two possibilities. When a mountain range is eroded down and the sediments thus produced are laid down on an adjacent plain, two kinds of adjustment take place: downward (as in the delta) where deposition takes place; upward where erosion is making the lithosphere lighter. A second possibility involves the advance and retreat of glaciers. Some tens of thousands of years ago (very recently, geologically speaking), thick ice covered the area of today's Great Lakes. That ice weighed heavily on the crust, and the lithosphere responded by adjusting as in the case of the delta. Then the ice melted, and the crust rebounded—it actually rose somewhat after the ice disappeared. The whole drainage pattern (and, thus, the regional landscape) of the Great Lakes area is affected by this movement, a movement that cannot be perceived by the eye but which has far-reaching consequences (Fig. 2.13).

The principle involved here has been called **isostasy**, and it is but one of several forms of shifts and movements in and on the crust. One important conclusion is that continental topography probably has a matching profile down deep: high orogenic mountain ranges have "roots" of sial that extend deeper into the asthenosphere than lower plains do. Look again at the cross-section on Figure 2.2. Note that the deepest "root" of the landmass lies beneath the highest mountains. What would happen if, during horizontal plate movement, that mountain range and its root became dislodged, out of balance? A major movement would take place in the lithosphere to restore the equilibrium. That movement would tilt plains, change the speed of rivers, and modify whole landscapes.

DIASTROPHISM

Movements in the crust that lead to deformation of rocks in the lithosphere are cases of *diastrophism*. These range from the gentle tilting or slight warping of a large segment of a tectonic plate to the folding (bending) and faulting (breaking) of rocks. When the plates move, they sometimes rise up slightly toward their leading edge, and this creates an equally slight tilt at the surface. It is suspected that the asthenosphere's convectional currents may not provide the same support under all parts of the plates; so, part of a plate may "dimple" just a little, perhaps temporarily. Such changes are instances of *epeirogeny*, the tilting of a large area but with very little crustal deformation.

Does epeirogeny have much effect on the landscape? At first glance, it may not seem so. But even a very slight tilt will increase the speed of flow of a river's water. That will increase that river's capacity to erode. This, in turn, will permit its tributaries (smaller rivers that flow into the main river) to erode more

53

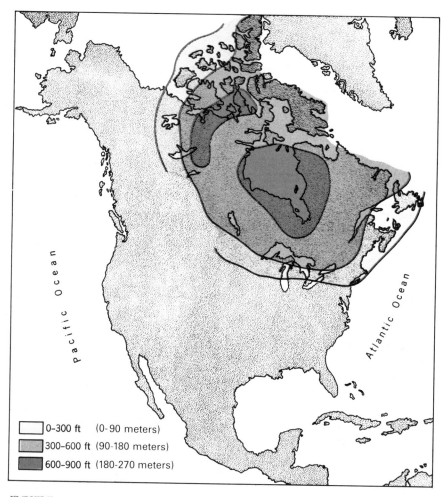

0-300 ft   (0-90 meters)

300-600 ft (90-180 meters)

600-900 ft (180-270 meters)

FIGURE 2.13

Map showing the rebounding of the earth's crust after melting of the ice of the last glaciation. (Melting began eighteen thousand years ago and was completed six thousand years ago.) The region of maximum uplift over Hudson Bay indicates the region of maximum ice depression of the crust and hence the region of thickest ice. (Generalized after W. Farrand and R. Gajda, *Geographical Bulletin*.)

actively, too. The rivers carve deeper; slopes become steeper. In time, the whole landscape may change from a nearly flat plain to rolling hills. Epeirogeny can do it.

If epeirogeny causes a "dimple" in the crust, that slight depression might fill with water. The water has weight, and isostatic adjustment begins to occur. When more water fills the basin, the crust yields still more. Eventually that lake can take on major proportions. Geomorphologists believe that Lake Victoria,

Africa's greatest lake, may have just such origins. At one time, local sediments prove, there was a slightly domed upland here; then its asthenospheric support failed, and a basin formed. Soon that basin filled with water and sediments.

But diastrophism associated with orogeny and plate collision is much more vivid. Rocks can be creased, wrinkled, pleated, and bent as though they were mere stacks of towels. When forces from two sides put pressure on horizontal layers of sedimentary

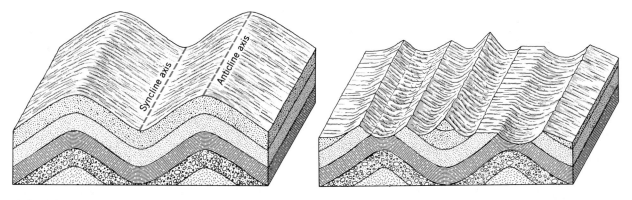

FIGURE 2.14

(a) Three-dimensional (block) diagram showing two uneroded upfolds or anticlines with an intervening downfold or syncline. (b) Erosion of the geologic structure results in a series of parallel ridges and valleys that form over the resistant and weak rock strata, respectively. Redrawn from a sketch in Donn, *The Earth*, 1972.

rocks, those rocks (e.g., sandstone, shale, limestone) will yield by *folding*. Part of the area will fold up, like an arch, into what geologists call an **anticline**. Other parts fold down into a basinlike **syncline**. Mountain ranges such as the Alps and the Andes have numerous series of anticlines and synclines, showing the force of the opposing pressures. Older mountains such as the Appalachians, where the orogenic activity has stopped, still retain their folded structures and thus give evidence of an earlier period of active diastrophism (Fig. 2.14).

What happens when rocks do not yield by bending or folding? Softer sedimentary rocks can fold in response to pressure, but harder, crystalline rocks—igneous rocks and metamorphics—are less able to do so (they can warp over large areas, but not bend into tight folds). Their response is to break, to fracture along faults, as we noted in connection with earthquakes (Fig. 2.15). Often you can see faults exposed in the landscape, because once the fault formed, the pressure was released and one side dropped down to create a scarp. And, of course, faults also can form

55

FIGURE 2.15

(a) Sketch of the relative motion and topography resulting from a normal or gravity fault. The resulting mountain is often called a *block mountain*. (b) Stream erosion quickly modifies the appearance of a fault mountain even as uplift occurs, which results in this more typical appearance. The mountain is characterized by a straight base that indicates the fault trend.

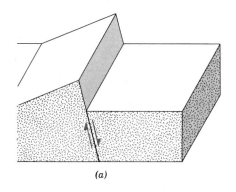

(a)

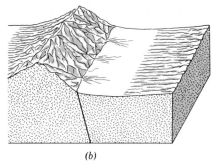

(b)

A Mobile Lithosphere

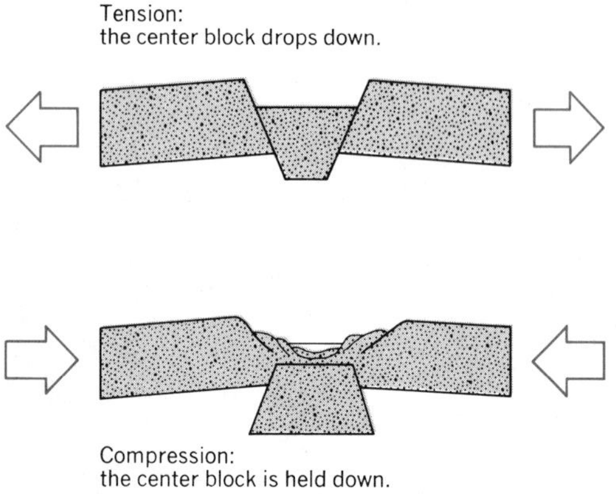

Tension:
the center block drops down.

Compression:
the center block is held down.

FIGURE 2.16

Cross section of tensional and compressional rift valleys in the crust.

when rocks are being pulled away from each other. When this happens, parallel faults sometimes develop, and the rocks between the faults slide down. This creates a valley not made by a river but by faulting—a **rift valley** (Fig. 2.16). The role of diastrophism in fashioning the landscape obviously involves many aspects. Such structures, as we will see, also can help in the interpretation of past geomorphic events, as in the case of the history of Africa's Lake Victoria.

• Lessons from the Continents •

Although the scientific developments that have produced information and theories on midocean ridges, crustal plates, and subduction zones result mainly from careful observation of recorded data, study of the configuration of the continents still serves a scholarly purpose. As we noted earlier in this chapter, the globe confronts us with some intriguing questions. One of these was noted more than a century ago: the way the continents, when cut out and reassembled, "fit" into a jigsawlike pattern. At the time, it was inconceivable that whole continents could move. Also, the physiography and geology of the continents were not well known. But geographers kept studying landscapes,

and geologists unraveled the rock alignments. Soon the issue (to many scholars) was not *whether* the continents had moved away from earlier relative locations, but *how*.

A marvelous example of the lessons landscapes can teach us relates to our earlier observation about the distribution of shield areas and orogenic belts. Among the continents, we noted that Africa has much shield and very little evidence of large-scale orogenic activity. Only in the Atlas Mountains in the north and in the Cape Ranges in the far south are there landscapes comparable, say, to the Andes and Appalachians. From Figure 2.17, we conclude that most of the continent is relatively flat or undulating country.

When we examine the map of Africa in greater detail, that conclusion turns out to be essentially correct, but with a qualification: Africa's shield areas are not low-lying as are most shields in the Northern Hemisphere, but plateaulike. In fact, much of the continent is rimmed by a scarp that leads rather suddenly (sometimes in a series of steps) from this plateau to a very narrow coastal zone (Fig. 2.17). In Africa, this is called the Great Escarpment, and in various degrees of prominence, it can be found from West Africa all the way around the continent to beyond Ethiopia. It is noticeable, though, that the escarpment is generally steeper and the plateau higher in the east than in the west. It is as if Africa were lifted up in the east and, perhaps, pushed down somewhat in the west. This is the process we encountered previously as epeirogeny—here it affects an entire landmass.

If Africa is a plateau continent, it is certainly no ordinary plateau. A textbook definition might describe a plateau as an area with a level surface, raised sharply above the surrounding country, or simply as a tableland. But look again at that map of Africa, *any* map that shows the continent's elevations. Note that the African plateau, the shield, has somehow been depressed into at least five major basins, depressions so large that they dominate the whole continent. The names given to these basins (Fig. 2.17) remind us that these are features of world significance. Such basins do not form simply because the crust collapses. Something drags or pushes the area down. What was it in Africa?

Still another feature of Africa's landscape provides us a clue. All continents have rivers, but Africa's rivers

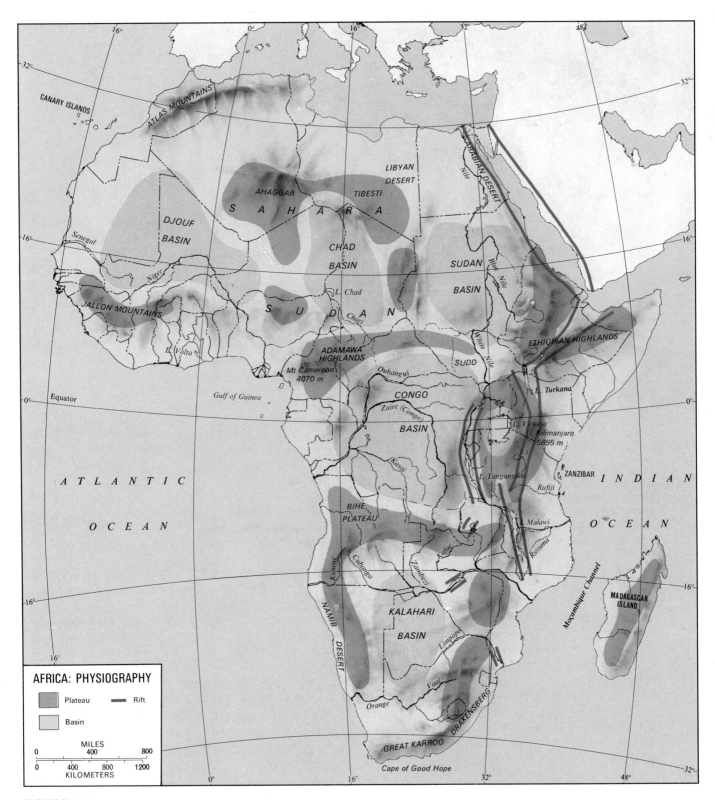

FIGURE 2.17

Major physiographic features of the African landmass.

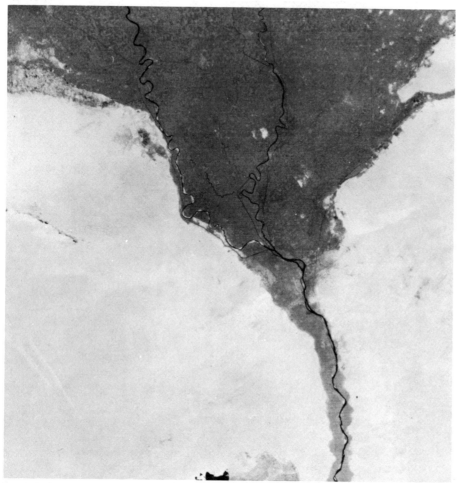

Two of Africa's major rivers, the Nile and the Niger, generate large deltas. This is a satellite view of the delta of the Nile River, showing the Nile as it emanates from its valley and divides into two major (and many minor) channels before entering the Mediterranean Sea.

deserve a second look. Take, for example, the Niger. It starts in the Futa Jallon Mountains, near the West African coast, and then proceeds straight into the Sahara. Near the edge of the Djouf Basin, the river divides into numerous channels and forms what is, in effect, an interior delta. Then the river gathers itself into one channel again and, in crossing the basin, turns sharply to sea. In its lower course, it plunges over several major falls, and then it forms another delta, this time on the Atlantic Ocean shore.

So the Niger has two deltas. What about that most famous of delta rivers, the Nile? It, too, has two deltas, though very different ones. The river spreads into almost innumerable channels in an area called the Sudd (Fig. 2.17), which happens to be near the edge of another of Africa's major basins, the Sudan Basin. Then it travels across that basin, turning south in the process, and flows over a series of falls and cataracts, finally to form its more famous delta on Africa's Mediterranean coast.

In equatorial Africa, the Zaïre (formerly Congo) River mirrors several of the peculiarities of the Nile and the Niger. The most obvious of these is its course. Rising in the southern part of the Congo Basin, the Zaïre River commences its course by flowing northeastward. It then proceeds northward for more than 600 miles (1000 km) before turning westward. Eventually it flows in a southwesterly direction, is joined by its major tributary, the Ubangi River, and cuts through the plateau-edging Crystal Mountains to reach the Atlantic Ocean.

The Zambezi–Cubango river system's upper reaches are oriented toward the Kalahari Basin, not the Indian Ocean coast toward which the Zambezi eventually flows. The remnant of a large delta on the edge of the Kalahari Basin still exists, but the Zambezi elbows northeastward, plunges over the Victoria Falls, and forms another delta on the Indian Ocean, in Moçambique.

If only a single river were to display the peculiarities of the Niger or the Zambezi, it might be ascribed to a particular, perhaps regional, set of circumstances. But the entire drainage pattern of all of Africa is involved, with several interior and exterior deltas, course reversals, and midcourse waterfalls. Any explanation, therefore, would have continentwide implications and must link Africa's Great Escarpment, its plateau character, the huge interior basins, as well as its drainage system. The obvious hypothesis is that, earlier during its geologic history, Africa did not possess the coastlines it has today. Surrounded on all sides by land, Africa lay at the heart of a much larger landmass; its rivers could not drain to the sea. They filled low-lying areas with water and sediment, and these areas were depressed isostatically as the weight of water and sediment increased. Eventually the rivers flowed into large interior seas, creating deltas on the coastlines. Then the huge landmass began to break up, and African coastlines were formed. The Great Escarpment marked the fracture, and rivers started to flow over the new African periphery. Before long, the great interior seas were drained and the interior deltas left high and nearly dry, but new deltas formed along the modern African coastline. Thus, today's tortuous rivers represent two drainage systems, not just one: the prebreakup and postbreakup segments are connected.

CONTINENTAL DRIFT

A South African geologist named A. L. du Toit was the first to observe African landscapes and to draw the conclusion just described. He was not the first to propose the idea that the earth's continents might not be permanently located where they are today, but he did produce the earliest well-researched, systematic study of African geology and landscapes in support of continental drift. Du Toit looked at more than just Africa's precipitous coastline, its basins, and its odd rivers. These gave the hint. Now what was needed was research into geology, the layering of the crust in Africa and the continents that once were supposed to have been adjacent to it, especially South America. Speaking of landscapes, who could escape noticing the amazing similarity between the bulges and indentations of the sides of the African and South American continents that face the South Atlantic Ocean? Du Toit wrote several books, including *A Geological Comparison of South America with South Africa* (1927) and his memorable *Our Wandering Continents* (1937). He meticulously dissected and compared what he found on opposite sides of the Atlantic Ocean, finding that there were fossils that strongly indicated a former union, that the downfolds and upfolds in the Cape Ranges' rock strata (the **synclines** and **anticlines**) were repeated in Argentina's Sierra de la Ventana, and that the proposed supercontinent was at one time glaciated.

These discoveries, indeed, provided tests for the hypothesis that the continents were at one time assembled approximately as shown in Figure 2.18. Any continent, large or small, undergoes changes during its lifetime. Rivers attack it and seek to erode the higher parts down to lower levels. Ocean waves erode the coasts. Loose material is ground down into fine sand and deserts and dunes are created. And climates also change. Thus, a supercontinent as large as Gondwana (Fig. 2.18) should, during its lifetime, have undergone glaciation. When the evidence for glaciation was found in southern Africa and similar evidence came from South America and India, the idea that drift might be reality received an enormous boost.

Certainly not everyone believed that continental drift could possibly be taking place. Even when du Toit published his evidence for similar geological for-

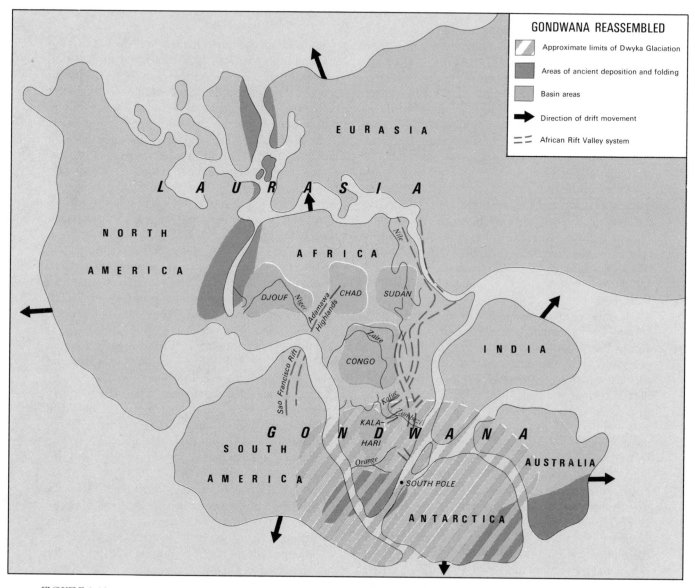

FIGURE 2.18

Relative location of the continental landmasses before the latest episode of fragmentation commenced.

mations on opposite sides of the South Atlantic Ocean, for fossil distributions, and for glaciation, doubts remained. This was partly because of the way the hypothesis had been proposed several decades earlier in *The Origin of Continents and Oceans* (1915) by Alfred Wegener (Fig. 2.19). Wegener had com- bined hunches, inadequate evidence, insufficient data, and a proposal for a mechanism that could easily be disproved—and he had been roundly crit- icized. Although his book was the first modern state- ment of the hypothesis, it probably set the idea back. But du Toit was not discouraged. In articles and mono-

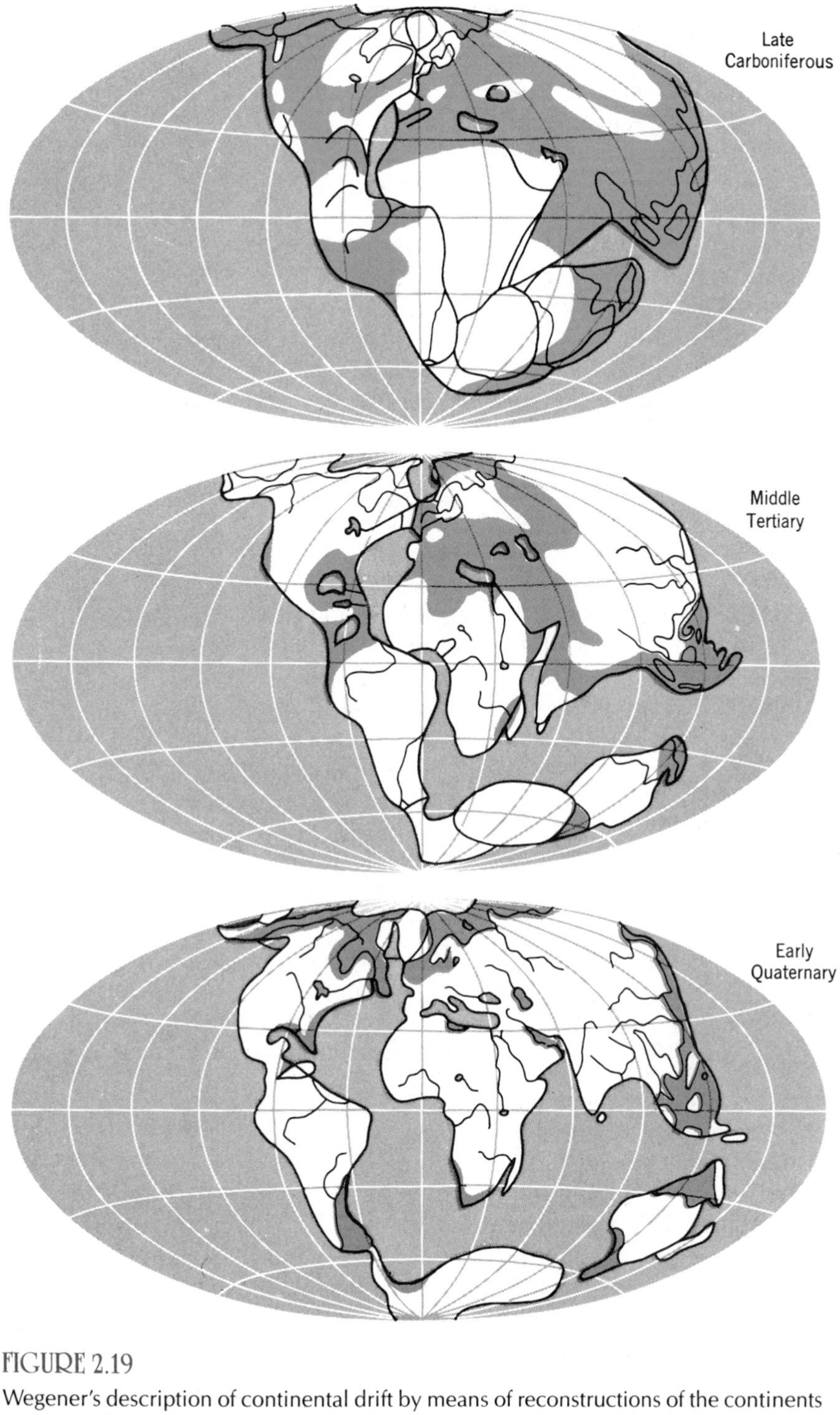

FIGURE 2.19

Wegener's description of continental drift by means of reconstructions of the continents at three stages in the drift process. The first stage shows the hypothetical predrift parent cotinent (*Pangea*). Continental areas covered by shallow seas are shaded. (A. Wegener, *The Origin of Continents and Oceans*, E. P. Dutton & Co., N.Y.; and Methuen & Co. London.)

graphs, he painstakingly built the geologic and geographic case for drift, reminding his readers that the world distribution of mountain ranges suggested that the continents had drifted away from Africa, crumpling and folding their leading edges into the great Pacific coast mountains. As they drifted away, the African continent attained its escarpment, the great basins were drained of their water by newly formed rivers (the enormous volume of water carved great waterfalls below the basins), and the great rift valleys were formed.

RIFT VALLEYS AND MIDOCEAN RIDGES

A closer look at the map of Africa's physiography (Fig. 2.17) leads us to examine the series of elongated lakes that extends from Ethiopia in the northeast (where Lake Rudolf, now called Lake Turkana, is the largest) to Moçambique in the south (where Lake Malawi approaches the Zambezi Valley). With the exception of massive, shallow Lake Victoria, the African great lakes are attenuated, oriented north-south (or nearly so), and contained between steep-sided walls that form deep troughs.

The lakes occupy parts of Africa's **rift valleys**, a system of trenches that extends from the Red Sea to South Africa over a distance of nearly 6000 miles (nearly 10,000 km). In East Africa, the system divides into an eastern rift and a western rift (Fig. 2.17). The Red Sea itself is the northernmost extension of this great system, which is marked, in Africa itself, by its uniformity. From Lake Turkana (Rudolf) southward, the rifts are between 20- and 60-miles (30- and 90-km) wide. Almost everywhere, the walls are steep and well defined, sometimes sheer, sometimes steplike. The landscape of the rifts is unmistakable. When the geologist J. W. Gregory first saw the rifts early in this century, he assumed that strips of the African shield had fallen down between parallel faults (Fig. 2.16). Later, it was found that some of the "floor" segments have not simply collapsed, but are being held down or pulled down. But not until very recently did it become clear that the rift valleys have a relationship with the midocean ridges. As Figure 2.7 indicates, the midocean ridge that forms the contact between the postulated Australian and African plates extends between India and East Africa and then bends into the Gulf of Aden—exactly where the Ethiopian rift begins.

It is difficult to avoid the conclusion that the East African rifts and the Red Sea are manifestations of the same divergent plate contact that marks the midocean ridges.

For this and other evidence derived from the physiographic map of Africa, it may be premature to assume that an African plate (such as that shown on Figure 2.7) exists. The rift valleys appear to signal a separation of parts of East Africa from the rest of the continent, just as the Arabian Peninsula is now separating (an ocean will form where the Red Sea now lies), and Madagascar has moved away. Because the plates diverge along midocean ridges, geologists have called the drift process seafloor spreading—but let us not forget what the map taught us. When Gondwana split up, the first fractures probably were like the rift valleys we see now in East Africa; they were anything but seafloor (Fig. 2.8). The better term undoubtedly is **crustal spreading**, and this is what the African rift valleys may signify. It is possible, then, to observe several stages of the process recorded on the map. The East African rifts may be the beginnings of the kinds of escarpment that ringed Africa when South America, India, and Antarctica broke away, starting about 125 million years ago. The Red Sea represents a more advanced stage—when ocean water has invaded the rift. Madagascar is a more advanced stage still, where rock and fossil data are needed to reconstruct the former link. And the ultimate stage shows Africa's old neighbors (South America, Antarctica, Australia, India), first affected by the spreading process, now thousands of miles away from the ancient core of the Gondwana supercontinent. Maps of landscapes tell more than first impressions might suggest.

•   Sculpturing the Landscape   •

It is clear that every feature, every aspect of the landscape that forms the crust's upper surface is in some way affected by what lies and occurs below. Continental landmasses drift, riding on the tectonic plates. They are subject to diastrophism of all kinds. Even if there were no rainfall, no streams, no wind or waves, the earth's landscapes would change; they would not be static. Forces, stresses, and pressures from inside the earth would be the causes.

Add to this the different degrees of resistance of various rock types to the forces from above, the forces of weathering and erosion, and we can see that the study of landscape poses many challenges. The physiographers who a century ago tried to find physical laws governing the evolution of landscape were unaware of many of the complications they confronted. and yet, if you read their papers and books carefully, it is impressive to note how some of the later scientific discoveries were anticipated. Long before epeirogeny could be attributed to the effect on landmasses of moving plates, the reality of it had been read from the landscape. W. M. Davis knew about it. So did other physiographers in the late nineteenth and early twentieth centuries.

Before we begin our study of the visible surface, it is necessary to introduce some terminology. As with any science, geomorphology has a rather extensive vocabulary. This makes the communication of ideas efficient. One important distinction should be made between the terms **landscape** and **landform**. A land-*form* is a single, discrete, physical feature: a dune is a landform, as is a volcano, a river valley, and a scarp. The land*scape*, on the other hand, is the total natural appearance of an area. A landscape is constituted by all the landforms in the area, by the relief, the soils, the vegetation. A landscape, thus, is an aggregate of features; a landform is a single feature.

When we try to reconstruct the sequence of events in the formation of a certain landscape, individual landforms often provide key bits of information. For instance, when an area is glaciated and the glaciers melt, they leave behind mounds, hills, small ridges, and other landforms so characteristic that there can be no doubt about that landscape's genesis. A trained physical geographer can tell almost at a glance that the landscape owes its origin to glacial conditions. Often geomorphology involves a kind of detective work, and a small piece of evidence can have large implications.

WEATHERING AND EROSION

Another set of terms with which we will become well acquainted is *weathering* and *erosion*. These were mentioned briefly previously in this chapter; now they assume great significance, because these are processes occurring at, rather than below, the surface.

Weathering goes on continuously almost everywhere, breaking down rocks and preparing them, in this way, for removal by erosional processes.

Mechanical, or physical, weathering destroys rocks through frost action when ice crystals form between grains or in joints and the alternate freezing and thawing wedge the pieces apart. In arid regions, the formation of salt crystals has a similar effect. The mineral crystals of which rocks are made have different rates of expansion and contraction in response to temperature changes, and rock surfaces exposed to

Biological weathering: the roots of this tropical plant are forcing apart the sections of rock between which they have taken hold. Enough nutrients can collect in exposed rock joints to permit seeds to sprout; once established, the plant becomes an agent of weathering.

63

the sun's searing heat during the day and rapid cooling at night also weaken.

Chemical weathering also causes rock destruction when **hydrolysis**, the absorption of water, causes expansion of certain grains and consequent loosening. Oxidation (mostly of iron and aluminum particles) and carbonation (e.g., of limestone in humid areas) contribute to chemical weathering as well. Obviously, mechanical weathering has a greater effect in dry and cold areas, whereas chemical weathering is more active in humid and warm regions.

Biological weathering contributes to a lesser extent to rock breakdown as the roots of trees and other plants penetrate cracks in rocks and open them farther.

Together these weathering processes contribute enormously to the disintegration of rocks and thus to the modification of landscape. Where there is comparatively little erosional action, for example, in the driest desert areas where streamflow is rare and wind action of limited power, the quiet work of weathering plays a major role in changing the surface.

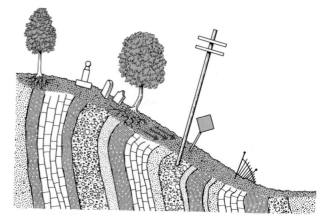

FIGURE 2.20

Soil creep slowly moves the soil downslope. The process moves the upper layer more rapidly than those below, so that trees, fence posts, and telephone poles begin to lean downhill. Redrawn from a sketch in Donn, *The Earth*, 1972.

MASS MOVEMENT

When weathering has loosened rock material (ranging from small grains to large joint blocks), the process of removal begins. The first force that moves the particle may not be a stream, but gravity. This movement of earth materials by the force of gravity is known as **mass movement** (sometimes **mass wasting**). One prominent product of rock weathering, for example, is soil (see Chapter 4), and soil is subject to just such gravity-induced movement. On slopes, soil slowly moves downhill in a process called **soil creep**, the slowest form of mass movement (Fig. 2.20). When a mass of earth material lying on a fairly steep slope becomes saturated by heavy rains, it may be lubricated enough to flow downward, a movement more rapid than creep. A mudflow, involving a thick layer of saturated mud, can be a major modifier of the landscape—and be very destructive of life and property. Under certain conditions, where soil and rock lie at a very high angle and become saturated or are disturbed in some other way, a **landslide** occurs. Landslides can take place without any lubrication of the

collapsed material, but saturation does contribute to many slides. They move much faster than flows and sometimes roar downslope with a thunderous sound and great, destructive force. Finally, some earth material, loosened by weathering from a scarp or steep slope, rolls and bounds downward in a nearly free fall. Signs posted on some highways saying BEWARE OF FALLING ROCK warn against this kind of condition along a slope or scarp overhead; again, gravity is the moving force.

Mass movement is a very consequential process in the modification of landscape. In a way, it is intermediate between weathering and erosion, because it involves rather longer distances of movement than weathered particles, but not as distant as erosional processes. A measure of its importance is its widespread occurrence and the large volume of earth material it displaces. When you stop virtually anywhere to observe a section of the landscape, you will be able to see some aspect of mass movement in progress: where a river is undercutting its bank; where trees and other plants, fences and posts lean downslope; where boulders and smaller rocks are rolling downhill. Gravity is a potent if silent partner in the sculpturing of landscapes.

A landslide in Alameda County, California, blocks Highway 24 near Oakland. Note the slumping in the soil and rock at the upper end of the slide.

EROSION

Important as weathering and mass movements are in the sculpting of landscape, they come nowhere near matching stream erosion in their total impact on the continents. Rivers and streams are the dominant agents of erosion, the major forces that attack mountains and destroy plateaus. In various parts of the world, glaciers, ocean waves, and even wind perform as erosional agents. But streams do more erosive work than all other agents combined.

River erosion is a complicated, intricate process. A river dislodges particles of soil and bedrock. When streams and rivers cut down and erode their channels, **degradation** occurs. The river has to carry its load, and so **transportation** is an important factor. And of course, rivers deposit their sedimentary load in levees and deltas near and at the coast. There they function not as erosional agents but as builders, and the prevailing process is **aggradation**. We can hardly overestimate the significance of rivers. Virtually all the materials accumulated in those thousands of feet of sedimentary rock on all the landmasses were, at one time or another, transported by rivers.

The valleys of all rivers are constantly changing. Rivers oversteepen their banks, causing rock and soil to collapse into the water. When a river's course bends and curves, the erosive action of the water is especially strong on the outside of the curve, so the valley wall against which the river bends undergoes the greatest erosion. Near the river's source and even near its mouth, cross-profiles often show the asymmetry that results from such differential erosion.

In the area where the river begins, where the valley shape tends most nearly to resemble a V, the process of **valley deepening** is usually most active. The narrowness of the valley, the volume and velocity of the water, and the large size of the rock fragments being moved along all contribute to this deepening. In the meantime, the valley is being lengthened, because the river is eating its way back into the countryside through a process called **headward erosion**.

Downstream, **valley widening** becomes much more important, and the valley opens up. The river is calmer, more adjusted. Change is also less perceptible, although it occurs all the time. The river begins to meander, the valley opens still wider, and rather than erode everywhere, the river actually begins to drop some of its load, filling the bottom of its valley with **alluvium**.

Stream erosion takes place in more than one way. We give the name *hydraulic action* to the force of the moving water that alone is enough to dislodge and drag away material from the valley bottom. These loosened materials, especially larger pieces, knock loose still other parts of the valley floor as they roll and bounce along in the water. In the process, they are ground down to finer sediment. This is **abrasion**, an important contributor to the valley-deepening process in the upper reaches of river systems. Certain

65

A landscape in youth: this slope, newly exposed following a glacial episode, now is attacked by running water. Small streams carve V-shaped valleys and drop their sedimentary loads as soon as they reach the base of the steep slopes. Near Mount Cook, New Zealand.

rock fragments of considerable size, even boulders, are carried along by the sheer force of the rushing water. Hikers know it is not safe to cross a mountain stream when they can hear boulders being moved along the stream bed. A river near its ocean mouth presents a very different picture. The water may be brown or gray with mud, but you can see no rocks or boulders. There the river carries only fine particles.

Where the river moves rock materials (and gravel and sand as well) by dragging them along the bottom, the process is one of **traction**. When the fragments are sufficiently reduced in size, they are carried along the channel floor for brief moments, as if bouncing: this is **saltation**. When the particles are reduced still further, part of the load is carried permanently along by the water itself, kept above the floor by currents and eddies. Now the load is in **suspension**. And, of course, rocks contain minerals that can be dissolved in water, so that solution is a part of the transportation work of rivers.

Imagine for a moment a situation in which a river does all this work over a long period of time on a section of the earth's crust that is completely undisturbed and in an area where the rocks are entirely uniform. Eventually this river will erode downward as far as it can go, and the bottom of its valley will show a smooth curve from its mouth to its source. This theoretical smooth curve represents a **graded profile**. Since the river cannot erode lower than the sea level at its mouth, its **absolute base level** forms the low end of this profile (Fig. 2.21).

Such a situation does not actually exist, of course. The crust is unstable and moves. River courses cross many types of rock strata that offer different resistance to erosion. Times of flood and drought cause varia-

FIGURE 2.21
Local and absolute base level.

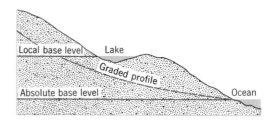

rocks that the river flows over are susceptible to solution. This chemical form of erosion is identified as **corrosion**, the least significant contributor to the overall erosion by rivers. Of course, we should not lose sight of the river's contribution to mass movement of the valley sides. Collapse of valley sides from the river's undercutting is a most significant process in the sculpturing of land surfaces.

Rivers erode in different ways; they also transport their load in different ways. Anyone who has ever watched a mountain stream in action has noted that

66

Three views of the life cycle of a river. In the photo at top left, the stream's upper reaches, its headwaters, lie in mountainous topography. Valleys are narrow, boulder-strewn; the water rushes; erosion is vigorous. In the upper right picture the river is meandering over a wide floodplain, traversing an area of lower relief. The third photo shows the stream braiding, its water volume large, slow-moving, and sediment-laden.

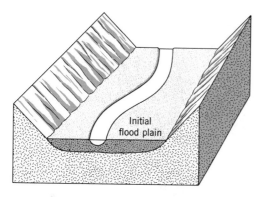

FIGURE 2.22

Relationships of the bedrock, alluvium, and stream channel in the floodplain of a river.

tions in the river's volume. Nevertheless, we may assume that the river system seeks always to establish **grade**, a balance among the volume of water, its speed of movement, the load it carries, and the valley's physical characteristics. By studying rivers' lengthwise profiles, physical geographers can sometimes reconstruct the stages of development of the regional landscape through which they flow. Those major African rivers described earlier are a case in point. Their present deltas represent their response to present absolute base level. But when the great interior lakes existed, the upper courses drained into those basins and formed deltas there in response to that **local base level**. Today the two course segments are connected, but they still carry the imprint of the two graded profiles—one from their source to the basins, the second from the basins to the coast.

Because rivers are degradational agents, and because they cut deep and sometimes spectacular valleys into mountains and plateaus, their aggradational (depositional) work is occasionally understressed. But rivers are builders as well as destroyers (Fig. 2.22). The rocks broken loose and ground down in upper parts of the valley are deposited in the lower course, where the stream sometimes becomes overloaded and sluggish. The river swings sideways, ever widening its reach, and fills its valley with alluvium. Eventually it flows over its own deposited sediment in a wide **floodplain**, so named because the river in times of heavy rain or meltwater may overflow its channel and flood the whole alluvium-filled valley (Fig. 2.23). But the river cannot stop the arrival of still more sediment. So, it creates **meanders** to lengthen its course; it sometimes divides into several parallel channels, or **braids**, to overcome obstacles; and when it reaches base level, it drops its remaining sedimentary load in a **delta**. If it were not for isostatic adjustment, this deposition at the river's mouth would eventually clog the river and back it up. But as the volume

FIGURE 2.23

The asymmetry of stream flow causes the valley to be eroded strongly where flow is greatest. Meanders thus migrate sideways and downstream, thereby developing a flat-floored valley as shown in the series of diagrams. (After Longwell, Knopf, and Flint, John Wiley & Sons.)

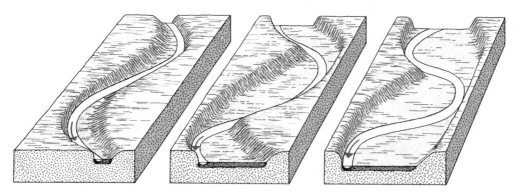

and weight of the accumulated sediments increase, the crust declines just enough to permit the river to continue its deposition. Isostatic adjustment and occasional heavy, channel-clearing floods combine to allow the river to continue reaching the sea.

Rivers and Plains

When William Morris Davis studied landscapes in the late nineteenth and early twentieth centuries and concluded that high-relief landsurfaces are reduced to peneplains during the geographic cycle, he devoted much of his attention to the behavior of streams and the downwearing of slopes. In 1889, for example, Davis published "The Rivers and Valleys of Pennsylvania" in the *National Geographic*, and later he wrote articles on rivers in New Jersey, New England, and Western Europe. The study of rivers is a key element in landscape study, as Davis often noted.

GEOMORPHOLOGY

Today research on slopes, streams, and the genesis of landscapes comes under the rubric of **geomorphology**. Sometimes this term is used as a synonym for **physiography**. By physiography, however, we mean the more general study of the total natural environment. It includes the identification and description of landforms and the study of relief, soils, natural vegetation, even the climate. A dominant concept that has come out of this work is the **physiographic region or province**. A physiographic province is a part of the earth's surface that displays a certain degree of physical homogeneity. In the United States, for instance, one could say that the eastern coastal plain, the Great Plains, the Colorado Plateau, and the central Rocky Mountains constitute physiographic provinces. The books by N. Fenneman and W. W. Atwood (see the bibliography for Chapter 1) represent this approach to physical geography.

Geomorphologists, on the other hand, are less interested in regions than in processes. When, for

example, they see a nearly flat surface that has obviously been eroded down over a long period of time, they want to know the age of that surface, whether it can be correlated with other surfaces elsewhere (even on other continents), what happened to the sediments that were removed, and why certain remnants still rise above the otherwise even plain.

Although the work of Davis was thought of as physiography in his day, Davis laid groundwork for geomorphology. As long ago as 1889, Davis defined some of the world's extensive, nearly flat surfaces as **peneplains**. As he envisioned it, mountains and hills would be lowered into very low, rounded, almost imperceptible elevations in the landscape, merging into equally faint valleys. Here and there a hill of especially hard rock (or one somehow protected from erosion) might still rise as a remnant above the peneplain. These were called **monadnocks**, after a prominent mountain in New England. The peneplains, he thought, would be created by long periods of erosion, toward the end of which almost all the higher areas would have been worn down by weathering and erosion. Davis was always concerned with processes, and he realized that such a plain could never become completely flat; rivers would stop flowing before that could happen. Thus he called those areas of gentle, broad slopes and wide valley floors *pene*plains, *near*-plains (Fig. 2.24).

This concept of peneplanation seems so obvious and easily understood that no one would dispute it. Certainly Davis thought so. As he pointed out, so long as areas are not disturbed by orogeny or by epeirogeny (which would interrupt the cycle), erosion has free reign and the peneplain will become evermore nearly perfect. And he noted that some very old peneplains had been preserved when lava flows covered them or ice sheets and their deposits buried them. Now erosion was exhuming them, so all could see that peneplanation had occurred throughout past geologic eras.

But Davis was challenged by a powerful opponent, the German geomorphologist Walther Penck. Peneplanation presupposes that the slopes of a mountain or plateau wear *down*ward and become lower and lower as the forces of weathering and erosion attack. A landscape's slopes will, therefore, become convex, until finally the whole peneplain consists of

69

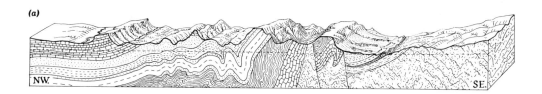

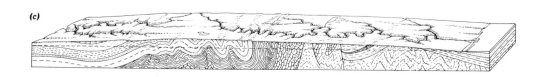

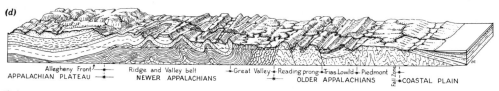

Allegheny Front | Ridge and Valley belt | Great Valley Reading prong Trias Lowl'd Piedmont | Coastal Plain
APPALACHIAN PLATEAU — NEWER APPALACHIANS — OLDER APPALACHIANS

FIGURE 2.24

Example of an uplifted erosion surface formed over the northeastern United States as exemplified
by the simplified history in (a) through (d). The deformed rock structures in (a) are shown eroded
to an essentially flat surface near sea level (peneplain) in (b). After uparching as in (c), renewed erosion
produced the topography of the regions labeled in (d). The summits of the ridges in (d) preserve
remnants of the uplifted erosion surface. (Modified after E. Raisz, from A. N. Strahler, *The Earth Sciences*,
John Wiley & Sons, Inc.)

convex interstream tracts. Penck argued that the landscapes he was studying in Germany had many con-
cave rather than convex slopes, and this suggested to
him that slopes do not wear downward but *backward*.
Think, for example, of an extensive mesa under erosion. Davis would suggest that it would become more
and more rounded and that the flat upper surface
would soon be worn away. Penck postulated that the
mesa's sides would retreat under erosion until only a
small butte was left; when this was ultimately de-

stroyed, only the slight rise around its base would
bear witness to the sequence (Fig. 2.25). This was the
pediment. Much later, the geomorphologist L. C. King
called the coalescing pediments a **pediplane** (from
pedi, foot of the retreating mountain, and *plane*, geometric flatness). The whole issue became one of the
most celebrated debates in the history of physical
geography as the two scientists exchanged articles
and books written to proclaim their respective positions.

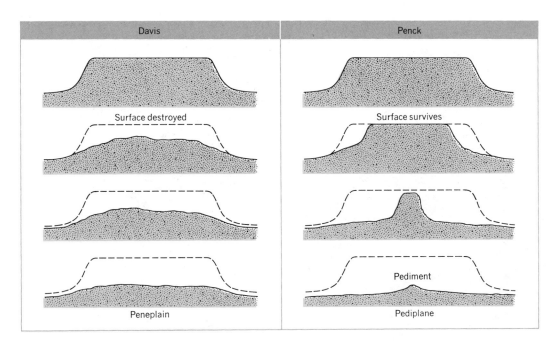

Davis	Penck
Surface destroyed	Surface survives
Peneplain	Pediment / Pediplane

FIGURE 2.25

Alternative erosion-cycle models. Peneplanation is at left, pediplanation at right.

Other Agents

The study of landscape planation has been a central concern in physcial geography, and it entails a wide variety of topics: the nature of streamflow and the formation of river valleys, the deposition (aggradation) of the eroded materials, the effects of rock hardness and underlying structures on planation, and much more. But some geomorphologists have concentrated their work on altogether different kinds of landscapes: those created or affected by ice, those related to wind action, those carved by waves and tidal action along coasts, and those involving the solution of rocks—a process that creates a unique terrain of caves and sinkholes. In this chapter, we can only sample the kinds of studies these geographers pursue. Again, you will note that one of the starting points is a basic vocabulary: a single term or name can summarize the significance of a landform or process.

LANDSCAPES OF GLACIATION

During its existence, the old supercontinent of Gond-

wana was glaciated at least once, and the landmasses that drifted apart all carry evidence of that glaciation. Thick sheets of ice formed in high-latitude (i.e., near-polar) areas and spread over the countryside, scraping and scouring the rocks below and piling up the rubble along their leading edges. Later, the ice age waned and only the ground-up material remained to bear witness to its occurrence. Lava flows and sediments were piled on top of the glacial rubble (called **till**) and buried it deeply, but an occasional exposure through erosion and a great deal of borehole evidence have confirmed the Dwyka ice age (Fig. 2.18) beyond a doubt.

That ancient glaciation was not the last of its kind the earth has experienced. In fact, we may be in the middle of another one right now. Between 2 and 3 million years ago, during the Pleistocene, occurred a new Ice Age that generated enormous ice sheets. These continental ice sheets pushed from Canada across the Great Lakes and into the Midwest as far south as the Ohio River (Fig. 2.26); in Europe they covered nearly all of Great Britain, Scandinavia, the

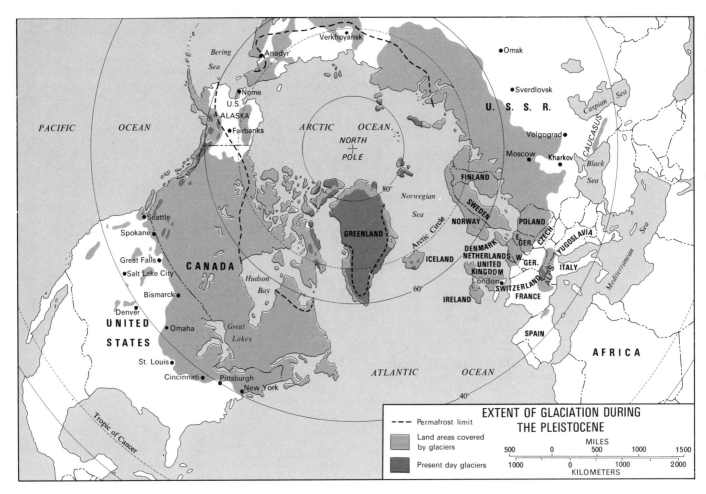

FIGURE 2.26
Areas of the Northern Hemisphere covered by ice during the Pleistocene glaciation.

Netherlands, and most of Germany as well as Poland and the whole northern part of eastern Europe. After much study of the till these glaciers left behind; the scratches, or **striations**, they made in the rocks over which they moved; and the topography that remained after they disappeared, it was clear that the **Pleistocene** glaciation has involved not just one, but several advances of the ice, the last phase ending only a few thousand years ago. Between these glacial advances came periods when the ice melted and somewhat warmer conditions prevailed, the **interglacials**.

An interesting debate presently involves the question of how many Pleistocene glacial advances may

have occurred. For a long time, physical geographers and geologists believed that 4 major advances took place, each named after a state where representative till is especially well developed (the Nebraskan, Kansan, Illinoian, and Wisconsin—from oldest to most recent). Further research has indicated that this interpretation, based on the deposits left behind by each successive glacial advance, may be in need of revision. As many as 14 advances of the ice may, in fact, have taken place, each followed by a warmer interglacial. At present, as just noted, we may be enjoying such a comparatively brief interglacial period. The ice may be back—and sooner than we have assumed.

Smoothed and striated surfaces reveal the presence of a glacier. This surface of quartzite acquired its telltale gouges and scratches during the most recent glacial advance; it lies exposed in the Dells area of south-central Wisconsin.

A glaciation, of course, is caused by a climatic change involving a drop in temperature. Exactly what causes this drop and what determines the timing when it occurs are still not known for certain. But there can be no mistaking the impact of glaciation on the earth's landscapes, and this is what interests geomorphologists. It is practical to distinguish between **continental glaciation** in which large areas are covered by major ice sheets and **mountain glaciation**. When the climate cools, mountainous areas also acquire large glaciers, but these then flow outward through valleys and are generally more confined by the underlying relief. Continental glaciers, on the other hand, are independent and larger. Fed by near-continuous snowfalls piling on enormous quantities of ice, these sheets of ice expand outward, weighing heavily on the underlying bedrock. It is not surprising that the bedrock below is scraped, shattered, and scoured; pieces of the rock become mixed with the lower layers of the ice. The ice sheets are huge; the quantities of rock dislodged and carried along are

When the great ice sheets of the Pleistocene reached their maximum extent, and mountain ranges such as the Rocky Mountains and the Alps were buried under snow, the landscape looked like this.

The Muir Glacier emanates from its mountain source area to reach the sea at Glacier Bay Monument, Alaska. Note the tributary glacier at left and the morainal material streaking the ice.

similarly great. A look at a section of the Canadian Shield shows hollowed-out areas now filled by water, forming the region's numerous lakes, and flattened, scraped-down areas between them.

What happens to the rock debris carried along the ice sheet? Ice sheets erode; they also deposit. **Glacial** deposits, or **drift**, consist of all kinds of rock debris. Sometimes when melting began and flowing water became involved in the process, the rubble was sorted. The resulting material is called **stratified drift**. More often, however, the glacier simply dropped its load of debris in the unsorted mass called till. An idea of the amount of rock carried along can be gained from the thicknesses of till that have been found in the Midwest. In some places, parts of Iowa, Illinois, and Ohio, for instance, it measures more than 300 ft (100 m) in depth; the average is probably about 50 ft (15 m). Along their leading edges, the ice sheets left belts of piled-up debris known as **terminal moraines**, which can be seen in many parts of the Midwest, diversifying an otherwise rather monotonous land-

scape. Speaking of landscapes, what happened to the original landscape of areas covered by the great ice sheets? It was buried by ice and covered by till brought from elsewhere. In one area, the so-called Driftless Area of southwestern Wisconsin, the glaciers for some reason (still uncertain) failed to cover the scenery. There we can get a glimpse of what this part of the Midwest may have looked like.

The continental glaciers covered enormous areas, but the mountain glaciers made by far the most spectacular landscapes. Mountain glaciers start the same way as ice sheets—that is, by the excess accumulation of snow and its transformation into permanent ice. This happens high up near mountain summits where mechanical weathering and streams did the erosive work before the Ice Age began. As more ice accumulates in the high valleys, the glacier begins to move, aided by the existing incline of the river valley.

When glacial erosion takes over from river erosion in a mountainous area, the overall topography changes dramatically (Fig. 2.27). As the glaciers grow

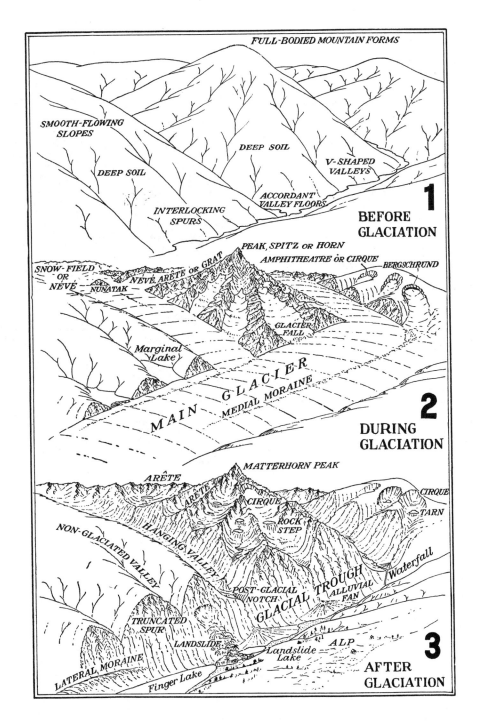

FULL-BODIED MOUNTAIN FORMS

SMOOTH-FLOWING SLOPES

DEEP SOIL

DEEP SOIL

V-SHAPED VALLEYS

ACCORDANT VALLEY FLOORS

INTERLOCKING SPURS

1

BEFORE GLACIATION

PEAK, SPITZ OR HORN

AMPHITHEATRE OR CIRQUE

BERGSCHRUND

SNOW-FIELD OR NÉVÉ

NÉVÉ

NUNATAK

ARÊTE OR GRAT

GLACIER FALL

Marginal Lake

MAIN GLACIER

MEDIAL MORAINE

2

DURING GLACIATION

ARÊTE

MATTERHORN PEAK

CIRQUE

CIRQUE

TARN

ROCK STEP

NON-GLACIATED VALLEY

HANGING VALLEY

POST-GLACIAL NOTCH

GLACIAL TROUGH

ALLUVIAL FAN

Waterfall

TRUNCATED SPUR

LANDSLIDE

Landslide Lake

ALP

3

AFTER GLACIATION

LATERAL MORAINE

Finger Lake

FIGURE 2.27

The transformation of a river-eroded, mature mountain landscape by alpine glaciation. (From *Geomorphology* by A. K. Lobeck. Copyright © 1932, McGraw–Hill Book Company. Used with permission of McGraw–Hill Book Company.)

and move down the valleys, they deepen, widen, and straighten them through their weight and erosive power. The source area or **cirque** grows larger as temperature-induced mechanical weathering breaks up the valley sides. Steep walls replace them, making the cirque look like an amphitheater.

Glaciers, too, are capable of headward erosion, and where two or more cirques intersect, they create a sharp jagged ridge called an **arête** or a steep-sided peak called a **horn**, like the famous Matterhorn in Switzerland. So, glaciers replace the roundness in the river-sculptured landscape with angular roughness.

Glaciers receive tributaries as rivers do, but the floors of the tributary valleys are not adjusted to those of the main glaciers. When the glaciers melt, there are waterfalls where those side glaciers entered the main valley. These **hanging valleys** on the side of the glacial trough are sure evidence that the glaciers existed and that they disrupted the drainage system.

The glacial valleys really do resemble troughs, because the glacier scoops them out and gives them a U shape that is quite unlike any river valley. Glacial valley sides are steep, sometimes nearly straight up, and the spurs that the river formerly twisted and turned around are simply sliced off. The famous Half Dome in Yosemite National Park is an example of the valley glacier's capacity to wipe out obstacles. **Truncated spurs** can be seen on the valley sides in Figure 2.27, "3. After Glaciation." The valley floor also shows evidence of the glacier's irregular action. Moving ice responds to differences in elevation and often excavates a series of steps that, after the glacier has disappeared, may accommodate standing lakes.

In high latitudes and even midlatitudes—for example, in Norway and New Zealand—the glaciers continued to erode below sea level when they reached the sea. Thus glaciers sometimes create gaping seaward valley mouths with characteristic steep sides. When the warming trend melted the glaciers, the sea occupied those troughs. **Fjords** formed, often with quite breathtaking scenery.

When mountain glaciers contain rock debris, this material is not distributed randomly throughout the ice, rather it is concentrated in zones that show up clearly as dark bands on the surface. The mountain glacier's **lateral moraine** consists of rock debris that has been acquired from the valley sides and is carried

A moraine-streaked glacier traverses the mountains of Alaska. Note the steep sides of the U-shaped valley it has carved.

along near the glacier's edge. When two glaciers with lateral moraines meet, however, their inner lateral moraines lie in the middle of the joined valley as a **medial moraine**. Add several other tributary glaciers and a whole series of such medial moraines forms. When the glacier begins to melt, this material is left in a series of ridges in the valley. When the mountain glacier stops advancing, rock debris is left as a pile of rubble across the valley floor. This terminal moraine marks the farthest advance of the glacier. If the glacier's melting is interrupted, **recessional moraines** form where there was a stillstand.

A natural bridge on the island of Aruba in the Caribbean. Earlier, a wave-cut platform developed on the upper surface, where the rocks are very resistant. Then this coastal zone was uplifted and wave erosion attacked the less resistant rocks beneath. Eventually the waves undercut the platform enough to collapse most of it, but this section of it remains.

The modern world's human cultures emerged and evolved in the wake of the last of the Pleistocene glaciers' withdrawal. Certainly there were communities and villages during earlier phases, but the great transformations of human society—plant and animal domestication, urbanization, mass migration, agricultural and industrial revolutions—have occurred during the past ten thousand years as the most recent glacial advance melted away. We are not yet sure why these momentous changes occurred when they did, but they have been accompanied by an unprecedented explosion of human numbers. This population explosion (see Chapter 6) is taking place at present, with the earth's available living space at a maximum. The implications of a return of the ice sheets and its impact on the globe's climates and habitable space stagger the imagination. Nevertheless, scientists warn that such a prospect lies in the world's future.

COASTAL LANDSCAPES

The landscapes of glaciers are distinctive and often unmistakable. Similarly, the landscapes and landforms created by wave action (and by currents and tides) have clear and obvious origins. Cliffs, beaches, terraces, and lagoons mark continental shorelines that are endlessly pounded by waves and stroked by currents. Gale-force winds at times drive the waves onshore with great force, causing rapid and destructive erosion; but elsewhere the waves reach the shoreline gently, bringing sand and gravel that build beaches and offshore barriers. Here the waves do not destroy except under the most unusual circumstances—a hurricane, for example.

As in the case of stream erosion, the behavior of the earth's crust in the coastal zone is a crucial element in any study of the evolution of the coastal landscape. The edge of the landmass may be sinking or rising or temporarily stable. The erosive power of the waves along coasts depends greatly on the configuration of the adjacent sea bottom, which is the product of the recent geologic history of the land margin. To appreciate this we need only compare the unstable

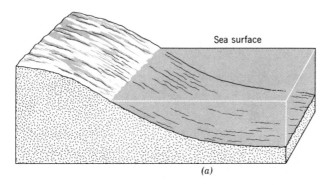

Sea surface

(a)

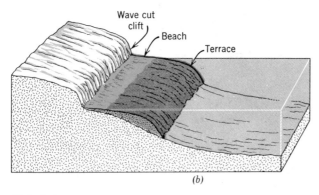

Wave cut
cliff

Beach

Terrace

(b)

FIGURE 2.28

78

Effects of wave action on an initially steeply sloping coastline.

and active crustal deformation of the western coast of the United States with the edge of the comparatively quiet coastal plain of the southeastern United States. The western coast is studded with cliffs, peninsulas, islands, and deep bays; the southeastern coast is comparatively smooth, straight, and shallow offshore.

Although currents and tides have some effect on the formation of coastal landscapes, most of the erosional work along shorelines is done by waves (Fig. 2.28). Persistent strong winds generate large, powerful waves whose effect is enhanced by rock fragments smashed against the shore: Like streams, waves erode by abrasion. The erosional process is affected, of course, by the resistance of the rocks onshore. Hard crystallines will stand up against wave erosion much longer than softer sedimentary rocks can. But the key factor is the topography along the coast. A gentle decline offshore and shallow waters, even far from the

beach, inhibit wave erosion as the waves lose their full force long before they reach dry land. When high relief marks the coast and offshore sea bottom, the water remains deep very close to the land; waves reach the land with full power and their erosional force is great. Coastlines are most practically classified, then, into high-relief and low-relief (sometimes called cliffed and noncliffed) landscapes, and their landforms are viewed accordingly.

The most characteristic landform of high-relief coasts is the **cliff**. Steepsided cliffs are especially likely to develop on exposed **headlands**, or **promontories**, that extend into the sea. Often the cliff retreats in such a way that sections of it become separated and stand apart as **stacks**. The erosive action of the waves often produces caves, especially where the cliff is made of soft rocks below and harder rocks above.

The rapid retreat of cliffs occasionally produces a feature also found in glacial topography: hanging valleys. Streams that once led all the way to the sea plunge off a cliff because the slope they once flowed down has been cut back by the waves.

At the foot of a cliff, the waves and the agitated rock debris tend to create a smooth, seaward-sloping surface called a **wave-cut**, or **abrasion platform**. If the area is uplifted (or the sea level drops) after the erosional phase, such a platform, unmistakably wave-cut, forms a terrace that gives evidence of the coast's complex history.

Low-relief coastlines are marked by wave-deposited belts of sand some distance from the original shoreline. These **offshore bars**, or **barriers**, eventually begin to emerge above sea level because deposition continues during high tides and even during all but the most destructive storms.

Between the offshore barrier and the original shoreline, a **lagoon** forms (Fig. 2.29). If rivers reach the coast in this area, they no longer spill their sediments into the open sea but into the relatively confined lagoon. Thus the lagoon may slowly fill up, developing mud flats that eventually begin to emerge permanently above the tidewaters. What the waves have done, therefore, is to extend the land rather than reduce it.

On occasion, an offshore bar will build from the tip of a peninsula or promontory and terminate some distance away. Such a partial bar is called a **spit**, and

A high-relief coastline in Oregon. Note the stack forming in the foreground and the uplifted wave-cut platform that forms the promontory (headland) bounded by a sheer cliff.

A low-relief coastline on Long Island, New York. The beach slopes gently beneath the waves; no cliffs or stacks mark this landscape.

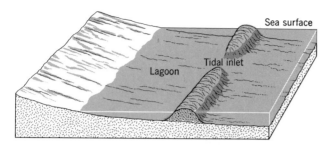

FIGURE 2.29

Offshore bar and lagoon on a low-relief shoreline.

like the offshore bar, it lies parallel to the shoreline except where strong currents or tides cause it to curve inward or outward.

In addition to these high- and low-relief coastlines, there are numerous special circumstances that do not fit into any simple scheme of classification. The coastlines of deltas are low-relief coastlines, but with the difference that they build on both the landward and the seaward side. In some areas, faults parallel the coast and modify it repeatedly. Elsewhere lava flows reach the sea and create unique landforms. Still another exception is the coral reef coastline, which is built from the skeletons of marine organisms. The physical geography of coastlines is endlessly varied and very complicated.

LANDSCAPES OF WIND ACTION

When we think of landscapes where wind action prevails, deserts come immediately to mind. But even in deserts, running water from rainstorms, however infrequent, has a great impact on the development of the landscape. Physical geographers have found it difficult to find landforms created exclusively by wind erosion, but there is no doubt that wind plays a significant role in the deposition and accumulation of loose materials.

When wind acts as an erosive agent, it does so by moving particles (from the tiniest grains to sizable pebbles) from one place to another and by hurling small grains against obstacles, wearing them down. Most of us have experienced the latter effect by being at a beach on a windy, dry day. Sand grains may be

swept along in sheets, hitting ankles and legs. Anything permanently in the way of such **abrasion** will eventually erode away.

Wind, like other erosional agents, lifts, transports, and deposits, The capacity of the wind to move loose particles depends on its speed. The higher the velocity of the moving air, the greater the size of the particles that can be moved and the more effective the erosion. Even on a nearly still day, there is some dust in the air, small specks of clay too light to fall to earth as long as there are the slightest upward air currents. When wind eddies pick up and remove clay and silt-sized particles, the process is called **deflation**. As the wind speed increases, sand grains are moved along. Sand never travels as high as dust does, but it may be a meter off the ground if the wind has storm strength. Unlike the dust particles that are, in effect, in suspension in the air, sand grains bounce along the ground, in a process called **saltation**. Larger particles such as gravel and pebbles are occasionally rolled and pushed by gale-force winds, but they do not move more than a few inches through the action of wind alone. It takes running water to transport these larger fragments over long distances.

The main erosional action of wind is a process of abrasion, the impact of the wind-driven particles against obstacles in the landscape. This process fashions such characteristic features as mushroom-shaped **pedestal rocks** and **deflation hollows**, shallow depressions excavated by deflation. But it is not the wind alone that does the work. Weathering and erosion by running water are predominant processes even in climatically dry areas. A single rainstorm modifies the landscape more than months of wind action. Still, the wind removes particles loosened by weathering or left behind in the valley of a dried-up stream and thus contributes to the formation of typical arid-area topography.

The landform feature most closely associated with wind action and deserts is the **dune**. Dunes are given their shape and form by prevailing wind activity, and fields of sand dunes are characteristics of many (but not all) deserts. Several kinds of dunes have been identified. Among the most typical is the **barchan**, a crescent-shaped dune whose ends are pointed downwind. When an area is completely and thickly covered by loose sand, the wind may throw the sand into

Coalescing barchans in an area of the Arabian Peninsula. These wind-fashioned landforms
are constantly (but slowly) moved in the direction toward the photographer; the prevailing wind
here is from the upper left.

a series of parallel ridges called **transverse dunes**, which lie at right angles to the wind direction. There are other places where the dune chains lie parallel to the wind. These are **longitudinal dunes**.

A very important feature associated with wind action is a very fine-grained, wind-carried silt deposit known as **loess**. Loess occurs in a large area in China, in extensive sections of the middle Mississippi basin, in Europe, South America, and elsewhere. Hundred-foot depths of loess are not uncommon, and it is believed that this material originated during dry periods of the Pleistocene glaciation. Strong winds probably swept the margins of the ice sheets, from which streams emerged to carry loads of silty sediment away from the glaciers. These winds dried out the riverbeds and picked up the finest silt and carried it to nearby areas, where it accumulated in quantity. For humanity, it turned out to be a most fortunate sequence of events. In China, the Midwest, Europe, and Argentina's Pampa, among other places, this loess has developed highly fertile soils capable of sustained production.

KARST LANDSCAPES

One of the world's most spectacular coastlines is that of Yugoslavia along the Adriatic Sea across from Italy. It is renowned for its numerous sheer-walled islands, its steep cliffs, deep caves, and mysterious underground passages. This unusual landscape results from an unusual process of erosion: the solution of rocks and their removal in dissolved form by running water.

The topography resulting from this process is referred to as **karst terrain**, a name that comes from the area in Yugoslavia where groundwater's role in dissolving limestone beds was first analyzed. For karst topography to form, soluble limestone must make up the underlying geology—either at the surface or beneath, say, a porous sandstone layer. When rain falls, it picks up a small amount of carbon dioxide present in the air and becomes a very weak acid. This acid dissolves the calcium carbonate in limestone and carries it away. Streams remove some of it, but (and this is where the process really differs from surface erosion) groundwater also transports it.

The resulting landscape is unmistakable. The solution process causes depressions or *sinkholes* to form, so that, from the air, a karst landscape has a pockmarked look. In central Florida, there are thousands of sinkholes, and some of them have filled with water, creating a pleasant scene. But there also are dangers. Solution may go on quietly underground without visible evidence on the surface. Suddenly the "roof" over the developing underground sinkhole (really a cave at that point) collapses, and whole buildings are sometimes swallowed up.

Karst terrain also has an underground dimension. Solution along the base of the limestone layer connects one sinkhole (or cave) to another, and there are networks of subsurface channels and streams. Rainfall over one area of lake-filled sinkholes causes water to flow below ground to other sinkholes, and strong currents develop through the network. Many swimmers have been lost this way.

Groundwater as an erosional agent has this in common with surface streams: it, too, deposits as well as removes. The best place to see this is in caves, for example, the Mammoth Cave in Kentucky or the Carlsbad Caverns in New Mexico. Here the groundwater, slowly trickling from the walls and ceilings of such caves, deposits its calcium carbonate as it evaporates. Some remarkable forms develop as a result: huge, iciclelike *stalactites* hang from the roof of these caves, and great columns, or *stalagmites*, rise from the floor. It is, indeed, a unique element of a distinctive landscape.

Regional Physiography

One general objective in physical geography involves the correlation of erosion surfaces (peneplains and pediplanes) in various parts of a continental landmass or even the world at large. Another goal (as just noted) is the study of particular landscapes and the processes that fashion them. A third aim is to dissect the regional physiography of an area—an area that may be as large as a whole continent or as small as a state in the northeast of the United States.

Regionalization does for geographers what classification does for other scientists (although other forms of classification are useful in geography as well). The regional concept in geography is discussed in more detail in Chapter 7, but in physical geography it involves the identification of areas within whose boundaries there is substantial uniformity and homogeneity of landscape. These areas are called **physiographic regions or provinces**, and they formed the basis for the books by N. Fenneman and W. W. Atwood referred to on page 00. In everyday language we use names to identify such regions: the Rocky *Mountains*, Great *Plains*, Colorado *Plateau*. But when it comes to drawing lines on maps to give substance to these mental images, complications arise.

One of these problems involves the criteria we employ in defining physiographic regions or provinces. The term **physiography** involves more than just the landscape and the landforms of which it is made. It relates (check your geographic dictionary or any dictionary, for that matter) to all the natural features on the earth's surface, including not only the landforms, but also the climate, soils, vegetation, hydrography, and whatever else may be relevant in the formulation of the overall natural landscape. For example, an area such as the North American Great Plains may be subdivided into several physiographic provinces because it extends across several climatic zones. The relief may remain generally the same, but the natural vegetation in one part of the area may be quite different from that of another section. This, of course, is a response to climatic transition, soil differences, and other regional contrasts.

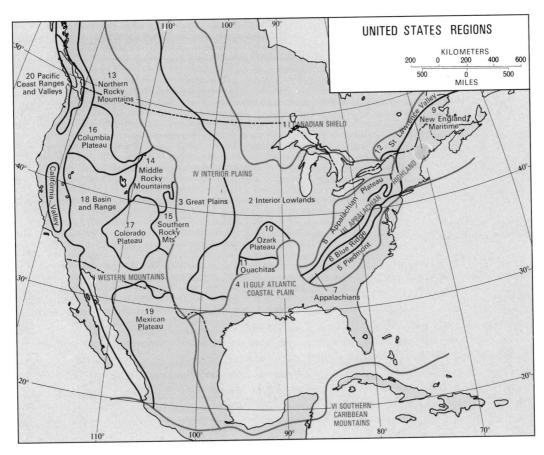

FIGURE 2.30

An outline of the physiographic regions of the United States and southern Canada.

This leads us to the crucial question: How are the boundaries of physiographic regions established? Occasionally there are no problems. Along much of the western edge of the Great Plains, the Rocky Mountains rise sharply, terminating the rather level surface of the Great Plains and marking the beginning of a quite different region in terms of relief, slopes, rock types, landforms, vegetation, and other aspects. But as Figure 2.30 indicates, the transition is far less clear and much less sharp in other instances. The eastern boundary of the Great Plains is a good case in point. Somewhere in the tier of states from North and South Dakota through Nebraska, Kansas, and Oklahoma, the Great Plains terminate and the interior lowlands (or central lowlands) begin. Exactly where this happens depends on the criteria we use to define the two adjacent regions and the method we use to draw the dividing line. Sometimes a persistent linear landform such as an escarpment may serve may serve effectively, overshadowing by its prominence the other transitions in the landscape. Where such a natural dividing line does not present itself, another solution must be found.

Nature rarely draws sharp dividing lines. Vegetation zones, climatic regions, and soil belts merge into one another in transition zones, and the lines we draw on maps to delineate them are the products of our calculations, not those of nature. In such a situation, we can establish dividing lines for each individual criterion (soil change, vegetative change, climatic change, etc.), superimpose the maps, and draw an "average" line to delineate the physiographic province as a whole. Still, another solution is to place a grid system over the area that the boundary is expected to lie in and to give numerical values to each of the criteria mapped. The physiographic region's boundary can thus be mapped by computer. But as always, everything ultimately depends on our choice of criteria and our ability to map them or attach values to them.

PHYSIOGRAPHIC REGIONS OF NORTH AMERICA

Twenty physiographic regions of the United States and southern Canada are represented by Figure 2–30. To simplify the picture some of these provinces can be grouped together. For example, the Appalachian Highlands would include not only the familiar ridge-and-valley region identified on the map as the Appalachians (7), but also the Appalachian Plateau (8) to the west, the Blue Ridge (6) and the Piedmont (5) to the east, and the New England Maritime Province (9) to the north. There is even some justification for including the Ouachitas (11) and Ozarks (10) in this eastern upland cluster; the Ouachitas's rock types and structures parallel those of the Appalachians and Ouachita landscapes resemble Appalachian scenery, except that overall relief is lower.

Thus the Appalachian Highlands group of physiographic regions lies between two plains: the Gulf–Atlantic Coastal Plain (4) and the Interior Lowlands (2), both of which lead into the Great Plains (3). The Appalachian Highlands are characterized by higher relief, higher maximum elevations, and greater internal variety. The ridge-and-valley topography of the Appalachians themselves (region 7) is formed on the long, persistent folds (anticlines and synclines) of the sedimentary rocks in this region. As the map shows, one of its prominent valleys is the Saint Lawrence Valley (12), which bounds the great North American core area, the Canadian Shield (1). Appalachian landscapes consist of lengthy, parallel, vegetation-clad ridges whose crests reach remarkably even elevations between 3000 and 4000 ft (about 900 and 1200 m). Along the eastern edge of the Ap-

palachians lies a wide lowland called the Great Valley, with fertile soils derived from limestone and slate, tilled by Pennsylvania Dutch farmers since before the Revolution.

This central, ridge-and-valley Appalachians region is also called the Newer Appalachians, to differentiate it from the also mountainous, but older, crystalline-supported Blue Ridge to the east. This is the country of the Great Smoky Mountains, higher than the ridges of the Newer Appalachians and, in the absence of linear folds, more jumbled and irregular. To the east of the Blue Ridge, which reaches 6000 ft (1800 m) in places, the Piedmont is still sustained by old, altered crystallines, but at lower elevations and more moderate relief. Eventually the crystallines of the Piedmont are covered by the sediments of the Gulf–Atlantic Coastal Plain, and the contact zone between the two provinces is marked by the Fall Line, the limit of navigation on many eastern rivers and the location of numerous towns.

West of the Appalachians proper, the Appalachian Plateau (8) has been carved by river erosion from horizontal or slightly inclined sedimentary rocks, so that the landscape differs markedly from the ridge-and-valley province. And to the north, the New England Maritime Province (9) presents a very distinctive landscape that is affected by glaciation, dotted with lakes, and streaked with moraines as well as diversified by high, crystalline-supported mountains and bounded by a high-relief coastline.

West of the plains regions, the **Rocky Mountains** also can be viewed as a cluster of physiographic regions. The Southern Rocky Mountains Province (15) lies across the heart of the state of Colorado. The mountains rise sharply out of the adjacent Great Plains along one of the most distinct physiographic boundaries anywhere on the continent; the easternmost range of the Southern Rockies is supported by crystalline rocks (it is called the Front Range as well as the Laramie Mountains and the Sangre de Cristo Mountains), and the sedimentary rocks of the plains are turned up against its flank. The Southern Rocky Mountains as a region are marked by a certain parallelism of the ranges, with a north-south orientation that is lost in the Middle Rocky Mountains (14). In the Middle Rockies, the ranges lie in various directions, with wide, open valleys and basins, and with a less congested landscape than in the Southern Rocky Mountains. As in the Southern Rockies, many of the major ranges are large anticlinal uplifts, but there are also fault blocks and lava flows. The Uinta Mountains of Utah have a massive anticlinal structure, but the Wasatch (also in Utah) and the Tetons in Wyoming are fault blocks. In the Northern Rocky Mountains (13) the structures reach their greatest complexity, and the landscape reflects it by the absence of persistent ranges and valleys and a jumbled topography. Gone are the conspicuous, high-crested ranges of the Middle and Southern Rockies. Exposed batholiths, folded and crushed sedimentary rocks, and extrusives, all heavily eroded and strongly glaciated, create an altogether different scenery.

The Rocky Mountains are higher than the Appalachians, reaching over 13,100 ft (4000 m) in several places. During the Pleistocene Ice Age, large glaciers formed on the uplands and occupied the valleys; some permanent snow still exists even during the current interglacial. The recency and intensity of the glaciation in the Rockies have produced some most spectacular scenery with deep U-shaped valleys, large cirques, horns, and many aggradational glacial landforms that set the Rocky Mountains apart as a physiographic province.

Between the Rocky Mountains and the Pacific Coast Ranges and Valleys (20) lies still another set of physiographic regions that can be viewed as a group: the intermontane basins and plateaus that include the Columbia Plateau (16), Colorado Plateau (17), and the Basin-and-Range Province (18). The **Columbia Plateau** lies wedged between the Northern Rocky Mountains to the east and the Cascade Range of Washington and Oregon to the west. The plateau itself is one of the largest lava surfaces in the world. We call it a plateau only because it lies as high as 6000 ft (1800 m) above sea level. The plateau offers few obstructions to travel. The relief is rather low and the countryside rolls, except where volcanic and other mountains rise above the flood of lava that engulfed them. The southern boundary of the province is determined by the limits of this basaltic lava field; in many places it is not clear in the topography at all. There is not much doubt that this region differs basically from the mountainous provinces that flank it. The two major rivers that drain the plateau, the Columbia River and

the Snake River, form deep canyons in the lava. The Snake River's canyon is one area well over a kilometer deep. Alluvial plains and loess accumulations date from the glacial period, when the rivers were blocked and when the winds piled dust in deep layers.

The **Colorado Plateau** (17), by contrast, is underlain chiefly by flat-lying or nearly horizontal sedimentary strata, weathered in this arid environment into vivid colors (*colorado* is Spanish for red) and carved by erosion into uniquely spectacular landscapes. The eastern boundary, of course, is the Southern Rocky Mountains, a very different topography. On the western side, a lengthy fault scarp bounds the plateau and separates it from the Basin-and-Range Region (18). To the south, the boundary is not marked by linear landscape features or by strong topographic contrasts. The southern boundary is traditionally taken as the divide between the plateau's major river, the Colorado, and the drainage area of the Rio Grande.

The landscapes of the Colorado Plateau are dominated by vast expanses; spectacular, steep-walled canyons; numerous mesas, buttes, and badlands; and countless multicolored rock exposures. The plateau surface reaches as high as 10,000 ft (over 3000 m) in elevation, broken by fault scarps that expose the variety of its sedimentary layers.

The **Basin-and-Range Province** (18) is not just one basin, but an entire region of basins, many of them internally drained and not connected to other depressions by permanent streams. Dryness is also the hallmark of this region.

A large number of linear mountain ranges rise above the basins. These mountain ranges are generally from 50 to 80 miles (80 to 130 km) long and from 5 to 15 miles (8 to 24 km) wide. They rise 2000 to 5000 ft (600 to 1500 m) above the basin floors, reaching heights of 7000 to 10,000 ft (2100 to 3000 m) above sea level. The mountains are mainly oriented north–south and have a steep side and a more gently sloping side, which suggests that they are structurally related. Faulting is the most likely explanation, because the Basin-and-Range Province looks like a sea of fault blocks thrust upward along parallel axes.

Today, with aridity prevailing, the basins are mostly dry, streams are intermittent, and vegetation is sparse. Wind action plays a major role in sculpting the landscape. This was not always so, because old,

85

The Colorado Plateau and the Grand Canyon. Horizontal or nearly horizontal layered rocks support a plateau landscape deeply carved by the Colorado River and its tributaries.

proximately parallel zones. In the interior, away from the coast, lies a zone of mountain ranges that includes the Cascade Range of Oregon and Washington and the Sierra Nevada of California. West of these ranges lie several large valleys, chiefly the California Valley and the Puget Sound–Willamette Lowland. And on the shores of the Pacific Ocean lie the Pacific Coast Ranges such as the Olympic Mountains of Washington and the Klamath Mountains of northern California.

Except for the plains and low hills in the valleys, the topography in this province is rough, the relief high, and the rocks and geological structures extremely varied. Mountain building is in progress, earthquakes occur frequently, and movement can be observed along active faults. The coastline is marked by steep slopes and deep water, pounding waves, cliffs, and associated landforms of high-relief shores. Elevations in the interior mountains rival those of the U.S. Rockies in places, reaching over 14,000 ft (4200 m) in the U.S. sector of the region and exceeding that in the Canadian Rocky Mountains along the Alaskan coast where some mountains are over 18,000 ft (5400 m) in height. This is the leading edge of the North American tectonic plate, and the landscapes of this coastal province reflect the forces at work there.

The physiographic regions just discussed represent the surface expression of the materials, forces, and processes described previously in this chapter. Streams, ice, waves, and wind have sculpted igneous, metamorphic, and sedimentary rocks into an infinite variety of landforms while the crust itself bent and buckled, crumpled and stretched. Glaciers came and went, lakes filled and dried. All of this is etched on the surface of our changing earth.

glacial-age shorelines of now-extinct lakes have been found along the margins of some of the basins in the province. Undoubtedly the most famous of the remaining bodies of water in the province is the Great Salt Lake in Utah. This lake's volume has varied enormously in recorded times, and its high salt content tells a story of contraction and withering for all the province's bodies of water.

The **Pacific Coast Ranges and Valleys Region** (20) forms the westernmost physiographic province in North America, extending all the way from Alaska to Central America. This region incorporates three ap-

Reading about Landscapes

The geography of landscapes and landforms is a part of physical geography, and a good place to start supplementary reading on this topic is in several of the standard introductory textbooks in the field. A very comprehensive and well-illustrated volume is A. N.

Strahler and A. H. Strahler, *Elements of Physical Geography* (3rd ed.; New York: Wiley, 1984). This book (like other introductory texts) includes sections on climate, biogeography, and soil geography, too. Another recent volume is T. McKnight's *Physical Geography: A Landscape Appreciation* (Englewood Cliffs, N.J.: Prentice-Hall, 1984), also well produced with good maps and illustrations. Easier reading is J. G. Navarra, *Contemporary Physical Geography* (Philadelphia: Saunders, 1981). Another current and well-presented physical geography book is R. A. Muller and T. M. Oberlander, *Physical Geography Today* (New York: Random House, 1984).

As we noted in Chapter 1, physical geography can be approached in various ways—purely as a natural science or in the context of human-environment relationships. In this latter connection, see two books by J. E. Oliver, *Physical Geography: Principles and Applications* (North Scituate, Mass.: Duxbury, 1979) and *Perspectives on Applied Physical Geography* (North Scituate, Mass.: Duxbury, 1977). Another work that relates physical, human, and environmental geography is C. F. Bennett's *Man and the Earth's Ecosystem* (New York: Wiley, 1975). Also see D. Greenland and H. J. de Blij, *The Earth in Profile: A Physical Geography* (San Francisco: Canfield Press, 1977).

Other sources include G. H. Dury, *An Introduction to Environmental Systems* (Portsmouth, Eng.: Heinemann, 1981); C.A.M. King, *Physical Geography* (Totowa, N.J.: Barnes & Noble, 1980); and W. M. Marsh and J. Dozier, *Landscape: An Introduction to Physical Geography* (Reading, Mass.: Addison–Wesley, 1981).

Books that concentrate on landscapes and landforms (not on the broader range of topics in physical geography) deal with specialized subjects such as river dynamics, slope behavior, and so on. One of the more readable is K. Butzer's *Geomorphology from the Earth* (New York: Harper & Row, 1976). More difficult to read is S. A. Schumm, *The Fluvial System* (New York: Wiley, 1977), a book about rivers, river dynamics, and river systems and their role in the sculpting of landscape. On the topic of slopes, see A.

Young, *Slopes* (New York, Longman, 1975). On glaciation, a fine source is D. E. Sugden and B. S. John, *Glaciers and Landscape: A Geomorphological Approach* (New York: Wiley, 1976). For more on coasts and wave action as an erosional force, see W. Bascom, *Waves and Beaches: The Dynamics of the Ocean Surface* (New York: Doubleday, 1980).

Good background for the sections on earth structure and plate tectonics can be found in J. P. Kennett, *Marine Geology* (Englewood Cliffs, N.J.: Prentice-Hall, 1982).

A. L. du Toit's *A Geological Comparison of South America with South Africa* (Washington, D.C.: Carnegie Inst., 1927) and his prophetic *Our Wandering Continents* (Edinburgh: Oliver & Boyd, 1937) continue to hold interest. A. Wegener's groundbreaking *The Origin of Continents and Oceans*, first published in German in 1915, was translated into English by J. G. A. Skerl (London: Methuen, 1924). An early speculation on the mechanism for continental drift that turned out to be very nearly correct is part of A. Holmes' *Principles of Physical Geology* (London: Nelson, 1944), a book still worth your attention.

The work of William Morris Davis was cited in Chapter 1. N. M. Fenneman's two great studies of U.S. physical geography, *Physiography of the Western United States* and *Physiography of the Eastern United States* (New York: McGraw-Hill, 1931 and 1938 respectively) continue to be useful more than a half century after their appearance. Another book that is likely to retain its relevance is W. D. Thornbury's *Regional Geomorphology of the United States* (New York: Wiley, 1965).

Articles on current research in physical geography abound in such journals as the *Annals* of the Association of American Geographers, in the *Professional Geographer*, the *Geographical Review*, and other major periodicals of the discipline. There also are several specialized journals for physical geography, including *Physical Geography* (Washington, D.C.: V. H. Winston) and *Geografiska Annaler Series A Physical Geography* (Stockholm: Almqvist & Wiksells). Also see *Quaternary Research* and *Catena*.

87

CHAPTER THREE

Climate:
Pattern
and
Process

When the Spanish conquerors sailed southward along the west coast of the Americas, they encountered ocean currents that were a blessing and a curse. A curse because the waters moved in the opposite direction: northward and offshore. A blessing because in the tropical heat those waters were cool, even cold. The Spaniards hung their wine flasks over the side of the sailing ships and enjoyed cellar-temperature wine cooled by Pacific waters.

Soon scholars were asking the obvious questions: why are the waters off Ecuador and Peru so cool under the heat of the equatorial sun? From the northward movement of the current, it was concluded that this was Antarctic cold, carried to the tropics. Alexander von Humboldt, the great explorer–geographer, measured water temperatures that seemed to confirm that idea; for a time, his name was associated with the phenomenon: the Humboldt Current. Today we know that something else contributes more significantly to the coolness of the ocean off tropical western South America: an upwelling of cold water from hundreds of feet below the surface. And the current is now called the Peru Current. Between them, the Peru Current and those upwelling, cold waters, produce the temperatures that Humboldt recorded.

But that is not the end of the story or the research. As time went on, those cold waters near Peru and Ecuador were found to contain huge amounts of marine life. In fact, they were the richest fishing grounds anywhere in the world, and from Peruvian and Ecuadorian ports, fleets of fishing boats brought back millions of tons of fish over the years. The boat captains noticed, however, that something happened every year around Christmas or very early in the new year: their catch shrank to only about 10 to 20 percent of the average for other periods. The fish seemed to disappear. Then, a couple of months later, the fish would come back and the catch would be back to normal.

Before long, it was known that the scarcity of fish, beginning in late December, had something to do with another phenomenon: the water that was usually so cool, warmed up. And the direction of the current changed. Warm water started to flow southward along the coast, always around Christmas time. So, the locals called this annual event *El Niño* (The

Little One), a name given to the Christ child whose birthday coincides very nearly with the onset of the warming trend.

Sometimes the people along the coast noticed that *El Niño* would come earlier and stay longer. This caused havoc with the fishing industry, changed the weather along the coast, and generally upset the area. Had this happened, say, along the coastline of Western Europe, or the Eastern United States, it would have attracted a lot of attention. But on the distant South American coast and in the vast eastern Pacific, these events remained mostly a matter of scientific curiosity for a long time.

Shortly after the beginning of the present century, scientists began to realize that this *El Niño* might be more than a local phenomenon affecting only a few coastal residents of two South American countries. One question soon answered was this: What causes the upwelling of cold water in the coastal area when *El Niño* is absent? The answer lay not in the ocean, but in the air above it. As more became known about the prevailing winds in the area, the mechanism was revealed: when winds blow from east (land) to west (toward water, the Pacific), surface water is moved from the coast into the farther ocean. Taking the place of this windblown water is water from the deep—cold, nutrient-rich water that rises to fill the now-available space.

Another question proved much more difficult to answer. How does *El Niño* overcome this mechanism, replacing the cold upwelling water with a warm southward-flowing current? And why is *El Niño* so much stronger in some years than in others?

As climatologists and meteorologists studied these problems, they realized that they had come upon something much more important than a mere local phenomenon. When *El Niño* was particularly strong, as in 1891 and 1925 and in 1982–83, it caused more than just a drop in the fish catch. Climatic events elsewhere in the world seemed somehow connected to it: monsoon failure in India, famine in Africa, typhoons in unusual Pacific locations. *El Niño* appeared to be a manifestation of global climatic irregularity, evidence that something was out of balance.

The search for solutions led in all kinds of directions. The impact of the 1982–83 *El Niño* was recorded in greater detail than any previous one, and it was staggering. In the early 1970s, the fishing boats were bringing back more than 12 million tons of fish annually; in 1983, the total was less than a half million tons. Coastal desert areas that normally received just a few inches of rain per year received 100 in. in one season. Biological effects on the mainland and on Pacific islands were catastrophic. Bird life on Christmas Island was wiped out because marine food disappeared. Coral beds died off over large areas. Droughts and floods in various Pacific locations appeared connected to whatever caused this giant disturbance.

Scientists collected data and, with the help of Landsat imagery and computers, tried to understand what might be the fundamental cause of these events. Their search led, first, to the relationships between ocean and atmosphere over the equatorial Pacific: the movement (circulation) in water and air. Next, they found reasons to go farther afield, to the Northern Hemisphere, to North America, and beyond. *El Niño*, they knew, was merely an especially intense manifestation of a global event, one that was (and is) capable of causing floods in one part of the world and drought in another. That research continues today, and *El Niño* is a constant reminder of a reality of our natural world: systems are interconnected, and an apparently local phenomenon may be a barometer of global forces at work.

This chapter deals with air and water, the energy that stirs them into motion and the patterns—global and local—that they develop. Complicated mathematical models and predictions begin with an understanding of the fundamentals of climatology, one of geography's most interesting fields.

The Life Layer

Planet Earth consists of a set of roughly concentric spheres, and the outermost solid sphere, the crust, was the major topic of study in Chapter 2. But the crust's surface does not mark the outer limit of our planet. On the crust lies the **hydrosphere**, covering more than 70 percent of it with water to an average depth of 13,000 ft. (nearly 4000 m). On both crust and hydrosphere lies the **atmosphere**, itself consisting of a set of layers. Elements of the thinning atmosphere persist more than 6000 miles (10,000 km) above the earth's surface.

Lithosphere, atmosphere, and hydrosphere interact in many ways. The French geographer Jean Brunhes remarked in his *Human Geography* (1952) that "every human establishment is an amalgam made up of a little humanity, a little soil, and a little water." We might add that without water, there would be neither soil nor humanity. Water is the critical agent in the breakdown of rocks into soil. Water sustains the plants that grow in the soil. But water would not reach the interior of the earth's landmasses if the atmosphere did not carry it there. Water from the life-giving rains that fall on wheatfields in North Dakota and rice paddies in northern India was carried from the oceans by the atmosphere's air in an endless cycle of replenishment. The surface of the crust, therefore, is an **interface**, a plane of interaction. The processes and products of this interaction have made possible the emergence of humanity and modern human society. In the following pages, we concentrate on the atmosphere and the hydrosphere, on climate, weather, and the oceans. In Chapter 4, we focus on soils and vegetation—the biospheres.

EARTH AND SUN

The spheres of the inner earth are mobile, their motion sustained by energy generated in processes of chemical change. Even the "solid" crust moves as oceans widen and landmasses separate to coalesce elsewhere. And so it is with the hydrosphere and the atmosphere. These layers, too, are in constant motion—but the energy that propels them comes not from the earth itself, but from a distant planet, the sun.

The earth receives solar energy from the sun. This **solar radiation** comes chiefly in two forms: shortwave, consisting of light rays we are able to see, and infrared rays, thermal rays whose wavelength is beyond the visible spectrum. As the earth orbits

91

around the sun, 93 million miles (150 million km) away, it receives a tiny fraction of all the heat energy the sun radiates in all directions. But that tiny fraction is enough to penetrate the protective atmosphere and heat it as well as the land and water of this planet.

When the sun's radiation reaches the earth's outer atmosphere, several processes begin. Molecules of nitrogen and oxygen, dust particles, water vapor, and other contents of the atmosphere interfere in various ways with the rays. Some of the radiation is reflected back into space. Some of it is scattered. Some of the infrared rays are absorbed by water vapor or carbon dioxide in the atmosphere. Clouds play a considerable role in reflecting radiation back into space. On a cloudy day with lots of moisture and a great deal of reflection, as much as two thirds of the radiation headed toward the surface of the earth may be lost through these circumstances and processes. And the reflective power or **albedo** of the earth varies greatly from place to place. Ice-covered polar areas reflect as much as 90 percent of incoming shortwave radiation, but forest-clad tropical areas reflect as little as 5 to 15 percent. In tropical areas, however, clouds add significantly to the albedo. On an annual average, tropical zones have an albedo of 15 to 25 percent (lower over low-reflecting ocean water), polar regions range from 45 to over 70 percent, and intermediate regions from 25 to 45 percent.

This latitudinal and locational range in the earth's albedo constitutes a significant factor in the global distribution of heat energy. The solar radiation that does penetrate all the way through the atmosphere (and is not reflected) reaches the surface of the earth and is there converted into heat. When shortwave rays reach an exposed surface, during a process called **insolation**, they are transformed into longwave (infrared) rays. This process is accompanied by rising temperature, and the heated surface now emanates these longwave rays. Most of the longwave radiation that is produced is absorbed by components of the atmosphere, and so the air temperature rises. In turn, the heated atmosphere emits longwave radiation as well. Some of this longwave radiation is lost through the upper atmosphere into space, but much more of it is radiated back again to the earth's surface. This **counter radiation** strengthens the atmosphere's role as a warming blanket for the earth's surface.

Anyone who has seen a distant section of asphalt road shimmer in the midday heat has seen these processes going on. Not only is the lowest layer of air heated by longwave radiation, but conduction, too, raises its temperature. Thus our atmosphere is actually heated from *below*, not directly by the sun. This is often referred to as the **greenhouse effect** because greenhouses function on the same principle: shortwave radiation passes through the glass, the interior surface is insolated, longwave radiation is generated and heats the air inside the greenhouse, and the same glass that let the shortwave radiation in prevents the heat from being lost. The same principle applies when we find our car's interior hot, even on a cool day, after it has been exposed to the sun for several hours.

Different parts of the earth receive different amounts of the sun's solar energy; herein lies the root of a ceaseless system of circulation that keeps the atmosphere in motion all the time. The ultimate effect of this contrast is familiar enough: equatorial regions are associated with heat, and polar regions display the effects of low-intensity insolation. What brings this about is a combination of factors: (1) the earth's spheroidal shape and (2) the inclination of the earth's axis of rotation.

The earth's spheroidal shape has the effect of reducing the intensity of radiation received at greater distances from equatorial areas. Figure 3.1 is a cross-section of the earth, showing only its surface and two parallel rays arriving from the sun. Note that rays *X*

FIGURE 3.1

Angle of the sun and intensity of insolation.

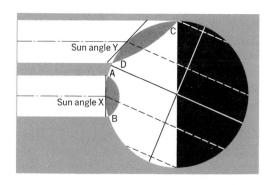

In parts of Western Europe, greenhouses virtually cover the countryside, creating a controlled environment for the cultivation of specialty crops that would not otherwise grow here. This particular (and quite representative) scene is from the island of Guernsey in the English Channel.

and *Y* have equal width. Ray *X* strikes the surface of the earth at right angles, and the sun angle where it is marked on Figure 3.1 would be 90°. The area this ray insolates is represented by *A–B*. Ray *Y*, on the other hand, strikes the surface at a low angle. Its area of insolation is represented by line *C–D*. Now, although the rays are drawn to equal width, *C–D* is much longer than *A–B* and, therefore, represents a much larger area. Thus, the same amount of radiation received by a rather small area at the equator is spread over a considerably larger area at higher latitudes. The larger area will, therefore, be insolated with less intensity. The earth's spheroidal shape has this effect: where the angle of the sun's rays is high, the intensity of insolation is greatest.

This brings us to the reasons that the equatorial areas happen to be the part of the globe to receive this maximum insolation. Our planet orbits around the sun at a distance averaging about 93 million miles (150 million km). While it does so, it rotates around its own axis, so that for about half of each 24-hour period, we are on the sunward side. For the other half, we are in darkness. If we visualize the earth's path of orbit and the sun as a surface, we thereby conceptualize the **plane of the ecliptic**, the name given to it by astronomers. Of importance is the fact that the earth's axis of rotation is not vertical to this plane: it sits at an angle of 23.5° (Fig. 3.2). If, indeed, the axis of rotation were vertical to the plane of the ecliptic, then the sun's rays would always fall vertically on the equator alone. But because of the 23.5° angle, the earth during its 365-day orbit is swept by the vertical ray between latitudes 23.5° north and 23.5° south.

Figures 3.2 and 3.3 help explain this. Figure 3.2 shows the earth in four positions during its orbit. Figure 3.3 shows in detail how the sun will be ver-

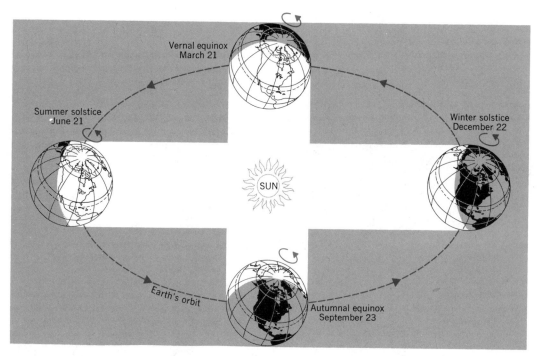

FIGURE 3.2

Sun–Earth relationships and the seasons.

tically above the Tropic of Cancer when the earth is on one "side" of it and how the ray falls vertically on the Tropic of Capricorn when the earth has moved to the opposite side. These are the two solstices that occur during orbit, during every year. Because of the inclination of the earth's axis, this is the farthest north and south the vertical ray reaches. Midway between the solstices, the vertical ray falls on the equator. That situation can be simulated by pointing a pencil straight down to the center of the sphere on Figure 3.3. It happens twice, as Figure 3.2 shows, once in late March and once in late September.

We have focused in some detail on this matter of the inclination of the earth's axis of rotation with good reason. It is perhaps the critical element in the whole system of heat reception and distribution over this planet. From our vantage point in the Northern Hemisphere, most of us see the sun angle rise each spring, reach a maximum on June 21 (the longest day), and then decline again. The warmest weather is still to come, during July and August. There is a lag because

the earth and its atmosphere need time to convert and distribute all the energy received in such enormous quantities in tropical areas. Such a lag, incidentally, can also be observed to occur each day. On a hot

FIGURE 3.3

Summer and winter solstices.

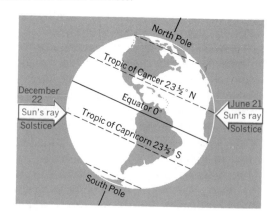

summer day, the hottest time will be about 2 P.M., although the highest the sun rises in the sky is as early as 12 noon. Again, the conversion of heat and its accumulation have the greatest effect sometime *after* maximum insolation. And there is another reason to look into the earth's axis of rotation and its effects. Anyone who looks at the globe will notice that maximum radiation areas, the tropical areas, consist of much water and relatively little land. Scientists have already wondered about the effects of moving the axis and artificially relocating the equator in a more advantageous position. Almost no one takes such a suggestion seriously, but do not be surprised if it becomes practical in the future.

One additional factor relating to the distribution of temperature on our globe becomes apparent from Figure 3.2. Again, if the earth's axis were vertical to the plane of the ecliptic and thus to the sun, all our days everywhere on this earth would consist of 12 hours of daylight and 12 hours of night. But our inclined axis changes that, as our long summer days prove. When the sun's vertical ray is at or near the Tropic of Cancer (i.e., near the June solstice), it continuously illuminates the northernmost latitudes, the North Pole, and as far south as the Arctic Circle. In these polar areas, then, there are months of continuous light and no darkness at all. But when the winter comes and the sun illuminates the southern polar areas, there is continuous darkness for months. Areas between the tropics and the poles experience intermediate effects, with long summer days but short winter daylight hours. In this way, the diminishing effect of low sun angles at higher latitudes is increased: not only is the insolation of *lower intensity*, but when the sun is lowest, the daylight hours, the *hours of insolation*, are shorter, too.

LATITUDE ZONES

Solar radiation and insolation diminish from what we have termed equatorial and tropical areas to polar zones of the globe. If our planet were made in such a way that it had a homogeneous surface, say, all water of equal depth, then earth temperatures would decline from the equator to pole along a steady curve. But the earth has a very varied surface, with the land made of different kinds of rocks at varying elevations

and waters not only of different depths, but also of different composition (e.g., salt and fresh). Thus, if we made a record of temperature conditions from the equator to the pole, say, from Africa through the Mediterranean Sea and Europe or from the interior Amazon Basin in South America across the Caribbean and eastern North America, our curve of poleward temperature decline would be anything but steady. Even so, we would be able to recognize some recurring patterns that are so pervasive that they have given rise to some common geographic terminology.

At the lowest latitudes lies the equatorial zone, known for its heat and (with so much water and so little land) its high humidity (Fig. 3.4). This equatorial zone is not bounded by any sharp line (none of the other latitudinal zones we will identify is either), but its width averages about 20 degrees, 10 degrees on either side of the equator. Here the sun angle is always very high, and there is not much seasonal variation in the length of days and nights, which are about equal. Beyond the equatorial zone lie the tropics. Here we can recognize the beginning of seasonal variations, although insolation is still very intense. As we will observe, one feature of this seasonal cycle is reflected by the precipitation regime: the steady, year-round rainfall of the equatorial zone gives way to a wet and dry season. Technically the tropical zone extends to the Tropics of Cancer and Capricorn, but again, we are dealing with generalized zones and not with specific

95

FIGURE 3.4

Latitude zones of the Northern Hemisphere.

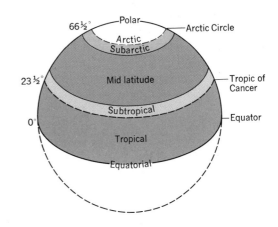

In order to make is possible to identify precisely the location of places and areas on the surface of the globe, a grid system has been devised. This system consists of a set of lines running east–west parallel to the equator and a set of lines running north–south connecting the poles.

The lines running east–west are called **parallels**, and they are measured by the angle from the plane of the equator, as shown in Figure 3–5. The poles, then, are at 90°. Positions of **latitude** are given as $x°$ north or south **latitude**. For greater precision, the degrees are, in turn, divided into minutes and seconds. One degree of latitude is about 69 miles (110 km) on the surface of the earth.

The lines connecting the poles are **meridians**. In this case, there is no convenient equator from which to measure, so a **prime meridian** was established to run through Greenwich, England.

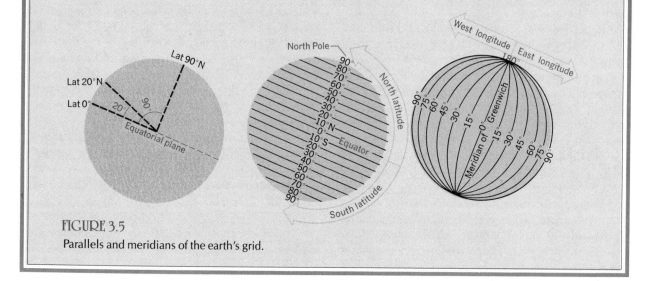

FIGURE 3.5
Parallels and meridians of the earth's grid.

latitudes. The tropical zone gives way to what is normally called the subtropical zone, where tropical conditions prevail during the high-sun season, but where the moderations of high latitudes are really beginning to be felt. In North America, Florida lies in this zone. Adjacent to the subtropical zone is the midlatitude zone, an appropriate term because it lies astride the middle parallel, 45°. It extends about 10° in either direction; in this zone, the seasonal contrasts brought by the movement of the sun's rays are strong: summer days are lengthened and winter days are shortened; summers can be balmy and tropical; winters can be laced with polar air.

Beyond the northern margins of the midlatitudes (about 55°) lie the subarctic (and subantarctic in the Southern Hemisphere), arctic (antarctic), and polar zones. Here we are at the opposite end of the curve we began at the equator. Variations in the length of days and nights are enormous, climaxing in the polar six-month day and six-month night, and seasonal contrasts in insolation are similarly immense.

• The Atmosphere •

We return now to the atmosphere itself, the gaseous layer that lies upon the liquid hydrosphere and the solid crust. The atmosphere of the earth consists of a mixture of gases that includes nitrogen (over 78 percent), oxygen (nearly 21 percent), and argon (less than 1 percent). The remaining fraction consists mainly of carbon dioxide, a very significant component of the atmosphere because it absorbs longwave radiation from the earth's surface, thus sustaining the atmosphere's warmth. It does this far more effectively than nitrogen or oxygen, so that the amount of carbon dioxide present in the atmosphere is an important factor in air temperature. In recent decades—indeed, ever since the onset of the Industrial Revolution—factories, automobiles, and other burners of coal, petroleum, and gas have been pouring carbon dioxide into the atmosphere from smokestacks and exhausts. The long-range effect of this on global atmospheric temperatures is unknown, but climatologists are concerned that a delicate balance of nature is being upset with consequences that could be serious.

The atmosphere's nitrogen, oxygen, argon, and carbon dioxide are identified as the **nonvariant gases**, because they prevail in approximately the same proportion from sea level to about 50 miles (80 km) into space. (The changing quantity of carbon dioxide is a human modification, not nature's design). Other gases, however, can exist in the atmosphere in varying amounts. Two important **variant gases** are water vapor—prevailing at under 1 percent by volume in desert areas but at 3 to 4 percent in humid equatorial zones—and ozone—a form of oxygen concentrated in very small but important quantities in a layer between 10 and 30 miles (about 15 and 50 km) above sea level. Water vapor, of course, is crucial in the atmosphere's capacity to absorb, transfer, and discharge moisture. Without water vapor there would be no clouds, no rain— and little agriculture. Ozone has the ability to absorb ultraviolet solar radiation, the kind that produces suntans but, in excess, blindness and skin cancers. The damage done to the ozone layer of the atmosphere by the engines of high-flying aircraft is another area of practical concern.

Like the other spheres of our multilayered earth, the atmosphere consists of several distinct layers (Fig. 3.6). This layering is *not* based on any change in the composition of the air, because this remains the same until 50 miles (80 km) out in space. Nor does the atmosphere's density decrease in stages. The key to the recognition of atmospheric layers lies in temperature changes.

It is reasonable to expect that air temperature will decrease with altitude. Permanent snow on the highest mountains, even in equatorial regions, appears to confirm this principle. The rate of decrease, 3.5°F/1000 ft (6.4°C/1000 m), has been long known and is referred to as the **environmental lapse rate** (sometimes the **normal lapse rate**). But when balloons carrying recording thermometers were sent higher than the highest mountains, they brought back some amazing data. At a height of 8.25 to 9 miles (13 to 14 km) temperatures stopped declining and, after briefly holding steady, they actually began to rise. At first, no one believed that this was actually the case. But eventually the fact was established: beyond about 9 miles (14 km) above sea level, where the temperature has dropped to about −76°F (−60°C), the mercury rises slowly until at about 30 miles (50 km) above sea level, it is back up to 32°F (0°C).

LAYERS

The lowest layer of the atmosphere, where temperature declines according to the lapse rate, is the **troposphere**. This layer is deepest at the equator, where it may extend as high as 10 miles (16 to 17 km) above sea level; over polar areas it reaches no higher than 5 miles (8 km). The troposphere is capped by the **tropopause**, where the temperature gradient changes. Above the tropopause lies the **stratosphere**, where the air is already very thin, water vapor and clouds almost completely absent, and vertical winds a rarity. Aided by persistent and predictable horizontal air currents, powerful jet aircraft perform well in the stratosphere. But the ozone layer (sometimes called the **ozonosphere**) lies within the stratosphere, and scientists have warned that the presence of increasing numbers of aircraft may threaten this fragile component of the atmosphere. At the **stratopause** the upward temperature gradient prevailing in the stratosphere changes again, and in the **mesophere**—between 30 to 50 miles (50 to 80 km)—it declines steadily. At this level, 50 miles, above the earth, the **mesopause** signals yet another reversal of temperature change, and in the **thermosphere**, it rises spectacularly (Fig. 3.6b). It is important to remember that the atmosphere at these upper levels is so thin (pressure is less than 1/50,000 of the sea-level value in the thermosphere) that temperature here refers to individual, widely separated

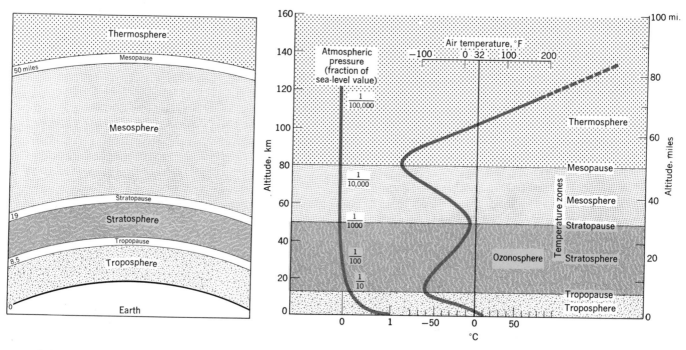

FIGURE 3.6
Layers of the earth's atmosphere.

molecules, not to the general environment. The surface of a spacecraft would not heat to 212°F (100°C) in its passage through the thermosphere.

THE TROPOSPHERE

For obvious reasons, the lowest layer of the atmosphere, the **troposphere**, interests us most. It is the layer of life, the essence of our existence on this planet. It constitutes a large part of the natural environment we experience as we sense its changing temperatures, its breezes and storms, its droughts and downpours. We may visualize the troposphere as an ocean of air resting on both land and hydrosphere, and at the bottom of this ocean—at the interface—humanity has emerged and thrived. Our life layer is very thin, however. The troposphere may average about 8 miles (13 km) in depth, but people can live and breathe comfortably only in the lowest 13,000 ft (about 4000 m). By the time mountain climbers reach the upper slopes of the earth's highest mountains,

they need oxygen equipment. The ocean of air is shallow, indeed.

The troposphere is warmed by the longwave radiation emitted by land and water beneath it. Earlier we noted that various latitudes of the earth do not receive shortwave radiation in the same amounts, so that the available longwave radiation also varies. The combination of high sun angles and low albedo contributes to the strong longwave heating in equatorial areas; low sun angle and high albedo inhibit the incoming radiation in polar areas, so that much less longwave radiation can be emitted there. But even within the same latitude, the conversion of shortwave radiation to longwave radiation does not take place everywhere at the same rate. This is really rather obvious: on a hot day the sand of a beach may become too hot to walk on, but the water will never become that warm. Neither by ground radiation nor by conduction could the water ever heat the air above it as fast as the adjacent land could. This is true on a large scale as well. Under a hot sun of summer, whole continental areas heat up much faster and to higher temperatures than the oceans of the same latitudes.

Why should this be so? In the first place, the water is constantly in motion whereas the land is not. No sooner is the top layer of water heated up than a wave pushes it several feet down and cool water replaces it. The heat is redistributed. In contrast, the radiation received by land surfaces builds up in the stationary upper layer. And water is transparent, which allows the sun's rays to penetrate to a considerable depth. Rocks, of course, are opaque. And although evaporation takes place both on sea and on land, the rate involved, also a cooling process, is much greater at sea than on land. Altogether, then, water requires nearly five times as much heat to increase its temperature the same amount as rock does.

But there is another side to this situation. Although the land surface heats up very rapidly, it also loses its heat much faster than the water does. Returning to that beach late at night, you may find that the sand is unpleasantly cold, but the water feels comfortably warm. When the solar radiation is ended by the coming of night, the rock and soil quickly lose much of their heat. This means, again on a larger scale, that temperature contrasts over land areas (seasonal as well as daily) are much greater than over water areas. Water tends to retain its warmth longer, partly for the reasons we identified before, the depth of heat distribution and mobility. In a short time, we will see that this land–water contrast has great implications for the world's distribution of climates.

AIR PRESSURE AND WINDS IN THE TROPOSPHERE

When, in 1783, two Frenchmen filled a large bag with air from a wood fire and watched it climb several hundred feet into the sky, they discovered something that is fundamental to the study of the atmosphere. What they realized is that when air is warmed, it expands. It becomes lighter than colder air and will rise.

On this principle, the entire atmosphere of the earth, especially the troposphere, is kept in constant motion. We know that different parts of the earth receive different amounts of solar energy and that differences in surface material such as between land and water create differences in rates of ground radiation. Let us reconsider for a moment that day at the beach when the sand became so hot. What happened to the air there? It was heated by intense ground radiation and by conduction, and it began to rise. All afternoon, the air over the land rose like the Frenchmen's balloon. What replaced the rising air? Air from the water, where it had remained relatively cool. Thus the bathers enjoyed a **sea breeze**, the cool air from over the water blowing landward to fill the place of the hot, rising air over the land. But that late night, when we went back and found the sand rather cool but the water pleasantly warm, we could observe the reverse. Now the air over the water was warmest and rising, though not as rapidly as the air over the land had earlier. Nevertheless, smoke from a beach fire would drift out over the water. Air from the cooler land was replacing the rising air over the water, and a **land breeze** was blowing.

The buoyancy of air is transplanted to maps and charts by measurement of the amount of **pressure** it exerts. Air heated to a high temperature (and rising as a result) exerts less pressure on the surface than air that is cold. Thus the differential heating of the earth we discussed before has the effect of creating not only temperature contrasts, but also pressure differences. And pressure differences lead to constant readjustment as the air moves to replace ascending (or descending) air.

This would lead us to assume that the equatorial zone, where so much heat is received, should be a zone of rather low pressure. And so it is. Here the process of **convection** prevails. On either side of this equatorial zone of low pressure, at about 25° to 30° north and south latitude, is the reverse, a zone of subtropical high-pressure belts. The **equatorial trough** (as the low-pressure zone is called) and the **subtropical high** (sometimes called the high **horse latitudes**) are parts of a global system of pressure zones and wind belts.

Thus air moves from areas of relatively high pressure to areas of lower pressure as a result of the **pressure gradient force**. When the pressure difference over a given distance is great, the air movement will be rapid; when the difference is comparatively small, the air flows more gently. You can

99

Atmospheric air has weight. The force of gravity acts upon air as it does on all matter, and this weight is expressed as **atmospheric pressure**. It is the force of gravity that keeps the atmosphere pressed against the land and sea surfaces.

Atmospheric pressure is expressed as the weight of a column of air on a unit area of the earth's surface, and it amounts to about 15 lb/sq in. (1 kg/sq cm) at sea level. Climb a mountain, and the pressure goes down, since less air remains in the column above. At 20,000 ft (about 6000 m), atmospheric pressure is well under *half* of what it is at sea level!

As long ago as 1643, the Italian scholar Evangelista Torricelli performed an experiment in which he filled a glass tube with mercury, put his finger over the open end, and inverted it in a dish of mer-cury. The mercury did not rush out of the tube: rather, it dropped a bit and then stood nearly 30 in. (76 cm) above the mercury in the dish.

Torricelli realized that the atmospheric pressure on the mercury in the dish kept the mercury in the tube elevated. He had invented the first **barometer**, and his observation became the standard: at sea level, the mercury barometer reads 29.92 in. (76 cm). Atmospheric pressure also is expressed in **milli-bars**, and the standard sea-level pressure on this scale is 1013.25 mb.

On a map, points of equal pressure are connected by lines termed **isobars**. The dimensions and properties of pressure systems can be interpreted from the location and spacing of such isobars.

read this off a weather map: when the isobars are close together, the wind will be strong; when they are farther apart, the wind will be weaker.

Earth's Rotation

But the pressure gradient force is not the only force that propels or affects moving air. The earth's constant spinning—its rotation—also has an important effect.

The earth rotates quite rapidly: it makes a complete turn every 24 hours. This means that a location at the equator travels nearly 25,000 miles (40,000 km) every day, moving constantly at 1050 miles (about 1680 km) per hour! The atmosphere, held to the earth by gravitational force, travels with the moving surface so we do not notice that we are constantly in motion. But this does not mean that the air of the troposphere is not affected by the earth's rotation. Indeed, without noticing it, we are affected, too, as are all moving objects on the globe's surface.

The earth's rotation causes all moving matter on the surface to be deflected from the path it is traveling. This force of deflection, the **Coriolis force**, deflects all moving objects in the Northern Hemisphere to the right of their direction of motion and all objects in the Southern Hemisphere to the left of their direction of motion. This is not just a bit of abstract or theoretical physics: river courses, people walking, automobiles moving, aircraft flying, and rockets during launch all are pushed from their original, intended direction of motion. The same is true for the air of the atmosphere. When air moves as a result of a pressure gradient, Coriolis force immediately deflects it.

Coriolis force affects global pressure systems and wind belts as well as smaller, local weather systems. If the Coriolis force did not exist, air would simply flow equatorward from higher latitudes in meridional (straight north–south and south–north) directions, and the global circulation pattern would look like Figure 3.7. But air flowing from north to south in the Northern Hemisphere will be deflected to the right, so toward the southwest. And the return portions of the cells, in the upper air, also are subject to the Coriolis force. Near the equator, then, surface winds in the Northern Hemisphere are the **northeast trades** (they flow toward the southwest, but winds are always named according to the direction *from which they come*) and in the Southern Hemisphere, deflection

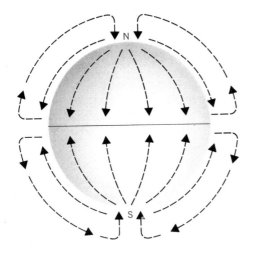

FIGURE 3.7
Air circulation without the Coriolis effect. Redrawn from a sketch in Donn, *The Earth*, 1972.

being toward the left, the surface winds are the **southeast trades**. The upper air ''return'' of the system would, theoretically, have an opposite or westerly flow.

Cells and Jet Streams

It has taken climatologists several decades to unravel the details of global circulation systems. As long ago as 1735, the English scholar George Hadley had explained the trade winds of the equatorial zone on the basis of differential heat and earth rotation (thus anticipating Coriolis's calculations), but his idea failed to account for the westerly wind belts that lie poleward of the tradewind zones. Not until 1941 did Carl Gustav Rossby produce a significant improvement by postulating a system consisting of several **cells**, among which the one generating the trade winds is named after Hadley (Fig. 3.8). Beyond the high-pressure

FIGURE 3.8

Global pressure and wind belts; cells in the atmosphere.

101

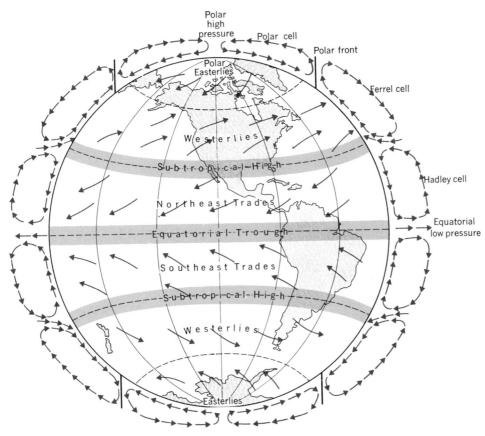

zones of the subtropical latitudes lies a second cell that has westerly winds at the surface and "return" winds with an easterly component in the upper air. Separating this cell, the Ferrel cell, from the Polar cells is the **polar front**, the contact zone between the frigid air of the polar regions and the more moderate air of the middle latitudes.

The Rossby model accounted for the observed surface circulation across the globe, but when more became known about the upper air—the "return" portions of the system—it proved to be inaccurate. For example, in the middle latitudes (where there are westerlies on the surface), no persistent easterlies were recorded in the upper air. In fact, westerly winds exist in these middle latitudes to the top of the troposphere! Moreover, the circulation in the Hadley cell, when carefully measured by scientists, proved to be much too slow to redistribute the heat from equatorial to higher latitudes. Something else had to be happening in the upper air. This turned out to involve

a great, continuous wave of high-altitude air in the middle latitudes, a tubelike westerly "river" of air that encircles the globe on the midlatitude side of the polar front (Fig. 3–9). This is the **polar front jet stream**, sweeping air from west to east in the zone of contact between polar and tropical air.

This polar front jet stream does not have a steady position around the polar zone; rather, it snakes irregularly and unpredictably, sometimes bending far equatorward and creating great loops, at other times withdrawing to the north. Its position varies with the seasons. During the winter of the Northern Hemisphere, the polar front jet stream comes to the south and the polar front lies to the south; thus much of the hemisphere is engulfed by cold polar air. When summer returns, the polar front moves north; but it can always generate severe atmospheric disturbances, because it is a contact zone between polar and tropical air masses.

The polar front jet stream is not the only zonal

FIGURE 3.9

Jet streams in the atmosphere.

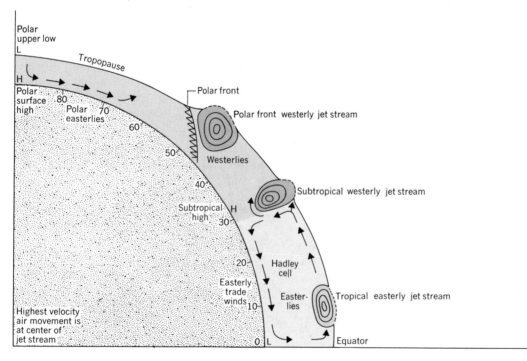

river of air to encircle the globe. At the poleward end of the Hadley cell lies another great westerly stream of air, the **subtropical jet stream**. These jet streams move air in great quantities at extremely high rates of speed (well over 220 miles/hour [350 km/hour] is not unusual) and can thus achieve the heat and volume transfers that could not be accomplished by the cells of the Rossby model alone. A third jet stream, the **tropical easterly jet stream**, lies near the equator in the Northern Hemisphere only (polar front and subtropical jet streams occur in both Northern and Southern hemispheres). This tropical easterly jet stream flows in an opposite, westward direction, and like other jet streams it migrates and bends as it does so, changing positions all the time. It is believed that the tropical easterly jet stream has much effect on weather in equatorial regions. In India, the annual monsoon can begin when the jet stream has "jumped" to the north of the Himalayan barrier, thus permitting the inflow of moisture-laden air.

It is important to remember that the preceding discussion necessarily conveys generalizations, not specifics. The global wind and pressure belts, for example, are not invariably made of the winds and pressures after which they are named. What does happen in the westerly wind belt is that the *prevailing winds* are westerly and that weather systems *tend* to move from west to east. In the subtropical high-pressure zones, high-pressure systems *prevail* at the surface. The equatorial low-pressure belt is marked by comparatively little horizontal air movement (in the past, this becalmed sailing ships, and the zone became known as the **doldrums**, but this same zone can experience storms and gales as well).

Figures 3.8 and 3.9 also should be seen in the context of the migration of the sun's vertical rays. Only twice during the year—shortly after the equinoxes in March and September—can we expect a symmetry such as depicted by Figures 3.8 and 3.9, with the equatorial low-pressure zone astride the equator. During the Northern Hemisphere's summer, the whole system moves to the north, so that the southeast trades actually cross the equator (and thus, coming under the opposite influence of the Coriolis force, bend to the northeast, becoming south-*westerlies*). During the Southern Hemisphere's summer, in December and January, all the global wind and pressure belts move southward; in the Northern Hemisphere, the area under the influence of polar air expands at the expense of the zone of westerlies.

PRESSURE SYSTEMS

So far, we have discussed pressure and wind zones on a global scale. But when we listen to weather reports, we are told that there is a "large high" over central

Canada or "a low-pressure cell" over the southern Great Plains. What this means is that centers of high and low pressure develop within the system of the global zones described earlier.

How does this happen? Several factors are involved. One phenomenon accurately suggested by Figure 3.8 relates to the persistence of air in specific parts of the troposphere. Note how air flows from the subtropical high through the tropics to equatorial regions at the surface and back again at upper levels; note also (on the vertical sketch) how polar air tends to get caught up in polar–arctic circulations. Thus **air masses** develop certain semipermanent characteristics, and we can recognize tropical air masses and polar air masses. They develop these characteristics by prolonged exposure to certain conditions (heat, humidity; cold, dryness) prevailing in a certain area, referred to appropriately as a **source area**. Northern Canada is such a source area. The cold, dry air that sweeps south into the heart of North America during the winter develops there. The Gulf of Mexico is another source area, in this case for warm, moist, tropical air well known by anyone who has spent the muggy days of a Midwestern summer in a sweltering city.

There is more involved. We know that cold air is heavier than warm air and exerts more pressure on the surface. That polar–arctic air, therefore, develops

104

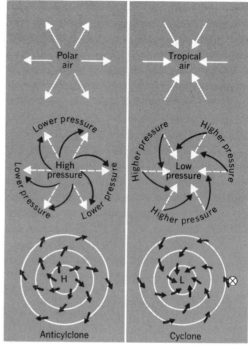

FIGURE 3.10

Air circulation in high- and low-pressure cells in the Northern Hemisphere.

• *Inter-Tropical Convergence* •

Since the northeast trades flow toward the southwest and the southeast trades travel toward the northwest, it is reasonable to conclude that they must meet. They do so along a narrow zone termed the **Inter-Tropical Convergence** or ITC (sometimes ITCZ [Z for zone]).

The ITC is a zone of mixing and rising air that has considerable influence on weather patterns in parts of the tropics. Over the oceans, it migrates northward and southward only little during the swing of the seasons. But over the landmasses of Africa, South Asia, and South America, it moves north and south over a zone of as much as 35° latitude. In West Africa, the rainfall brought by the inflow of the southeast trade winds cannot begin until the ITC has moved north into the Sahara interior. Elsewhere, too, the ITC functions as a barrier that controls the influx of moist air. Recent severe droughts in the Sahel and Ethiopia were directly related to the movement of the ITC.

A cyclonic storm system in the Northern Hemisphere. The streaked clouds reveal the counterclockwise direction of air movement in this low-pressure system, photographed from the Apollo spacecraft.

high-pressure characteristics. This means that it is, in effect, a heavy pile or bulge of air and that the air is busy flowing outward from it in all directions to places where it can push lighter air away. But now the Coriolis force comes into play. Imagine the outflowing air by drawing a series of arrows fanning straight out from the center of the polar air mass. The Coriolis force begins to affect those arrows, and (in the Northern Hemisphere) they are deflected to the right. As Figure 3.10 shows, that results in a closed cell of circulation in which the air moves clockwise. This is called an **anticyclone**.

Low-pressure tropical air can produce the reverse effect. Imagine a mass of such warm air has flowed into the southern Great Plains where it is additionally heated by the high-angle July sun. Air in the center of the mass rises, heated by ground radiation, and air from adjacent parts of the air mass flows into the center, replacing the risen air. Again, the arrows would point straight along the pressure gradient (Fig. 3.10, right top), but the Coriolis force deflects the movement of the air, again to the right. The result of this is a counterclockwise circulating system known

as a **cyclone**. (In the Southern Hemisphere, where the Coriolis force deflects to the left, the cyclone's circulation is clockwise and the anticyclone counterclockwise).

The weather map that is part of a good daily weather report, thus, has lines connecting points of equal pressure (*isobars*) closing around cyclones and anticyclones, and it shows wind directions associated with these systems. Weather forecasters look for changes in these systems as they traverse the country, and some of these changes will shortly be discussed. In the meantime, let us not lose sight of this important reality: even though a cyclone may control the whole central United States and winds in this cyclone blow from all directions (as Fig. 3.10 shows), the whole cyclone lies in the westerly wind belt (Fig. 3.8), and it will slowly travel an eastward path. This is why, if your home were located at point X on Figure 3.10, you could predict with fair certainty that winds in the next few days would change from south to southwest, then to west, and eventually to northwest. All you have to do is move the cyclone across point X and read the arrows!

The Atmosphere

The Hydrosphere

We now interrupt our study of the troposphere to consider the world's oceans, because here the air of the atmosphere acquires its moisture. Like the troposphere, the **hydrosphere** is constantly in motion, and some of the forces that cause the air to move also affect the water. But the most important influence on oceanic circulation patterns is the atmosphere itself. The persistent low-pressure systems over the Atlantic Ocean to the north and south of the equator combine with the easterly winds near the equator and westerlies in the middle latitudes to stimulate a great clockwise circulation in the North Atlantic and a counterclockwise circulation in the South Atlantic (Fig. 3.11). The water is set in motion by the surface friction of the wind, and once it moves it comes under the influence of the Coriolis force. These forces plus temperature and density contrasts caused by differential heating produce a global circulation system that is essentially (but not exclusively) clockwise in Northern Hemisphere oceanic zones and counterclockwise in Southern Hemisphere waters. The oceanic circulation cells or **gyres** vary in size from huge, ocean-filling systems to small whirls no more than a few hundred miles in diameter.

The worldwide system of ocean currents is perhaps best explained by a close look at the situation in the Atlantic Ocean. Once we understand the system of circulation there, the worldwide mechanics will not be difficult to grasp. Figure 3.11 represents (in generalized form) the circulation systems of the Atlantic. Note, first, that some of the water movements are called **currents** (such as the Benguela Current), others are identified as drifts (the West Wind Drift), and still another is referred to as the Gulf Stream. These terms do not represent sharp differences between the various kinds of water bodies in motion. A current may be regarded as a rather more defined, fairly rapidly moving "river" of water in the ocean; a drift is slower, less confined. The Gulf Stream is simply another name for a current in that area. As it crosses the North Atlantic, it loses its definition and becomes a drift.

In equatorial latitudes, the effect of the persistent easterly winds is to drive water from east to west as the **North and South Equatorial Current**. On reaching South America, some of the water piling up in the western Atlantic turns southward, carrying warm tropical temperatures along the coast in the **Brazil Current**. Other water moves on along the north coast of North America, eventually entering the Caribbean Sea or directly merging into the Gulf Stream, also a warm current. The Gulf Stream and Brazil Current flow poleward along the North and South American coasts, respectively, the Gulf Stream fading into the **North Atlantic Drift** and the Brazil Current turning into the **West Wind Drift**. Here the water is cooling, especially in the South Atlantic; by the time the current called the **Benguela Current** develops, the water is quite cold. Now, the water is moving from higher, near-polar latitudes toward tropical zones; in the process, it carries the coldness of high latitudes to the continental coasts. In the North Atlantic, the **Canaries Current** is something of an equivalent to the Benguela, but it emerges from the North Atlantic Drift still a lot warmer than the West Wind Drift is. As the water nears the equatorial zones it warms up; by the time it becomes the Equatorial Current, it has reached its temperature of 80°F (27°C) or even higher once again.

This broad, regular system, whereby east coasts have warm offshore currents and west coasts have cold currents nearby, is disrupted in various places. In the North Atlantic Ocean, the North Atlantic Drift crosses the entire ocean and brings warmth to European ports, keeping them ice-free when they would otherwise freeze up; the water then turns back along Greenland and between Greenland and Labrador, from where the **Labrador Current**, a cold stream of water, emerges to round Newfoundland and cool the Gulf Stream. In the South Atlantic Ocean, the **Falkland Current**, another cold-water body, moves around Tierra del Fuego and heads off the southward-flowing Brazil Current.

Broadly outlined, this pattern—a westward movement in equatorial latitudes and circular systems of circulation, which are clockwise in northern waters, counterclockwise in southern oceans—is repeated in

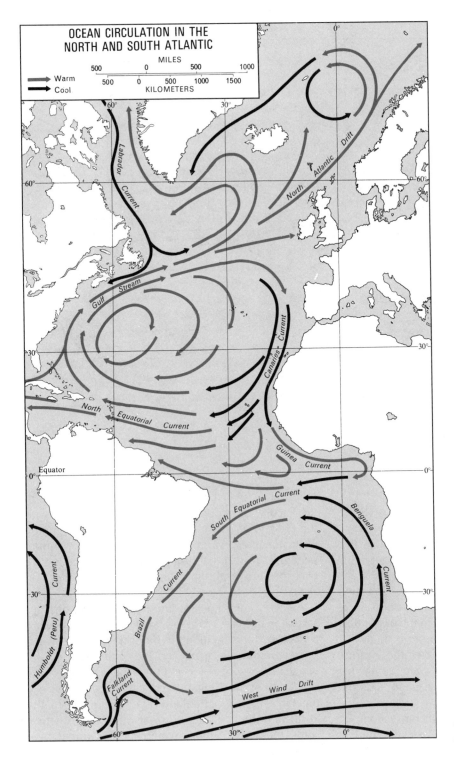

107

FIGURE 3.11

Movement of ocean water in the
Atlantic Ocean.

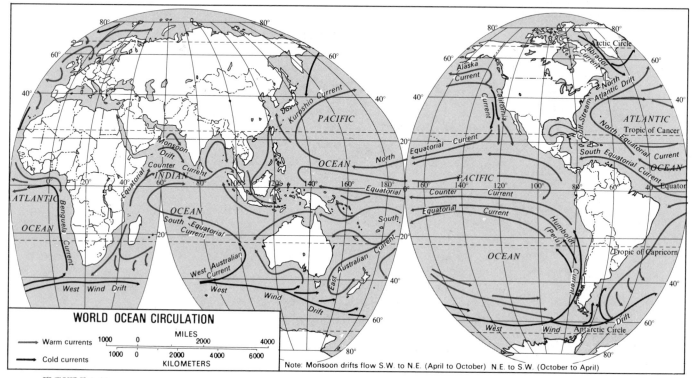

FIGURE 3.12

Major currents and drifts in the world's oceans.

the Pacific and Indian oceans as well (Fig. 3.12). These oceans are less confined than the Atlantic, though, and so the patterns are less clearly defined.

The Indian Ocean has much more area in the Southern than in the Northern Hemisphere, but there is enough space in the north to permit the development of full-scale gyres with essentially clockwise circulations: one to the west of the Indian peninsula in the Arabian Sea, the other to the east in the upper Bay of Bengal. South of the equatorial currents there develops a large counterclockwise gyre, the (warm) Moçambique Current moving southward along the African coast and the West Australian (cold) Current northward near Western Australia.

The Pacific Ocean, largest of all, has a complicated pattern. Both the Indian and especially the Pacific oceans have a strongly developed, eastward-flowing countercurrent between the westward-moving North and South Equatorial Currents. In the case of the Pacific, this countercurrent has much to do with the *El Niño* phenomenon described at the beginning of this chapter. Warm water from the countercurrent penetrates the normal system and, flowing southward, overrides the cool, upwelling water off Ecuador and Peru.

North of the Equator, the Pacific Ocean displays several major clockwise-moving gyres (Fig. 3.12). Important currents include the Kuroshio Current, which brings warmth to Japan, and the California Current, which brings cool waters to California's coasts.

South of the Equator, the Pacific Ocean develops the largest of all circulations, extending from the northward-flowing Peru Current off South America, across the ocean in the South Equatorial Current, then southward in the East Australian Current, and then eastward in the south. This enormous counterclockwise-moving gyre has several spinoffs, minor circulations such as that partially developing between Australia and New Zealand. But the overall system moves water as shown in Figure 3–12.

All three major oceans, the Atlantic, Indian, and Pacific, merge into the ocean that surrounds Antarctica, the Southern Ocean. As the map of world oceanic circulation shows, this Southern Ocean has a continuous eastward movement, a motion set up by westerly winds that prevail here. This giant Antarctic circulation is called the West Wind Drift. Here the warmer, southward-flowing currents are swept up and chilled, and from here comes the cold water for the Benguela and Peru Currents, among other northward-flowing currents.

This outline of the oceanic circulation is relevant to our study of the troposphere, because the temperature of the water offshore and the direction of air movement across ocean and land are important elements in the total climatic environment. Air drags the water into drifts and currents; water, in turn, can affect the temperature and moisture content of the air.

THE HYDROLOGIC CYCLE

The water of the oceans and the air of the atmosphere combine to deliver enormous quantities of moisture to the landmasses in an unceasing system called the **hydrologic cycle**. The hydrologic cycle could not take place if water could not change from the liquid state to the vapor state and back again to the liquid state (and even, as ice, to the solid state). When the surface of the ocean is in contact with an air mass, **evaporation** takes place, and water in the vapor state becomes part of the atmosphere (we might say that the air has high humidity). Evaporation rates are higher where temperatures are high, and warm air can contain much more moisture than cold air can, so that air masses in equatorial zones are typically humid, whereas polar air masses are dry (we might say crisp).

Once the water vapor has entered the air mass and the air continues its movement in the trades, westerlies, or some other pressure system, it may reach a landmass and overspread it. By various processes, which we will discuss shortly, the moisture in the air now condenses and falls on the land as precipitation (Fig. 3.13). On reaching the land surface, some of this precipitation evaporates back into the air again—from the leaves of vegetation, from the soil, and from the surfaces of lakes and rivers. Part of it seeps into the soil to become ground water, but much of this eventually drains into lakes and streams and even back into the ocean itself. And part of the precipitation becomes runoff, flowing directly into streams that carry it back to the ocean as well. As the water drains back into the ocean it mixes with the passing current, and eventually it may evaporate back into the air again. Then the whole circulation system is renewed.

In this way the hydrologic cycle serves as a giant, worldwide pumping system that brings life-giving water to even the deepest interiors of the continents.

The hydrologic cycle is a global system, and it is difficult to measure its components. When climatologists began to record precipitation, humidity (moisture in the air), seepage in soils, runoff in streams, evaporation, and other measurable processes in various areas on land and sea, they made an important discovery: some areas have what may be called a water surplus, others have a deficit. This in itself is not surprising—the landscape and vegetation give strong indications of this. But what *was* surprising was that large areas that would seem to have a surplus actually do not or have it only seasonally. This gave rise to the concept of **water balance**, the annual (or seasonal) water budget of a locale.

Equatorial zones are best supplied with water and have a favorable water balance. But moving north and south from the equatorial zone into the tropics, we find large areas between 10° and 40° latitude that have a negative balance—and not just desert areas. There may be quite considerable rainfall in such areas, but when various losses are measured (including evaporation in the high heat of the low-latitude "summer"), the loss often exceeds the gain. This means that vegetation in such areas is under stress, which can make it vulnerable to destruction. Even areas that have a positive water balance during part of the year may record an overall (annual) loss.

EVAPOTRANSPIRATION

One aspect of the loss of moisture and negative water balances has to do with a combination of processes

109

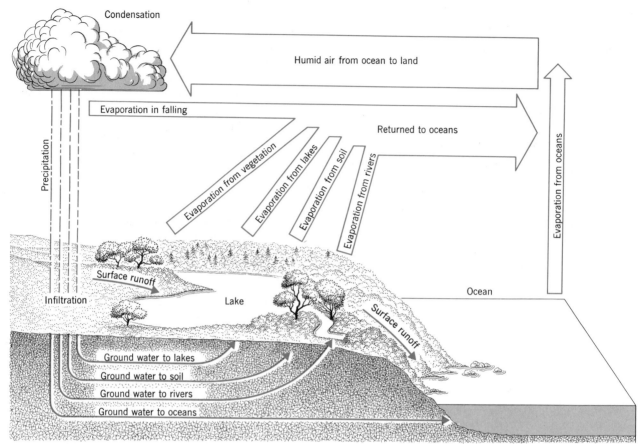

Condensation

Humid air from ocean to land

Evaporation in falling

Returned to oceans

Precipitation

Evaporation from vegetation

Evaporation from lakes

Evaporation from soil

Evaporation from rivers

Evaporation from oceans

Surface runoff

Infiltration

Lake

Surface runoff

Ocean

Ground water to lakes

Ground water to soil

Ground water to rivers

Ground water to oceans

FIGURE 3.13

The hydrologic cycle.

called evapotranspiration. Moisture is lost in several different ways, as we have noted. Evaporation from the soil is one process whereby moisture is returned to the atmosphere during the hydrologic cycle's circulation of water; when there is no rain, the soil gives up its moisture to the air and becomes progressively drier, and the dry-soil zone becomes deeper as the rainless days wear on.

At the same time, the roots of plants absorb soil moisture and transfer it through trunks and branches to the leaves. There this moisture collects in the leaf pores and is lost to the atmosphere during a process known as transpiration.

The two processes of direct evaporation and plant transpiration are combined as **evapotranspiration**, since they account for the bulk of the water loss in any unit area (not counting runoff). In combination, the two processes provide a ready measure of the state of the water balance, because it is possible to measure the *actual* amount of evapotranspiration taking place and to calculate the *potential* amount of it that would be taking place if ample moisture were available. Obviously, the rate of evapotranspiration is faster where it is warm and slower where it is cool, just as we perspire more profusely when it is hot. When a certain area receives as much as 40 in. (100 cm) of rainfall annually, it is important to know how this relates to the rate of evapotranspiration there. In warm subtropical regions, 40 in. may be far below the need indicated by measures of potential evapotranspiration. But in the cooler middle latitudes, 40 in. of precipitation may be ample. When we consider the world distribution of precipitation later in this chapter and soil distributions in the next, the patterns should always be viewed in light of actual and potential evapotranspiration.

• Moisture in the Troposphere •

The troposphere (we noted previously) has properties of temperature, pressure, and moisture. Earlier, we studied temperature and pressure in the troposphere, and now, after examining the hydrosphere (where the moisture originates), we are in position to approach this third property.

Water vapor is one of the troposphere's variant gases, constituting as much as 4 percent of the total volume of air in equatorial zones but a fraction of 1 percent in cold polar regions. When air at a given temperature contains all the water vapor it can contain, that air is said to be **saturated**. If it contains only a part of the moisture it could contain, say half of it, the air would have a **relative humidity** of 50 percent. Saturated air, therefore, has a relative humidity of 100 percent. We notice the difference in our experience during the summer: a temperature of 86°F (30°C) is much more bearable when the relative humidity is 50 percent than when it is over 90 percent.

Air has a temperature level at which the particular amount of water vapor present fully saturates it. To take the example just given, air that has a temperature of 86°F (30°C) and a relative humidity of 90 percent needs very little cooling for that 90 percent to saturate it completely. When that happens, the relative humidity has reached 100 percent and the air is said to have reached its **dew point**. If the temperature of the air continues to drop, some of the water vapor must be removed, and this produces **condensation** of the water vapor (in the gaseous state) into drops of liquid water. The term *dew point* is certainly appropriate, because dewdrops on the grass and other objects (e.g., car windshields) in the morning indicate that overnight cooling of the air has taken the temperature below that critical level.

Thus the air of the atmosphere will release the moisture it contains only when it is cooled. How does this cooling take place in sufficient degree and involve enough air to produce precipitation, not just a few dewdrops on vegetation? The kind of nighttime cooling that produces dew can occur even during a clear, cloudless night. But dew does not saturate the soil and moisten plant roots. This requires rain or some other form of precipitation. Hail and snow also result from the condensation of moisture in humid air.

Air in large quantities can be forced to rise (and therefore expand and cool substantially) in three ways. First, air can be heated by longwave radiation from the ground, supplemented by conduction, and rise just as heated air in a balloon would. Second, air is forced to rise when it is pushed up against a mountainside that lies in its path of movement. And third, air rises when it collides with colder, denser, heavier air that wedges underneath it. Respectively, these three processes produce convectional, orographic, and cyclonic (or frontal) precipitation; each will shortly be examined in some detail.

Whatever the means by which the air rises, the rate of cooling is about 5.5°F/1000 ft (10°C/1000 m) so long as no condensation is taking place. This is the **dry adiabatic lapse rate**, not to be confused with the temperature decline or **environmental (normal) lapse rate** in stationary air, discussed earlier. When the temperature of the rising, cooling air drops below the dew point and condensation begins, latent heat of vaporization is released. This energy, which was locked up in the air when it absorbed the water vapor over the ocean, has the effect of slowing the rate of cooling during condensation to between 2° and 3°F/1000 ft (3° and 6°C/1000 m), the **wet adiabatic lapse rate**. Thus, a rising air mass initially cools quite rapidly, then reaches its dew point, condenses its excess water vapor, and cools more slowly. As the air rises still farther, it continues to shed its water vapor as condensed water (or as ice crystals), thus generating large amounts of precipitation.

111

• Clouds •

Clouds form when tiny droplets of water (or ice) collect around extremely small dust particles in the atmosphere. These small dust particles are essential, because they serve as the nuclei of condensation. This is true of fog also. A dense city fog often is generated by the high pollutant level in the lower air layer and by radiation cooling.

Cloud types are classified according to their appearance and the altitude at which they occur. *Stratiform* clouds are sheetlike, of wide areal extent, but thin. *Cumuliform* clouds are billowing, tall clouds.

Altitude is signified by prefixes. The lowest stratiform cloud is simply known as *stratus*, or *nimbostratus* when precipitation falls from it. Above

6500 ft (2000 m) or so, a stratiform cloud is known as *alto*stratus. Above about 20,000 ft (6000 m), it is *cirro*stratus. The lowest cumuliform cloud is **strato**cumulus, or simply *cumulus*, the middle form is altocumulus, and the highest, *cirro*cumulus. Very-high-altitude streaks of thin cloud are called *cirrus*.

Cumuliform clouds tend to develop to great height. Cumulus clouds, caught up in a convection cell, may develop a base at 1000 ft (300 m) and a head, or **anvil**, at 33,000 ft (10,000 m) or even higher. When this happens the cloud is identified as a ***cumulonimbus***, signifying that it produces rainstorms and often thunder and lightning along with squalls of high wind.

112

Cumulus clouds over a Midwest landscape. These puffy clouds, in contrast to *stratiform* clouds, develop vertically; on a hot day, some will evolve into cumulonimbus clouds.

CONVECTIONAL PRECIPITATION

The process of convection involves two critical elements. The first is the heating of a particular area during a period of uninterrupted strong insolation and, consequently, the rapid warming of the parcel of air lying over this area. The area need not be large, perhaps a few square kilometers; usually it is an exposed area, unprotected by leafy vegetation or surface water. Here the heated air begins to rise in a column termed a **convection cell**, and before long the dry adiabatic lapse rate has the effect of dropping the temperature below the saturation level. Condensation begins, and a cumulus cloud forms. Now the second critical element in the process starts to play its role: the release of 600 cal of heat for every gram of water that is condensed. This heat was stored as latent heat of vaporization when the air mass absorbed its water vapor from the ocean surface; it is released again when that same water vapor returns to the liquid form. Thus the updraft that began as a convection cell now perpetuates itself by its own energy; as condensation continues, the cumulus cloud builds into a huge cumulonimbus cloud to a height of 33,000 to 40,000 ft (10,000 to over 12,000 m). Not all convection cells reach such dimensions, of course; in many the amount of water vapor is insufficient to sustain development. But when a full-scale cumulonimbus cloud surges skyward, a series of updrafts push forward to its anvil-shaped top, which is not far from the roof of the troposphere where temperatures fall below freezing, thus producing the nuclei for hailstones that are carried down in vicious downdrafts capable of destroying aircraft. Hail and rain strike the ground along the base of the cloud in squall winds that can uproot trees, damage buildings, and destroy crops. All this occurs amid lightning and thunder as the enormous energy packed in a cumulonimbus cloud is released.

Not every conventional rainstorm is attended by this scenario. Some convection cells develop into cumulus and cumulonimbus clouds that produce some precipitation, but then peter out as the supply of water vapor ends. But in moist areas in equatorial and tropical latitudes, huge cumulonimbus clouds develop almost daily. It happens wherever the air is **unstable**—capable of propelling itself upward through condensation following initial convection. Over the Florida Everglades from June to September, late-afternoon and early-evening skies are streaked by the lightning from dozens of towering cumulonimbus clouds.

OROGRAPHIC PRECIPITATION

When a mountain range lies in the path of a moving air mass, the air is forced to rise over the mountains' crests and, in the process, may be cooled below its saturation level. If the mountain range lies near the coast and the air mass has passed over warm ocean water (thus increasing its humidity through evaporation), not much upward movement may be needed to cool the air to its dew point. When this happens, precipitation begins and the rain (or snow) continues to fall until the air has crossed the summit of the range. During the moist ascent, the air cools at the lower wet adiabatic rate of between 2° and 3°F/1000 ft. On the other side of the mountain range, the air mass descends, warming at the dry adiabatic lapse rate of about 5.5°F/1000 ft, so that it warms up on the lee side much faster than it cooled on the windward side. The result is a strong contrast on the two sides of the mountain range: a cool, moist, well-watered windward slope and a dry, perhaps desertlike, warm lee

side. Since it was the mountain slope that wrested the moisture from the approaching air, the rainfall is termed **orographic precipitation**.

Again, the situation is not always so clear-cut. But even mountain ranges located deep in the interior of the continents can draw some moisture from the air, and many of the world's deserts begin in the rain shadow of mountain barriers. In the United States, the **chinook winds**—flowing down the lee (eastward-facing) sides of mountain ranges of Montana and Wyoming and warming at the dry adiabatic lapse rate—can raise the adjacent plains' winter temperature by nearly 50°F (30°C) in a few hours. There are instances when the temperature in South Dakota rose from about 20°F (–7°C) to 70°F (21°C) when chinook winds descended during a winter day.

CYCLONIC PRECIPITATION

The third process that generates substantial precipitation involves air masses in low-pressure cells called **cyclones**. Earlier we discussed the formation of these lows, but it is necessary now to look more closely at the air-mass concept.

In large areas of the world such as the icy Arctic or the warm Caribbean Sea and Gulf of Mexico, a layer of air may remain stationary for several days. As it hovers over such regions, the air takes on the characteristics prevailing there, for example, extreme coldness in Arctic zones, or warmth and high humidity in the Gulf and Caribbean. Thus those areas are the **source area** for the air masses. And the air masses become extensive bodies of air in the lower troposphere with relatively uniform properties of temperature and humidity.

Climatologists use code letters to identify air masses according to their sources. Two letters, (m) for maritime and (c) for continental, indicate whether the air mass developed over water or on a landmass, and hence its moisture content. Then follows a letter for the latitudinal position of the source region: whether Arctic (A) or Antarctic (AA), signifying the highest possible latitudes; Polar (P), between about 45° and 70° north and south latitude; Tropical (T), between 20° and 35° north and south; or Equatorial (E), the lowest latitudes astride the equator. Thus an mT air mass has its source region over tropical waters, and we can expect it to be warm, humid, and unstable. A cP air mass, on the other hand, will be dry, heavy, and stable, having originated over a landmass in polar latitudes.

Now consider as we did previously a low-pressure cell that occupies the central part of the continent, lying between the Rocky Mountains and the Appalachians and extending from southern Canada to northern Texas. We know that this cyclone has winds converging on the center and that these winds emanate from central Canada as well as southern Texas and the Gulf of Mexico. Imagine that this cyclone prevails during the fall, say, in late October. The air from the north, with its near-Arctic source area, is cold, heavy, stable. It flows into the system's northern half and is caught up in the circulation. (On the weather map, this air may be identified as cP, continental polar in character.) From the south comes air from the Gulf, warm, moist, buoyant, unstable. It flows into the southern half of the system. (The map shows this as mT air, maritime tropical in nature.) Now our low-pressure cell has two air masses in it, air masses with very different properties.

How do these air masses interact? Do they mix and eliminate their differences? No. Rather, the air masses make contact with each other along a surface called a **front**. If this term suggests wartime combat, that impression is correct. Just as ground armies face each other along a front and seek to push each other back, so the front between air masses is a zone of contact and combat. Figure 3.14 indicates where these fronts might be positioned in our hypothetical low-pressure cell. In the southwestern quadrant, the cP air is pushing back the warm air along a front that looks like a line on our map but is in reality a plane of contact, as the cross section *A* shows. In the southeastern quadrant, warm mT air is pushing back the cool air (cross section *B*). The small arrows indicate the wind directions in this particular system.

Here, then, we have the third process whereby air can be forced to rise to higher altitudes, generating rainfall or other forms of precipitation. Along the cold front, the heavy cP air wedges beneath the more buoyant mT air and forces the latter to rise quite rapidly, lowering its temperature below the dew point. Such a cold front is marked by billowing cumulus-

114

• Tornadoes •

When a cold maritime polar air mass and a moist, unstable maritime tropical air mass are in contact along an active cold front, it is not unusual to see a line of dense, dark-based cumulonimbus clouds ahead of the advancing cold front. All too frequently, the extreme turbulence associated with the squalls from these clouds leads to one of the most destructive of weather phenomena: a tornado.

A **tornado** is a small, local cyclone in which air spirals at speeds ranging up to as much as 250 miles (400 km)/hour. The tornado is funnel-shaped, and its base usually has a diameter of 100 to 200 yards (some are larger). In the center of the funnel, the pressure is so much lower than elsewhere that, when it passes over a house with closed windows and doors, the structure may explode. Tornadoes sometimes form in clusters associated with a severe

squall line and may travel many miles across a plain, the base alternately touching the ground and then lifting up, only to touch down again farther along the track. As the giant funnel snakes and writhes along, its base is blackened by the soil, dust, and debris it has picked up. Eventually the squall line weakens and the tornadoes dissipate.

No description could convey the terror and destruction of a tornado, whose power can lift entire sections of buildings and smash them to the ground elsewhere. In the United States, the area of greatest tornado risk lies in the heart of the Great Plains, especially from northern Texas through Oklahoma and Kansas into southern Iowa, but the whole region between the Rocky Mountains and the Appalachians—where air masses of contrasting properties interact—is vulnerable.

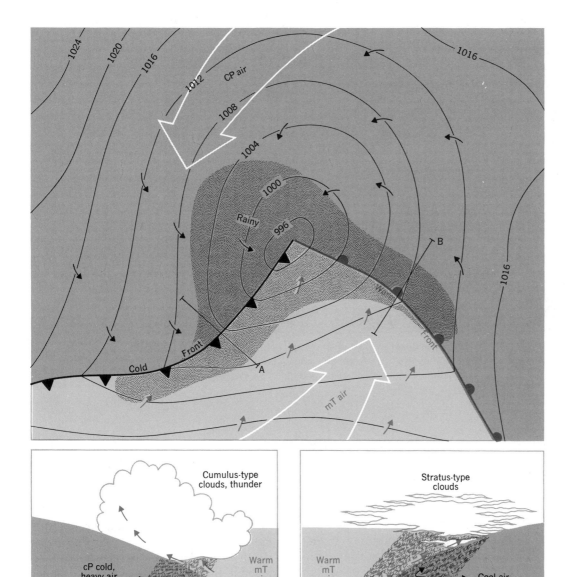

FIGURE 3.14
Cyclonic systems and associated fronts.

type clouds, intense rainstorms, and, often, lightning and thunder. Along the warm front, the updrafts are less rapid and the clouds more likely to be of the stratus variety, with rather persistent, gentle rain. Here the warm mT air overrides the cooler air (cross section B) and cools below condensation temperatures; in the cool air, too, there is some updrafting. Along the warm front, mixing of air is more pronounced than along the cold front.

In our middle latitudes, these frontal cyclones traverse parts of the North American continent almost daily, traveling eastward at rates varying from a few to

The source cloud and funnel of a tornado at Greenwood, Indiana (south of Indianapolis). The funnel moves across the countryside, sometimes in a straight line, sometimes in zigzag fashion; the base of the funnel may ''hop,'' touching the ground in places and lifting above it elsewhere.

several dozen miles per hour. Their paths are difficult to predict, for reasons we have discussed. They may move to the northeast into Canada, stay on a due-easterly track, or turn to the southeast. In our example, much of North America between the Rockies and the Appalachians was dominated by the cyclone pictured on Figure 3.14, but it will move eastward, and soon another system will take its place. Not all systems that cross the continent are cyclones. Sometimes, especially in the fall, winter, and early spring, an uncontested high-pressure area will overspread the region, bringing stable air, dryness, and blue skies. Fronts do not develop in such anticyclones as they move eastward; they do not draw in air, but rather disperse air outward. A careful look at tonight's weather report might reveal some high- as well as low-pressure areas in control of parts of the country.

TYPES OF PRECIPITATION

The term *precipitation* refers not only to rainfall, but to other forms of precipitation as well. Rain itself may arrive at the surface as a drizzle in which the individual drops are just large and heavy enough to fall to the ground. But rain can also come in drops so large that they cause **splash erosion**, the dislodging of soil particles by the impact of individual raindrops. This happens during severe rainstorms that fall on areas unprotected by vegetation. In equatorial zones, raindrops are likely to develop, grow, and fall in the liquid form throughout. But in middle and higher latitudes, rain often starts as small ice particles (snow, literally) that melt as they fall through warmer air at lower altitudes.

Precipitation also reaches the ground as snow. Snow begins as small ice crystals in a cloud in very cold air, and these ice crystals form nuclei that attract a film of water. This water also freezes, and soon clusters of the larger crystals adhere to each other and form snowflakes. When the snowflakes fall earthward through air at subfreezing temperatures, they are preserved and accumulate on the surface. A warmer layer of air melts the snow and rain falls. The density of snow on the ground varies considerably, depending on the shape of the crystals and flakes as well as the weight, temperature, and "wetness" of the snow. On an average, 1 in. of rainfall is equivalent to 8 to 10 in. of snow. In many areas, snow is an important form of water storage, with late spring and summer melting periods that provide dependable supplies of water during dry seasons. California's farmlands depend considerably on such meltwaters.

Occasionally raindrops fall earthward but cross a low layer of very cold air before reaching the ground; snowflakes may melt as they come down, only to be refrozen in such a low-lying cold air layer. This produces sleet, in which the precipitation reaches the earth in frozen form but is not crystallized as snowflakes are.

In high, dense cumulonimbus clouds where much condensation takes place, powerful updrafts can take water drops, freeze them as they penetrate colder layers, and coat them several times with new films of water, each of which freezes. Finally, the growing balls of ice become too large to remain suspended and fall earthward as hailstones. Hail as large as baseballs has been recorded; golfball-size hailstones are not uncommon. Even marble-size hail can be extremely damaging to crops.

• World Precipitation Patterns •

Precipitation, in its various forms, results from a complex set of processes and circumstances. We have studied fundamental systems and their results; now it is time to look at the global pattern. Plate 2 displays, in generalized form, the distribution of precipitation across the land areas of the globe during an average year. The equatorial and tropical areas immediately draw attention because they receive comparatively high amounts of rainfall. South and Southeast Asia, equatorial Africa, and northern South America generally receive more than 40 in. (100 cm) of rain, and large areas within these regions record more than 80 in. (200 cm) annually. And some locales such as coastal southwest India, parts of Indonesia and New Guinea, a small area in West Africa, and some Pacific islands receive as much as 120 in. (300 cm) and even more. The significant exception is East Africa, which appears to lie in the path of Indian Ocean trade winds and should have far more rainfall than it does. We do not know exactly what interferes with rain-bringing systems here, although there are theories that the trade winds are broken (by Madagascar from the southeast), that the layer of moisture-producing air is thin and the upper air is dry, and that East Africa gets leftover rainfall from the west, from equatorial Africa, rather than from the Indian Ocean. Even on a scale like this, the world map still poses unsolved problems.

It is clear, however, that the abundance of rainfall in the equatorial and tropical areas relates to the limited size of the land areas, the warmth of the air, and its capacity to carry huge amounts of moisture. Most of the rainfall comes by convection, though orographic processes also play a role, as in the case of the Indian monsoon and the escarpment of eastern Madagascar. And the apparently anomalous dryness of the west coast of South America and Western Africa south of the equator also can be explained. Here subtropical high-pressure systems extend somewhat inland, and since high pressure means subsidence and warming air, there is little or no rainfall.

Poleward from the equatorial and tropical zone of abundance, rainfall declines, a precipitous decline in Asia north of India and in Africa north and south of equatorial Zaïre. Here we are in the belt of subtropical highs. Australia, southern Africa, and South America all have desert and near-desert (steppe) areas at these latitudes, and in the Northern Hemisphere, the vast Sahara–Arabian Desert lies astride the Tropic of Cancer. In Asia, the zone of persistent high pressure shifts somewhat northward and lies to the north of the Himalaya Mountains. Note how dry much of interior Asia is. Remoteness from the moisture-bringing air masses and mountainous obstacles to air flow have much to do with this.

In the middle latitudes, the effect of the westerlies is shown clearly in North America and in Europe from about 30° northward and in South America from the same latitude southward. In North and South America, the areal extent of orographically induced precipitation is rather limited because coastal ranges (the Rocky Mountains and the Andes) confine the rainy

117

areas to coastal strips, but in Europe there is less impediment to the flow of the moist air. Note how, in North America, the trend of the Rocky Mountains coincides with the zone of annual precipitation below 20 in. (50 cm) and how, in Europe, the moist coastal zone extends as far eastward as the Soviet Union. East of the Rocky Mountains, much of the rainfall originates in the unstable air masses drawn from the Gulf of Mexico and the Caribbean Sea.

North of the middle latitudes, polar and arctic areas receive comparatively little annual precipitation, as Plate 2 indicates. In these areas the air is cold, containing as little as 0.5 percent of the moisture carried by a warm equatorial air mass. It is appropriate to describe the Arctic and Antarctic regions as polar deserts, because annual precipitation totals resemble

those prevailing in the world's driest areas, amounting to under 12 in. (30 cm) almost everywhere and declining to below 4 in. (10 cm) in some Arctic areas.

The map (Plate 2) thus shows enormous contrasts, but we should not read more into it than it shows. This is a map of averages, compiled from long-term records. As we know all too well from our own experience with the weather, averages are by no means guarantees that "expected" conditions will occur. When it comes to rainfall, there is a general rule that variability increases as the annual average decreases. Thus, the moist, equatorial area that records an average of 100 in. (240 cm) of rain is likely to get it, more or less, year after year. But the steppe area where the average is just 12 in. (30 cm) may experience a series of years with 4, 6, 5, and 3 in.—and then suddenly a

• Hurricanes, Typhoons, Cyclones •

Every year, in the tropical oceans between the approximate latitudes of 5° and 20° north and south, tropical depressions form that develop into tropical storms and, under certain favorable conditions, into **hurricanes** (as they are called in the Atlantic Ocean and in the Pacific east of 180°), **typhoons** (their name in the western Pacific), and tropical **cyclones** (in the Indian Ocean, especially in the Bay of Bengal).

In one way or another, these storms all begin as cyclones as we know them: low-pressure circulation systems. When such a pressure cell develops into a hurricane, it is almost perfectly circular in shape. It has a steep pressure gradient that leads to a central (core) low where pressure sometimes drops below 950 mb, when standard sea level pressure is 1013.2 mb. This generates an intense air circulation around the hurricane's core, with surface winds reaching 125 miles (200 km)/hour and even higher. The system grows to a diameter averaging 300 miles (500 km), but hurricanes more than twice this large have been recorded. Wind strength increases toward the center, but the core of the hurricane, its eye, is calm. The eye measures 3 to 9 miles (5 to 15 km) across; here, the air descends from high altitude, is warmed

adiabatically, and rejoins the circulation. Temperatures in the eye can be as much as 18° F (10° C) higher than in the other segments of the hurricane.

Hurricanes tend to move slowly and often erratically, although they drift westward and northwestward in the North Atlantic Ocean's trades and then curve northward and even northeastward as they penetrate the westerly wind belt. This path takes them through the Caribbean Sea; occasionally they curve into the North Atlantic and threaten the U.S. East Coast, but at other times they enter the Gulf of Mexico and strike the coast of Mexico or the Gulf states. In 1965, Hurricane Betsy crossed the Gulf, traversed Florida, and entered the North Atlantic. Suddenly the hurricane stopped, then reversed its course and struck Florida just south of Miami, reentering the Gulf of Mexico to batter New Orleans.

The wind damage done by hurricanes is aggravated in coastal areas by the storm surge swept up by the storm. Often the water does more damage than the wind itself, although hurricanes are normally attended by other dangers as well, including tornadoes set off by the incredible turbulence in the system.

Wind and wave action from Hurricane David threaten Fort de France, capital of the Caribbean island of Martinique, in August 1979. Hurricane paths are difficult to predict and their incidence varies by season.

year with 22 in. High variability, of course, means low predictability. And to people who depend on every drop of rain, it also means high risk. Again: the map is a generalization of world patterns, nothing more.

A Fragile Balance

Our study of the earth's atmosphere has taken us through a complicated framework of interrelated systems ranging from wind-induced ocean currents to the hydrologic cycle. This is the life support for humanity on the planet, a life support that depends on equilibrium and balance among innumerable elements. More than 5 billion people consume its oxygen continuously, as do factories, furnaces, fires; yet there has been no measured change in the atmosphere's oxygen content. Cities, farms, and manufacturing plants use immense quantities of water, but the system keeps replenishing the supply. Automobiles, smokestacks,

and coal burners pour tons of dirt into the atmosphere. But even over the dirtiest city, a brisk breeze still manages to produce a beautiful, blue sky. It appears as though nothing can permanently injure the troposphere.

So it seems. But it may be that while we are busy seeking solutions to the energy crisis foreseen by some experts, while we drill bore holes into the continental shelf, while we search for new alloys, we are risking permanent damage to the most vital part of our legacy of abundance. We can survive without food for several weeks, without water for several days. But we can survive without air for only a matter of minutes. This earth may still conceal many resources, but the thin ocean of air at the bottom of which we live is all the air there is, and all there is ever likely to be.

POLLUTION OF THE ATMOSPHERE

By our growing numbers and our drive for progress, we are making increasing—and excessive—demands on the air and the oceans. Somehow the atmosphere seems to be able always to cleanse itself, even when we blacken the skies with soot. But in the end, that may prove to be an illusion. True, the air disperses even the densest smoke and the most acrid chemical fumes. Some of the waste pouring into the atmosphere, however, is producing irreversible changes in the troposphere, changes that may ultimately affect the radiation balance and thus the whole life-support system on this earth.

It is true, again, that people are not the only source of atmospheric pollution. Volcanic eruptions spew millions of tons of dust into the atmosphere, and some scientists believe that periods of especially active volcanism during earth history may have produced so much volcanic pollution that surface environments were directly affected, and thus animal and plant life. Lightning can start grass and forest fires whose smoke forms a kind of natural pollution, and plants send pollens into the air that also constitute pollutants. Dust storms are still another form of natural pollution, although human activity often contributes to such storms through the destruction of natural vegetation. Always nature has been able to overcome these threats, and no permanent damage was done to the troposphere.

Humanity is a new (and very recent) factor in this situation. Never before has the atmosphere had to absorb and cleanse itself of so voluminous and so wide a range of **particulate** and **gaseous pollutants** as produced today. We experience pollution of the air and its consequences in our personal lives (e.g., when someone smokes in our presence), at work (in an office or a factory), and in our community in general. In our North American society, automobiles have been major contributors to the volume of pollution in the air. Before devices to reduce this pollution became mandatory, cars contributed about 60 percent of all pollutants by volume, factories about 20 percent. Automobile-generated pollution still is substantial, as carbon monoxide, (mainly) along with hydrocarbons and nitrogen dioxide. The last two are the main elements in a now-familiar urban phenomenon, smog.

Industries such as steel mills, refineries, and chemical plants produce another gas, carbon dioxide, among many pollutants. As will be noted later, carbon dioxide in the atmosphere plays a major role in increasing the greenhouse effect in heavily industrialized urban areas. Thus the outpouring of carbon

dioxide may have an impact on global climate also. Many scientists are expressing concern over the possibility that an enhanced greenhouse effect could cause global warming, the melting of ice in polar areas, and consequent flooding of low-lying coastal areas where major cities and densely populated farmlands are located.

Notwithstanding the rising production of nuclear energy, much of the electricity produced today is still derived from burning coal and oil.

Although this is so, power generators spew huge quantities of sulfur dioxide into the atmosphere, over 2 million tons annually in the New York urban area, for one example. Sulfur dioxide has played a prominent role in several pollution disasters of the past decades and undoubtedly will do so again.

There are other types and sources of air pollution. Who has not seen a municipal refuse dump sending a column of smoke into the air or city buses emitting ugly, black exhaust fumes? Even the winter fire in the fireplace or the summer barbecue contribute pollutants—multiply it by millions and the cumulative pollution figures are staggering.

POLLUTION AND WEATHER

The mechanisms that force air in the troposphere to rise (convectional, orographic, and frontal) are the systems by which pollution in the air is dispersed. When the air is forced to rise, the pollutants that have accumulated near the bottom of the troposphere are scattered through the whole atmosphere. No matter how dirty the air over a city, the passage of a cold front can clear the skies.

There are times, however, when this clean-up mechanism takes a long time to arrive and atmospheric conditions actually favor a buildup of pollutants. It is not difficult to recognize such conditions. We have already discussed the nature of high-pressure cells, in which the air has a tendency to subside, to press down on the surface. When this occurs, there is obviously no rising of air and no quick removal of pollution. We have to wait until the anticyclone has been pushed eastward and a new system (we hope a frontal cyclone) has taken its place. Summer high-pressure systems, then, can be a mixed blessing. They bring rainless, sunny days and clear nights, but those days and nights may turn smog-ridden and hazy.

Inversion Risks

There is a meteorological condition that is especially hazardous when it develops and lingers over a big, industrial city. This is called an *inversion* (Fig. 3.15). The term is not difficult to interpret. We know that under normal circumstances temperature in still air decreases with altitude about 3.5°F/1000 ft (6.4°C/1000m). When the temperature fails to decrease this way but actually *increases* with height, the situation is upside down, and there is an inversion. How can such a condition develop? There are several ways. First, we have seen that under clear skies the surface loses heat rapidly at night through ground radiation. The layer of air adjacent to the ground is the first to be cooled, too, through contact with the cold surface and through its own loss of radiation upward. Thus a cool layer of air perhaps several hundred feet thick lies on the ground, but the air farther up has not yet been cooled. It has not lost as much radiation outward. Hence, the lowest layer may be cooler than the layer above it, which means that the temperature increases with height. The result often is the develop-

ment of a dense fog, a dirty fog full of pollutants. Most commonly this eye-searing fog is broken up by the sun at midmorning, but there have been cases where the sun failed to penetrate the fog, could not heat the lowest air layer, and could not insolate the ground. The inversion intensified, and more pollutants continued to be poured into it. In London in 1952, hundreds of people died as a direct consequence of a situation such as this.

What has just been described is a **ground inversion**, but an inversion can also develop at higher altitudes. Under certain circumstances the environmental lapse rate can change in such a way that a rising parcel of air will remain warmer than the still air for several hundred feet, but then its adiabatic cooling carries its temperature below that of the still air—with the result that the air parcel stops rising and sinks back to a lower layer. This means that there is a "roof" on the convectional process and that pollutants contained in the rising air cannot be wafted away. This situation occurs rather frequently over central cities, which serve as **heat islands** because of their concrete and asphalt buildings, streets, and rooftops, and their automobiles and factories. The heat island causes air near the surface to rise, but then the **upper-level inversion** prevents it from being carried higher into the

121

FIGURE 3.15

The development of a strong inversion between the California coastal mountains and the Pacific Ocean from the flow of cool air off the ocean. Smog is concentrated in the stable cool air layer below the inversion. (After Donn, *The Earth*, 1972).

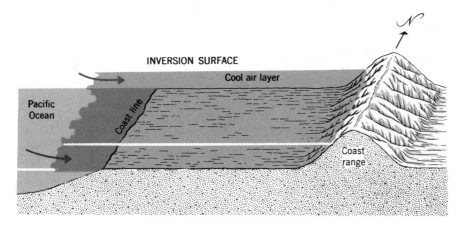

troposphere to be dispersed and replaced by fresher air. Smokestacks and automobile exhausts continue to pour pollutants into the air, and a persistent **pollution dome** develops. This is the kind of condition often associated with the presence of an anticyclone, a high-pressure cell, over an area. The wait for an air-clearing front is an anxious one.

The great cities of the world experience the most serious pollution problems today. Some cities are more susceptible than others, especially those where air movement is impeded by a topographic barrier. Among these are Los Angeles, between cool water offshore and mountains inland; Denver, flanked by the high Rocky Mountains; and Mexico City, in a basin that frequently fills with still air and acrid pollutants. Many of the worst-afflicted places lie in climatic zones where atmospheric conditions frequently favor the development of high-pressure cells with their sub-

siding air.

But air pollution is not the monopoly of Western industrial cities. It would be difficult to find a more pungent example of the point source than the house of an African family in the rural countryside where wood burns right in the living place and the smoke collects under the roof, seeping into the air but bringing streams of tears to uninitiated eyes before doing so. For the children born and raised here, glaucoma is one of Africa's serious diseases. An evening scene in rural savanna Africa is swathed in the smoke of countless indoor and outdoor fires that hangs in the valleys and gives the air an unmistakable fragrance.

The dangers of air pollution were recognized long ago. An English king, in more authoritarian times, forbade the burning of coal within the city of London on penalty of death. But never before has it

An inversion over Denver, Colorado traps a mass of dense smog. Urban pollution domes such as this constitute a health hazard to many city residents whose sole defense is to remain indoors.

122

been the global phenomenon it is today. The worsening spiral seems to have begun about 20 years ago. In July 1970, cities as far apart as Rome, Tokyo, Sydney, and New York all reported that their air was so polluted that people with breathing problems were suffering by the thousands. Shortly thereafter, a significant event was reported from southern Florida. Although there is little heavy industry in that area, the skies of June 1972 were colored by a yellow haze that blocked out much of the sun's summer rays. Meteorologists found that a high-pressure cell with a rather easterly location was wafting to Florida the pollutants generated by industries of the Northeast, and it was on a path that would take it all the way across the western Atlantic. We now produce enough pollution to span the world.

Pollution from the exhausts of thousands of automobiles enters the atmosphere of major cities every day. This is *linear source* pollution, and the photograph shows daily commuter traffic near Oakland, California on the approach to the San Francisco Bay Bridge.

POLLUTION AND CLIMATE

There are various ways of appraising the impact of air pollution on the human environment. We will dispense here with its effects on our health; it is hardly necessary to emphasize that air pollution is bad for people. Carbon monoxide is a lethal gas. Sulfur dioxide in excess makes people ill. Tobacco smoke has been linked to lung cancer. Our present interest, however, is in the way air pollution affects the climate, our life-support system, locally and globally.

It is noteworthy that some textbooks of climatology now include chapters on urban climate. Indeed, by their considerable production of heat, by their emission of smoke, gases, and dust into the atmosphere, and by their relief and its effect on air movement, urban areas modify their own weather and climate. It has been estimated that some European cities lose between one third and one half of their incoming winter radiation because the sun's rays cannot penetrate the smog dome over them. Some of the loss, however, is made up for by the smog particles' interception of outbound ground radiation after dark, which makes night temperatures usually somewhat higher than in the far suburbs and beyond. In Chicago, for example, the night temperature in winter at outlying O'Hare International Airport is sometimes as much as 10°F (5.5°C) below that of the Loop. Overall, the annual mean temperature in a large city can be expected to be between 0.9° and 2.7°F (0.5° and 1.5°C) higher than in the surrounding countryside. In terms of fog, this has become a prominent urban climatic phenomenon. With so many dust particles in the air around which water droplets can form, cities experience from 60 to 70 percent more foggy days than the countryside, where the air is less polluted. And it has been proved that cities, contributing as they do to the uplift of air both by their physical mass (tall buildings) and by the heat they generate, produce excess rainfall as well. Estimates range from 5 to 12 percent; not surprisingly, cities also have more cloud cover than the nearby countryside.

All these are local urban climatic changes brought on by the cities and recorded in the cities themselves. But is it possible that the cumulative effects of the air pollution of many cities might be changing the climate of the earth as a whole? Al-

123

Meuse Valley, Belgium, 1930   The valley of the Meuse River became a death trap for dozens of people in December 1930 when an inversion confined a sulfur dioxide fog produced by steel mills, factories, and chemical plants. Hundreds more fell seriously ill.

Donora, Pennsylvania, 1948   Donora is in the valley of the Monongahela River. Its air is frequently polluted by numerous factories and plants. In October 1948, an inversion interrupted the valley's natural ventilation system and pollutants, chiefly sulfur dioxide, made the air acrid and almost unbreathable. Half of Donora's 12,000 people fell ill, and many probably lost years of their lives. Some were even less fortunate: 20 people died from the smog that began on a Tuesday and held on until the following Sunday night.

London, England, 1952   Cool, stagnant air under high pressure, a temperature inversion at low altitude, hundreds of thousands of coal fires in private homes, and industrial emissions combined to produce a smog consisting of smoke, sulfur dioxide, and other pollutants that lowered visibility to a few feet. The fog remained for four days, and death rates from respiratory diseases rose dramatically. At least four thousand people died as a direct result; an estimated fifteen thousand died from complications during the next three months.

though there is no agreement among scientists, there are those who argue that the answer is yes and that far-reaching changes will come sooner than most of us expect. One problem is that world temperatures fluctuate in a cyclic manner in any case, and it is difficult to determine whether recorded changes are due to this natural oscillation or are pollution induced. We know, for example, that the mean world temperature from 1880 to 1940 rose by about 0.7°F (0.4°C) and that after 1940 it dropped about 0.4°F (0.22°C). This decline after 1940 is especially interesting since sunspot activity, meaning an increase in the intensity of solar radiation, has continued to grow. Is the pollution in the air responsible for this apparent anomaly?

That question has no direct or simple answer. One possibility is that we are, indeed, causing a cooling trend, in part, through the increased outward reflection of incoming solar radiation by all those pollutants floating in the atmosphere. A decrease in temperature of only 0.4°F (o.25°C) or so in 30 years does not seem much until you realize that a full-scale ice age could be brought on by a drop of just 3.5° to 5.5°F (about 2° to 3°C). Some scientists warn that we may be bringing back the ice of the Pleistocene in just a few hundred years, perhaps less.

But there are observers who interpret the facts differently. True, they say, the temperature is presently dropping just under 0.4°F/30 years (since 1940), but it would have been more had it not been for all that atmospheric pollution. These scientists worry that we are upsetting the balance inherent in the cyclic changes of temperature that have been recorded before and when a warming trend begins again, it will warm up so much that the ice caps will melt and the oceans will drown large areas of all the continents. Their arguments are based on calculations that involve chiefly the carbon dioxide we are pouring into the air. Carbon dioxide, a by-product of combustion processes, has little effect on incoming solar radiation, because it is nearly transparent to those rays. But it does absorb ground radiation and radiates energy back to the surface. Calculations have shown that increased carbon dioxide content in the atmosphere results in a warming of the lower troposphere. The amount of warming depends on the quantity of water vapor also present in the air, but it is estimated that a 10 percent increase in carbon dioxide concentration results in a warming of 0.35°F (0.2°C). At present, we are causing a carbon dioxide increase in the atmosphere of about 2 percent a decade, but the rate of increase is speeding up. Again, it does not take much extrapolation to arrive at the conclusion that we may be warming up the atmosphere enough to melt the

124

polar ice caps.

Another concern among scientists studying the earth's outer layers has to do with the ozone layer. This layer consists of a special kind of oxygen, and its molecules are concentrated in a zone about 10 to 30 miles (15 to 50 km) above the surface in the stratosphere. The ozone layer forms, in effect, a protective shield around the earth, warding off most of the sun's damaging ultraviolet rays. Without the ozone layer, skin cancers would multiply, bacteria would be destroyed, ocean life would be severely damaged and crop yields would be reduced. In the 1970s, it was realized that Freon, which is used in aerosol spray cans and in refrigerating and air-conditioning equipment, can damage the ozone layer. Once released, these halocarbons (as Freon compounds are also called) float upward through the atmosphere and stratosphere, where they combine with ozone molecules, reducing them to ordinary oxygen. The use of halocarbons in spray cans was prohibited by the government of the United States in 1976, but many Freon compounds still pour into the air from other sources and from other countries.

In 1986, new fears were voiced by researchers. Color-enhanced satellite imagery showing the ozone layer revealed a hole in the layer over Antarctica—the first gap in the ozone layer ever recorded. The possibility that this gap would grow and that this would have catastrophic impact on the human environment on earth seemed real. Fragmentation of the ozone layer also would greatly speed the warming trend postulated by many scientists, bringing the prospect of melting ice caps into the realm of imminent peril.

Are these estimates matters of distant concern, that is, fit for the geologic time scale but not for the calendar? Many experts do not think that it will be long before we come to face the consequences of our unconcern. Some have even suggested that the in-

FIGURE 3.16
Coastal areas that will be submerged when sea level rises about 300 ft (90 m).

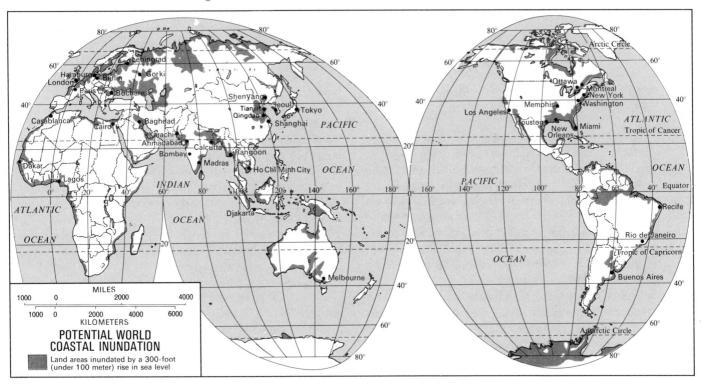

undation of low-lying continental areas will start in 80 years or so. A map of continental outlines following such a rise in sea level has begun to make its appearance (Fig. 3.16). But, in fact, we cannot predict either nature's course or the effect of our own intervention. As *El Niño* proved, there are times when the global systems themselves go through oscillations. Over the longer term, we should remember that we live in what may prove to be an interglacial and that an overpowering meteorological force may dominate environmental change in decades and centuries ahead. By comparison, our actions may serve merely to intensify or, perhaps, to lessen slightly what nature has in store for the earth.

Still, of all the natural wealth with which humanity was endowed on this planet, the atmosphere and the life-support system of which it is a part has probably been treated with less caution than any other resources. Although we have been concerned over the future of energy sources, of alloys, and of metals, we have allowed our vast reservoir of fresh air to deteriorate. To millions of people on this globe who enjoy the benefits of material abundance, one thing has become scarce: fresh, safe, clean air.

Climates of the World

It is one of the primary objectives of geography—physical as well as human—to distill a regional synthesis from a mass of data and thereby discover a spatial order that will serve as a basis for further study much as any classification system does. This is an especially difficult challenge in climatology, because of the almost infinite variety and complexity of the world's weather patterns. (The **weather** at a given location is the record of the temperature, pressure, relative humidity, and other environmental conditions at a particular instance; the **climate** is a generalization of all the accumulated data.) Several climatologists have devised regionalizations of the world's climates, but none of the results has gained universal acceptance.

One problem relating to any map representing the regionalization of world climates is obvious enough: weather and climate change from place to place gradually, not suddenly. Yet the map demands that we draw lines, with one kind of climate on one side and another climate on the other side. Transition zones cannot really be shown at the scale being used here, but there are not many places in the world where one can walk from one climate into another within 100 or 200 ft or so!

Another problem has to do with the criteria to be used in the regionalization of climates. Some geographers have preferred to base their systems on known temperature and precipitation records and on established world patterns of soil and vegetation, the latter a clear response to climatic conditions. More recently, the dynamics of evapotranspiration and water balances have been built into regional systems. But no matter what criteria are given emphasis, a world climatic regionalization must necessarily be the product of compromise. It is possible, of course, to establish a system for certain specific purposes, for example, to reflect the sensible climates of the world, that is, how climate in each established region *feels* to the human body. Another classification outlines the soil moisture and plant growth potential in each region. But for our purposes, the system must remain reasonably simple and yet reflect the diversity of world climates.

A Swiss botanist, Alphonse de Candolle, produced in 1874 the first comprehensive classification and regionalization of world vegetation based on the internal functions of plant organs. A German climatologist, Wladimir P. Köppen (1846–1940), recognized the significance of this effort, since plant associations in given areas represent a synthesis of the many weather variations experienced there. Köppen compared de Candolle's map with his own map of world temperatures, which he had prepared in 1884, and in 1900 he published the first version of a regionalization of world climates. Temperature, precipitation, and plant life were the key elements in the Köppen system; as more information became available, the system was revised numerous times. Köppen himself produced a major revision in 1918; after his death, two of his students, R. Geiger and W. Pohl, published an updated version that has become one of the most widely used systems in physical geography.

The Köppen–Geiger regionalization of world climates of 1953 is reproduced on Plate 3. As the map shows, the classification system that forms its basis employs three-letter symbols. Each of these letters has a numerical value based on monthly averages or seasonal extremes of temperature and moisture. The first (capital) letter is the critical one; the *A* climates are humid and tropical, the *B* climates are dominated by dryness, the *C* climates are still humid and comparatively warm, the *D* climates reflect increasing coldness, and the *E* climates mark the frigid polar and near-polar areas.

Humid Equatorial Climates

The **humid equatorial** climates (*A*), also referred to as the humid tropical climates, are marked by high temperatures all year and by heavy precipitation. No month in any *A* region records an average temperature below 64.4°F (18°C). This indicates just how warm these equatorial climates are. But not all *A* climates are the same. In some, rainfall comes all year-round. In others, rain is strongly concentrated in one or two seasons.

Rainfall comes throughout the year in the *Af* subtype. No month has less than 2.4 in. (6 cm) of rain. The result can be seen not only on the climate map we are considering now, but also on the precipitation map (Plate 2). Appropriately, the *Af* climatic regions are named after the vegetation that prevails there: these are the *rain forest* areas. When we discuss the rain forest as a vegetation region in Chapter 4, we will see it on the map again—showing a distribution very similar to the climatic type *and* the precipitation pattern.

But rainfall does not come evenly distributed in all *A* regions. In the monsoon zones, the rainfall is concentrated in a comparatively brief period (three or four months), leaving a pronounced dry season at other times. This is the *Am* subtype, showing up on the map in peninsular India, in parts of Southeast Asia, and in a corner of West Africa. Again, the vegetation mirrors the type; the natural forest is more open, the trees more able to withstand a long dry season.

A third tropical climate, the savanna, also has pronounced moist and dry seasons. This is the *Aw* subtype on the map. Here the pattern begins to reveal

a wider daily and seasonal temperature range, and rainfall is concentrated in one or two wet seasons. Overall, the total precipitation in *A* regions is lower than in rain forest or **monsoon** areas (Plate 3 underscores this). The seasons often show an especially interesting pattern, with rain coming in concentrated periods twice during the year and with two (not just one) dry seasons separating these. Climatologists refer to this as a double maximum of precipitation. In many savanna zones, the people refer to the "long rains" and the "short rains" to identify these seasons, and a persistent problem in these regions is the unpredictability of the rain's arrival. Savanna soils are not among the most fertile; thus when the rains fail, the specter of hunger arises. Savanna regions are far more densely peopled than rain forest areas, and millions of residents of the savanna subsist on what they manage to cultivate. Rainfall variability under the savanna regime is their principal environmental problem.

THE DRY CLIMATES

The dry climates are shown on the map as the *B* climates, the climates of low precipitation and prevailing aridity. The dividing line between the *BW* (true desert) and *BS* (slightly moister steppe) varies according to several criteria, but in the simplest terms, it may be taken to coincide with the 10-in. (25-cm) **isohyet**, or line of equal rainfall. As Plate 2 shows, parts of the central Sahara receive less than 4 in. (10 cm) of rainfall and, as Plate 3 shows, much of West Africa's **Sahel** is steppe country. This is a very tenuous, difficult environment for farmers and herders. A pervasive characteristic of the world's dry areas is the enormous daily temperature range: it may be burning hot in the daytime and frigid at night. Such a range may exceed 60°F (35°C) in low-latitude deserts. Recorded instances exist where the maximum daytime shade temperature was 120°F (49°C) followed by a nighttime low of 48°F (9°C). The monthly averages that form the basis of the Köppen–Geiger regionalization do not reflect such extremes, but there are two more letters in use for the B-type climates: an *h* for very hot deserts, where the mean annual temperature is higher than 64.4°F (18°C), and a *k* for cooler deserts, where the average falls below that level. The cooler deserts, of course, lie at higher latitudes.

HUMID TEMPERATE CLIMATES

The humid temperate climates, shown on the map as C climates, show the effects of higher latitudes as well. Here the coldest winter month has an average temperature below 64.4°F (18°C) and may be as cold as 26.6°F (−3°) but no colder than this. These climates are also referred to as the **mesothermal** (middle temperature) and as the midlatitude climates; as the map shows, almost all the areas of C climate lie just beyond the Tropics of Cancer and Capricorn. This is the prevailing clime in the southeastern United States from Kentucky to Florida, in California and the coastal Northwest, in Western Europe and the Mediterranean, in southern Brazil and northern Argentina, in coastal South Africa and Australia, and in eastern China and southern Japan. None of these areas is associated with climatic extremes or severity, but the winters can be fairly cold. These areas lie about midway between the winterless equatorial climates and the summerless polar zones.

The humid temperate climates range from quite moist, as along the densely forested coasts of Oregon, Washington, and British Columbia, to relatively dry, as in the so-called Mediterranean (dry summer) areas that include not only coastal southern Europe and northwest Africa, but also the southwestern tips of Australia and Africa, middle Chile, and southern California. In the Mediterranean areas, the scrubby, moisture-preserving vegetation creates a natural landscape very different from that of more lushly clothed Western Europe.

The second and third letters of the symbols for the humid temperate climates denote moisture and temperature patterns respectively. The f denotes an adequate amount of precipitation, 1.2 in. (3 cm) all year-round; the w indicates that winters are seasons of low precipitation, and the s identifies areas where the summers are comparatively dry seasons (as in the case of the Mediterranean subtype). In each case, a simple formula determines the classification. The third letter refers to monthly temperature patterns: a if the warmest month exceeds 71.6°F but at least four months of the year average above 50°F (10°C); and c if only one to three months average over 50°F. There is a d for areas with an especially cold winter month, but this happens rarely. The distribution on the map reveals the prevalence of cool (Cfc) conditions along Norwegian and Alaskan coasts, milder Cfb climes over much of Western Europe, and warm Csa conditions along the Mediterranean Sea coast. That Csa pattern is especially interesting, because it breaks a pattern that exists almost everywhere else on earth. Note that the s indicates a dry summer climate, so that the season of the highest heat is also the time when there is least rain. This unusual climatic type has been named the Mediterranean climate, and it occurs not only along Mediterranean shores, but also in parts of California, in Chile, South Africa, and Australia. Again, this very special climate is marked by a particular vegetation association.

HUMID COLD CLIMATES

The humid cold climates, the D climates, have several other names: they are called the continental climates, the microthermal climates, and the snowy-forest climates. The letter D, means that the average temperature of the warmest month is over 50°F (10°C), but the coldest month averages under 26.6°F (−3°C). The second and third letters are assigned on exactly the same basis as for the C climates. These humid cold climates may also be called the continental climates, for they seem to develop where there is a lot of landmass, as in the heart of Eurasia and in interior North America. No equivalent land areas at similar latitudes exist in the Southern Hemisphere, and no D climate occurs there at all.

Great annual temperature ranges mark the humid continental climates, and very cold winters and relatively cool summers are the rule. In a Dfa climate, for example, the warmest summer month (July) may average as high as 70°F (21°C), but the coldest month (January) only 12°F (−11°C). Total precipitation, a substantial part of which comes in the form of snow, is not very high, ranging from over 30 in. (75 cm) to a steppelike 10 in. (25 cm). Compensating for this paucity of precipitation are the cool temperatures, which inhibit loss by evaporation and evapotranspiration.

Some of the world's most productive soils lie in areas under humid cold climates, including the U.S. Midwest, parts of the Soviet Union's Ukraine, and north China. The period of winter dormancy, when all water is frozen, and the accumulation of plant debris during the fall combine to balance the soil-forming and enriching processes. The soil differentiates into well-defined, nutrient-rich layers, and a substantial store of organic humus accumulates. Even where the annual precipitation is light, this environment sustains extensive forests.

COLD POLAR CLIMATES

The cold polar climates, *E* on the map, are differentiated into two regions: those under tundra conditions and those even colder, where permanent ice and snow prevail and no plants can gain a foothold. The tundra, like rain forest and savanna, is a vegetative term; it stands for an association of grasses, herbs, shrubs, mosses, and lichens—all that can survive under these harsh conditions. The tundra climate is marked by one to four months of above-freezing averages, but the "warmest" month still remains below 50°F (10°C) average temperature. When you com-

pare the climatic map (Plate 3) with the vegetation map (Plate 4), you will note that the climatic boundary between *D* and *E* climates corresponds fairly closely to the boundary between the pine forests and arctic tundra.

HIGHLANDS AND MOUNTAIN SLOPES

Any climatic map will show cold regions not only in high latitudes, but also at high elevations. Where the elevation dominates, such regions are marked *H* on the map. Even in equatorial latitudes it is possible to ascend through subtropical, temperate, cold, and even polar climates as high mountains rise above the snow line. In East Africa, Mount Kilimanjaro stands near the equator, its permanent snow glistening in the tropical sun. In these highland regions altitudinal zonation replaces latitudinal progression.

The highlands of Middle and South America also include a series of altitudinal zones that have been given regional names: **tierra caliente** (hot land) up to an elevation of about 2500 ft. (750 m), **tierra tem-**

plada (temperate land) reaching up to approximately 6000 ft (1800 m), **tierra fria** (cold land) extending to around 10,000 feet (3000 m), and above this the **tierra helada** (frozen land), also called the **paramós.** Each of these altitudinal zones presents a different living environment, a different combination of temperatures, perecipitation patterns, and exposures. So varied are the highlands regions that our map (Plate 3) could not convey the complexities. That is why they are generalized into a separate category in the Köppen regionalization.

The Köppen-Geiger regionalization of world climate is only one of several systems that have been derived from the available data. It has the merit of simplicity and precision, but you should remember that it constitutes a gross generalization of a very complex world environment. Each of the regions in Plate 3 is itself varied and divisible; there are no constants on the world's climatic map.

• Reading about Weather • and Climate

Several different topics are discussed in Chapter 3, and you can choose your supplementary reading from a long and interesting list, only part of which can be suggested here. For additional details on technical matters you may wish to consult A. N. Strahler, *The Earth Sciences* (3d ed.; New York: Harper & Row, 1984). The physical geography books mentioned as outside sources for Chapter 2 have extensive discussions of weather and climate. Also see J. R. Mather, *Climatology: Fundamentals and Applications* (New York: McGraw–Hill, 1974) and H. J. Critchfield, *General Climatology* (4th ed.; Englewood Cliffs, N.J.: Prentice–Hall, 1983). Both these volumes follow introductory sections on general climatology with chapters on climate and agriculture, health and comfort, industry, architecture, and related topics. A straightforward climatology is J. J. Hidore's *Geography of the Atmosphere* (Dubuque, Iowa: W. C. Brown, 1972). An excellent concise volume is R. G. Barry and R. J.

Chorley, *Atmosphere, Weather, and Climate* (4th ed.; New York: Methuen, 1982). A very personal book, full of anecdotal material and well worth an evening is D. M. Gates's *Man and His Environment: Climate* (New York: Harper, 1972). Other volumes in the *Man and His Environment* series are also relevant to our study, especially L. R. Brown and G. W. Finsterbusch's book on food (see Chapter 6).

Other useful works in this area are J. F. Griffiths and D. M. Driscoll, *Survey of Climatology* (Columbus, Ohio: Merrill, 1982); J. E. Oliver, *Climate and Man's Environment: An Introduction to Applied Climatology* (New York: Wiley, 1973), and J. E. Oliver and J. J. Hidore, *Climatology: An Introduction* (Columbus, Ohio: Merrill, 1984). Climate, food production, and population growth are combined in a book by W. O. Roberts and H. Lansford, *The Climate Mandate* (San Francisco: Freeman, 1979). A nontechnical treatment of the topic is G. T. Trewartha and L. H. Horn, *An Introduction of Climate* (5th ed.; New York: McGraw–Hill, 1980).

Topics discussed briefly in our chapter form the topics of whole volumes, for example H. E. Landsberg, *The Urban Climate* (New York: Academic Press, 1981); and S. Nieuwolt, *Tropical Climatology: An Introduction to the Climates of Low Latitudes* (New York: Wiley, 1977). G. T. Trewartha discusses *The Earth's Problem Climates* (Madison: Univ. of Wisconsin Press, 1981).

One of the most interesting topics in climatology concerns long-term changes. Two volumes by the renowned British climatologist deal with this topic: H. H. Lamb, *Climate: Present, Past, and Future* (London: Methuen: Vol. 1 [1972] is *Fundamentals and Climate Now*; Vol. 2 [1977] is *Climatic History and the Future*). An earlier collection of Lamb's writings on climate and environment, *The Changing Climate* (London: Methuen, 1966) is now dated but still holds interest.

Papers given at the Second Carolina Geographical Symposium, which met at the University of North Carolina in 1975 to discuss atmospheric quality and climatic change, were edited by R. J. Kopec and published by the university's Department of Geography in the *Studies in Geography Series* (No. 9). The range of opinion on what is happening to the atmosphere and what the long-range effects will be is reflected by this interesting set of articles. For excellent photography and accompanying narratives, see E. O. Barrett, *Climatology from Satellites* (London: Methuen, 1974). Also see an unusual collection of papers edited by W.R.D. Sewell,, *Modifying the Weather* (*Western Geographical Series*, No. 9; Toronto: Victoria Univeristy, 1974). And if you wish to hark back to Ellsworth Huntington, spend an evening with N. Winkless, III, and I. Browning, *Climate and the Affairs of Men* (New York: Harper & Row, 1975). Or go directly to the source: E. Huntington, *Civilization and Climate* (3rd ed.; Hamden, Conn.: Shoe String Press [Archon Books], 1971).

Understanding the oceans and their influence over global climates is important, and a major part of our chapter refers to oceanic conditions. For additional reading, see W. A. Anikouchine and R. W. Sternberg, *The World Ocean: An Introduction to Oceanography* (Englewood Cliffs, N.J.: Prentice–Hall, 1981). A basic reading on the oceans is edited by J. B. Ray, *The Oceans and Man* (Dubuque, Iowa: Kendall/ Hunt, 1975). For other background on the physical

geography of the oceans, see J. M. McCormick and J. V. Thiruvathukal, *Elements of Oceanography* (Philadelphia: Saunders, 1976).

Many articles on climatology continue to appear in the geographic journals and reflect current areas of interest. In the *Annals* of the Association of American Geographers, for example, see R. W. Pease *et al.*, "Urban Terrain Climatology and Remote Sensing" (Vol. 66, No. 4; December 1976); D. Morgan *et al.*, "Microclimates within an Urban Area" (Vol. 67, No. 1; March 1977); and P. J. Robinson and J. T. Lutz, "Precipitation Efficiency of Cyclonic Storms" (Vol. 68, No. 1; March 1978). D. W. Gade discusses "Wind Breaks in the Lower Rhône Valley" in the April 1978 *Geographical Review* (Vol. 68, No. 2). There is always something interesting, too, in the *Professional Geographer*, for example, G. H. Dury's "Likely Hurricane Damage by the Year 2000," in the August 1977 issue (Vol. 29, No. 3).

Jean Brunhes's *Human Geography* (London: Harrap, 1952) was translated by E. F. Row from the 1947 French edition. The quotation in this chapter is from page 40.

131

Biospheres:
Geography
at the
Interface

◄══►

I s there a more satisfying way to view the landscape than from the gondola of a balloon drifting slowly over the countryside? Ballooning has, in recent years, become a major attraction for tourists visiting places as different as the wildlife reserves of East Africa and the wine country of California. For geographers, there is an added attraction. Height gives perspective, but too much height blurs detail. A balloon trip is a perfect compromise. It permits observation that is otherwise difficult to achieve. Anyone who has sat under the roar of a helicopter engine or flown—always too fast—in a small airplane will know the difference.

Geographers, in any case, like to view the earth from above. When a geographer visits a city he or she has not seen before, a trip to the top of the tallest building, if there is a lookout there, will be first on the agenda. Height is not an advantage only in the countryside.

From the balloon's gondola, one can see river courses, escarpments, the face of the land. But much of the underlying geology and its structures is covered by soils. Trees, grasses, and croplands create a cloak over the countryside. And these, like the design of geomorphology and the pattern of climate, also have their spatial plan. Relationships become quickly evident: sparser vegetation on steeper slopes, stands of trees in watered valleys. We are looking at the geography of biology.

What the perspective from above reveals is the topic of this chapter: that critical zone of contact between lithosphere and atmosphere where rocks weather, with the help of the atmosphere, into soils and where plants and animals are sustained. This is the layer of life, the biosphere. Much of the exposed surface of the earth's crust is covered by a cloak of soils, and the soil itself is alive with organisms, ranging from microscopic bacteria and fungi to worms, insects, snakes, and even burrowing animals such as ground squirrels and gophers. The soil sustains the earth's mantle of vegetation, from the smallest tuft of grass in a desert area to luxuriant forests in moister zones. Plants and animals (**flora and fauna**) thrive even in the seas and oceans. It is therefore appropriate to consider the earth's life layer not just as one sphere, but as a set of biospheres. This chapter focuses on two

of these biospheres: the soil layer and the mantle of vegetation. The fauna are discussed briefly.

On the landmasses, soils and vegetation exist at the **interface** between lithosphere and atmosphere, at the plane of interaction between rocks and their mineral constituents, on the one hand, and the air with its moisture and heat, on the other. The soil is the key to plant life: it produces and preserves plant nutrients and absorbs and stores water. But the vegetation itself contributes to its own sustenance by adding fallen organic matter to the soil surface, to be absorbed and converted to reusable nutrients. It is an intricate and complicated system of interaction in which all elements of the environment play a role.

The study of the geographic aspects of plant and animal distributions is part of the field of *biogeography*. This geographic field has been described as lying in the meeting ground of the earth sciences and the biological sciences. And certainly biogeographers are concerned with aspects of both: they are especially interested in the relationships between plant and animal communities and their natural environments. Biogeographers pursue these interests from their spatial perspective. They want explanations to account for plant and animal distributions: why certain species are found in some areas but not in others; why species exist cooperatively in some places but competitively in others. The study of biogeography was begun by Alexander von Humboldt, who gave the best example: biogeographers must know about soils and climates as well as plants and animals. In addition, biogeography involves genetics, energy exchanges, and other complicated topics.

Biogeography, so defined, deals with the world of plants and animals. The character and distribution of soils are so closely related to these topics that the study of soils and their spatial distribution is often regarded as a part of biogeography. But the main branches of biogeography are plant geography, or *phytogeography* (the strongest subfield today) and *zoogeography*, the geography of faunal distributions, which is full of opportunities but not pursued as much as it was formerly.

The concept of the *ecosystem* has much importance in biogeography, and in the preceding chapters we have been working toward this notion. An ecosystem is defined as an assemblage of organisms (animals and plants) and the natural environment with which this assemblage interacts. The implication is that an ecosystem exists in a certain area on the earth's surface and that this area has boundaries. Those are geographic properties, and so it is not surprising that biogeographers are concerned with this notion. But before we discuss plants and animals and their environments, let us concentrate on a crucial element of ecosystems: the soil.

• The Living Soil •

In Chapter 2, we noted that life as we know it would not exist on this planet without the presence of water. One reason is that without water there would be no soil. The moon's lifeless surface of rock fragments and pulverized rubble holds no water—and therefore no soil. In its life cycle, soil depends on water. Chemical and physical weathering of rocks begins the process of soil formation, with water playing a crucial role. As the soil develops or matures, water sustains the circulation system, promotes the necessary chemical reactions, transfers nutrients, helps decompose organic matter, dissipates excess heat, and ensures the continued decay of the rocks below. Soils capable of supporting permanent vegetation develop (on an average), over a period of between one hundred and two hundred years, but then it takes thousands of years for the soil to mature fully and for a permanent (equilibrium) condition to develop among soil, vegetation, and climatic regime. All this time, the soil absorbs and discharges water, breathes air in and out through its pore spaces, takes in organic matter, and dispenses nutrients. The soil is, indeed, alive.

But the soil is not always alive and well. Humanity has been on this earth for only 1/20th of 1 percent

This famous photograph dramatically illustrates the power of wind erosion (see Chapter 2). It was taken during the 1930s, when dust-bowl conditions affected large areas of the Great Plains. The picture shows an approaching dust storm, prevalent during that period, driven by a strong wind (50 km per hour). The storm lasted for nearly three hours. It removed thousands of tons of fine material from the upper horizon of the soils over a widespread area.

135

The Living Soil

of the planet's lifetime, and in this moment of earth history, our human numbers have taxed the soils that had taken not just thousands but millions of years to develop—and in places overtaxed them. Soils are fragile, and once the equilibrium of which they are a part is disturbed, an unstoppable cycle of destruction begins. Damage the natural vegetation, and roots will wither and die. The upper layer of the soil (the topsoil) is loosened, and wind and runoff carry it away. Soon patches of bedrock begin to protrude through what remains of the soil. The soil is dying, and only time and protection can restore it. All too often, neither is available as hungry livestock consume what little vegetation still manages a foothold. In the section of the Great Plains that lies in Texas, Oklahoma, and Kansas in the 1930s, the expansion of wheat cultivation, the destruction of natural grasslands, and a succession of dry years led to severe wind erosion as huge dust storms carried away the upper layer of the soil—and the area became known as the Dust Bowl.

SOIL DEVELOPMENT

The study of soils is the science of **pedology**, from a Greek word, *pedon* (ground). Today the term *pedon* is applied to a six-sided column of soil that might be dug from a soil layer so that the **soil profile** (the soil's vertical cross section from upper surface to bedrock) can be studied. The soil profile consists of a series of layers called **soil horizons**, which can be distinguished by their differences in color, structure, texture, and consistency. A fully developed soil is likely to have an O horizon on top (O for organic), followed by an A horizon from which soluble minerals and other matter have been removed by downward-percolating water, a B horizon marked by the accumulation of these downward-moving minerals and particles, and a C horizon consisting of weathered bedrock. Below the C horizon lies the unaltered bedrock, sometimes labeled the R horizon (R for **regolith**). Each of these soil horizons may be subdivided; for example, the A horizon may have an upper (A_1) layer with high organic content and an (A_2) layer below it that contains less organic material and displays transitional characteristics to the B horizon (Fig. 4.1).

When a soil profile displays both an A and a B horizon, we may conclude that this reflects an advanced stage of soil development, because this differentiation takes much time to evolve. In less mature soils, the upper (A) horizon rests directly on the C horizon of weathered bedrock as there has not been time to develop the necessary thickness to accommodate the intermediate horizon. But whether mature or immature, the soil consists of four components: minerals from bedrock, organic matter from decomposing plant material, water in the form of weak solutions from mixing with minerals and gases in the soil, and air marked by high carbon dioxide content and lower amounts of oxygen and nitrogen than in the atmosphere.

The soil profile of a mature grassland soil, a *mollisol*, in South Dakota. Mollisols have a thick, dark-colored (brown to black) upper horizon called the *mollic epipedon*, clearly visible and extending to a depth of about 45 centimeters (1.5 feet; the measures at the left are in feet). The A horizon extends to a depth of about 15 centimeters (6 inches) and the B horizon lies from 15 to 45 centimeters (6 to 18 inches). The parent material is a glacial till.

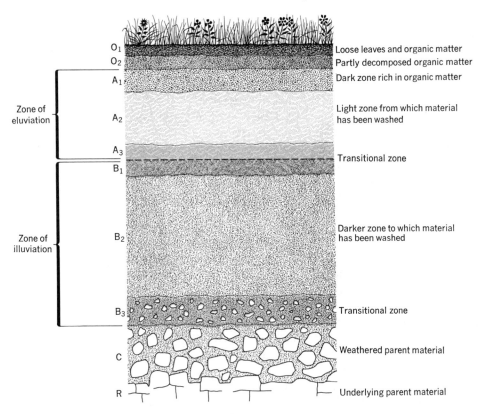

Zone of
eluviation

Zone of
illuviation

O₁ — Loose leaves and organic matter
O₂ — Partly decomposed organic matter
A₁ — Dark zone rich in organic matter
A₂ — Light zone from which material has been washed
A₃ — Transitional zone
B₁
B₂ — Darker zone to which material has been washed
B₃ — Transitional zone
C — Weathered parent material
R — Underlying parent material

FIGURE 4.1
The profile of a mature soil.

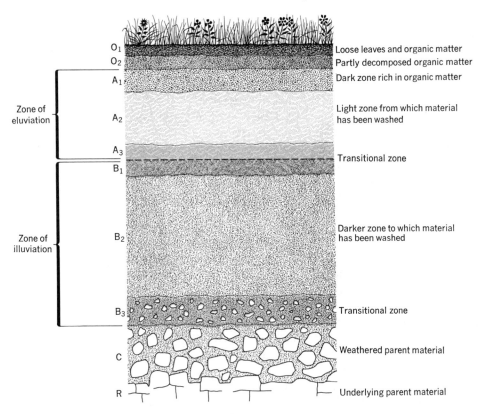

Zone of eluviation

Zone of illuviation

O_1 — Loose leaves and organic matter
O_2 — Partly decomposed organic matter
A_1 — Dark zone rich in organic matter
A_2 — Light zone from which material has been washed
A_3 — Transitional zone
B_1
B_2 — Darker zone to which material has been washed
B_3 — Transitional zone
C — Weathered parent material
R — Underlying parent material

FIGURE 4.1
The profile of a mature soil.

• Eluviation and Illuviation •

A soil develops when there is less material lost to erosion at the surface than is gained by weathering at the base. As soil matures, it will normally also develop **horizonation**, the layering seen in the soil profile.

The development of horizons depends considerably on the work of percolating water. Water moving downward through the pore spaces in the soil will remove from the upper (A) horizon soluble minerals and microscopic, colloidal-size particles of organic matter, clay, and oxides of aluminum and iron. This removal process is called **eluviation**, thus the upper layer of the soil is said to be eluviated.

As the soil solution carrying the dissolved and suspended matter moves downward, it begins to deposit its content in the pores and against the surfaces of the grains below. When this process of **illuviation** takes place, we are in the soil's B horizon. A mature soil displays horizons of eluviation and illuviation, and below this there is a horizon (C) consisting of weathered parent material.

In a mature soil, the A and the B horizons together are termed the **solum**.

The formation of a mature soil depends on a number of conditions and circumstances. Obviously, the minerals in the underlying bedrock, the **parent material** of the soil, strongly influence the kind of soil that will ultimately develop. We can recognize two kinds of parent material. The first is the permanent bedrock such as an area of granite or a zone of folded and eroded sedimentary rocks. Weathering of these rocks produces a soil that contains constituents derived directly from them, so that we may describe the parent material as **sedentary** (sometimes the term residual is used, but it is really the soil that is residual, not the bedrock). Elsewhere the bedrock below is buried beneath material deposited there by water (such as alluvium), or by ice (glacial till), or by wind (loess). When soil develops on this material brought in from elsewhere, such soil bears no relationship to the bedrock deeper down. This soil, then, is developed on **transported** parent material. But whether soil is derived from sedentary or transported parent material, you should not expect to be able to correlate the two invariably. Over time, as the soil matures, the impress of the parent material may diminish and other factors become more important in the constitution of the soil. A granite will not yield exactly the same soil wherever it is a parent material. If it did, pedology would be a less complicated science than it is!

Much depends, for example, on the *topography* where the soil forms. On a steep slope, soil will have difficulty forming because weathered material tends to be removed by efficient erosion. Soil has a better chance to develop on a gentler slope, and some of the best-formed soils occur in valleys of gently undulating country, especially when there is good drainage and an ample supply of warmth. In a poorly drained, clogged valley, a thick layer of organic material (**humus**) is likely to accumulate at the top of the soil, and the usual A and B horizons will be missing. And the orientation of the slopes matters, too. In the middle latitudes of the Northern Hemisphere, a northward-facing slope receives much less insolation than a southward-facing one, and this has the effect of reducing evapotranspiration; soils on the northward-facing slopes are normally more moist as well as cooler.

This leads us to still another factor in soil development, the role of the available *water* and temperature. These are climatic factors, of course, but we should be careful not to apply the conditions in the Köppen climatic regions (Plate 3) to all the soils within each of those regions. This is why C. W. Thornthwaite developed his water-balance system of climate classification, but even this more practical system conceals small-area variations. The same hill can have a slope facing sunward where moisture conditions differ significantly from the opposite slope, that faces away from the sun's rays. Water's critical role in soil development extends from the continuation of weathering processes near the bottom of the soil to the support of vegetation at the top. It varies from season to season: during some months, a soil's pores may be almost completely filled with water and a water surplus exists (the excess goes to the groundwater and the water table rises), but then the heat of summer, the growth of plant foliage, and the demands of the vegetation empty the pores and a water deficit develops. Eventually the growing season ends, the leaves fall to the surface, temperatures decline (and so evapotranspiration losses diminish), and the soil's water budget swings to the plus side again.

The role of *temperature* is already evident because plant growth and evapotranspiration rates are directly related to heat. But there is a temperature regime in the soil itself as well. Chemical processes and biological activity cease when the temperature falls below 32°F (0°C) in the soil. Low temperatures (below 41°F [5°C]) inhibit many of the processes of soil development, although water is not frozen and can move through the pore spaces. In general, the higher the temperature, the more rapid the processes of organic decay, plant growth, and chemical reaction—but not all soil horizons have the same temperature. The O and A horizons' temperatures are likely to reflect the daytime and nighttime (**diurnal**) swing of temperature and the annual range as well, but deeper horizons reflect the transitions of weather and climate less.

The functions of *biological* processes in soil development are also critical. Again, we must not generalize here and refer to vegetation alone as a soil-forming factor; vegetation, like weather and climate, changes from place to place and from hillside to adjacent valley, thus generalization is really impossible. It is, nevertheless, possible to identify plant-related processes going on in the soil and affecting its develop-

138

ment. These include the nurturing of plants and their roots; from this vegetative growth the soil later derives its organic matter on the upper surface as leaves and stems and within lower horizons as roots. The conversion of dead plant matter to humus and the extraction of water and carbon dioxide constitute one set of biological processes in the soil. Another process involves the recycling of plant nutrients from dead plant matter back to live plants without these nutrients being subject to downward percolation. This process is extremely important in tropical rain forests, where the nutrient cycle sustains the trees. When the rain forest is cut down and replaced by other plants, the nutrient cycle is interrupted and the soil alone cannot support the new vegetation. In addition, the animal life in the soil plays an important role in its development by creating passageways, by ingesting soil and passing it through their systems, by eating and passing organic matter, and by contributing to the soil's circulation systems in general.

Soil requires *time* to develop, but soils in colder environments need more time than soils in equatorial and tropical zones. After the Indonesian volcano Krakatoa exploded in 1883 and poured huge volumes of lava over what remained of its island, more than 1 ft of soil (35 cm) developed on the new rock in only 45 years. But this soil formed under equatorial conditions, where high temperatures and high humidity combine to speed the processes of soil development. In higher latitudes, the same soil depth might not be reached in hundreds of years. Soil geographer James G. Cruickshank reported in his *Soil Geography* (1972) that after a Swedish lake was drained, the first signs of a new soil profile's development could be observed in a hundred years; calculations showed that this soil would probably mature in a thousand years. Some soils in Alaska contain organic material that is still not completely decomposed, although it has been dated as twenty-nine hundred years old. Undoubtedly the full development of soils in higher latitudes involves thousands, not hundreds, of years.

It is surely appropriate to add *humanity* to this brief enumeration of factors affecting soil development. It was the soil that sustained the first large human communities when people learned to harvest wild crops and then to cultivate and irrigate. With irrigation and fertilization began the large-scale modi-

fication of the soils of huge regions of the world. Today the soils of millions of acres of the earth's surface are watered, fertilized, plowed, and planted—and thus modified drastically and permanently. Humanity ranks as one of the major factors affecting soil development in major areas of the world.

PROCESSES AND PROPERTIES

Within the soil layer, numerous processes go on continuously as matter is transferred from one horizon to another: added to the soil from above, lost to plant roots from below. These processes involve the movement and exchange of microscopic particles, called **colloids** and **ions**, as soil matter is also being transformed from one state to another. For example, the organic material that falls on the surface of the soil becomes humus and as such goes into the nutrient cycle, but continued decomposition and alteration of organic matter leads to two products, one a liquid (water) and the other a gas (carbon dioxide).

Soils that develop in areas of low precipitation (and thus with a negative water balance most of the time) undergo **calcification**. Here the limited rainfall, mixed with carbon dioxide to produce a very mild carbonic acid, penetrates the upper soil layer, dissolves some calcium, and percolates downward. But there is not enough rainfall to leach the soil effectively, and soon the available water is evaporated or absorbed. Now, the calcium carbonate is deposited, and over time a so-called calcic horizon develops: 6 in. (15 cm) or more thick. This is a hard, slablike layer (called **caliche** in the southwestern United States) whose depth below the surface is related to the local water balance. In very dry areas, this layer lies in (or even on) the A horizon, but where it is moister, the layer develops lower down in the C horizon. As we will see later, calcification is a process that occurs especially in soils termed **aridisols**.

In cool, moister climates, soil development is influenced by a set of processes collectively termed **podzolization** (*podzol* means ash in Russian). Here the water supply is more ample and evapotranspiration losses are lower, so that the arriving rains are sufficient to leach the A horizon quite thoroughly, thus removing most of the minerals except silica. This gives the upper horizon a gray, ashlike appearance to

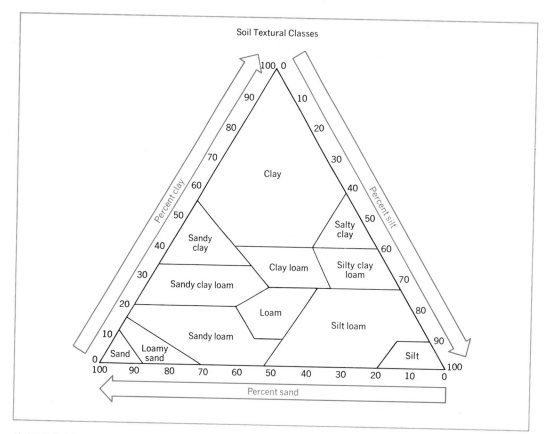

FIGURE 4.2

Percentages of clay, silt, and sand in soil texture. (USDA).

which the mixed-in organic matter contributes. The effectiveness of the percolating rainwater as a leaching agent is enhanced by the acids derived from the humus in the O horizon; so, the mineral contents of the A horizon are transferred downward and accumulate by illuviation in the B horizon. With low temperatures inhibiting bacterial action, the podzolized soils extend from the zone of boreal evergreen forests, where they are most strongly developed, into humid midlatitude areas, where their fertility increases. Podzolization processes prevail, especially in soils known as **spodosols**.

Warm, moist equatorial and tropical regions produce soils that undergo **laterization**. Here the precipitation is plentiful, so that large amounts of it percolate deeply into the soil, leaching it thoroughly and promoting weathering at depth. The high temperature promotes rapid bacterial action; organic matter that falls on the surface of the soil is rapidly destroyed and little real humus is left. Since it is humus that gives acid strength to percolating soil water, the water in tropical areas leaches the soil but leaves oxides of iron and aluminum behind. This gives rise to the characteristic reddish yellow color of tropical soils; sometimes the concentration of iron and aluminum oxides in an upper layer of the tropical soil is so great that a hard **oxic horizon** develops—termed **hardpan** in areas where it interferes with agriculture. Not surprisingly, the soils that are susceptible to laterization are called **oxisols**.

These three processes occur among many other variations, depending on changing moisture conditions and temperature regimes. These processes inevitably leave telltale marks on the soil in the form of color, texture, structure, and consistency—all clues to what is happening to the soil and what its capacities are. **Soil texture** refers to the proportions of sand, silt, and clay in each horizon. By definition, sand particles are the largest, silt particles are medium sized, and clay constitutes the finest material in soil (Fig. 4.2). A **loam** contains about the same amount of each. Since eluviation affects the A horizon, we may expect soil texture there to be coarser than in the B horizon, because the finer particles are removed from the A horizon and deposited in the B horizon.

Texture is an important soil property, because it indicates the amount of water a soil can hold. Sandy soils hold very little water, allowing it to percolate downward; silt retains an intermediate quantity of water; and clay holds the most water, in fact, so much that it is often waterlogged and too wet for most plant roots. The ideal combination for farming is loam, which permits reasonable drainage and air circulation, yet contains enough water to support plant growth.

Soil structure refers to the way soil materials tend to cluster together in larger units called **peds**. These lumps of soil separate naturally from others, thus revealing a property that plays an important role in the movement of water and air in the soil: the provision of channels for root penetration and the soil's ability to withstand erosion.

Pedologists identify four types of soil structure; the terms are self-explanatory. In **platy structure**, the peds are layered, like flat plates positioned horizontally. They are easily recognized from a soil sample in the palm of your hand. The individual plates are as much as 0.4 to 0.8 in. (1 to 2 cm) across, sometimes even larger. In **prismatic structures**, the soil forms prismlike, vertical columns from less than 0.5 to 4 in. (to 10 cm) across. Other soils display a **blocky structure**, soils in which the peds have angular shapes, usually about 0.4 in. (1 cm) or less per side but sometimes as large as 2 in. (5 cm) per side. In this angular arrangement, the peds tend to fit into each other, giving the soil more cohesion than when it has **granular structure**. In granular soils, the peds are usually small and nearly round, the soil is very porous, and there is not much strength.

Soil consistency manifests itself in several ways

• Permafrost •

The role of temperature in soil development is especially pronounced in areas of extremely cold climate (the tundra of E climate). There the entire depth of the soil freezes solid during the long winter; in summer only, the upper 3 ft (less than 1 m) or less thaws. Since there is solidly frozen ground below this upper layer, water percolating downward from the soil surface cannot go very far and thus the soil tends to be poorly drained, especially in flat-lying areas. Where this situation prevails permafrost exists and soil development is inhibited. **Permafrost** refers to the permanently frozen ground below the soil that alternately freezes and thaws.

The area subject to permafrost conditions is quite large (Fig. 2–26); permafrost does more than affect soil development, it also poses serious problems in construction. The annual freeze–thaw cycle contracts and expands the soil, so that structures are not adequately supported; the alternative is not simply to drive supports into the permafrost below, because it will not support heavy structures. Many roads and railroads in permafrost country have buckled and broken.

When the Alaska pipeline was built to carry oil from the North Slope to the Pacific coast, it was necessary to raise the pipeline above the ground on pedestals anchored deeply into the permafrost. This prevented the oil from melting the permafrost and distributed the pipeline's weight in such a way as to reduce the risk of fracture. About one quarter of the total length of the pipeline is raised above the permafrost zone.

and, although difficult to define and measure exactly, is a valuable indicator of soil properties. It is initially tested in the field by hand: one squeezes or rolls bits of the soil between thumb and forefinger to gauge the soil's stickiness, plasticity, hardness, and degree of cementation.

Soil Fertility

A soil is not automatically fertile because it happens to contain chemical nutrients that are essential to plant growth. Soil fertility is determined by its *capacity* to provide these essential ingredients to plants. This depends on the presence, nature, and contents of the **soil solution**: the often-complex mixture that fills pore spaces in part of the soil profile.

The soil solution consists of water that has been converted into a mild acid through the dissolution of carbon dioxide, which produces a weak carbonic acid. This acid, in turn, interacts with soil minerals to enrich the solution with nitrates (an important source of nitrogen for plants), phosphate, potassium, sulphur, calcium, and magnesium, as well as small quantities of such trace elements as copper, zinc, and manganese.

Texture, structure, and biological components also affect the fertility of soil. When a soil contains adequate humus and sufficient clay particles, its fertility will be greater than that of a sandy, less humus-rich soil. Structure influences penetrability and circulation; texture affects the rate of exchange of ions between soil, soil solution, and plant roots.

When soil and vegetation are allowed to interact over time without interference, a cycle of consumption and production develops in which mineral nutrients pass from soil to plant roots to foliage (and to animals), then return to the ground as waste that is, in turn, broken down and absorbed by the soil solution—becoming available nutrients once again. In this cycle, the fertile soil is the great converter and supplier.

Soil Classification and Regionalization

Scientists in the Soviet Union and the United States have, during the past century, taken the lead in the effort to classify and regionalize soils. the Russian scholar, Vasily V. Dokuchayev (1846–1903), was the first pedologist to demonstrate that soils of the same parent material develop differently under different environmental conditions, and he began the description of soils according to their horizon-layered profiles. In 1914, Dokuchayev's student Konstantin D. Glinka

142

FIGURE 4.3

Soil distribution on a hypothetical Northern Hemisphere continent. After USDA.

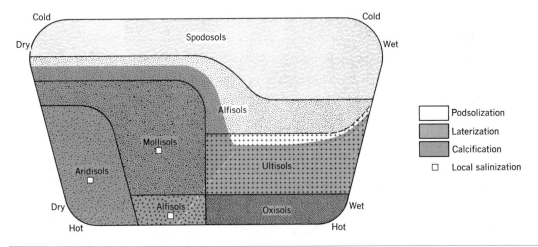

(1867–1927) published a book that summarized the achievements of the school Dokuchayev had started. Thus the knowledge of the Russian school was transmitted to the United States. The American pedologist, Curtis F. Marbut (1863–1935), laid the groundwork for a genetic soil classification (first published in 1938), using Russian ideas and terminology. Marbut was a director of the U.S. Department of Agriculture's (USDA's) Soil Survey Division, and throughout the 1940s, revisions were made to improve the Marbut classification system as his team of soil scientists gathered new information. But as time went on, it became obvious that the Marbut system had fundamental shortcomings that could not be remedied by revisions. What was needed was a totally new system.

THE SEVENTH APPROXIMATION

The effort to build a new system was begun during the 1950s by the USDA Soil Conservation Service. The Soil Survey staff had assembled an enormous mass of data on soils in the United States, and now a first attempt was made to produce an all-encompassing classification system based on the soils themselves, *not* the soil-forming factors and processes believed to be operative. This first attempt went through a series of adjustments until, in 1960, the seventh revision was adopted. This is why the *Comprehensive Soil Classification System* (CSCS) is also called the Seventh Approximation.

The world's soils are grouped into 6 categories in this new classification. We can concern ourselves only with the 10 orders, since these are divided into 47 suborders, 185 great groups, about 1000 subgroups, 5000 families, and 10,000 series.

Perhaps the best way to approach the regionalization of the 10 soil orders of the CSCS is to consider their spatial relationship on a hypothetical Northern Hemisphere continent (Fig. 4.3). This map should look familiar, because it contains elements of the Köppen climatic arrangement. Thus the southeastern region would be dominated by mT air masses, the southwest by cT air, the northwest by cP air, and the northeast by mP air. In very broad terms, the hot moist southeast experiences laterization, and soils in that corner of our continent would be laterized oxisols. In the dry, hot southwest, temperatures are high, moisture is deficient, and calcification prevails; as noted

earlier, this is a quality of aridisols. The north is cool to cold, moister in the east than in the west, and podzolization occurs; these types of soils are spodosols. We have considered these three soil orders in the context of the major processes affecting them. It is noteworthy that no oxisols occur in the United States, where no area receives sufficient heat and moisture to generate the amount of laterization required to produce them (Fig. 4.4). Oxisols (*oxidé* is French for containing oxygen) are not generally useful for farming; although they are often extremely thick soils, their relationship with the tropical vegetation they sustain is fragile and easily disturbed.

Spodosols (*spodos* comes from the Greek for wood ash) do occur in the United States, although not over a wide region. They are confined to a sector of the Northeast, where they form an extension of the broad belt of spodosols that supports the needleleaf forests of Canada. Lumbering is a major industry in these areas, but spodosols can be farmed when limestone is used to offset the soil's acidity; root crops, especially potatoes and sugar beets, do well in the soil as it is.

Aridisols, a soil so named for obvious reasons, prevail over much of the dry U.S. Southwest. Indeed, these soils cover a larger area of the world as a whole than any other soil order (Fig. 4.5), nearly 20 percent. Calcification prevails, and aridisols contain horizons rich in calcium, gypsum, and various salt minerals (**salinization** of the B horizon also occurs).

Calcification processes also affect the **mollisols** (*mollis*, Latin for soft) that extend over much of the western Great Plains and large areas of the Northwest. The mollisols possess a dark, humus-rich upper layer and never become massive and hard, as aridisols do, even under dry conditions. As the map suggests, mollisols are located in intermediate positions between dry and moist climates; these are the grassland soils that support large livestock herds or, when farmed, sustain vast expanses of grain crops.

On the moist side of the mollisols lie the **alfisols** (they used to be called ped*alf*ers, hence the new term), located between arid and subhumid soils on one side and more humid ultisoils on the other (Fig. 4.4). In the United States, they occur in the physiographic province known as the Interior Lowlands (page 00), where they have developed on the glacial debris deposited during the Pleistocene Ice Age.

143

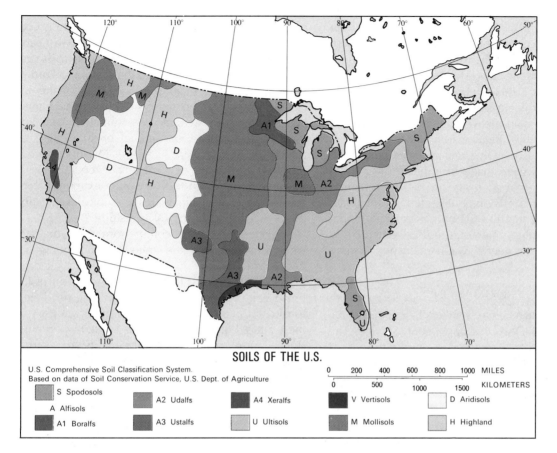

SOILS OF THE U.S.

U.S. Comprehensive Soil Classification System.
Based on data of Soil Conservation Service, U.S. Dept. of Agriculture

| 0 | 200 | 400 | 600 | 800 | 1000 | MILES |
| 0 | | 500 | | 1000 | | 1500 | KILOMETERS |

S Spodosols
A Alfisols
A1 Boralfs
A2 Udalfs
A3 Ustalfs
A4 Xeralfs
U Ultisols
V Vertisols
M Mollisols
D Aridisols
H Highland

144

FIGURE 4.4
Generalized map of the distribution of soil orders in the United States.

There is also a zone of alfisols in the southern Great Plains. They occur in a broad zone in interior western Canada as well (Fig. 4.5), reflecting the wide range of environmental conditions under which they can evolve. These are soils that are subject to neither calcification nor podzolization, nor do they show a horizon where oxidation has occurred. They do have a marked lower horizon of clay accumulation. Alfisols are generally fertile and capable of sustaining intensive agriculture.

The southeastern United States is largely a region of **ultisols** (*ultimus*, Latin for last). Unlike the alfisols farther north, these ultisols are very old, not having been covered by Pleistocene glaciers. Many ultisols are deeply weathered, therefore, but they are not

particularly fertile. It is possible to see the transition to the tropical oxisols, because ultisols develop where there is a pronounced wet season (during the summer) when tropiclike leaching can occur, followed by a water-deficient dry season. The moist season is accompanied by the same kind of removal of solubles that goes on continuously in the oxisols, so that ultisols often display the characteristic reddish yellow coloration in the B horizon due to the concentration of iron oxides. Although ultisols are not very fertile— in much of the subtropics, they support shifting cultivation—they can be made productive with the aid of fertilizers and the addition of lime.

So far we have identified 6 of the 10 soil orders of the CSCS. Each is broadly associated with a particular climatic regime. It is less easy to generalize about the remaining 4, because they can develop almost anywhere and sometimes occur in small patches within, or surrounded by, other soils. There are the entisols, inceptisols, vertisols, and histosols.

Entisols (*ent* from recent) can be grouped with *inceptisols* (*inceptum*, Latin for beginning) and *vertisols* (*verto*, Latin for turn) because they do not display the horizonation expected of a mature soil under their particular climatic environment. Soils formed on the alluvium deposited by major rivers, for example, are likely to be **inceptisols** (Fig. 4.5), so that they are found in the Mississippi Valley, in the Amazon Basin, and in India's Ganges Valley—but they also prevail in the highest latitudes in the region of tundra climate (compare Fig. 4.5 and Plate 3), where quite a different moisture-temperature regime nonetheless produces the incomplete development characteristic of inceptisols. In the United States, areas of entisols extend from the tip of the Florida Peninsula to the northwestern Great Plains, but larger zones of entisols occur in Africa (on Sahara and Kalahari sand), in Australia (on sand accumulations in the interior), and in China (on the alluvium and coastal-plain sediments of the lowlands in the east). Again, entisols either have not existed long enough to develop mature horizonation or lie on parent material (e.g., quartz sand) that does not readily evolve into horizons. Persistent erosion also can lead to the development of entisols by constantly removing surface material that might have generated an A horizon. The **vertisols** are clayey soils (more than 35 percent of the volume consists of clay particles) characterized by deep, wide cracks that develop during the dry season. These cracks close again when the available moisture increases and the clay swells, but before it does so, a portion of the surface material has washed into the cracks—hence the soil's name (from *invert*, turn over). In the United States, vertisols occur in limited areas of the Gulf coastal plain in Texas, but larger areas exist in Australia between the western desert and the eastern mountains, in western India, and in the Sudan immediately west of the Ethiopian Plateau. These soils, then, are confined to tropical and subtropical locales where a pronounced dry season is part of the climatic regime.

The **histosols** (from *histos*, Greek for tissue) differ from all the other orders in the CSCS because they are primarily organic, not mineral, soils. These occur in bogs, moors, or as peat accumulations, and so they are waterlogged most of the year. They can develop in almost any climatic region, so long as the water supply exists, from equatorial to tundra latitudes. Although histosols exist in limited areas in many parts of the world (too small to be mapped at the scale on Fig. 4.5), the largest contiguous areas lie in northern Canada south of Hudson Bay and in the far Northwest just below the permafrost limit.

Although the CSCS consists of 10 soil orders, maps of the world distribution of soils according to the CSCS include one other category—the soils of mountainous areas. These soils vary from valley to slope to crestline and cannot, on a global scale, be represented accurately; mountainous areas also include patches of permanent snow and barren rock. Hence, mountain zones, including the great orogenic belts and also lesser mountains such as the Appalachians and the Australian cordillera, are mapped separately as "Soils in Mountainous Areas," signifying a wide soil variety that includes many of the orders discussed.

SOIL AND VEGETATION

Notwithstanding the shortcomings of the Köppen climatic regionalization and the sweeping revision of the classification of soils, it is possible to discern patterns of similarity between climatic regions and soil zones (Plate 3 and Fig. 4.5). The east–west orientation of climatic belts in Eurasia and their north–south direction in North America is mirrored by the broad latitudinal arrangement of mollisols and alfisols in Asia and the longitudinal zone of mollisols in North America. And something survives as well from the old association between vegetation regions and soil areas. Before the CSCS was introduced, the soil classification in use referred frequently to vegetative associations. There were, for example, prairie soils (now part of the mollisol order), tundra soils (inceptisols and entisols), and brown forest soils (partly alfisols). In the new soil regionalization, such vegetative terms have been dropped—but the oxisols are still the soils of the tropical rain forest and the near-equatorial savanna, the spodosols still prevail under the zone of boreal needleleaf forests, and the aridisols are still the soils of the dry steppe and desert.

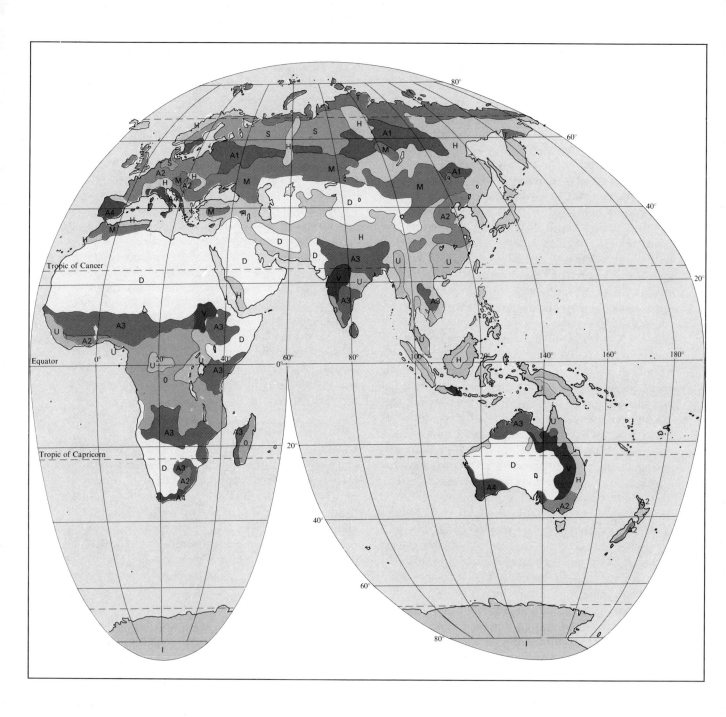

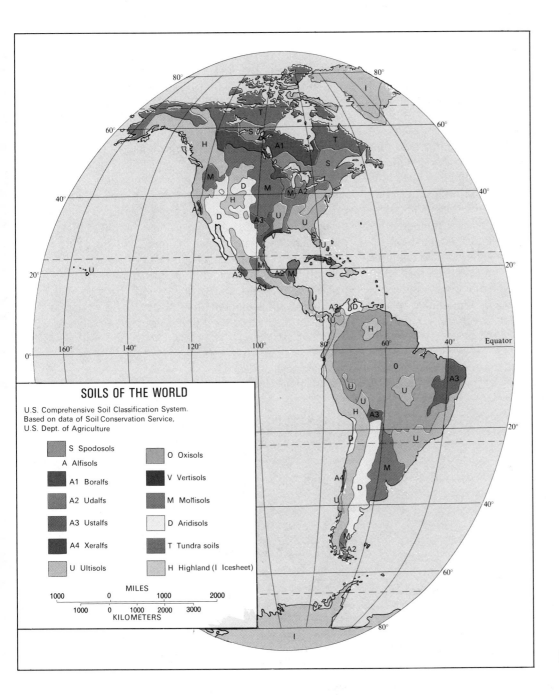

SOILS OF THE WORLD

U.S. Comprehensive Soil Classification System.
Based on data of Soil Conservation Service,
U.S. Dept. of Agriculture

S Spodosols

A Alfisols

A1 Boralfs

A2 Udalfs

A3 Ustalfs

A4 Xeralfs

U Ultisols

O Oxisols

V Vertisols

M Mollisols

D Aridisols

T Tundra soils

H Highland (I Icesheet)

MILES
1000   0   1000   2000
1000   0   1000   2000   3000
KILOMETERS

147

FIGURE 4.5

We turn now to the study of the vegetative biosphere, the plant associations that have developed at the interface between soil and atmosphere. The distribution of soils is one of the factors that affect the distribution of vegetation, the **edaphic factor**, and we are already familiar with its role in the provision of moisture, nutrients, and other essentials for plant development. Another influence is exerted by landform conditions, including steepness of slope, altitude of the land surface, and exposure to sunlight. Naturally, temperature is a powerful factor in limiting the distribution of natural vegetation. Some plants have developed the capacity to withstand low temperatures (the **microtherms**), whereas others are adapted to heat (the **megatherms**); many are intermediate, neither very heat- nor cold-resistant (the **mesotherms**). Still another factor affecting plant distributions is the availability of water. As in the case of temperature, plants have in some cases adjusted to varying conditions.

Certain plants are able to cope with very little moisture (the **xerophytes**); these, of course, exist in regions of dry climate, but they also occur in places where drainage is rapid and moisture availability is locally low, as on sand dunes and in small cracks against bare rock surfaces. Other plants are adapted to large amounts of moisture (the **hygrophytes**), for example, the vegetation growing in shallow lakes or on lake margins, in swamps, marshes, and bogs. Vegetation that grows in areas of ample moisture and good drainage (the **mesophytes**) does not adjust well either to excessive dryness or to excessive moisture; severe droughts as well as saturation rains can affect these plants adversely. In tropical climates where a pronounced dry season occurs and in wet-summer, cold- and dry-winter humid cold (continental) climates, trees have adapted to the dry periods by dropping their leaves, thus reducing their water loss. Such trees and other plants that shed their leaves seasonally are

Xerophytic (moisture-preserving, adjusted to limited moisture) plants in the Angeles National Forest, California. Xerophytic vegetation gives the natural landscape an unmistakable appearance wherever it occurs.

148

called **deciduous**; those that retain their leaves all the year round are **evergreen**.

As in the development of mature soils, a critical factor in the evolution of a vegetative association that is in a condition of equilibrium with the prevailing climatic regime and the soil is time. During this evolution, the vegetation (and the fauna as well) goes through an ecological succession in which biotic communities succeed one another, becoming ever more complex until a stable end point, a **climax vege-. tation**, is achieved. When we refer to a vegetative association such as the tropical rain forest or the tall-grass prairie, we refer to the climax vegetation that could exist everywhere in the region of occurrence—but that does not occur everywhere by any means. Natural and human causes can interfere with the ecological succession or destroy the climax vegetation, and the cycle must start over again. As in the case of climates, it is necessary to generalize.

Biomes

When vegetation and animal life reach a stable adjustment to the natural environment, they constitute an ecosystem. Such an ecosystem consists of a large number of plant and animal species, interacting with each other and with elements of their environment. Ecosystems develop in the oceans as well as on the continental landmasses, but our interest is in the terrestrial ecosystems—particularly in the vegetative assemblages that have developed on the earth.

The world's land vegetation includes an enormous variety of species, but these can be grouped into trees, shrubs, grasses, and mosses. These major groups do not include several specialized plants such as lianas (woody vines that climb the trunks and limbs of trees in tropical and midlatitude forests) and epiphytes (plants that live off other plants by attaching themselves to tree limbs).

The dominant forms (trees, shrubs, grasses, mosses) give the ecosystem its particular visual character, however. **Trees** are perennial (enduring) plants, normally with a single main trunk and few branches in the lower part, but spreading by means of several limbs into a crown in the upper part. At times, these crowns meet, creating a layer of vegetation high above the ground that effectively reduces the amount of sunlight reaching lower layers and the ground below. **Shrubs** are smaller and lower, but they are also woody plants that develop several stems branching off close to the ground, so that the foliage spreads out at a much lower level than it does in trees. **Grasses** (technically combined under the term **herbs**) are non-woody, thin-stemmed plants that may be perennial or annual. In an ecosystem that consists of all four dominant forms, grasses occupy a lower layer, although mosses and similar vegetation forms exist in the lowest zone.

Biogeographers have classified the earth's land vegetation into five great **biomes**; these, in turn, are subdivided into **formation classes**. This has permitted the regionalization shown in Plate 4, where the biomes are identified as forest, savanna, grassland, desert, and tundra (see the map's legend). This map should be viewed in conjunction with Figure 4.5 and Plate 3, and some interesting relationships will become apparent.

FOREST BIOME

Forests occur in equatorial, middle, and high latitudes, although there are major differences between individual formation classes even in the same general latitude. Forests are, of course, dominated by trees, but in different ways. In some forests the tree crowns interlock, blocking sunlight and casting permanent shade on the ground below. Elsewhere the trees stand sufficiently far apart to permit the sun to shine on patches between them. In still other locations trees lose their leaves every autumn. Forests grow on high mountains and on monsoon-drenched coastal plains.

Rain Forests

The **equatorial, tropical,** and **monsoon rain forests** can be grouped together, although some plant geographers (phytogeographers) prefer to treat these assemblages as separate formation classes. These low-latitude forests reach their fullest development in equatorial areas of high heat and high humidity. The tallest trees reach sunward to a height of as much as

Equatorial and tropical forests develop under conditions of warmth and ample moisture. In drier zones, the climax vegetation may be a savanna—except where elevation wrings enough water from the air to sustain a different kind of forest, a *montane* forest. In the savanna-encircled Chyulu Hills of Kenya in East Africa, such a forest still exists. Note the range of tree sizes, the gaps in the canopy, and the comparatively dense undergrowth.

130 to 200 ft (40 to 60 m), forming broad crowns that rise above the nearly continuous, thick canopy of foliage created by the trees of medium height (50 to 100 ft or 15 to 30 m). Thinner, less-leafy trees stand below this canopy, losers in the constant struggle for sunlight. This three-layer aspect of the equatorial rain forest is one of its dominant properties; so is the virtual absence of foliage near the ground. Contrary to popular images of impenetrable jungles, there are few obstructions between the tree trunks; there is too little sunlight to sustain ground-level vegetation. Of course, the rain forest does contain numerous lianas and epiphytes, giving it a cluttered appearance. But other than a layer of leaf litter on top of the soil, the ground is quite clear.

The bark of trees in the rain forest of equatorial and tropical regions is usually quite smooth, and the leaves are large and broad. Green foliage is rapidly replaced, so that there is always a supply of leaves to add to the layer of leaf litter on the ground—a layer that is important in the forest's nutrient cycle.

Another prime characteristic of the equatorial rain forest is the enormous number of species that exist even in a small area. Hundreds of species thrive in a patch as small as half a square mile, and this has had much effect on lumbering operations in these forests. A square mile may contain several thousand trees, but only a dozen may belong to the species being sought for harvesting. This can render their extraction unprofitable, so the variety of the rain forest has helped protect it against exploitative destruction. But not, unfortunately, against destruction from encroachment.

Away from equatorial zones, the slight climatic changes (a moister, slightly less humid season and a mild annual temperature range) result in a transition from true equatorial rain forest to tropical rain forest. The tropical rain forest has somewhat fewer species,

Equatorial rain forest in Brazil's Amazon Basin. High temperatures and high humidity prevail here, and a nearly continuous, thick *canopy* of foliage develops. The people living in the forest must clear patches of it to cultivate their crops. See "Shifting Cultivation" in Chapter 8.

and there are also fewer lianas, although epiphytes continue to thrive. In Plate 4, the equatorial and tropical rain forests are combined because no clear dividing line could be drawn between equatorial and tropical rain forests. As Figure 4.5 indicates, the low-latitude rain forests grow on oxisols and ultisols, heavily leached soils where nutrients are quickly returned to the plants from the decaying leaf matter; no humus layer develops here, and the soils (as many farmers know) are not themselves nutrient rich or fertile.

The stress of the slightly drier season that imposes a reduced number of species in tropical (compared to equatorial) areas has a much greater impact in Am (tropical monsoon) regions, producing a distinct assemblage in the monsoon rain forest. Although the general aspect is that of a tropical rain forest, the tree trunks are less straight, limbs develop at lower levels, the bark is coarse and rough, the leaf litter is thicker, and shrub vegetation stands between the trees, so that the forest is quite dense and often difficult to penetrate. Most significant, many of the tree species are deciduous, and the period of dormancy between the monsoons is marked by a pronounced thinning of the forest's foliage. Plate 3 provides an indication of the distribution of the monsoon rain forest in Asia, Africa, and Middle and South America.

The **temperate rain forest**, also called the broadleaf evergreen forest, prevails under a climatic regime quite different from equatorial and tropical zones, the humid temperate (Cf) clime with cooler temperatures and a more pronounced seasonal temperature range. But there is moisture all year round and many trees are evergreens, as in tropical forests. The temperate rain forest contains far fewer species and thus many more individuals of particular species, and trees are less tall, stand farther apart, and have smaller leaves. The foliage does not form a continuous canopy, so that sunlight can reach the forest floor where shrubs, ferns, and other low-level foliage develop.

Regions of temperate rain forest are widely distributed, from a large area in southeastern China and southern Japan to much smaller zones in the southeastern United States, southwestern South America, eastern Australia, and New Zealand (Plate 4). The forest's aspect differs considerably from region to region as particular species dominate. In the southeastern United States, species of evergreen oak, laurel, and magnolia trees mixed with beech and pine prevail, but in New Zealand the kauri pine, a resinous timber conifer, which reaches 150 ft (45 m) in height and as much as 23 ft (7 m) in thickness, dominates a forest of podocarp conifers and large tree ferns.

Figure 4.5 indicates that temperate rain forests develop on ultisols, especially in China and the southeastern United States; alfisols and mountain soils support rain forests in Australia and New Zealand.

Deciduous Forests

The **midlatitude deciduous forest** extends across the eastern United States, Western Europe, and eastern

Deciduous forest in the United States. Tall oak, maple, beech, and other trees dominate these midlatitude forests, the ground covered by a sometimes thick layer of leafy matter decaying and contributing to the fertility of the soil below.

China (Plate 4), and with the exception of small areas in southern South America, it is a vegetative assemblage that occurs only in the Northern Hemisphere. In the United States, this is the familiar forest of tall oak, beech, hickory, walnut, maple, ash, and elm trees. When the forest is fully developed, the summer canopy of foliage is almost continuous, but in the autumn the leaves drop. In the spring, grasses cover the ground thickly before the new leaf growth begins, but soon the canopy develops again, the sunlight is shut off, and the grasses thin out markedly.

Midlatitude deciduous forests occur among the world's most densely populated areas; here is one of those instances where the map shows theoretical (potential) climax vegetation, not the real situation. Much of this forest, adaptable to moist, cold- or cool-winter climates, has been cut down, and some of what we see today is actually a regeneration, not the original primeval forest. This destruction resulted not only from the value of the wood and the comparatively small number of useful species, but also from the fertility of the soils beneath, including alfisols and some mollisols. Compare the map of world agriculture, Plate 8, with the map of world vegetation (Plate 4).

Needleleaf Forests

The **needleleaf forest** extends from Alaska across Canada to Newfoundland as well as across northern Eurasia from Scandinavia to the Soviet–Japanese island of Sakhalin. This is the zone of Dfc (cold, humid) climate, and the vegetative response is a forest of trees with needle-thin leaves, which are comparatively short (40 to 65 ft or about 12 to 20 m), mostly evergreen, and with a small number of species. Although this is technically a year-round humid climatic region, the winter cold and freezing, in effect, constitute periods of prolonged drought; some trees such as the larch prevailing in parts of northern Siberia shed their leaves during this time. But the dominant spruce, fir, and pine are evergreens, clustered so closely over large areas that the ground is permanently shaded. Mosses cover the soil here, but these northern forests are not quite so full as their tropical counterparts. Open patches do exist where grasses and hardy shrubs survive. Poorly drained depressions (bogs) are occupied by hygrophytic vegetation.

• Jungle •

If the equatorial and tropical rain forests consist of tall, well-spaced trees with well-developed trunks without low limbs and branches and with little undergrowth, how did these forests acquire the name **jungle**, implying high surface density and impenetrability? In truth, the rain forest can be traversed on foot with relative ease. Many other forests are much more difficult to cross on foot, especially the thorn-tree thickets of drier zones.

It is human activity that transforms rain forest into jungle. When the trees are cut down and the forest destroyed, the tropical vegetation will regenerate—but the well-ordered layering of a tropical rain forest takes many years to restore. First a mass of disorganized second growth arises, a clutter of trees with low crowns, shrubs, lianas, and other plants all competing for the sunlight that now penetrates the broken-down canopy. Every square foot of soil carries its candidates for primacy in the newly developing forest, and only a machete can achieve a path through the nearly impenetrable ''jungle.''

Destruction of the rain forests has taken place as long as people have occupied tropical areas. By fire and by ax, the trees have been felled to make way for a year or two of patch farming; then the exhausted soil is left behind and another area is cleared. This shifting cultivation has done its share of damage, but modern machinery does it faster—and more permanently. The shifting cultivators leave the forest to restore itself, but now there are other objectives in the tropics: the wholesale clearing and reforestation of huge tracts. The true climax vegetation we know as equatorial and tropical rain forest may be on the verge of destruction.

The needleleaf forest is a comprehensive class formation. It includes not only the high-latitude boreal forests, but also the forests on higher mountain slopes in lower latitudes, where conditions resemble those of the cold northlands; and it also incorporates the great coastal needleleaf forests of British Columbia (Canada) and California. These western coastal forests contain the giant redwoods that grow to heights of more than 330 ft (100 m) and attain diameters of as much as 65 ft (20 m).

Sclerophyll Forest

The **sclerophyll forest** differs sharply from the class formations discussed earlier. Its distribution (Plate 4) suggests why: this plant assemblage fringes the Mediterranean Sea and occurs in smaller areas in southern California, central Chile, southern South Africa, and western Australia, thus coinciding rather closely with the dry summer humid temperature climate (Csb, Plate 3). This is a low-precipitation climatic regime, and the sclerophyll forest consists of low, knotty, low-branched trees with leathery, drought-resistant leaves. Thick-barked shrub or scrub covers much of the Mediterranean countryside, but there are also live oak, cork oak, olive trees, and various species of pine. The sclerophyll forest has a most characteristic and unmistakable aspect that has been given various names in different parts of the world: **maquis** in Mediterranean areas, **chaparral** in southern California, and **fynbos** in South Africa. Almost everywhere it occurs, the sclerophyll forest has been damaged or destroyed by human activity, and what we observe today is regeneration in areas that may have been covered by much more luxuriant Mediterranean-type vegetation in the past.

SAVANNA BIOME

Between the tropical rain forest and the desert lies a vast region of Aw (savanna) climate (Plate 3), a climatic regime that derives its name from a vegetative assemblage. This is the tree-studded grassland of the **savanna**, also called (as it is in the legend of Plate 4) the rain-green vegetation of low latitudes. It is so named because over large parts of the region its grasses and trees desiccate for 8 to 10 months, turning gray and frequently burning from lightning strikes and

153

human fires; then they are revived by the life-giving rains of the wet season. When this happens, the savanna turns green, the animal herds return, and the area briefly seethes with life. Soon, however, the rainy season ends, and the long drought resumes.

As noted earlier, parts of the savanna have a distinct double maximum of precipitation each year, a long rains and a short rains—always separated by pronounced dry periods. Even here, the countryside changes from dusty grayness to new green each time the rains return. But the rainfall is variable, not nearly so dependable as it is in the equatorial and tropical rain forest zones. Rainfall in the savanna—especially on the high-latitude side—is a scarce commodity.

In Africa, a type area for the savanna, this zone is mostly an open woodland with widely spaced, flat-topped trees standing in vast fields of thick grassland. On the moister margins, savanna trees are deciduous. But the typical savanna country has trees of medium height, many with several trunks branching from near the ground with thorny stems and branches and small leaves. The xerophytic character of the vegetation signals the transition into the adjacent (and even drier) steppe, as acacia trees and baobabs dominate the landscape.

Savannas presenting different aspects exist in Australia, southern and southeastern Asia, and in South and Middle America. In Australia, the eucalyptus tree dominates in a dense savanna woodland. In South America, much of Brazil's plateau is savanna country, with tall grasses and widely spaced trees such as mimosa, varieties of acacia, and palms. In India much of the savanna has been destroyed by the human explosion, but where it remains it is shrub country that changes into the monsoon rain forest toward the moister margins and into thorn bush in the drier fringe. The savanna biome develops on oxisols and alfisols; it is also found on ultisols and in Australia and India on vertisols.

The role of fire in the maintenance of the savanna biome should not be underrated. Whether set by nature or by humans, savanna fires prevent the encroachment of trees and shrubs at the expanse of the grasslands. The fire does not destroy the roots of the grass, but it does kill off many weaker woody plants above the ground; only the fire-resistant species can survive the annual fire cycle. This leaves the ground open and the grasslands can regenerate when the rains return. Toward the end of the dry season, long ribbons of fire cross the African savannalands, and a smoky haze hangs over the countryside; shortly after the first rains, the new shoots of green appear from the blackened, ashy grass tufts.

GRASSLAND BIOME

Huge expanses of the continental interiors in the Northern Hemishpere are covered by grasslands. Smaller but still substantial areas of southeastern South America and plateau South Africa also are grasslands. These grasslands range from thick, tall grass where trees are virtually absent and confined to the valleys of streams to sparser, shorter grasses interspersed with shrubs and low trees. As always, the critical commodity is water: the moister zones support denser, treeless grasslands, but declining precipitation (toward steppe and desert margins) brings sparser, shorter grass and patches of open ground.

The tall-grass **prairie** occupies a large part of the midsection of North America (Plate 4) and a region in southeastern South America extending from southern Brazil to northeastern Argentina. The prairie consists of perennial grasses (annual grasses are unusual here) that do well under humid temperate and humid continental climatic regimes where no serious water shortage develops. As Figure 4.4 shows, they develop especially on mollisols.

In South America, the prairies are known regionally as the **Pampa**, the type area in Argentina that supports large herds of livestock. Here, as in North America, a humid climatic regime prevails that yields westward into a drier zone. The vegetative class formation changes to the **steppe**, or short-grass prairie; grasses are bunched and shorter, and there are patches of exposed soil as well as shrubs and stunted trees that dot the countryside. In addition to the large region of short-grass prairie in North America, there are extensive steppelands in interior Asia and in South Africa where this vegetative assemblage is known as the **veld**. Soils remain mollisols but under a somewhat drier regime, although aridisols make their appearance on the drier margins of the steppes.

Every vegetative assemblage supports a varied and often large animal population. In this chapter, we concentrate on the interrelationships that exist among climate, vegetation, and soil, but a complete ecosystem also includes the regional fauna.

The savannalands sustain the earth's remaining great herds of wildlife and their predators. Just two centuries ago, huge herds of wild animals occupied the grasslands of North America and Asia. Bisons by the millions roamed the prairies from Canada to Mexico, their numbers controlled by the vagaries of the natural environment and by packs of wolves, coyotes, and bears—and by a fairly stable population of Indians. In the plains of Asia and the savannas of Africa, a variety of wildlife, ranging from the Arabian oryx to the greater kudu and including buffalo, lions, elephants, and hundreds of other species, formed part of local and regional ecosystems.

But the pressure of human numbers and the advent of mechanized means of soil and forest exploitation have dramatically reduced these once-vast herds. Africa's savannalands now remain the last great refuge for millions of migrating wildebeest and other grass eaters, their movements made possible by large national parks and other means of protection. But Africa's wildlife, too, is under stress. The search is on for ways to domesticate eland, zebra, and other species, and research is being done to counter the tsetse fly, the insect that spreads sleeping sickness in domestic livestock and thus renders much of the savannalands useless for ranching. If both efforts succeed, the days of Africa's wildlife may be numbered.

DESERT BIOME

The desert biome includes the whole transition from the short-grass prairies to the driest environment. It includes areas where there are some perennial plants, areas where plants are annual (completing their entire life cycle in just one year), and places that are almost completely without plant life. The most convenient division is into two regions: semidesert and true desert.

The **semidesert** prevails over vast areas of the world, including southwestern North America, interior southern South America, much of central Asia, large regions of northern and southern Africa, and about half of Australia (Plate 4). Here the xerophytic vegetation consists of thorn trees (some biogeographers recognize a **thorn tree semidesert** as a separate class formation) and shrubs that have adapted to a lengthy dry season followed by brief sometimes heavy rains. In South Africa, this vegetation is called the **doornveld** (thornbush), and in places it can develop into dense, thorny, almost impenetrable thickets. The climate, of course, is prevailingly the BS (semiarid) regime of the desert climates (Plate 3); the soils are aridisols.

The true **desert** occupies the interior regions of North Africa, south-central Asia, and Australia; smaller areas exist in southern Africa, coastal western South America, and southwestern North America. Here vegetation is either very sparse or completely absent, and most plants that do exist are annuals; they spring from seeds that may have lain dormant for years, awaiting a period of rainfall. Thus the desert presents a totally different aspect after an occurrence of rain when hundreds of plants emerge to give it an almost luxuriant appearance. But soon the sun takes its toll, the vegetation withers and dies, and the landscape again is one of desiccation and aridity.

Deserts are not always vast expanses of sand and dunes, as is sometimes believed. Much of the Sahara, for example, is rocky desert, and plants that do remain alive manage to find a means of survival in the cracks of joints in the rock surfaces where moisture is retained a little longer and where some nutrients collect. But in the Sahara, as in other deserts, there are places where one can stand on a hilltop and see not even a hint of plant growth as far as the eye can reach.

155

A view over the Canadian tundra near Hudson Bay in Manitoba. Excess water and low temperature mark the tundra environment; few trees manage to survive under these circumstances.

TUNDRA BIOME

Except for a small patch of tundra on Tierra del Fuego at the southern tip of South America and the high-altitude tundra on scattered mountains, the **tundra** biome consists of a broad zone in the northernmost land latitudes of the Northern Hemisphere. This region extends along the northern margins of North America, Greenland, and Eurasia, and it exists also on islands in the Arctic region.

The tundra prevails under harsh, extremely cold climatic conditions. During the long winter, high winds and blizzards attack any plants that protrude above the snow cover, the soil freezes solid, and the vegetation lies dormant. The short summer melts only the upper layer of the soil; the ground below remains frozen as permafrost. Meltwater thus cannot percolate downward very far, and much of the soil in tundra areas is waterlogged in summer, although soils on low hills and ridges fare somewhat better. The tundra vegetation, therefore, must be adjusted to low temperatures (microthermal plants) and in many cases to ex-cess water (hygrophytes). These circumstances have generated quite a wide range of flora, and the dominant aspect of the tundra is of a lichen- and moss-covered ground with sparse stands of grasses and sedges. An occasional stunted tree survives (e.g., the dwarf willow), but this is a difficult environment for tree growth, the fairly substantial humus content of the topsoil notwithstanding.

It is no surprise that the soils beneath the tundra vegetation are mainly inceptisols; full development cannot take place here. This is true also of the **alpine** tundra that develops above the limit of tree growth but below the snow line on high mountains. Apart from the contrast in landforms and overall scenery, the vegetative aspect of alpine tundras is quite similar to that of the high-latitude tundra.

BIOSPHERES

The close interaction of climate with the development of vegetation and the formation of soils is reflected by the similarity of the distribution patterns we have just examined. But it is important to remember that we have dealt with these patterns on a global scale, so that it has been necessary to generalize. Soil-forming processes and vegetative assemblages often change quite drastically over short horizontal distances and we cannot, at this scale, account for these changes. Also, the lines on the maps must not be seen as sharp dividers but rather as transition zones. For example, the zone of contact between northern needleleaf forests and tundra is actually a wide belt where both forest and tundra exist intertwined, one affecting the other. The trees of the forest prevent the winter snow from drifting as it does over the open tundra, and they also act as a windbreak, so the tundra in the transitional zone has an aspect quite different from that farther north. In more temperate climes, the grasslands of the prairie interdigitate with the adjacent midlatitude deciduous forest; of necessity, we have been viewing **type areas**, but the boundary line on Plate 4 represents a zone of transformation.

Another important context relates to the earth's constantly changing environments. Even if Plate 3, Figure 4.5, and Plate 4 were totally accurate and infinitely detailed representations of the globe's climatic and biospheric patterns, they would only constitute still photos representing a moment in a sequence of unending change. Just a few thousand years ago, the tundra existed across the heart of Western Europe, and there was permafrost in France; northern needleleaf forests occupied Spain and Italy. The climatic, vegetative, and pedologic belts have shifted poleward, and they have not stabilized. Our picture of the biosphere is a snapshot; a thousand years from now such a snapshot is likely to look quite different.

•   Zoogeography   •

If you compare definitions and descriptions of biomes as they appear in biology, botany, and geography textbooks, you will note that there are various ways to delimit these. The great biomes as represented in Plate 4 are based primarily on vegetation, but we know that animals are closely associated with plants and with the soils that sustain them. A biome, therefore, is actually an interacting set of ecosystems that extends over a relatively large area of the earth. Among the participants in the establishment and maintenance of a biome are the fauna—animals ranging from the tiniest organisms to large mammals.

Zoogeographers study spatial aspects of animal populations. This field emerged during the nineteenth century following the work of Alexander von Humboldt and, of course, Charles Darwin. But the best-remembered work of that period was a two-volume book by Alfred Russel Wallace, *The Geographical Distribution of Animals*, published in 1876. Southeast Asia and Australia were realms that especially interested Wallace because their fauna differ so strongly. Australia is the last major refuge of the earth's marsupials (animals whose young are born very early in their development and then carried in a pouch on the abdomen). Some marsupials remain elsewhere such as the opossum in America. But Australia has a wide range that includes the kangaroo, koala, and wombat; Australia also is the home of the only two remaining egg-laying mammals, the platypus and the Australian anteater. Wallace drew a line across the islands of Indonesia, a line that was to separate Australia's fauna from that of Southeast Asia. Wallace's line went between Borneo and Sulawesi (Celebes), and between the first (Bali) and second (Lombok) islands east of Djawa (Java). Wallace's line became one of the most debated zoogeographic delimitations ever drawn (in fact, one of the most intensely discussed boundary lines in all of geography). Other zoogeographers and zoologists drew alternate lines to show, for example, how far Southeast Asian animals such as the rhinoceros, the elephant, and the tiger had progressed along the island stepping stones between mainland Southeast Asia and continental Australia. Various substitutions for Wallace's line were suggested, all of them proving how difficult the problems of regional zoogeography can be. It is one thing to draw maps of stands of rain forest, expanses of desert, regions of tundra vegetation, but to do the same for African elephants, American jaguars, or Indian tigers is another matter. Animals move and migrate, their range (or region of natural occurrence) fluctuates, and detection may be a problem, too.

Among zoogeographers' interests are the past migrations of animals, the routes by which they reached their present homes, and their present migration habits—for example, by birds as they fly seasonally between summering and wintering grounds. At the present time, much of this zoogeography is done by biologists, not geographers. But for geographers interested in wildlife, there are many opportunities to join this research. This is emphasized by P. J. Darlington, Jr., in a book that is now more than 30 years old but still has relevance, *Zoogeography: The Geographical Distribution of Animals* (1957). This book states four key questions in zoogeography: (1) What is the main pattern of animal distribution? (2) How has this pattern been formed? (3) Why has this pattern developed as it has? (4) What does animal distribution—past and present—tell about lands and climates? Another interesting approach involves what G. G. Simpson calls ecological zoogeography, the study of animals as they relate to their environment, which is a major theme in his book, *The Geography of Evolution* (1965). This work, too, devotes much attention to what the author calls historical zoogeography, that is, questions (2) and (3) above.

Zoogeography also has a more current, theoretical dimension. Biologists believe that there may be as many as 30 million species of organisms on our planet, of which only about 1.5 million have been classified. One obvious question has to do with the rules of nature that govern the habitats of all these species. What determines where species will live and how many can be accommodated in a certain region? Part of the answer has come from a specialized field called island zoogeography. Much observation and some experimentation have led to several conclusions, one of which is that the number of species living on an island is related to the size of that island. Another is that there is a cap on the number of species that can occupy an island of a given size. Many new species keep arriving on islands, brought by birds from afar or washing up on the shoreline. These species cannot succeed (if an island has a stable population) unless another species becomes extinct, thus making room. This notion was actually tested on some small islands off the coast of Florida where biogeographers counted the species of insects, spiders, crabs and other arthropods. They then sprayed the experimental islands, wiping out the entire popu-

lation. After a few years, the islands were re-inhabited—by the same number of species, although not in the same proportion. This indicates that the number of species is characteristic of an island in a certain environmental zone but that the kinds of species making up this number may vary and depend on which arrived earliest to fill the available niches.

Thus biogeographers watch with great interest when new land is formed—for example, when an island emerges above the ocean surface through a volcanic eruption—to see if their predictions about species numbers and kinds will hold true. The repopulation and reforestation of Krakatoa, where all life was destroyed in the great explosion of 1883, has been followed in detail. From these observations, zoogeographers can predict what populations might be in other isolated places, for example, on the tops of buttes and mesas with their steep slopes and flat upper surfaces.

Thus, biogeography (and, by extension, the geography of soils) is a field with many opportunities for students who are interested in the spatial dimensions of the biosphere, the layer of life.

• Reading about Biogeography •

Biogeography is a wide-ranging field, involving not only phytogeography (plant geography) and zoogeography (geography of animal life), but also aspects of climatology and soil geography. This diversity is reflected by the available readings, some of which range beyond the limits of geography itself. Thus a very readable and interesting introductory work is R. F. Dasmann's *Environmental Conservation* (5th ed.; New York: Wiley, 1984); this book contains chapters on soils, vegetation, animal life, and the human impact. Also see the introductory chapters in R. H. Wagner's *Environment and Man* (2d ed.; New York: Norton, 1974).

On soils, it is important to remember that some books do not reflect the new soil classification, although their discussions of soil-forming processes and other topics are very helpful. Undoubtedly the best available work is D. Steila, *The Geography of Soils* (Englewood Cliffs, N.J.: Prentice-Hall, 1976). This volume explains the Seventh Approximation in understandable terms. If you want to see the full background, consult a volume prepared by the Soil Survey Staff of the USDA's Soil Conservation Service, *Soil Taxonomy: A Basic System of Soil Classification for Making and Interpreting Soil Surveys*, Agriculture Handbook No. 436 (Washington, D.C.: U.S. Govt. Printing Office, 1975).

Another readable work is *Soil Geography* by J. G. Cruickshank (New York: Wiley, 1972); our reference to the role of time in soil formation comes from page 63. Also see R. M. Basile, *A Geography of Soils* (Dubuque, Iowa: W. C. Brown, 1972), part of the Foundations of Geography Series. A well-illustrated and very practical volume is E. M. Bridges, *World Soils* (London/New York: Cambridge Univ. Press, 1970). A more recent work is that of J. W. Batten and J. S. Gibson, *Soils: Their Nature, Classes, Distributions, Uses, and Care* (University: Univ. of Alabama Press, 1977). Also see N. C. Brady, *The Nature and Properties of Soil* (9th ed.; New York: Macmillan Co., 1984), a very comprehensive book to keep in your library. Another useful volume in this area is E. A. FitzPatrick, *Solids: Their Formation, Classification, and Distribution* (New York: Longman, 1983). A book of relevance in geography and related fields, one that deals with soil classification problems, modern (computer) applications, and ecological matters is *Principles and Applications of Soil Geography* (London: Longman, 1982) by E. M. Bridges and D. A. Davidson.

A classic book on phytogeography is R. Good's *The Geography of the Flowering Plants* (4th ed.; London: Longman, 1974). Also see a work edited by S. R. Eyre, *World Vegetation Types* (New York: Columbia Univ. Press, 1971), a book that includes a lengthy and interesting article by L. Cockayne, "The Subtropical and Subantarctic Rain Forests of New Zealand." Also see A. S. Collinson, *Introduction to World Vegetation* (Boston: Allen & Unwin, 1977), M. C. Kellman, *Plant Geography* (New York: St. Martin's Press, 1980),

and J. Tivy, *Biogeography: A Study of Plants in the Ecosphere* (New York: Longman, 1982).

On general biogeography, consult D. Watts, *Principles of Biogeography* (New York: McGraw-Hill, 1971) and C. B. Cox, I. H. Healey, and P. D. Moore, *Biogeography: An Ecological and Evolutionary Approach* (New York: Wiley, 1973). First published in the United Kingdom, this book ranges more widely than most biogeographies and includes sections on the biogeography of Gondwana and the impact of continental drift on the distribution of world vegetation. For a fascinating view of the layers of life, see *The Biosphere, a Scientific American Collection* (San Francisco: Freeman, 1970). Also on biogeography, see J. H. Brown and A. C. Gibson, *Biogeography* (St. Louis: Mosby, 1983); I. G. Simmons, *Biogeography: Natural and Cultural* (London: Arnold, 1979); and P. A. Furley and W. W. Newey, *Geography of the Biosphere: An Introduction to the Nature, Distribution, and Evolution of the World's Life Zones* (Stoneham, Mass.: Butterworth, 1982).

And, as always, check the recent periodical literature. In June 1976 P. J. Gersmehl published "An Alternative Biogeography" in the *Annals* of the Association of American Geographers (Vol. 66, No. 2), an article that focuses on nutrient cycles. The August 1977 *Professional Geographer* includes an article by V. Meentemeyer and W. Elton, "The Potential Implementation of Biochemical Cycles in Biogeography" (Vol. 29, No. 3), and in the September 1977 *Annals* of the Association of American Geographers Professor Gersmehl writes about the new soil-classification system in "Soil Taxonomy and Mapping" (Vol. 67, No. 3). P. J. Darlington, Jr.'s work, *Zoogeography: The Geographical Distribution of Animals* (New York: Wiley, 1957) still is of interest. Also still relevant is the volume by G. G. Simpson, *The Geography of Evolution* (Philadelphia: Chilton Book Co., 1965).

The field of biogeography is served by a prestigious journal, the *Journal of Biogeography* (Oxford: Blackwell Scientific Publications). The journal *Catena* (Cremlingen-Destedt: Rohdenburg) has its focus on soil science and related topics.

159

Resources:
Consumption
and
Conservation

lanet Earth may be nearly 5 billion years old—5000 million years that began in the gases and dust of the solar system and saw the formation of continents and oceans, atmosphere and biosphere. Early in its turbulent history, a large body of matter from space (one of many that bombarded the planet) struck the earth, and from that collision the moon was born. Much later the planet, now more stable and endowed with a core, mantle, and crust, became the stage for the evolution of life. The evidence of early beginnings is found in rocks 3 billion years old, but vertebrates (animals with backbones) did not invade the landmasses until about 500 million years ago. And the rise of humanity is the story of the past 5 million years or so . . . the last 1/10 of 1 percent of the planet's lifetime.

But this last 5 million years was no ordinary phase in the earth's existence. Since about halfway through its lifetime, or about 2.5 billion years ago, the planet has periodically been affected by great ice ages. There was such an ice age when Gondwana still was a unified supercontinent, the so-called Dwyka ice age of Permian times, more than 220 million years ago. And again during the past several million years, there has been the Pleistocene Ice Age. There is (as we have noted earlier) reason to believe that this latest ice age is not yet ended and that we are enjoying a respite between cold periods.

The Pleistocene Ice Age may have had a precursor, a forewarning. Between 5 and 6 million years ago, the evidence indicates, there was a time of catastrophic environmental change on the earth. A long period of warm climate came to an end with a sudden cooling. The ice sheets in polar areas grew larger and thicker. Cold air masses invaded normally warm latitudes. Moist regions became dry. Forests disappeared. Sea level dropped, and large parts of formerly submerged continental shelf became exposed. Animals became extinct, but other animals managed to adjust to the new conditions, and thrived.

Among animals forced to cope with the looming ice age were primates, and some did better than others. Habitats changed, and adaptation was necessary for survival. It may not be a coincidence that chimpanzees and hominids (ancestral humans) diverged from their earlier lineages at about that time. Again, the hominids were the more successful. *Homo habilis*, the survivor, gave way to aptly named *Homo erectus*, the walking, bipedal hominid. With a larger brain, arms free for purposes other than locomotion, and the ability to make needed tools from rock fragments, *Homo erectus* multiplied in numbers and spread from his original home—Africa—into southern and southeastern Asia, China, and probably also southern Europe.

Let us not lose sight of the environmental circumstances under which these momentous events took place. After those early portents of the Pleistocene glaciation's coming, the Ice Age arrived in full force. By the time *Homo erectus* was finding new habitats, about five hundred thousand years ago, the glacial age was in full swing. Colder periods alternated with warmer phases (see Chapter 2), and environments changed time and again. It was a challenge that could be met by migration, by moving as familiar environments changed locales, and . . . by making tools to enhance the chances of survival. But that very capacity to adapt and adjust may have created another problem: population growth. Scholars who study the diffusion of the early hominids suggest that population pressure may have contributed to early migrations as much as environmental changes did.

That means, of course, that the better toolmakers, our more inventive ancestors, had the best chance. They learned that stones could be chipped to make sharp edges that could serve the purposes of a knife, to cut skin and carve meat. *Neanderthal* people (*Homo sapiens Neanderthal*) lived through cold glacial times in southern Europe and in the Middle East. They had inherited the controlled use of fire and were good hunters. But still more capable humans, modern *Homo sapiens*, replaced them. By about thirty thousand years ago, modern humans held center stage.

All this—from *Homo habilis* to *Homo sapiens*—took place in just 5 million years, in the last 1/10 of 1 percent of the earth's existence! Still, the great changes that brought modern civilization had yet to begin. Just 30,000 years ago, after nearly 5 million years of development, the tools known to humans were made of stone and wood and bone. The raw

162

materials of survival were limited in scope and number. Shelter remained basic and minimal. The concept of an earth with bountiful resources had yet to emerge.

Momentous Warming

Time and again, Pleistocene cold gave way to warmer interglacials. It happened once more about fifteen thousand years ago; at the time, there was little to suggest that, on this occasion, it would be different. Again, there was melting of polar ice and rising of sea level. Gently sloping continental shelves flooded, driving back human communities perched near the water's edge. Areas long dry and cold became moist and warmer. And something changed in the way of life of Stone Age humanity. Our ancient ancestors had been predators and makers of simple stone tools, but now they began to gather shellfish in the shallow, warming waters offshore; they learned to fish; and they gathered berries, fruits, nuts, and other plant foods. Thus the natural vegetation became a major source of sustenance in this new (Mesolithic) era, with the important consequence that human communities became more sedentary and permanent. The idea that vegetation could be expected to provide food during every annual growing season and the knowledge that certain plants produce fruits earlier during the year than others, undoubtedly contributed to the concept of organized farming and the formation of settled farming communities for the first time in human history. This happened between ten and twelve thousand years ago, and the introduction of agriculture in its most rudimentary forms marks the beginning of the Neolithic era. For the first time, the soil became a resource for human survival, just as wild animals and natural vegetation were a means of survival. These developments did not, of course, take place simultaneously wherever human communities existed. Peoples in the warming Middle East and its river valleys took the lead.

Even as our earliest human ancestors led a hunting and gathering existence, they began to make use of the earth's rocks and minerals. From those beginnings, when spear- and arrowheads were chipped from fine-grained flint (a kind of quartz), the human search for this planet's mineral wealth has continuously intensified. Hundreds of thousands of years ago the glassy lava obsidian became a valued commodity, and communities that did not have access to it bartered eagerly for it. Obsidian could be fashioned into sharp-edged hand axes that could skin animals, carve meat, even cut human hair. In those ancient days it was a vital resource.

METALS

Eventually people learned to use metals. Among the first metallic minerals to be used was copper that occurs in the native state. In that condition, copper can be fashioned rather easily into any desired shape. Historical geographers report that copper and gold (another mineral that can be found in pure or nearly pure form) were prized commodities in the Middle East more than ten thousand years ago. But not until bronze was made by combining copper and tin was the real usefulness of metals for toolmaking understood. This occurred between six thousand and seven thousand years ago, and it heralded the beginning of the Bronze Age.

The peoples of the ancient civilizations in Southwest Asia and North Africa found uses for a growing number of such **nonferrous (noniron) metals** as copper, tin, zinc, and lead. They devised smelting methods to purify ores that needed it, and their search for additional deposits intensified. For thousands of years they had been taking gold from the stream waters in the region, and it is likely that the quest for new deposits of other minerals first followed the banks of these productive stream courses into the mountains surrounding the populated river basins. There mineral deposits were found, and systematic mining and distribution began. But the metal of the future, iron, escaped the ancient cultures for a long time. During the Bronze Age, what little iron was known to exist came, it appears, from scattered mete-

163

orites found during the exploration for other ores. Iron was so scarce that it was valued even more highly than gold in this period.

The spreading and improving knowledge of metallurgy had a major impact on the development of the civilizations of the ancient Middle East. By 5000 B.P. (before the present) the use of copper was spreading into still-Neolithic (Late Stone Age) Europe; between four thousand and three thousand years ago, the Bronze Age reached its strongest development, stimulating the growth of cities in several parts of the Southwest Asian–North African realm. At that time, the peoples of the Middle East viewed their environment in very different ways than their Stone Age predecessors had done. Not only had minerals become useful resources, but soils were being farmed and irrigated, and the new tools made possible the exploitation of the natural vegetation because wood could be fashioned for the construction of houses, vehicles, and boats.

From Bronze to Iron

By about 3200 B.P., the discovery of high-grade iron ores and successful experiments in the smelting and purification of iron initiated the Iron Age and brought the Bronze Age to an end. Now, tools and weapons that had been made of softer and relatively scarcer bronze were created from iron, their effectiveness increased. Initially the technology of iron metallurgy was carefully monopolized by those in power because it gave them an insurmountable advantage over adversaries. But inevitably this knowledge diffused to other peoples and other regions (Europe still remained well behind the Middle East), and the Iron Age brought the first mass production of weapons and, hence, the first of a series of power struggles in Asia, Africa, and Europe that were to last for many centuries.

Iron, the **ferrous metal**, eventually became fundamental to modern industrialization, and the Iron Age still continues, although other metals have also come into widespread use. In the form of steel, iron remains an industrial mainstay, but other metals with special qualities (e.g., light aluminum) have replaced iron in many functions. In any case, our age may better be called the Energy Age (the Hydrocarbon Age).

Nothing characterizes our use of the earth's resources so much as our voracious consumption of petroleum, natural gas, and coal. The ancient peoples of the Middle East lived right on top of the oilfields that have fueled the modern age, but even if they had found them, the oil deposits would not have constituted a resource. Power in the Bronze and Iron Ages was provided by people and animals, not by fuel-driven machines.

Nonmetallic Minerals

The ferrous and nonferrous metals that came into use during the Bronze and Iron Ages to supply the ingredients for the developing metallurgical industries were not the only earth materials to be perceived as resources. The ancient Middle Easterners also used a variety of nonmetallic minerals. Clay, for example, was used in buildings and artworks that have been dated as more than twenty thousand years old. Salt appears in the very oldest records of organized commerce, and undoubtedly it was in use long before. Stone was quarried for building foundations and fortifications. Gemstones have been prized as long as people have appreciated beauty. And like the ancient Greeks, the Babylonians and Egyptians of the Middle East were fond of bright colors and painted their buildings and statues with ocher and other pigments. As human societies became larger and more complex, their demands for commodities and services rose. This rising demand expanded the range of items that were viewed as resources and intensified their exploitation. From stone to tin and copper to iron to oil, the earth has proved to contain an enormous variety and huge volume of resources. But our planet is not inexhaustible, as we are about to discover.

Natural Resources

The ancient Babylonians, Mesopotamians, and Egyptians undoubtedly had a notion of just what a resource is and means. The raw materials needed for building, boatmaking, metal smelting, and other industries were not always in easy reach. Some had to be transported from afar, others were scarce—scarce enough to be fought over. There were advantages to possession of a source of iron ore or gold. Some of the ancient scholars and philosophers certainly viewed the soil that sustained annual crops and the life-giving waters that irrigated them as resources too. To this day, soil and water are critical resources of our planet.

Defining the term *resource* is not, however, as easy as it might seem. Economic geographers often classify the resources of a national or regional economy into three groups: land, labor, and capital. Our concern in this chapter is with the first of these categories—the land, under which is subsumed the soil and vegetation, metallic and nonmetallic minerals, fuels, and all other resources that are part of the natural composition of the earth. Thus we concentrate on the **natural resources**, those that occur in nature.

This still does not answer the question of what is a natural resource and what is not. To our hunting and gathering ancestors, copper and tin were not resources. Copper (as noted) did not become a resource until it was deemed to have value as a malleable substance for the creation of various implements; its worth was heightened when, in combination with tin, it produced bronze. To the ancient Egyptians petroleum was not a resource; our incredibly fast consumption of mineral fuels attended the emergence of the Industrial Age. Uranium did not become a resource until the twentieth century. There may be substances in the earth that will become resources in the future but that are not resources today. In other words, human development determines and defines the nature of resources; what is and what is not a resource is a cultural matter. Each civilization in the history of humanity has depended on its own set, or complex, of resources. In ours, at present, the power resources dominate, and the earth's mineral fuels (coal, petroleum, natural gas) are the critical elements of the complex. Although it is true that iron is also fundamental to our Industrial Age, numerous substitutes for iron have already been developed. Substituting for our current power supply is quite another matter, and although scientists are searching for alternatives (nuclear fusion is one such possibility), no one is sure where the energy the world will require, say, a hundred years from now will be coming from.

RESOURCES AND TECHNOLOGY

This brings us to another aspect of resource use: changes brought about by necessity, technological progress, or both. A prominent example is that of iron ore. One of the world's largest sources of iron for a long time was the Mesabi Range, not far from the western end of Lake Superior. The Mesabi deposit is the largest of six such deposits that extend from Minnesota and Wisconsin into the Upper Peninsula of Michigan. These deposits are of sedimentary origin, and circulating waters for hundreds of millions of years have been removing silica and other impurities from them. The iron content, when mining of these ores began, was as high as 56 percent on the average and even went to 62 percent in places. The rock, called **hematite**, became the leading source for iron ores used in the steel mills on the shores of the Great Lakes and beyond. But it could not last long in the face of massive mining operations, and as it became exhausted, steelmakers had to look for other iron ores. Since major ores were located in West Africa and South America, it was sensible to build new plants on the East Coast so that the cost of importing those overseas ores could be minimized. Certainly there were other ores around Mesabi, but they were so low grade that it was not practical to exploit them. One of these was a rock called **taconite**, just a rock with 30 percent or so iron. Taconite was no more a resource than sandstone.

But then technology changed things. It was no accident, and indeed a matter of necessity, that scientists had been working on methods to extract iron from low-grade ores. In the 1950s, they succeeded, and suddenly it became feasible to refine taconite's iron at costs that were competitive. Just as suddenly, taconite, which had been merely another rock, became a valued resource, and the iron workings at Mesabi and its surroundings had a new lease on life.

The taconite story is not unique. In South Africa, where gold has been mined on a large scale for nearly a century, the waste from the mining operations was piled in big heaps, the so-called mine dumps that dominate the urban scene of parts of Johannesburg and nearby towns. Gold-extracting techniques have improved to the point that the gold left behind in the waste material of the old days can be profitably extracted. The mine dumps themselves have now be-

166

The destruction of the natural landscape is well illustrated by this photograph of an open-pit iron mine in the Mesabi area, Minnesota.

If a resource is defined as a means, a source of supply or support, then *relative location* can be a resource. A place may be almost totally devoid of natural resources—the raw materials of industry, good soils, rich fishing grounds—and yet have a great advantage over other places. This advantage arises from its situation, perhaps on a natural waterway, on the shore of a busy lake, or at the head of a bay. In each case, its relative location puts it at a focus of activity, a hub of interaction. The city-state of Singapore is an example. At the tip of the Malaya Peninsula, at the head of the Strait of Malacca, opposite Indonesia, on the route between Japan and the Middle East, Singapore has translated its relative location into economic success. Once a place where transshipment was a major industry, Singapore now is an industrial center itself, and the wealthiest country in Southeast Asia.

Circumstance also can constitute a resource. At the surface of the earth, gravitational attraction is high. But in recent years, technical experiments under conditions of weightlessness have become common. Such experiments are best carried out where weightlessness prevails, for example, on a spacecraft; and research organizations have spent millions of dollars to secure the opportunity to perform them aboard the space shuttle. The moon proved to have no extraordinary mineral raw materials at or near its surface when the first exploration took place. But it does have a low-gravity environment that might constitute one of its potential resources.

come a resource. The same thing has happened with copper. Not only can ores long considered too low in copper content be mined, but also the mining wastes of decades ago contain extractable amounts of copper.

Thus it is reasonable to identify **potential** resources, deposits that cannot under present conditions of technology and economics be exploited but that await the necessary developments. In this respect, our modern age of science differs from preceding periods of human history. We *seek* ways to make resources out of presently unproductive sources, whereas for centuries our predecessors relied chiefly on accident or trial and error in this context. People were smelting iron in pre-Christian times; they were doing essentially the same thing as late as the nineteenth century. The Industrial Revolution brought with it changed (and still changing) perceptions of the earth's resource base.

RENEWABLE AND NONRENEWABLE RESOURCES

A very important distinction must be made between resources that can be exhausted, leaving none available, and those that are replenished by nature, or renewed. These are called, respectively, the **nonrenewable** and **renewable resources**.

The distinction is easily understood. There is, on this planet, a finite, limited amount of, say, platinum, copper, or uranium. Not all of the deposits may have been discovered, but the quantity is a constant. Use up all of a nonrenewable resource, and there will be no more. Renewable resources, on the other hand, can be used even as they are replenished. The hydrologic cycle ensures that rivers will flow indefinitely (barring climatic change, in which case a dying river in one area will be replaced by a new stream in another). Soil can be used year after year; nature ensures that it is renewed. Trees in a forest can be felled, and the forest will regenerate.

From this it would seem that the only concern lies with the nonrenewable resources: that the earth will fail to provide enough (or any) petroleum, natural gas, or some other vital resource at some point in the future. But there is more to the issue. Renewable resources can also be overused to such an extent that the processes of regeneration cannot match the rate of exploitation. When too much water is taken from a river for irrigation purposes, the river may respond by

167

failing to clear its lower course and its mouth; sand and silt may close its mouth off altogether, and permanent damage may be done. If forest exploitation does not proceed in a planned manner, regeneration may never occur, and the forest may disappear altogether, as has happened in much of Europe and in large areas of the Americas already. When fishing fleets harvest the oceans' bounty, they must adhere to catch limitations or the whole resource (in this case, the world's continuously renewed supply of fish) is endangered. Renewable resources are *not* inexhaustible.

• The Soil as Resource •

In Chapter 4, we studied the earth's soils as part of the biosphere, and we viewed their structure, texture, and distribution. But the mantle of soil is one of the world's most important resources—perhaps, with water, the most important of all. Again the ancient civilizations in the Middle East, where organized farming probably evolved earliest (it may have happened in China very shortly thereafter), were the first to realize that certain soils were more productive than others. The irrigable alluvial soils could be depended on to produce, undoubtedly more so than some other soils in what we have come to call the Fertile Crescent. The better soils constituted a greater resource. But the notion that soils are subject to destruction, can be lost forever, and need conservation developed much later.

The burgeoning, mushrooming of the planet's human population over the past century has put unprecedented demand on the earth's soils to produce food. River basins with their fertile soils have filled to—and beyond—capacity with human numbers. Farmlands have crept up steep mountain slopes, into desert margins, onto flood-prone coastlands. Soils once anchored by protective natural vegetation have been exposed and put under hoe and plow. As a result, soils in many areas of the world are being destroyed, exploited for the present without regard to conservation and the future. They have become, in those areas, nonrenewable resources.

In recent years, geographers and other researchers have alerted governments and agencies that the rate of soil loss has reached crisis proportions and that this resource is under severe stress. The problem is that a growing soil shortage does not have the same immediate impact as, say, an energy shortfall. When, in the 1970s, there was an oil crisis, governments set to work solving the problem, and (temporarily at least) the shortage disappeared. A destructive earthquake, volcanic eruption, or flood will attract immediate remedial attention. But the loss of soil is much less dramatic. It is much more difficult to mobilize a response.

We should remember that it is the valuable A horizon, the topsoil in ordinary language, that is most vulnerable to overuse and eventual loss. When erosion by natural forces takes place, it is the topsoil that is most directly exposed. But nature also rebuilds. The problem is that human intervention is exceeding what we may call the soil's tolerance level. A study done in the early 1980s suggested that about 44 percent of the total soil area of the United States is beyond this tolerance level, and soils are—every year—degenerating.

When scientists calculate the loss of soil resources, therefore, they measure the *excess* loss, that is, the loss above and beyond what nature would be removing. After all, a great river such as the Mississippi or the Ganges will carry millions of tons of sediment to the ocean, and would do so even if no farming took place in their catchment areas. It is this excess loss that has scientists worried.

To ask the geographic question: Where is the problem most serious? Many observers believe that the losses are greatest in countries where data are difficult to obtain. Lester C. Brown and Edward C. Wolf, in their work, *Soil Erosion: Quiet Crisis in the World Economy*, say that the Soviet Union, which has the world's largest cropland area, may be losing more topsoil than any other country. Soviet scientists have sounded many warnings, but the government faces pressures to increase farm production. So the scientists are all too often ignored. As a result, an estimated 1.25 million acres (500,000 ha) of cropland are abandoned every year because they have been overused and are severely eroded. The winds now blow away what remains of a topsoil that once yielded an annual harvest.

Another country with severe soil erosion problems is China. The volume of silt that enters China's major rivers and is carried out to sea represents a huge excess loss of topsoil. In addition, large areas of farmland have been opened up in China's marginal northern areas where wind erosion has begun to take an enormous annual toll. It is believed that China's soil loss exceeds that of India by as much as 30 percent.

And India, too, suffers severely. Unlike the Soviet Union and China, India has compiled an official national estimate of soil loss from all its states. Using the same methods as in the United States (i.e., calculating the excess loss above the tolerance level), India is measured to have more than twice the rate of loss prevailing in the United States.

So the four major food producers (the United States, the Soviet Union, China, and India) all experience a net loss of topsoil. Other countries probably fall somewhere in the range exhibited by these countries. Estimating the global situation on this basis is not encouraging. "At the current rate of erosion," state Brown and Wolf, "this [soil] resource is being depleted at 0.7 per cent per year—7 per cent each decade." It is no overstatement to describe such a situation as a crisis.

• Natural Vegetation as Resource •

Forests and other forms of natural vegetation are as vulnerable to overexploitation as are soils. For centuries, wood was the chief fuel for domestic and industrial purposes, and forests diminished—first in Europe, later when the Europeans reached these shores, in the Americas. Wood was used for the building of houses and the construction of boats. When the technique of making charcoal was developed, the raw material, again, was wood.

And so, long before the present cycle of exploitation of forests, large stands of trees disappeared. Spain's plateau, the *meseta*, once was covered with forests, but these disappeared during the Spanish rise to world power and colonial expansion. Then, when Spain colonized Mexico, huge stands of forests there fell to the invaders' ax.

Today much more is known about forests, their role in the environment, their requirements, and their commercial value than in days past. Forests are crucial in the maintenance of river catchments and soils, they provide a refuge for wildlife, they provide timber, and they serve recreational and tourist purposes. Especially in equatorial and tropical latitudes, forests, according to many scientists, contribute to the maintenance of oxygen levels by releasing oxygen during photosynthesis. In short, forests are vital elements of the global environment.

This is not to suggest that forests cannot be used and managed. Specific objectives can be met by careful planning: logging, for example, requires the planting of replacement seedlings at a certain rate. Forests in recreational areas may have their undergrowth cleared. Old trees, which might be removed under other circumstances, are left to decay in forests serving as wildlife (including bird) reserves.

All this knowledge and the fact that wood no longer serves as the world's main source of fuel would suggest that forest resources are secure for the future. Unfortunately this is not the case. The map of biomes (Plate 4) gives a rather favorable impression of the real situation: many areas that *should* carry a certain forest association as their climax vegetation have been damaged or destroyed. Phytogeographers describe forest associations that have been damaged as disturbed and those completely felled as destroyed. The U.N.'s Food and Agriculture Organization (FAO), in the early 1980s, estimated that as much as 44 percent of the earth's rain forests already were disturbed and that more than 1.1 percent of the remainder was being logged each year, with no concern for conservation or regeneration. At this rate, all the rain forests would be gone within a century from now. Geographically, only three areas seem temporarily secure from assault for about 40 years or so: western Brazil, the interior Guyanas, and the Congo (Zaïre) Basin of Africa. In these areas, the comparatively small human population (at least, up to the present) will not have an impact as severe as elsewhere. The overall situation can be looked at another way: about 35 acres of tropical rain forest (14 ha) are being logged every minute of every day and another 35 acres are being disturbed. An area about half the size of the state of Iowa is being lost every year.

The Developed Countries of the world (DCs) often chide their neighbors for poor conservation

169

efforts; in recent years Brazil, for example, has been denied funds from international agencies to assist in projects that will further damage its rain forests. But what about one prominent DC, the United States, and its own rain forests? The United States has rain forests in Hawaii, Puerto Rico, and controls several rain forested islands in the Pacific Ocean. In Hawaii, two thirds of the area of native rain forest has been cut away to make room for plantations, ranches, and settlements. These forests once supported 57 bird species and subspecies. Today only 9 species remain unthreatened; 23 species are extinct; and another 25 species are threatened. As far as plants are concerned, there were about 1250 native flowering species in Hawaii two centuries ago. About 10 percent of these are already extinct; another 40 percent are threatened or endangered, with many on the brink of extinction. The less prosperous countries of the world have no monopoly on poor conservation practices.

Of course, loss of vegetation and loss of soil go hand in hand. China's soil degeneration was coupled with such a loss of forest that Mao Zedong, the former ruler, ordered every citizen to plant at least one tree, thus achieving the addition of more than 800 million new trees on the denuded Chinese landscape. The effort did not meet its goals, but the order was evidence on the severity of China's deforestation.

Reforestation, as conservationists have learned, is much more difficult than forest maintenance. Once a forest is lost (or any natural vegetation is destroyed), replanting and sustaining a substitute is very expensive and often a failure. The destruction of forest is not the only case in point. Environmentalists are warning that a combination of human and natural forces is laying waste much land that formerly carried grasses, shrubs, and other vegetation that bound the soil and retarded erosion. Severe **desertification** of this kind has hit Africa hardest of all the continents, but none has escaped it. In Africa, the margins of the Sahara—both along the Sahel in the south and along the Mediterranean north—are being denuded by the goats and cattle of human communities in search of survival. In Ethiopia, drought has contributed to the denudation of millions of formerly grassed-over areas. In northeastern Brazil, overuse and variable rainfall have combined to lay waste a large corner of the country. In

northern Mexico, southeastern Spain, central Turkey, coastal Pakistan, southern Madagascar, western Australia, interior China, and in many other areas of the world, the cycle of destruction goes on. It has been estimated by scientists of the U.N. Environment Program that about 67 million acres (27 million ha) each year fall from productive (even marginally so) to unproductive land. As a result, it is calculated that arid lands now must sustain 850 million people or about 17 percent of the world's human population, and this number is growing continuously.

Whether as forest, grassland, or coastal wetland, the natural vegetation is under stress. As a resource, the natural vegetation may be designated as renewable, but only when properly managed to ensure its maintenance. What is happening in many areas of the world reminds us that renewable resources, when overused and poorly managed, are irreversibly exhaustible.

• Wildlife as Resource •

When the natural vegetation is threatened, so is the wildlife for which it forms a habitat. Wildlife, from the smallest organisms to large mammals, forms part of the total ecosystem. Wildlife also is one of the earth's resources. When our ancestors still were hunter-gatherers, they followed migrating animal herds, a process that undoubtedly contributed to early human dispersal.

The domestication of animals (see Chapter 8) created a new relationship between human communities and livestock; as this innovation spread, dependence on hunting declined. But wildlife continued to be a resource. Animals were hunted for their skins, for decorative feathers, for antlers and horns, for tusks and, simply, for trophies. All this still goes on, and the refuges of wildlife dwindle.

In what way is wildlife a resource? In the first place, wildlife can serve the same purpose as forests and other kinds of wilderness: as a source of recreation and inspiration. Second and more important, wild animals have played a crucial role in medical and biological research, and the loss of any species is a serious matter. Yet literally countless species of animals are being lost every year, some tiny and obscure, others large and well known. The Tasmanian tiger, for example, a large mammal of unique characteristics, became extinct in the 1930s. The list of endangered species includes part of the earth's great faunal heritage and includes the Florida panther, the African cheetah, the Asian snow leopard, and many other magnificent animals.

Today Africa south of the Sahara contains the earth's last great wildlife populations. The idea that African wildlife can be domesticated and ranched has led to numerous experiments to exploit, control, and maintain this potential resource. Although some antelopes have proved unsuitable for controlled ranching, others, notably the eland, have been herded and penned with comparative ease. The eland is a large, heavy animal whose meat is quite acceptable for human consumption; in protein-poor areas, this alternative may help improve dietary balances while the breakthrough against the tsetse fly is awaited. But the creation of large domestic eland herds is not a simple matter. As with the tree species in the tropical rain forest, the eland is only one of numerous species of antelope in the African savannas; the catching and securing of each animal is a laborious and difficult process. The pasturelands must be fenced (but still animals are lost as fences are damaged) and systems for feeding and watering laid out. The wait for natural reproduction is a long one. In this sense, Africa's wildlife is a potential rather than an actual living resource.

Marine Resources

So vast are the world's oceans and so large the hauls of fish that have been made year after year that the oceans began to be viewed as the hope for the future to supplement food supplies and to balance diets that could not be adequately provided by the world's farm-lands. All that seemed necessary was more efficient fishing equipment, better tracking methods, and improved distribution systems. During the 1950s and 1960s, after all, the oceans yielded a harvest that had more than doubled to some 50 million metric tons; by 1970 it was more than 65 million metric tons. The limit seemed far higher, and the developed countries built more trawlers and canning ships.

But the oceans are not inexhaustible, and the fish fauna can be damaged, set back, and even destroyed by overexploitation. The world map of huge oceans and extensive fishing grounds induces a false optimism, and the warnings of scientists have gone unheeded. In the first place, those vast oceans include areas almost as devoid of fish as there are deserts on land nearly devoid of people. Second, fish are part of a **food web**. Small fish live on plankton, drifting plant and animal matter composed of minute organisms; somewhat larger fish live on smaller fish; still larger fish depend on the medium-sized and smaller varieties. Break this hierarchy somewhere, by fishing out the smaller- or medium-sized fish, and the whole balance is destroyed. And third, plankton happens to develop best in shallower parts of the sea, principally the continental shelves adjacent to the continents. Fish, therefore, tend to concentrate on and near these shelf areas. Here they are susceptible to harvesting in great numbers—and vulnerable to overexploitation.

The oceans themselves can also turn against the schools of fish they have nurtured so long. A shift in the path of an ocean current, a change in temperature or salinity can make a once-hospitable zone that teemed with fish a barren stretch of empty water. It happens quite suddenly, and scientists are not always sure just how such a cycle of change begins.

Certainly the oceans always gave warning signals. During the 1920s and 1930s, the annual harvests of sardines off the west coast of North America rose spectacularly, from under fifty thousand to nearly eight hundred thousand metric tons. Until the mid-1940s, the harvests remained above half a million tons, but danger signs abounded: it took ever-more boats and longer hours of fishing to keep the harvest high. The fishers could not be persuaded to limit their catches to save the small, younger fish, and the inevitable happened in a few short years. The

171

catch fell to a few dozen tons in the early 1950s and the industry collapsed. There is evidence that changes in the ocean water itself contributed to the disaster, and it is possible that the sardines would have been depleted even without commercial overfishing. What is clear is that under the circumstances of unlimited exploitation, they had no chance.

In Chapter 2, we studied the impact of the *El Niño* events on the Peruvian anchovy fishery. True, the de-

Anchovy fishing off the Peruvian coast has long constituted one of Peru's leading industries, but warning signals appeared during the 1970s here as elsewhere: the oceans' resources were being strained by overexploitation. Peru's anchovies do not contribute to improved diets in that country. They are mostly ground up into fish meal and sold to the developed, wealthy countries for consumption by domestic animals.

cline in the catch was related to this atmospheric and oceanic aberration, but there had been severe overfishing earlier. That overfishing probably led to the failure of the anchovy population to reestablish itself when normal conditions returned.

How important is fishing in the economic geography of the world? The answer lies in employment figures as well as consumption totals. In the late 1980s, more than 12 million people supported themselves by fishing or *aquaculture* (the farming of fish in artificial ponds). Millions more work in related industries: the preparation, handling, transportation, and marketing of fish. But even more significant is the contribution made by marine resources to the diets of people nearly everywhere. In the late 1980s, the annual fish harvest averaged 75 million tons, slightly more than in the early 1970s, when world population was more than 1 billion less! Still, the 1987 harvest, by weight, was larger than world beef production. Even more important, fish reach the dinner plates of many millions in the less developed world; beef rarely does.

What is evident from the statistics, however, is that world fish production *per person* has been falling since 1970 when it was nearly 40 lb (17.9 kg); today it is 33 lb (15 kg). That loss reflects the rapid growth of the world's human population, but it also results from a slight decline, during the 1980s, of actual harvests. This decline is evidence of the troubled state of several fisheries, a consequence of overfishing and a lack of restraint. Geographers who study the world's fishing industries do so from two principal viewpoints: the regional and the specific. Regionally, the fishing industries of the world concentrate in several marine areas where a favorable combination of circumstances (nutrient-rich waters, shallow depths) produces a clustering of fish. Specifically, there are several dozen world fisheries of individual species: herrings, sardines, and anchovies (the largest annual catch); cods, hakes, and haddocks (second ranking); jacks and mullets (third ranking); and so on through the tunas, shrimps, lobsters, and krill. Many of these species are caught worldwide, for example, the tuna; others such as the krill (a small, shrimplike shellfish) is found mainly in one marine region.

Both regionally and specifically, the world's marine resources are stressed. We have already noted the collapse of South America's anchovy fisheries, a regional example. In the North Atlantic Ocean, one of the world's most productive fishing grounds, some species are holding their own, but others such as the hake are not. The Food and Agriculture Organization of the United Nations (FAO) in the mid-1980s listed six species in the Atlantic and five in the Pacific as depleted to the point of biological collapse. Some, for example the Atlantic cod fishery, has been recovering as a result of international agreement limiting catches. But others were not. And in 1986 marine biologists announced that the seemingly unlimited krill grounds near the Antarctic were not so bountiful after all. The marine resources of the planet can only be preserved (and thus remain renewable) through an international conservation effort.

Energy Resources

Our image of what resources are tends to focus on the raw materials of heavy industry—on iron and aluminum ores, on alloys, and on fuels. And, indeed, those have been the mainstays of the Industrial Revolution, the age of machinery and technology. In Chapter 10, we will have an opportunity to view the distribution of these resources. Among them, however, the energy resources have a special role. The issues of supply and its cyclic character, politics and its role, and pollution and its hazards as well as the need for management and conservation are matters of constant concern where energy resources are concerned.

COAL AND INDUSTRY

Coal was used as a fuel in Europe in Roman times, and perhaps even earlier in China. Small quantities of it were traded in Europe after about the thirteenth century, but for thousands of years it was wood, not coal, that served as fuel for heating, cooking, and the making of charcoal.

In the eighteenth century, the potentials of coal were discovered, and coal lit the fires of the Industrial Revolution long before petroleum and natural gas began to fill the ever-rising need for energy in the industrializing world. And coal as an energy source may well outlast petroleum. The earth's crust carries enormous quantities of coal, and coal is a versatile source of energy. Progress has even been made in the conversion of high-grade coal into gasoline.

The usefulness of coal for burning (and therefore as a resource) was recognized thousands of years ago, perhaps as early as the Bronze Age in the Middle East. Coal was used as a heating fuel by the ancient Romans; there is evidence that coal was systematically extracted and used for heating and industrial purposes in China a thousand years ago. In North America, Indian peoples as early as the twelfth century were using coal on their cooking fires, in pottery making, and for burning during ceremonial events. In England, the use of coal increased after the thirteenth century since coal could be mined at the surface from exposed seams (it could even be gathered on the beach, washed ashore from seams exposed and eroded on cliffs). But the real stimulus for increased use of coal came during the Middle Ages when the excessive exploitation of Europe's forests for building material and firewood led to their exhaustion, and alternate sources for heating fuel were needed. This led to the first organized, commercial mining and distribution of coal, at first from the most accessible deposits nearest to major population centers but later from underground sources farther away. A network of coal transportation existed in Europe long before the Industrial Revolution commenced; in England, coal was being mined from shafts as deep as 330 ft (100 m) during the seventeenth century, at which time there were already hundreds of mines. But coal extraction was obstructed in many places by the groundwater that would fill the shafts; at first it was lifted out of the mine shafts with buckets hauled by hand or by teams of horses. Not until reliable pumps were invented (during the eighteenth century) could deeper mining proceed. Even then, the coal was raised in baskets carried to the surface on human backs or pulled upward by horses. The Industrial Revolution created an unprecedented demand for coal and produced the machinery that could extract it in far greater volume from deeper, less accessible sources. Coal became the essence of the industrial transformation.

173

Grades and Sources

Coal has its origins in swamp or bog environments, usually associated with forests. Under stagnant water conditions, the bacterial activity that destroys plant matter is slowed down and the falling vegetation accumulates faster than it decays. The initial product of this process is **peat**, a soft mass of vegetative tissue used in some parts of the world as a household fuel. During the Mississippian and Pennsylvanian periods (see the geologic time scale, Chapter 2) the continents were eroded down to very low relief, and vast expanses of plains with enormous swamps existed. Great accumulations of peat developed, hundreds of feet thick in many places. Then these peat layers were buried when new sediments were laid down on top, and the sediments began to weigh down on the fibrous mass below. This caused the peat to change into a somewhat higher grade of fuel known as **lignite**.

Although there are large lignite deposits in many parts of the world, this fuel is not very useful because there is too much water in it (30 to as much as 70 percent) and too great a quantity of volatiles (e.g., hydrogen and oxygen). But compaction continued as more sediments were laid down over the original peat areas and the weight of the sediments started to drive off almost all the water. The coal became harder and darker in color, a much better-quality fuel. This, **bituminous coal**, is the variety of coal mined in most of the areas shown on Figure 5.1. It is the coal that proved usable in the smelting of iron and the production of steel. There is a still higher grade of coal. When layers of coal between sedimentary rocks became caught up in orogenic movements, complete with intense folding, enormous pressures, and rises in temperature, practically all the water and all the volatiles were driven out and what remained was the purest coal of all,

FIGURE 5.1

World distribution of known coal reserves.

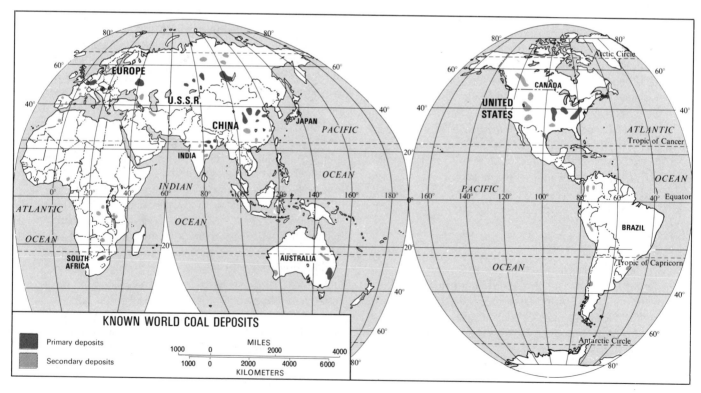

KNOWN WORLD COAL DEPOSITS

Primary deposits

Secondary deposits

MILES

KILOMETERS

consisting of over 90 percent fixed carbon, **anthracite**. This, the cleanest of the coals, long was a fuel in great demand for heating homes.

The world map of the distribution of known major coal deposits (Fig. 5.1) indicates that coal-bearing sedimentary strata are widely distributed across the earth. Comparing this map to the schematic representation of world fuel reserves (Fig. 5.2), we note that the United States is believed to possess more than one third of all the coal reserves in the world, with the U.S.S.R. and China about one fifth each. These figures should be seen as estimates, however, because new reserves of coal are still being discovered and the balance may shift somewhat. But the overall pattern is well established, and it is reflected by the production figures. The United States usually leads the world in the annual production of coal, although China's output is slightly larger in some years; the U.S.S.R. produces nearly as much, and East and West Germany and Poland also rank among the world's top 10 coal producers.

The world distribution of coal reserves emphasizes their concentration in sedimentary rock and their paucity in shield areas. In the Northern Hemisphere, a zone of coal deposits lies across the Western and Eastern United States, Western and Central Europe, Eastern Europe, the U.S.S.R., and China. The position of this belt coincides approximately with the contact zone between the Laurentian and Eurasian shields and the adjacent noncrystalline strata. In South America, the known coal reserves lie in the Andean zone of the west and on the margin of the Brazilian shield in the east. Africa, the shield continent, is not well endowed with coal, although substantial reserves do exist in South Africa, where they are part of the great sequence of deposition known as the Karroo System. And Australia's coalfields lie mainly on the eastern margin of that continent's shield along the flanks of the Great Dividing Range.

Reserves and Output

When coal reserves and coal production are discussed, it is important to take account of the fact that coal comes in several grades and that countries do not necessarily report their production in the same equivalents. Total world reserves of all grades of coal in 1986 were estimated to be about 2.5 trillion metric tons, but in reality they probably are much greater. Production of all countries combined in the mid-1960s averaged 3 billion tons, certainly a comfortable ratio. At current rates of consumption, there would be enough coal for about eight hundred years—compared to the decades in which imminent oil exhaustion is measured.

In the past, there were two undisputed leaders in the world of coal production: the United States and the U.S.S.R. Today, however, there is a Big Three: China's output now rivals that of the United States and the U.S.S.R.; in 1984 China actually outproduced its competitors for the first time. In that year, to provide an idea of the huge quantities mined, China produced 715 million tons, the United States 712, and the U.S.S.R. 675. No other country produced even half this quantity; East Germany ranked next with 278 million tons. Another interesting first in the mid-1980s involved Australia, which became the world's leading exporter of coal, much of it to energy-poor Japan.

Production and consumption patterns for coal are quite different from those of petroleum. Petroleum flows in large volume from countries that are major producers but only minor consumers. In the case of coal, the major producers possess the largest reserves *and* are the largest consumers. The United States exports nearly 10 percent of its coal production and has the security of large coal stockpiles; the U.S.S.R. is the world's next largest consumer as well as producer. In Africa, industrializing South Africa produces and consumes more than 90 percent of the entire continent's output. Both Australia and India are able to export coal in excess of their own consumption.

One significant change since the beginning of the Industrial Revolution involves the availability and production of coal in the United Kindgom, the early hearth of industrialization, and in western Europe. In the United Kingdom, nearly two centuries of intensive exploitation of the most accessible coal reserves took its toll. The peak year of coal production in the United Kingdom was 1913, when more than 290 million metric tons were mined, which was in excess of 20 percent of world production at that time. About one third of this volume was exported; the United Kingdom could afford to do so because production was at

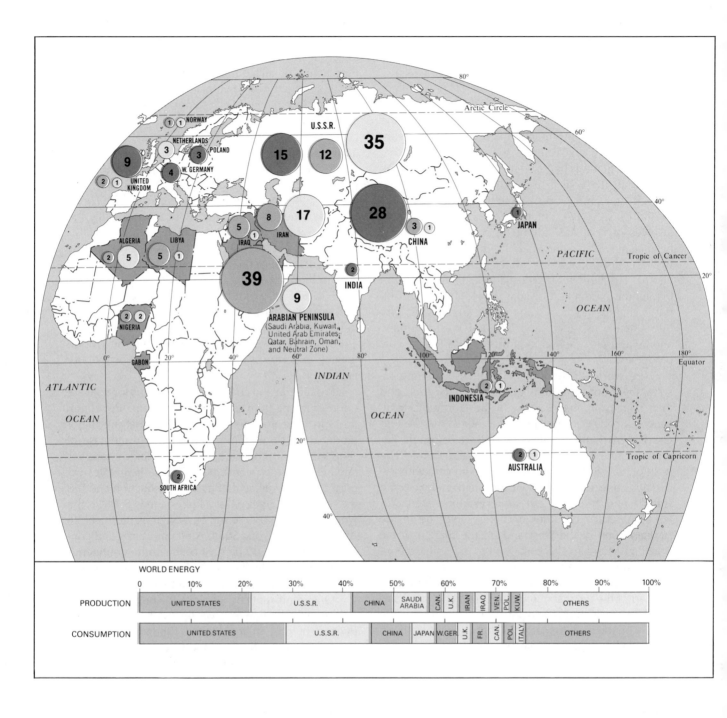

NORWAY ① ①

U.S.S.R.

NETHERLANDS ③
POLAND ③

W. GERMANY ④

UNITED KINGDOM ② ①

9

15   12   35

17   28

8   CHINA ③ ①

IRAQ ⑤ ①   IRAN

JAPAN ①

ALGERIA ② 5   LIBYA 5 ①

39

INDIA ②

9

ARABIAN PENINSULA
(Saudi Arabia, Kuwait,
United Arab Emirates,
Qatar, Bahrain, Oman,
and Neutral Zone)

NIGERIA ② ②

GABON

ATLANTIC
OCEAN

INDIAN
OCEAN

PACIFIC   Tropic of Cancer

OCEAN

INDONESIA ② ①

Equator

SOUTH AFRICA ②

AUSTRALIA ② ①   Tropic of Capricorn

WORLD ENERGY

	0	10%	20%	30%	40%	50%	60%	70%	80%	90%	100%
PRODUCTION		UNITED STATES		U.S.S.R.		CHINA	SAUDI ARABIA	CAN. U.K. IRAN IRAQ VEN. POL. KUW.		OTHERS	
CONSUMPTION		UNITED STATES		U.S.S.R.		CHINA	JAPAN W.GER. U.K. FR. CAN. POL. ITALY		OTHERS		

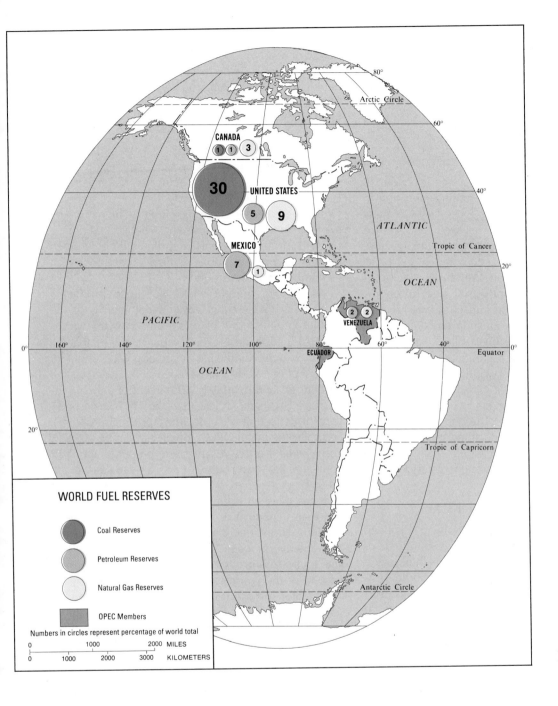

WORLD FUEL RESERVES

Coal Reserves

Petroleum Reserves

Natural Gas Reserves

OPEC Members

Numbers in circles represent percentage of world total

| 0 | 1000 | 2000 MILES |
| 0 | 1000 | 2000 | 3000 | KILOMETERS |

FIGURE 5.2

a maximum and reserves were strong. But as time went on, coal became more difficult and expensive to mine, and alternative sources of energy became available. In the 1950s, the availability of cheap, plentiful petroleum gave the coal industry severe competition; the gas industry reduced its use of coal as North Sea natural gas flowed to the British market. And so coal, the essence of the industrial surge, declined in importance—and in output. In 1960, the United Kingdom produced somewhat over 180 million tons of coal; in 1970 the output was about 150 million tons, and in 1985 it was down to 120 million tons. West Germany registered production declines as well, and France embarked on a program of investment in the development of coal reserves overseas to compensate for its shrinking domestic production. Western Europe's coal mines, mainstays of the Industrial Revolution for nearly two centuries, felt the impact of extended reserve exploitation, rising production costs, and competition from other energy sources.

The British and Western European experience is not, however, typical of the world situation. In most areas where coal is present, it remains accessible and production is rising, as it is in countries as far apart as Poland, China, and Australia. The position of coal as an energy source may have declined in Western Europe, but for the world as a whole the giant coal reserves are a guarantee against an energyless future.

PETROLEUM: PROMISE AND PROBLEM

Although there is evidence that petroleum has been used for some specialized purposes for hundreds of years, the systematic exploitation of oil is little over a century old. In some localities in the Middle East, North America, and elsewhere, oil seeps to the surface from underground sources: ancient Mediterranean seafarers used it for stopping leaks in their boats. An associated product, natural asphalt, was used by the builders of cities to cover streets and walls. American Indians sometimes used oil for fuel. But petroleum became a full-scale resource only in the second half of the nineteenth century when the first wells were drilled. From those beginnings, a single well drilled in 1859 to a depth of just over 69.5 ft (21 m) that produced a few barrels a day in Pennsylvania, world production has risen to over 60 *million* barrels a

day in the mid-1980s. This amounts to a staggering output of nearly 22 billion barrels a year. In 1985, the latest year for which statistics are available, proven petroleum reserves from around the world totaled 707 billion barrels. This would mean that, even at present rates of consumption, all petroleum would be used up in less than 35 years. But new reserves will probably be found in the future in areas yet to be explored, so that the situation may not be quite so bleak. On the other hand, consumption could also rise again and cancel out the gains of newly discovered reserves. One thing is certain: there is a limit to the petroleum available in the earth's crust, and whether that limit is reached in 30 or 50 years, the day will come when the mainstay of our mobile civilization no longer can be pumped from the ground.

A map of present world petroleum-producing areas (Fig. 5.2) emphasizes the dominance of the Middle East in terms of proven reserves as well as production. The most recent estimates suggest that the countries of the Middle East together possess about 56 percent of the world's proven petroleum reserves and account for approximately 20 percent of annual production. This is much less than the 1977 share of 37 percent, reflecting price declines, political problems, and a weakening of OPEC, the oil cartel. The U.S.S.R. has 9 percent of the reserves and produces about 22 percent of the yearly output. Compared to these quantities, other reserves are much less promising: even with the Alaskan oilfields, U.S. reserves amount to less than 7 percent of the world total, and those of Western Europe (recently discovered under the North Sea) about 4 percent. The reserve picture in Middle and South America (formerly estimated to comprise 4 percent of the world total) has been changed by recent, major discoveries that may place Mexico's reserves among the world's largest; the reserve share of Middle and South America is 12 percent. African reserves, including significant fields in Nigeria and Libya, constitute just under 9 percent of the world total, and China may have between 3 and 4 percent.

A comparison between Figure 5.2 and Figure 2.6 (the map of world landscapes and shield areas) indicates that petroleum occurs in nonshield zones. Petroleum is found in sedimentary rocks that have been folded or faulted into reservoir-forming structures (Fig. 5.3). The formation of petroleum involved

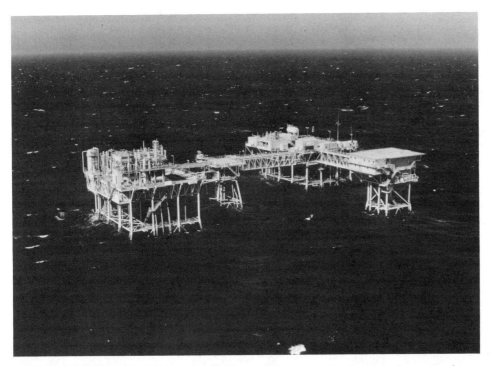

The continental shelves are likely to contain the world's major yet-undiscovered reserves of petroleum. The great reserves of Mexico, for example, recently found, extend under the Gulf of Mexico. Production of oil from beneath the sea floor requires a complex technology that has required ever-larger offshore structures. The production platform shown here lies off the coast of Louisiana in the Gulf of Mexico.

FIGURE 5.3
Two oil traps: an anticline and a thrust (reverse) fault.

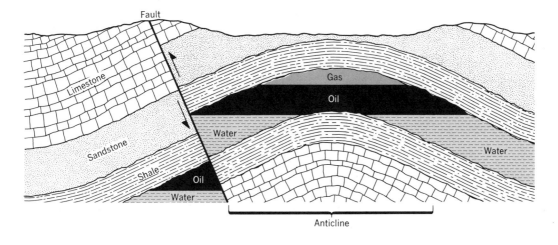

With the bulk of the world's known oil reserves, a comparatively small group of countries—the majority of them in Southwest Asia and North Africa—can control the supply and price of vital energy to the powerful developed nations. This was not always the case. In the 1950s, the large international corporations that actually exploit the oilfields had so diversified their operations that a large oil surplus developed and oil prices declined. With production coming from new fields in Nigeria, Libya, and even Australia, the oil companies could resist the pressure for higher prices by governments of longer-established oil-producing countries. The oil companies could threaten to cut back on production and exports in favor of other areas where governments were more cooperative.

In 1960, the governments of established oil-exporting countries decided to face this issue jointly to prevent the oil companies from playing this divisive economic game. They founded the Organization of Petroleum Exporting Countries (OPEC). Initially this had only limited effect, but time was on OPEC's side. Political changes brought countries that had formerly served as oil-company alternatives, including Libya and Nigeria, into the OPEC fold. At this point, the governments of the OPEC countries began to deal with the oil companies from a position of increasing strength. By imposing limits on what they would allow the companies to produce and export, OPEC governments reversed the price decline. The surpluses of the 1950s soon were a thing of the past. In the early 1970s, the OPEC countries could precipitate a full-scale energy crisis, underscoring their collective strength in the energy field. OPEC governments bought controlling shares in the operations of oil companies on their soil and no longer lacked the power to use their principal natural resource for political and economic gain.

In 1987, the membership of OPEC consisted not only of the regional countries of Algeria, Iraq, Iran, Kuwait, Libya, Qatar, Saudi Arabia, and the United Arab Emirates, but also included Ecuador, Gabon, Indonesia, Nigeria, and Venezuela.

The organization's fortunes, however, declined severely during the 1980s. This was due to several factors. First, world oil consumption declined about 6 percent from 1979 to 1986. This was related to the efforts by the developed countries to reduce their consumption, practice conservation, and develop alternate energy sources. Second, non-OPEC countries could produce as much as they wanted; when the OPEC countries limited their output to keep prices high, non-OPEC countries produced even more. This led to a collapse in oil prices. Now the OPEC countries not only lost their market share, but saw their oil-derived incomes drop precipitously. OPEC lost its strength and its unity as some of the member countries refused to adhere to OPEC production and pricing policies. But OPEC's day may come again. So long as major industrial powers depend on foreign sources for a substantial part of their fuel, the OPEC cartel holds a trump card.

---

large bodies of water where, scientists believe, microscopic plant forms (e.g., diatoms) contained minute amounts of it. On death, these tiny plants released this substance, so that it became a part of the sediments accumulating on the seafloor. Millions of years later a thick layer of sediments might contain quite a large quantity of the petroleum; then, when the sediments were folded and compressed, the accumulated oil would be squeezed into a reservoir. Such a reservoir might be an upfold (anticline) in the rock layers, or a dome, capped by an impermeable stratum. There it remains under pressure until its existence is discovered by exploration, then a well is drilled, and the oil rises to the surface.

Another aspect of the map of the distribution of petroleum-producing areas is the coastal or near-coastal location of most of the reserves. Given the origins of petroleum deposits, this is not surprising, but it does mean that the continental shelves are especially likely to contain the world's major yet-undiscovered reserves. In recent years, the exploration of the continental shelves has intensified enormously, not only off the coast of the United States but also in East and Southeast Asia, Europe, India, and elsewhere. Improving technology has brought ever deeper portions of the continental shelf, slope, and ocean floor within the reach of exploration; offshore boreholes have been drilled to depths as great as 26,000 ft (8000 m). This has implications for the future: the prospect of reserve exhaustion will raise the level of competition for the remaining supplies, and the ownership of the continental shelves in some areas of the world still remains a matter for contention. It is possible, for example, that petroleum reserves will be found in the continental shelf that surrounds Antarctica. Although an Antarctic Treaty was signed by 12 countries in 1959 limiting their activities to scientific research and exploration, the treaty is not likely to prove strong enough to withstand the pressures of national needs should major petroleum reserves be found there. On the other hand, the countries of Western Europe were able to agree on boundaries on the North Sea continental shelf so that the newly discovered oil- and gasfields in that region could be allocated and exploited. A period of delicate negotiation concerning the ownership, exploration rights, and exploitation of other shelf areas still lies ahead.

The map (Fig. 5.2) also indicates where the majority of the world's petroleum is consumed. The United States continues to lead, but its annual petroleum consumption, which approached 30 percent of the entire world's total in the late 1970s, declined to about 25 percent in the late 1980s. Still, the United States in the late 1980s required as many as 15.2 million barrels of petroleum per day to keep its power plants, automobiles, aircraft, ships, and other machinery running. But U.S. oil production averaged only about 16 percent of the annual world total, and U.S. reserves amount to less than 7 percent of the world total, emphasizing this country's present and future dependence on foreign oil supplies.

Other major petroleum consumers include the U.S.S.R., with just under 20 percent of the world total

(less than its contribution to annual production); Japan, with about 8 percent, practically all of it imported; and the largest countries of Western Europe, now benefiting from the developing North Sea oilfields: West Germany, nearly 5 percent; France, over 4 percent; and the United Kingdom, with well over 3 percent of annual consumption.

If you were to compare the just-quoted figures to those given in the Second Edition of this book, some interesting changes would become apparent. In the late 1970s, Middle East production was much higher than today, U.S. consumption was larger, and oil prices in 1977 were about double the price in 1986. In 1977, OPEC was strong and feared; in 1987, the oil cartel is weak and divided. What the experts of the 1970s said could never happen again—gasoline for less than $1 at the pump—in the 1980s became a reality once more.

The problem is that the underlying causes of the energy crisis of the 1970s have not really disappeared, and certain risks are greater now than they were a decade ago. Western governments worked toward a decline in the per barrel price of oil, and it dropped from around $30 to as low as $12. This had the effect of splitting OPEC and weakening the once-powerful Middle Eastern producers, but there were other consequences here at home. Wells that had produced modestly but just barely profitably at $30 a barrel could not be kept active at much less than that price and were, therefore, permanently closed off. Research and exploration, so strongly stimulated when oil was expensive and scarce, lagged. Should a political crisis cause a disruption of oil imports to the United States, this country will be less prepared than it was previously. The low price of oil does have its costs.

NATURAL GAS: EXPANDING OPTION

Petroleum is not the only energy source in which the United States leads world demand and consumption. Accumulations of **natural gas** often occur in association with oil deposits (see Fig. 5.3), and the use of natural gas as an energy source has increased enormously since the Second World War. In the late 1980s, the U.S.S.R., with the largest proven reserves (39 percent of the world total), and the United States

181

were the two giant consumers of natural gas. But U.S. reserves are estimated to be only one sixth as large as those of the U.S.S.R., and there is concern for the long-range future. The U.S.S.R. produces more than it requires and exports natural gas to Eastern and Western Europe through a network of pipelines. The United States consumes about 5 percent more natural gas than it produces and needs some imports of this commodity. Mexico's substantial sources are best positioned to provide what is required.

After the U.S.S.R., the Middle East and North Africa have large reserves (about 30 percent of the world total). Smaller but important reserves also lie in the Netherlands, West Germany, and under the North Sea. Canada's reserves account for about 3 percent of the world's proven gas resources.

As a result of the distribution of these reserves and the intensification of gas consumption by developed countries, a growing system of pipelines carrying natural gas now supplements the oil-carrying pipelines (Fig. 5.4). In North America in 1986, there were more than 2.5 million miles (4 million km) of pipeline carrying gas from source to consumer. In Western Europe, pipelines have been laid across the bottom of the North Sea from the producing gas- and oilfields to West Germany, England, Scotland, and Norway. Another pipeline runs across Saharan waste and Mediterranean Sea to connect the important Algerian gas reserve at Hassi-R'Mel to Bologna in northern Italy.

The growing consumption of natural gas has caused an intensification of the search for additional reserves of this resource, and has helped change the ways in which the commodity is stored and transported. Like oil, gas moves most efficiently through pipelines, but it can be transported in liquefied form by tanker. Although expensive, during the most recent energy crisis such transportation became a profitable proposition. Most natural gas is consumed comparatively near its production source—or at least within pipeline range—but some continues to travel to distant markets in liquid form.

Conservation of this resource also has improved. Gas that would have been allowed to flare, or burn away, at the wellhead just a few years ago is now being trapped. A good example of the changed circumstances is the natural gas production associated with the oilfields in Nigeria. Until as recently as 1974,

almost all the gas that came to the surface was flared at the wellhead, but then its potential as an export in liquefied form led to conservation. Of course, Nigeria still lies beyond pipeline range to major consumers, but both Western Europe and the United States were prepared to buy liquefied natural gas as far afield as Nigeria.

Natural gas is a clean and efficient source of energy. Although this fuel has somewhat reduced the pressure to produce more petroleum, the reserve situation is uncertain. Substantial new discoveries continue to be made, yet the overall picture is far from clear. The latest available figures, for example, suggest that reserves in the United States amount to about 198 trillion cubic feet, but these are **proven** reserves. Considering that U.S. consumption recently has averaged 20 trillion cubic feet per year, the situation appears rather bleak until another figure, the **potential** reserves, is taken into account. Experts calculate that further exploration may yield another 900 trillion cubic feet of natural gas if their estimates and speculations prove to be anywhere near correct. One of these guesses—that natural gas would be found in commercial quantities off the U.S. East Coast—bore fruit in August 1978 when successful strikes were made in the Baltimore Canyon area of the continental shelf. Even if the projected figure of 900 trillion cubic feet for the potential reserves is a gross overestimate, as some specialists believe it to be, it is certain that the United States contains much more than 198 trillion cubic feet of proven reserves. But the United States consumes more natural gas each year than all other countries of the world combined, so we have an enormous stake in exploration for this resource wherever potential exists.

PATTERNS IN NORTH AMERICA

No single country consumes as much energy—in total or per capita—as the United States does. From Mexico to the south and from Canada to the north flow energy resources to the great U.S. market—and from countries much farther away come additional supplies.

North America is endowed with very large coal reserves, but comparatively modest-sized oil- and natural gasfields (Fig. 5.4). Three major areas of coal deposits exist: (1) Appalachia produces about half of

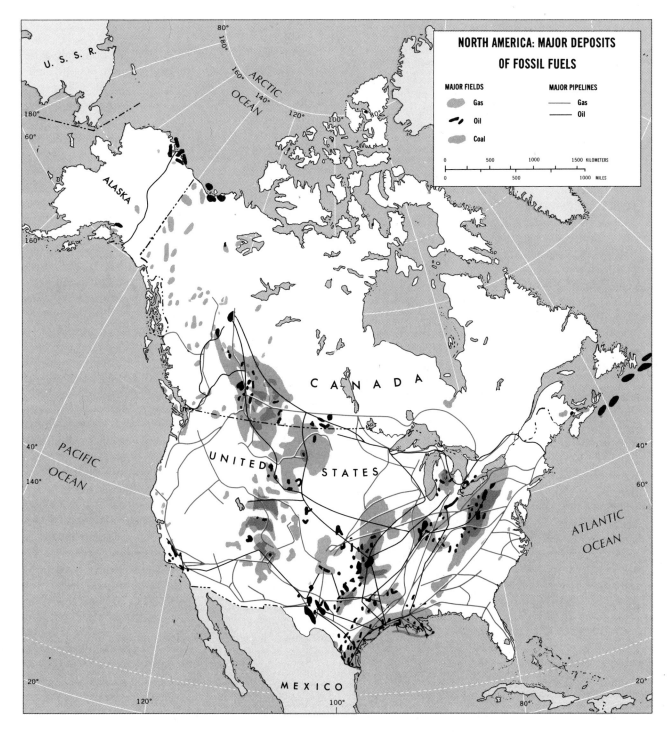

NORTH AMERICA: MAJOR DEPOSITS OF FOSSIL FUELS

MAJOR FIELDS

Gas

Oil

Coal

MAJOR PIPELINES

Gas

Oil

| 0 | 500 | 1000 | 1500 KILOMETERS |

| 0 | 500 | 1000 MILES |

FIGURE 5.4

all U.S. coal; it is still the largest, but it is declining because its sulphurous coal must be expensively filtered to meet the standards set by federal clean-air statutes; (2) the Western coal area, which is centered on the Great Plains within 500 miles (800 km) north and south of the U.S.–Canada boundary, is expanding vigorously thanks to the deposition of its vast low-sulphur supplies in thick near-surface seams—ideal for lucrative strip mining. This has even prompted the construction of the first long-distance railroad line in the United States in over 50 years (into Wyoming's Powder River Basin); and (3) the Midcontinent coalfields—a large arc of high-sulphur, strippable deposits centered on southern Illinois and western Kentucky—are also declining as the West surges ahead. North America's major oil-production areas are located along and offshore from the Texas–Louisiana Gulf Coast; in the midcontinent district that extends through western Texas, Oklahoma, and eastern Kansas; and along the front of the Canadian Rockies in central and northern Alberta. Lesser oilfields also exist in southern California, west-central Appalachia, the northern Great Plains, and the Southern Rockies. Both the United States and Canada have recently turned their attention to the Arctic, where ample supplies exist along Alaska's North Slope and adjacent territory in northwestern Canada. In 1977, the Trans-Alaska Pipeline was opened—a triumph over the harsh Arctic environment—enabling the pumping of North Slope petroleum south 800 miles (1300 km) to the warm-water port of Valdez for transshipment by tanker to the lower 48 states. Both countries have also agreed to jointly build another pipeline to carry North Slope natural gas all the way south overland through western Canada. The distribution of *natural gas* deposits resembles the geography of oilfields because both petroleum and gas are usually found in similar geologic formations (the floors of ancient shallow seas). Accordingly, major gasfields are located in the Gulf, midcontinent, and Appalachian districts. However, when subsequent geologic pressures are exerted on underground oil, the liquid is converted into natural gas; this has happened frequently in mountainous zones, so that western gas deposits in and around the physiographic province of the Rockies tend to stand apart from oilfields.

ALTERNATIVE ENERGY SOURCES

Coal, natural gas, and petroleum are nonrenewable resources; they exist in the earth in finite quantity and can be used up. Over the hundreds of millions of years of geologic time that preceded the emergence of human societies on this globe, vast amounts of these resources developed, but humanity has an enormous appetite. Some experts estimate that presently known reserves of zinc, lead, and tin ores will be exhausted early in the twenty-first century, that the world's usable forests will not last much longer, and that even iron ore—the taconite experience and its implications notwithstanding—will soon follow (see Chapter 10). Still, human history is one of technological innovation, and somehow the prospect of alloy shortages does not seem awfully alarming. As for iron ore, other metals have already been substituted for iron in many instances, including in automobile engines. In any case, low-grade iron ore, lower even than taconite, exists in vast quantities.

This, however, is an age of energy, and it is the energy picture that has resource specialists worried. It is not that we will run out of oil, gas, and coal suddenly some summer (although some observers have even forecast such a doomsday); the problem is that, over the long term, demand is skyrocketing and supply appears to be on the verge of falling seriously short. The most accessible reserves have been worked out, costs are rising, and technological developments are not coming as fast as they are needed. The effect of all this is felt especially strongly on the form of power for which demand has risen most rapidly of all: electricity.

Electric Power

Electric power is manufactured from oil, coal, and gas; it is produced by hydroelectric plants; and increasingly it comes from the alternative of the future, nuclear fission. Today, the United States still uses about 30 percent of its fossil fuels for the generation of electricity. Industry uses most of the output (more than 40 percent), and private homes and commercial establishments consume most of the rest. The problem with the conversion of these fuels into electric power is that it is a very dirty process. Just over one third of

Icebergs and pack ice float in the Southern Ocean off Antarctica. Icebergs form when glacial ice flows (or ice shelves) reach the ocean and segments break off (or "calve"). In northern seas, icebergs are a hazard to navigation; in the Southern Ocean, the zone of pack ice surrounding Antarctica expands and contracts with the seasons. There, too, ships have been trapped and crushed by the ice.

the fuel is made into electricity, and the rest becomes wasted smoke and heat. An increase in electricity supply, therefore, requires a disproportional increase in fuel supply and consumption, which means more landscape-destroying strip mining, ocean-polluting oil spills, and atmosphere-befouling soot and noxious gases. Is our technology finally failing us?

Not yet, but it looks like a close call is at hand. Even while industrial countries—including the United States, the United Kingdom, West Germany, and others—are converting much of their power-generated equipment to oil and natural gas, progress is being made in the switch to nuclear reactors, the apparent alternative when it becomes uneconomical to exploit oil and gas. Such nuclear plants already exist and are operating, but today they supply a mere 15 percent or so of U.S. electricity requirements.

Nuclear Fission and Fusion

If we already know the source of future power, why not give up on oil and gas (as we are giving up on coal before it is exhausted) and convert entirely to nuclear power? The answer lies in the exponential growth of human numbers and needs and, to some extent, in a growing concern over the future of the environment. On the one hand, technologists are having trouble planning, designing, and locating nuclear plants as fast as desired. On the other, individuals and groups of people are blocking construction because they fear that lethal radiation pollution will occur just as surely as soot emanates from coal-using plants. And certainly they have reasons. The uranium-powered nuclear fission process, apart from dangers of pollution or explosion, produces radioactive wastes whose storage and disposal constitute problems not only for today, but for future generations as well. This is one reason why the so-called breeder reactors, now emerging as the likely future generation of nuclear plants, are opposed by conservationists. Their case was strengthened by the accident involving the Three Mile Island (Pennsylvania) nuclear power plant in 1979.

The future of nuclear power was further clouded by the death-dealing explosion at the Chernobyl power plant north of Kiev, in the Soviet Union, in April 1986. Not only did the accident cause casualties among workers at the plant, but radiation pollution spread across a vast area of countryside, requiring the immediate evacuation of as many as one hundred thousand persons from homes to which many may never be able to return. And elevated levels of radiation were recorded in the air over much of Europe. Vegetables, dairy products, and other foods could not be safely consumed. There were fears over water pol-lution, soil contamination, and other hazards. On the day of the explosion and fire, the winds were blowing northwestward, carrying the lethal radiation away from the large city of Kiev; scientists could only imagine what would have happened if the winds had blown southward. Nevertheless, dozens of persons died and tens of thousands must now live with an increased risk of various forms of cancer.

In the United States, the nuclear industry has already been slowed by rising costs of construction and operation, by greater competition from (recently cheaper) traditional energy sources, and by public opposition. In the mid-1970s, experts were predicting that, by 1990, as much as 50 percent of electricity generated in the United States would come from nuclear sources. At that time, there were some 70 operating nuclear reactors, with another 96 in various stages of construction and planning. By 1985, there

---

• Alternative Energy Sources •

When shortages in fuel supplies develop, as was the case during the most recent energy crisis, the search for alternative sources of energy intensifies. During the shortage of the 1970s, prototype electric cars made their appearance. When the shortage disappeared, little more was heard of this technology. Its time may come again.

Some alternative sources of energy have served for a very long time. Hydroelectric power stations as recently as the early 1950s contributed as much as 30 percent of all electricity consumed in the United States, but that share has dropped to a mere 7 percent, reflecting the enormous increase in total consumption over the past 40 years. In other countries, such as Japan and Norway, the role of hydroelectric power is much more important in the overall energy picture.

Solar energy (energy from the sun) is another option. Harnessing that energy in quantities that would really help reduce dependence on fossil fuels is a difficult proposition. It can be done directly through glass panels or reflecting (ray-concentrat-ing) mirrors, but this remains a proposition for individual buildings or homes, not for general use. Energy cells capable of collecting and storing solar energy and converting it to electricity remain too expensive for large-scale application.

Wind power also is a long-used source of energy. The Western European landscape once was dotted with thousands of windmills that raised water, drove textile-making equipment, ground wheat, and performed other functions. Huge modern versions of these old windmills have produced electricity, but not as dependably as had been hoped. Again, wind energy is likely to remain a mainly local, supplementary source.

The earth provides energy in the form of tidal power, which is harnessed experimentally in a few places, and as geothermal heat in areas where ground water is heated by volcanic conditions in the upper crust. Geothermal development has met with some success, but none of the alternative sources promises to significantly reduce our dependence on the fossil fuels.

186

were just 82 functioning nuclear power plants, generating only 16 percent of all electricity.

The nuclear industry's slowdown in the United States has not been matched in other countries. In 1987, as much as 68 percent of France's electricity came from nuclear power plants; in Belgium, 60 percent; in Sweden, 50 percent; in West Germany, 33 percent. The Soviet Union's plans for expanded nuclear power production were not reversed by the Chernobyl accident. Soviet planners expect that 50 percent of their country's electricity supply will come from nuclear sources by the year 2000. But it has become clear that an accident in one country can have disastrous consequences for other nations. The Chernobyl accident could have been much worse than it was, locally and meteorologically.

And so the search is on for still other alternatives for the generation of electric power. At present, the most attractive prospect seems to lie in **nuclear fusion**. In this process, the joining of atomic nuclei sets off an enormous amount of energy. The high temperatures involved (this, in fact, is the process that powers the hydrogen bomb) and problems of storage will delay the appearance of fusion reactors, but when they emerge, the prospect of an energy crisis may disappear. Hydrogen, the fuel, is so common on this planet (e.g., in seawater) that the development of the nuclear fusion reactor will guarantee perhaps millions of years of power. And the process is clean. If it can also be made safe, it will solve the problems that fission poses. Another potential advantage may be the greatest of all: it may be possible to develop cells that convert fusion power directly into electricity. This may mean the end of huge power plants for many purposes. Cells may be installed in private homes; thus, when you buy a television set, for example, it will contain its own power source.

From all this, it is obvious that the world is in a state of energy transition; before one substitute technology is in place, another, with greater promise, appears. In the meantime, shortages in traditional energy resources are followed by times of comparative plenty—all in the face of political and economic uncertainties. Wood was the chief source of fuel for millennia; coal for centuries; now oil and gas have prevailed for decades. The time of nuclear energy may have come, but it is still measured in years. The outcome of our current energy adjustments is still in doubt.

Conservation

This brings us back to our theme of the earth as a storehouse of accumulated natural wealth from which human societies have, since their beginnings, chosen and exploited resources. It is not true that the specter of short supply (as in the case of petroleum) is now confronting us for the first time. History is replete with local and regional crises: when Europe's forests were severely depleted, there was an acute shortage of firewood, charcoal, and building material. But it *is* true that our human numbers today are straining the earth's resources more than ever in world history and that we are altering our natural environments in unprecedented ways. The extent to which the environment is affected is sometimes revealed in small but significant ways. In July 1972, four drums of water, buried in the sand of the Sahara since the Second World War, were accidentally discovered. This created great excitement, because these four drums were thought to contain the only water known to have been unaffected by radiation pollution. They were treated with the kind of care and curiosity usually reserved for moon rocks, and newspapers around the world reported on the careful recovery. It was a reflection on our time that so common a substance as water should be so universally affected that its pure form had become a rarity.

GEOGRAPHY AND CONSERVATION

It is a matter of record that the conservation of the natural environment and the reasoned exploitation of natural resources were areas of interest and concern for geographers long before they became popular public issues. Courses under the rubric of conservation were part of the geography curriculum at many

187

colleges and universities in the 1940's, and one of the most widely used geography texts, edited by Guy-Harold Smith, was entitled *Conservation of Natural Resources,* (1950). Since those days, when such issues as atmospheric fouling, noise pollution, radiation threats, pesticide danger, and oil-spill damage were not yet major concerns, the emphasis has changed considerably. But the fundamental idea, that we should educate ourselves to protect our natural heritage while recognizing the need for its use in the human interest, is still the same.

It is interesting to compare the major issues in conservation 30 to 40 years ago with those of today. Remember that the world's population in the mid-1940s was less than *half* of what it is today; one of the striking contrasts is the much greater urgency in today's writings on conservation. Not only have our growing numbers increased and intensified the exploitation of the earth's natural resources, but humanity's impact on the global environment has also changed qualitatively. Following the Dust Bowl experience of the 1930s, conservationists focused on land-use planning and soil-improvement programs. During the Roosevelt administration, called by some the golden age of conservation, the U.S. government created several offices and agencies to translate newly passed conservation legislation into productive action. The Soil Conservation Service was established, as were the National Resources Board, the Civilian Conservation Corps, and the Tennessee Valley Authority. Not all of those agencies have retained their original name and identity, but many conservation-oriented offices of the present federal government trace their origins to that period of national concern and commitment to the conservation idea.

Much happened during the 1950s and 1960s to divert the national interest from conservation to apparently more urgent issues: the Korean War, the growing tensions with the Soviet Union that culminated in the Cuban missile crisis, and the rising involvement that led to the Indochina War. The United States also was urbanizing rapidly, and many small farmers sold their land to large-scale agricultural corporations and went to work in towns and cities. Still, there were farsighted people who realized the conservation philosophies and programs must be supported,

thus several privately financed organizations were founded to promote resource management and environment protection. The Conversation Foundation, established in 1948, has sponsored research and conferences that have resulted in significant publications on specific and practical conservation problems. Another organization, Resources for the Future, founded in 1952, filled a gap by supporting long-range and large-scale studies on resource exploitation and management. Meanwhile, the issues involving conservation were widening, and the protection of environments, whether natural or artificial (e.g., urban living conditions) became a pressing concern.

While the public interest was somewhat diverted from the conservation idea, the problems were multiplying. Rapidly growing populations required more food, so pesticides were developed to reduce losses of agricultural crops. Success was considerable, and farm yields increased—but the pesticides, it was later found, entered the plants and endangered the consumers. Livestock that had eaten contaminated fodder had to be destroyed, their milk and meat were unsafe. Urban populations were exposed to contaminated drinking water; the air over large, densely populated areas was fouled by toxic fumes from car exhausts and factory smokestacks. Radioactivity from test explosions of nuclear bombs in the atmosphere diffused around the entire world, affecting the air almost everywhere, not just over the immediate testing ground. *Fallout* became a term in common usage. The introduction of jet aircraft on passenger routes raised concern over the impact on the ozone layer, and people near airports began to agitate over noise pollution. When the British–French Concorde, the first supersonic jetliner, was first scheduled into New York's Kennedy Airport, crowds protested by blocking roadways to the terminal. The fouling of California beaches by oil seeping from offshore boreholes and scenes of dying bird life brought demonstrators to the headquarters of oil companies.

Such issues served to revive public interest in environmental matters; in 1969, the U.S. government passed the national Environmental Policy Act. In 1970, a master agency, the Environmental Protection Agency, was created to take control over all the states, municipalities, and industries in areas of air and water

Contour plowing in Wisconsin. Although the slopes in this area are not very steep, this farm's fields are nevertheless carefully plowed along the contours. This practice reduces soil erosion by impeding the downslope rush of rain water; gulleying is prevented and the topsoil protected.

quality, pesticide use, waste disposal, and radiation hazards. Another agency, the Council on Environmental Quality, was established to coordinate and monitor the actions of all government departments that might have an impact on the environment in the United States. This was badly needed, because numerous U.S. government offices and agencies had been involved haphazardly with various conservation efforts without adequate planning and coordination.

PRACTICAL OBJECTIVES

The new era in conservation and environmental pro-

tection did not outdate the practices of old. The conservation of soil remains a primary objective, but more is now known about the effect upon the soil of insecticides used to protect the crops. The liberal use of DDT, for example, has been discontinued; research proved that this poison was washed into the soil and then ingested by the next generation of plants. Fertilizing of soil continues, but chemical fertilizers are used with greater insight into their long-term effect on the soil. Crop rotation practices also continue, because certain plants demand one particular set of nutrients from the soil, whereas others extract a different combination. In some areas, the soil is left uncultivated (fallow) during one year out of three or four so that it may rest and regenerate. Various methods to combat soil erosion have been employed for many years—such as **contour plowing**. When the furrows made by plows follow the contours of the land, each furrow helps stop the rainwater from running downhill and carrying the topsoil with it. Downward percolation of the rainfall is also enhanced, the water loss by runoff diminished. Where conditions of slope do not permit contour plowing as a means of soil conservation, **terracing** is an alternative. Terracing lends itself less well to mechanized farming than does contour plowing, and terraces are usually indicative of steep-slope conditions such as prevail in East and South Asia's ricelands. In the United States, the same principle is applied when efforts are made to stop soil damage from gulleying. Earthen dams across the incipient drainage lines help reverse the erosional process by interrupting the flow of water and leveling the land.

Such conservation efforts have the effect of modifying the landscape, of changing a physical (natural) landscape into a landscape of human purpose. Cultures, we will note in Chapters 7 and 8, make their own, distinct imprints on the earth's surface when they transform the landscape.

How can nonrenewable resources be conserved, or the rate of their consumption slowed down? The conservation of these natural resources involves several options. One of these is **beneficiation**, the use of lower-grade ores than have previously been mined. Taconite, discussed earlier, is an example of this alternative; improving technology can bring lower-grade reserves into production. In the case of the mineral fuels, large amounts of petroleum are locked up in oil shales and tar sands in various parts of the world, including the United States, but the exploitation of these potential reserves must await the exhaustion of more readily accessible reserves presently supplying world consumers. Another form of conservation of nonrenewable ores is their **recycling** following initial allocation. In recent years, the practice of collecting aluminum cans, glass bottles, and the steel from junked automobiles has gained currency and contributed to a reduction in consumption. Recycling, at the same time, has the effect of reducing pollution in our throwaway society. Still another conservation measure involves **substitution**, the use of more readily available substances in order to save scarcer resources. The use of copper in previously all-silver coins is an example, as is the mixing of plastics and ceramics with smaller amounts of metal for many purposes. We have previously alluded to the versatility of aluminum as a substitute for iron; aluminum also can replace copper and steel in wiring, and its rust-free quality makes it a favored roofing material over iron-based sheeting.

CONSERVATION OF LIVING RESOURCES

The conservation of living resources presents quite different problems. These include the management of marine fauna, which must be guarded against overexploitation; the protection of forests, where renewal programs must accompany extraction; and the preservation of terrestrial wildlife (including not only mammals, but also reptiles and birds). It also involves the

The last of the great herds of wildlife are under pressure in Africa, but fauna in such locales as Madagascar, Sumatra, and Tasmania also are endangered. The wombat is among native animals of Australia, part of a unique assemblage found nowhere else in the world. Australia's human population is small (15 million) but already major species (such as the Tasmanian "Tiger") have become extinct.

190

management of grasslands as well as the livestock that may share the land with wild animals.

The wildlife and nature conservation idea has had practical expression in European countries and in the United States for little more than a century. In the United States, the Yellowstone National Park was established as early as 1872, and Yosemite National Park was created in stages over a 40-year period that ended with its proclamation in 1905 when the Yosemite Valley was added to the park segment first created in 1890. But these were parks intended to serve people as well as protect landscapes and animals, and sometimes these objectives are not compatible. In 1916, the National Park Service was formed as an agency of the U.S. Department of the Interior, a major step in support of conservation; in 1929, the National Forest Service was established under the aegis of the Department of Agriculture. Over the years, areas totaling about 385,000 square miles (nearly 1 million sq km) in the United States have been set aside as national forests, with another 116,000 square miles (300,000 sq km) as national parks and other conservation areas under the Fisheries and Wildlife Service. As additional land is acquired, these totals continue to increase (Fig. 5.5).

By the time protective measures were taken, much of the wildlife that had been present in North America when the Europeans first arrived was already destroyed. The twentieth-century conservation effort has concentrated on the salvageable remnants of this realm's wildlife, and there have been successes as well as setbacks. The protection of wildlife poses complex problems. Special-interest groups such as hunters sometimes demand the right to shoot particular species, endangering what may be a delicate balance in the ecosystem. Others object to the closure of wilderness areas to the public for any purpose, even sightseeing, when this is necessary to allow an endangered area or animal species to recuperate. And wildlife is mobile, so that it may migrate beyond the boundaries of a wildlife refuge or national park, thus becoming endangered. The protection of birds, whose migratory habits may carry them across the length of a continent, is made especially difficult by their mobility.

Conservationists also learned a great deal about the most effective methods of wildlife preservation. This is not simply a matter of fencing off an area where particular species exist, but involves the management of the entire habitat, the maintenance of a balance among animals and plants—and, sometimes, people. A series of congressional actions and commissions during the 1960s and 1970s served to facilitate and implement the newly developed conservation practices, and there have been several successes, including the revival of the bison, the wild turkey, and the wolf as well as the survival of the grizzly bear and the bald eagle.

In European countries and in the Soviet Union, also, the story of the conservation effort is largely one of remnant preservation. Europe's varied wildlife fell before the human expansion; it was lost even earlier than America's wildlife. Conservation is an expensive proposition, afforded most easily by the wealthier, developed countries. European colonial powers carried the concept to their colonies where they could carve wildlife refuges from tribal lands with impunity. In many areas of Africa and Asia, the Europeans found the indigenous population living in balance with livestock and wildlife. The invaders disturbed this balance, introduced the concept of hunting for profit and trophies, and then closed off huge land tracts as wildlife reserves (and as controlled hunting zones). That the African wildlife refuges have so long survived the period of decolonization (some are now in danger) is testimony to a determination not present when orgies of slaughter eliminated much of America's wildlife heritage. We may have advanced knowledge of conservation theory and management practices, but we are in no position to proselytize.

The African Wildlife Heritage

In time, Africa's wildlife may suffer the same fate as that of America, Europe, and—more recently—India. Not long ago, parts of India teemed with herds of wild buffalo, many kinds of antelope and deer as well as their predators, lions and tigers and large numbers of elephants and rhinoceros. The Indian population explosion plus a breakdown in the conservation effort have combined to destroy one of the earth's great wildlife legacies. The Indian lion is virtually extinct;

192

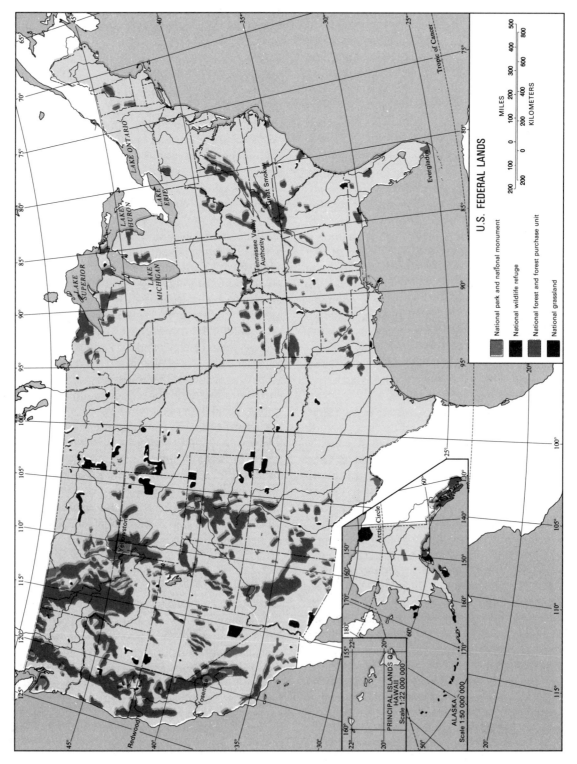

U.S. FEDERAL LANDS

National park and national monument

National wildlife refuge

National forest and forest purchase unit

National grassland

MILES
200   100   0   100   200   300   400   500
200   0   200   400   600   800
KILOMETERS

Tropic of Cancer

Everglades

LAKE SUPERIOR
LAKE MICHIGAN
LAKE HURON
LAKE ERIE
LAKE ONTARIO

Great Smokies

Tennessee Valley Authority

Yellowstone

Redwood

Yosemite

Arctic Circle

ALASKA
Scale 1:50 000 000

PRINCIPAL ISLANDS OF
HAWAII
Scale 1:22 000 000

FIGURE 5.5

the great tiger is endangered and survives in small numbers in remote forest areas. Other wildlife has been decimated by systematic eradication campaigns (to protect farmlands and domestic livestock) and by Europeans and other whites on hunting expeditions.

The future of Africa's wildlife hangs in the balance. Although the conservation effort begun during the colonial occupation has been sustained and even expanded in such countries as Kenya, Tanzania, and Zambia, the breakdown of order elsewhere has had the opposite effect. Angola's civil war severely damaged its national parks and their contents; wildlife in Zaïre, Uganda, and Ethiopia also suffered when political crises occurred. The elephant in Ivory Coast is being poached out rapidly. The gorilla is in danger in eastern Zaïre and other parts of equatorial Africa as a result of human encroachment on its mountain and forest habitat. Remote areas in Sudan and Botswana have been penetrated by hunting safaris and are about to be opened up; the results of such a sequence of events are all too familiar.

The survival of as much of Africa's wildlife as does remain can be attributed principally to four factors: (1) the value of tourism as a producer of foreign money; (2) the ability and willingness of African governments to withstand pressures to open wildlife reserves to human occupation; (3) the commitment of African governments to the conservation idea, and their recognition of the value and uniqueness of the regional fauna; and (4) the remoteness and lack of land-use alternatives that characterize major wildlife reserves. In Kenya, the tourist industry produces as much as 40 percent of the country's annual income; although Kenya is Africa's leader in this respect, tourism contributes substantially to the economies of other countries as well. African governments have resisted the invasion of wildlife reserves by squatters, notwithstanding the political risks involved. As population pressure grows and periodic droughts reduce the carrying capacity of populated lands, the natural objective of subsistence farmers is to occupy land set aside for wildlife. Only the most determined national policies can resist such movement. As for the recognition of wildlife's fundamental value, the Kenya government in 1977 banned all hunting and the sale of all trophies and products made from the skins, hair,

bones, and horns of wild animals. No other country ever implemented so far-reaching a policy, depriving itself of a major source of income, in the interest of long-range conservation (Fig. 5.6).

But the major ally of Africa's wildlife undoubtedly has been the tsetse fly, which is the carrier of sleeping sickness. As we noted in Chapter 4, much of Africa is grassland savanna country, potentially good grazing land for domestic livestock. But livestock that is introduced into areas infested by the tsetse fly soon is infected by sleeping sickness and the animals die. On the other hand the wildlife over hundreds of years has developed a state of balance with the disease; it is endemic (see Chapter 6) and the wild animals survive. When a tsetse fly bites a wild animal, it sucks up its blood containing **trypanosomes**, the agents of sleeping sickness. In the fly's body, these trypanosomes reproduce and eventually reach the insect's salivary glands. When the fly bites a cow, the animal is infected with trypanosomes and develops sleeping sickness. Thus Africa's wildlife acts as a giant reservoir that sustains this disease. Any program to turn the continent's grasslands into ranchlands would involve either the total elimination of all wildlife (as was ruthlessly done in northwestern Zimbabwe—then colonial Rhodesia) or the eradication of the tsetse fly. Progress against the fly is being made, and when sleeping sickness is conquered—as may happen before the end of the century—Africa's wildlife may go the way of India's.

When the Europeans entered East Africa, they found cattle-herding peoples already there, the livestock sharing the pastures with the great migratory herds of wildebeest, zebra, buffalo, and other wildlife. The cattle-herding Masai managed to protect some (but not all) of their livestock against the tsetse fly by keeping their cattle away from areas and places known to be fly infested, but they still lost animals to sleeping sickness. Other diseases and predators also acted to keep the Masai cattle herds limited in size. In time, the Masai were ousted from many of their better grazing lands, but their pastoral way of life still survives: the people and their livestock are a part of the overall ecosystem in southwestern Kenya and northern Tanzania.

The savannalands and forests of tropical and

193

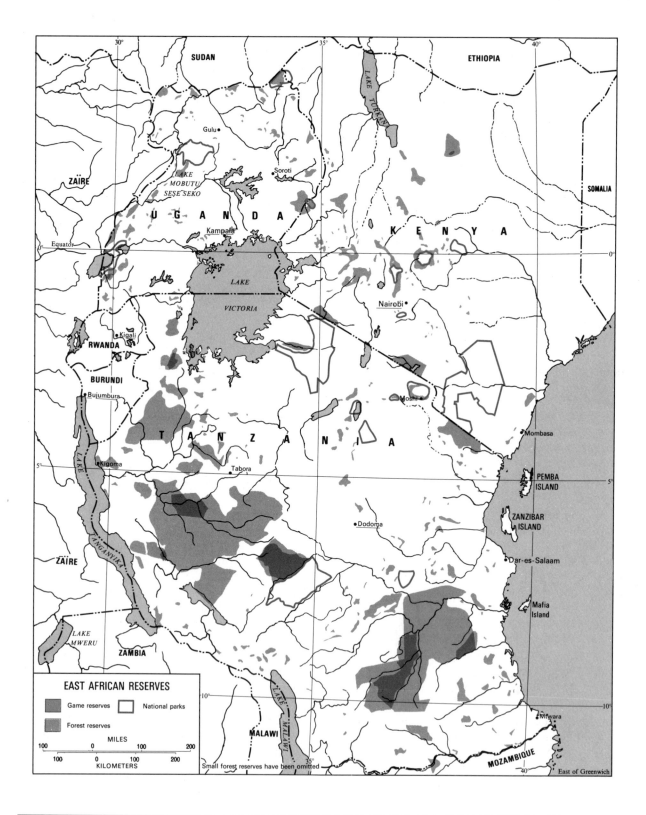

EAST AFRICAN RESERVES

- Game reserves
- National parks
- Forest reserves

MILES
100   0   100   200

KILOMETERS
100   0   100   200

Small forest reserves have been omitted

194

East of Greenwich

Chapter 5   Resources: Consumption and Conservation

Ranching East Africa's eland: an experimental station in Kenya. Africa's great antelope herds have not been successfully ranched on a large scale, but research continues in this direction. These eland are captured when young, and in captivity they prove to be quite docile and easily managed.

equatorial Africa now suffer the fate of midlatitude America and subtropical India: ecosystems that for centuries were in balance are disturbed and destroyed. North American Indians hunted the bison but did not exterminate it: that was very nearly achieved by the European invaders. For better or worse, India's human population increased at unprecedented rates following the onset of the colonial era, and a rich native fauna and flora fell victim. Now, African countries exhibit the world's highest birthrates, and wildlife sanctuaries stand against an ever-rising demand for soil and space. In the end, the rising tide of human numbers will engulf this last refuge of nature.

East Africa is the last refuge of the great wildlife herds and their predators. Many of the reserves shown on the map at left are now under pressure from human encroachment.

Past and Future

This chapter has painted a worrisome picture of the state of our earth and world in the context of one crucial element of our global life-support system: the resources that affect the *quality* of life. Four points stand out. First, surely, is the brevity of time, the short period it has taken humanity to exploit and, indeed, ravage so much of the earth. The whole sequence of events we have covered here has taken just $\frac{1}{10}$ of 1 percent of the total lifetime of the planet. What took hundreds of millions of years to develop was used up in just centuries and, in the case of certain resources, decades. And the rate of exploitation—of soils, plants, fuels, minerals—continues to rise. It is all happening so fast that we are unable to get it under control.

The second conclusion relates to the interconnected nature of our earth and our divided human world. To confront common problems we need concerted action, but such action seems impossible to achieve. If, for example, the depletion of the earth's ozone layer can bring on a 300-ft (90-m) rise in sea level in a matter of decades, as scientists in the late 1980s are warning, then surely a global program of response is needed. But individual countries must pass the laws that will control the damage. If protective measures are taken in Western Europe and North America, but not in Japan, the Soviet Union, and other countries, the risks will continue to rise. In this and in other urgent environmental concerns, our divisiveness is our greatest disadvantage. We have learned that fallout from nuclear weapons tests and from reactor accidents can literally encircle the globe, but we have no effective international mechanisms to reduce the risks.

The third reality has to do not only with our depletion of resources, but also with the political dangers this entails. Powerful states that depend on foreign sources for raw materials may feel compelled to take military action to protect those lifelines. Big-power military action in a nuclear age puts the entire world at risk and, indeed, endangers the future of all humanity. Our tame discussion of resource exhaustion should be seen against this background.

Finally, there is the matter of environmental degradation. To many of us this is a matter of occasional concern over the quality of drinking water, a high smog reading on a summer day, or a disturbing view of strip-mined countryside from a highway. But to hundreds of millions of people in Africa, Asia, and Middle and South America, environmental destruction means hunger and starvation, the loss of soil and livestock. Theoretical studies that attempt to project the future of our world invariably produce pictures of a twenty-first century world even bleaker than that of today: with even more rapid environmental degradation in larger areas than those now afflicted and greater human populations putting more pressure on less productive land. Of course, there is the hope that technological innovations and genetic engineering (to create ever-more productive crops) will overcome these ills and that human population numbers will somehow stabilize. But we do not have the time to

await these potentials. So we must work toward a world whose limits are acknowledged, whose environments are a joint concern, whose economics are based less on power and more on fairness—in short, a world with a future.

Reading about Resources and Conservation

The topics of Chapter 5 extends into several fields of geography. One of these fields, conservation, became a cornerstone of the discipline as recently as only two decades ago when most undergraduate course programs began to include at least one required course in this area. That was the time when some of the most interesting books in the geography of conservation were published. Two of these were G-H. Smith (ed.), *Conservation of Natural Resources* (4th ed.; New York: Wiley, 1971), and *Conservation in the United States* (2nd ed.; Chicago: Rand McNally, 1969) by R. M. Highsmith, J. G. Jensen, and R. D. Rudd. Today, most books on conservation issues are written by nongeographers. One of the best remains R. F. Dasmann's *Environmental Conservation* (5th ed.; New York: Wiley, 1984).

One of the fields that have emerged in this area is environmental science, and a book geographers will appreciate is that by G. T. Miller, Jr., *Living in the Environment: An Introduction to Environmental Science* (4th ed.; Belmont, Calif.: Wadsworth, 1984), which ranges from elementary ecology to relevant spatial-economic issues. The problems of resource adequacy are addressed by H. S. Brown in *The Challenge of Man's Future* (Boulder, Colo.: Westview, 1984); first published in 1954, this book has proven to be of lasting interest. Another relevant volume has a title that will ring familiar to many geographers: *The Human Impact: Man's Role in Environmental Change* (Cambridge, Mass.: MIT Press, 1982), which also presents a survey of the literature in this field.

Another good place to start your supplementary

reading in this general area is in the economic geography textbooks listed at the end of the chapters on farming and industrialization (Chapters 8 and 10). These books include good sections on resources and conservation problems and issues. In addition, see L. C. Brown and E. C. Wolf, "*Soil Erosion: Quiet Crisis in the World Economy*" (Washington, D.C.: Worldwatch Institute, 1984), a warning of substance; the quote is from page 47. In another context, see R. Goodland and S. Irwin, *Amazon Jungle: Green Hell to Red Desert?* (Amsterdam/New York: Elsevier, 1975), a volume with a dramatic title but a solid basis for the concern it expresses. A recent contribution by a geographer is D. Greenland's *Guidelines for Modern Resource Management* (Columbus, Ohio: Merrill, 1983). Additional books to be sampled include V. Smit, *The Bad Earth: Environmental Degradation in China* (Armonk, N.Y.: Sharpe, 1984) and J. Walls, *Land, Man, and Sand: Desertification and Its Solution* (New York: Macmillan, 1980). Among older works that date from the time when geographers were more active in this area, there is the classic volume edited by W. Thomas, *Man's Role in Changing the Face of the Earth* (Chicago: Univ. of Chicago Press, 1956), and T. Detwyler (ed.), *Man's Impact on Environment* (New York: McGraw–Hill, 1971).

The topic of wildlife conservation is addressed by D. W. Ehrenfield in a dated but still valuable study, *Conserving Life on Earth* (London/New York: Oxford Univ. Press, 1972). Also see P. Ehrlich and A. Ehrlich, *Extinction: The Causes and Consequences of the Disappearance of Species* (New York: Random House, 1981).

The damage that can be inflicted by development projects is the subject of another older but still thought-provoking book edited by M. T. Farvar and J. P. Milton, *The Careless Technology: Ecology and International Development* (New York: Natural History Press, 1972).

Energy issues have attracted the attention of many scholars. A good source is G. A. Daneke (ed.), *Energy, Economics, and the Environment: Toward a Comprehensive Perspective* (Lexington, Mass.: Health [Lexington Books], 1982). Also useful is J. T. McMullan, R. Morgan, and R. B. Murray, *Energy Resources* (2nd ed.; London: Arnold, 1983). A book by nongeographers with many implications for geography is *Food, Energy, Society* (New York: Wiley–Halsted, 1979) by D. Pimentel and M. Pimentel. When you reach Chapter 11, keep in mind the book by M. A. Conant and F. R. Gold, *The Geopolitics of Energy* (Boulder, Colo.: Westview, 1978). Also see S. W. Sawyer, *Renewable Energy: Progress, Prospects* (Washington, D.C.: AAG, 1986).

The first conservation textbook published in the United States was written by G. P. Marsh and was entitled *Man and Nature: Physical Geography as Modified by Human Action* (New York: Scribner's, 1864): A classic that offers a fascinating insight into a time when this was one of geography's central concerns.

No journal in geography is exclusively devoted to conservation or environmental issues, but articles by geographers on these topics continue to appear in the discipline's leading publications. For an example of one contribution to the resource issue, see the *Water Resources Atlas of Florida*, prepared by E. A. Fernald and D. J. Patton (Tallahassee: Florida State Univ., 1984).

197

COLOR PLATES

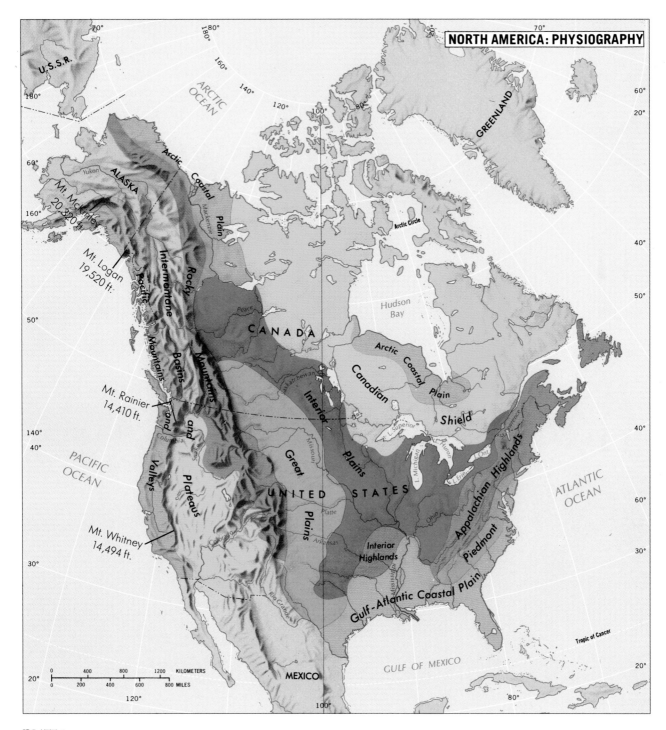

NORTH AMERICA: PHYSIOGRAPHY

U.S.S.R.

ARCTIC OCEAN

GREENLAND

ALASKA

Arctic Coastal Plain

Mt. McKinley 20,320 ft.

Yukon

Mt. Logan 19,520 ft.

Pacific

Intermontane

Rocky

Mackenzie

CANADA

Peace

Arctic Circle

Hudson Bay

Canadian

Arctic Coastal Plain

Shield

Mountains

Basins

Mountains

Saskatchewan

Interior

Plains

L. Superior

St. Lawrence

Mt. Rainier 14,410 ft.

Columbia

and

Missouri

L. Michigan

L. Huron

L. Erie

L. Ont.

Appalachian Highlands

PACIFIC OCEAN

Valleys

Plateaus

Great

UNITED   STATES

Plains

Platte

Ohio

Piedmont

ATLANTIC OCEAN

Mt. Whitney 14,494 ft.

Arkansas

Colorado

Interior Highlands

Mississippi

Gulf-Atlantic Coastal Plain

Rio Grande

Tropic of Cancer

MEXICO

GULF OF MEXICO

0   400   800   1200   KILOMETERS
0   200   400   600   800   MILES

PLATE 1

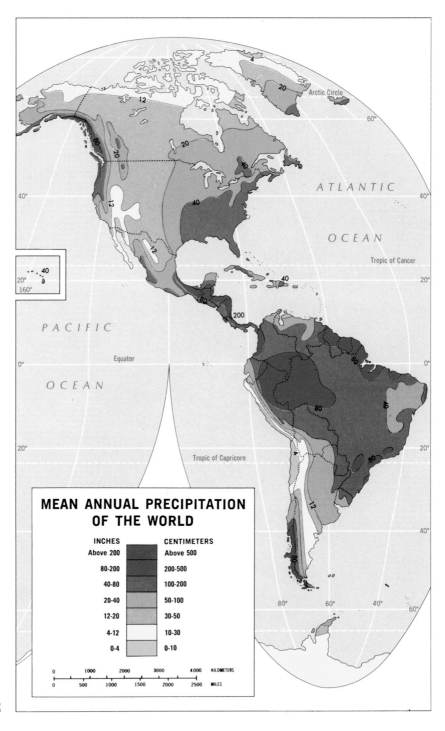

MEAN ANNUAL PRECIPITATION
OF THE WORLD

INCHES		CENTIMETERS
Above 200		Above 500
80-200		200-500
40-80		100-200
20-40		50-100
12-20		30-50
4-12		10-30
0-4		0-10

PLATE 2

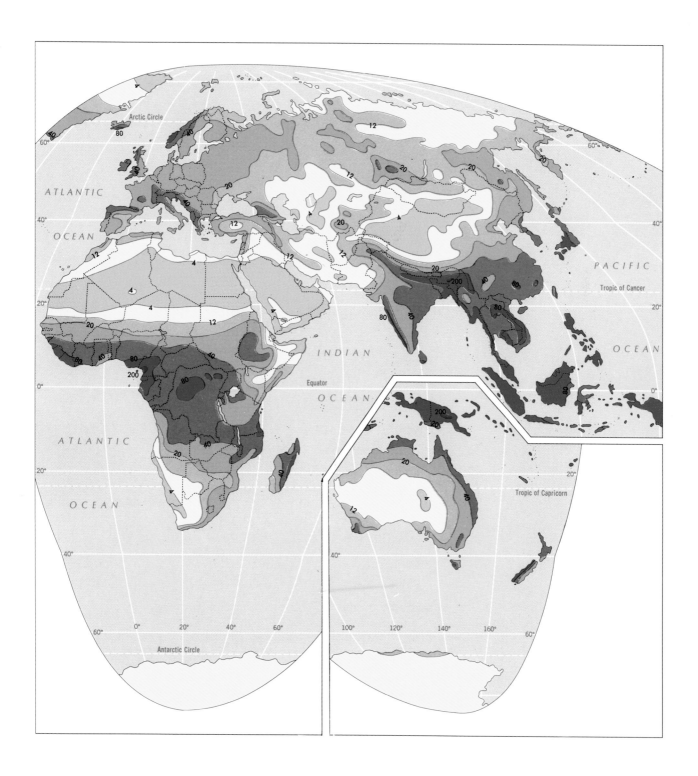

ATLANTIC

OCEAN

Arctic Circle

60°

40°

20°

0°   Equator

20°

ATLANTIC

OCEAN

40°

60°

Antarctic Circle

0°   20°   40°   60°

INDIAN

OCEAN

PACIFIC

Tropic of Cancer

OCEAN

20°

40°

60°

Tropic of Capricorn

40°

60°

100°   120°   140°   160°

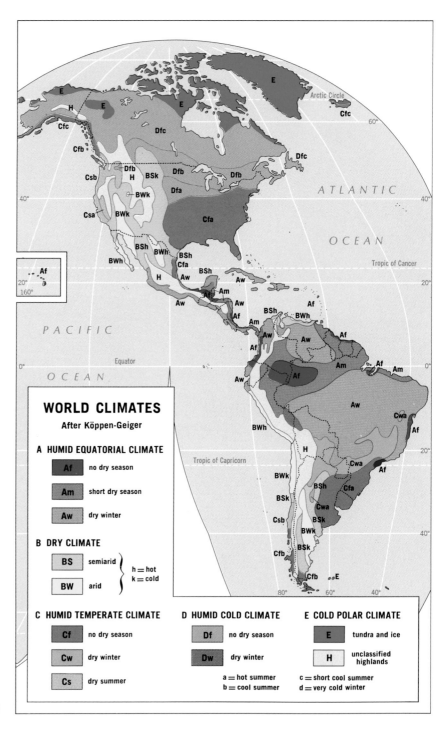

WORLD CLIMATES

After Köppen-Geiger

A HUMID EQUATORIAL CLIMATE

Af	no dry season
Am	short dry season
Aw	dry winter

B DRY CLIMATE

| BS | semiarid | } h = hot |
| BW | arid | } k = cold |

C HUMID TEMPERATE CLIMATE

Cf	no dry season
Cw	dry winter
Cs	dry summer

D HUMID COLD CLIMATE

| Df | no dry season |
| Dw | dry winter |

a = hot summer
b = cool summer

E COLD POLAR CLIMATE

| E | tundra and ice |
| H | unclassified highlands |

c = short cool summer
d = very cold winter

PLATE 3

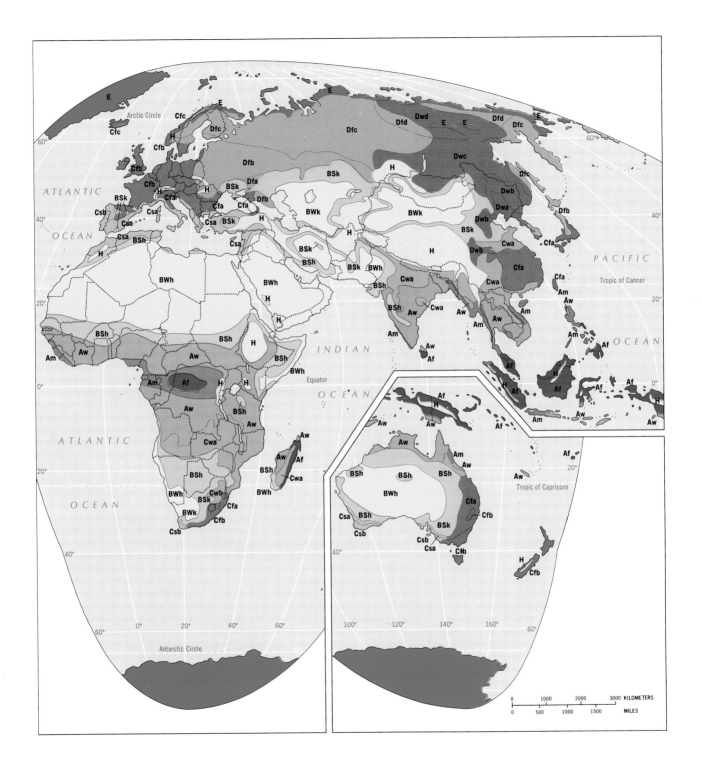

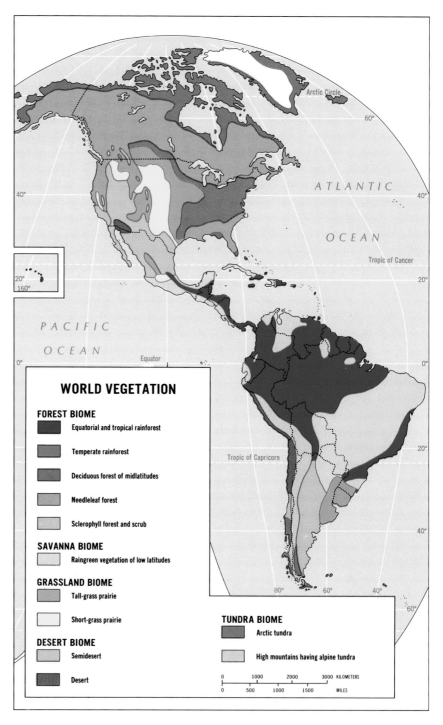

WORLD VEGETATION

FOREST BIOME

Equatorial and tropical rainforest

Temperate rainforest

Deciduous forest of midlatitudes

Needleleaf forest

Sclerophyll forest and scrub

SAVANNA BIOME

Raingreen vegetation of low latitudes

GRASSLAND BIOME

Tall-grass prairie

Short-grass prairie

DESERT BIOME

Semidesert

Desert

TUNDRA BIOME

Arctic tundra

High mountains having alpine tundra

0	1000	2000	3000 KILOMETERS
0	500	1000	1500	MILES

ATLANTIC

OCEAN

Tropic of Cancer

PACIFIC

OCEAN

Equator

Tropic of Capricorn

Arctic Circle

PLATE 4

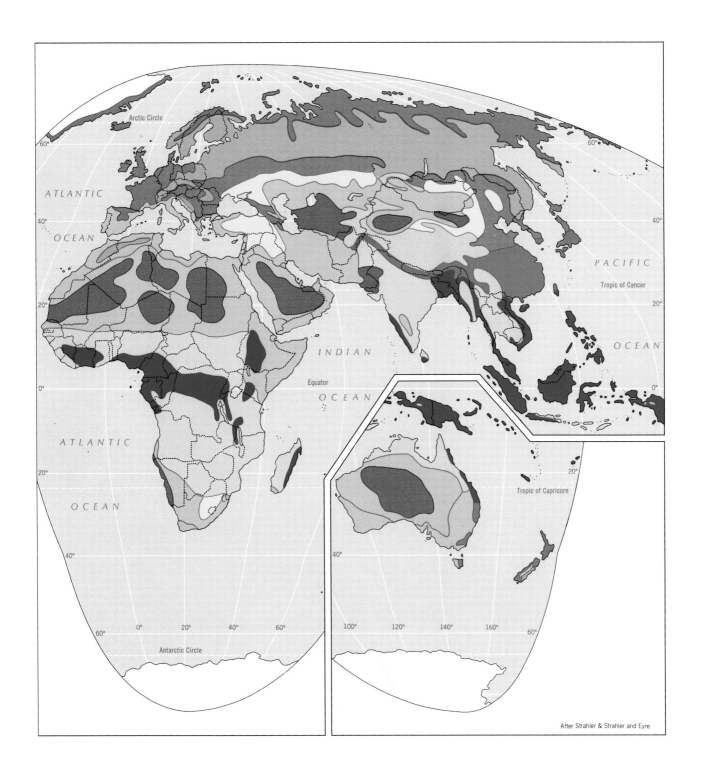

After Strahler & Strahler and Eyre

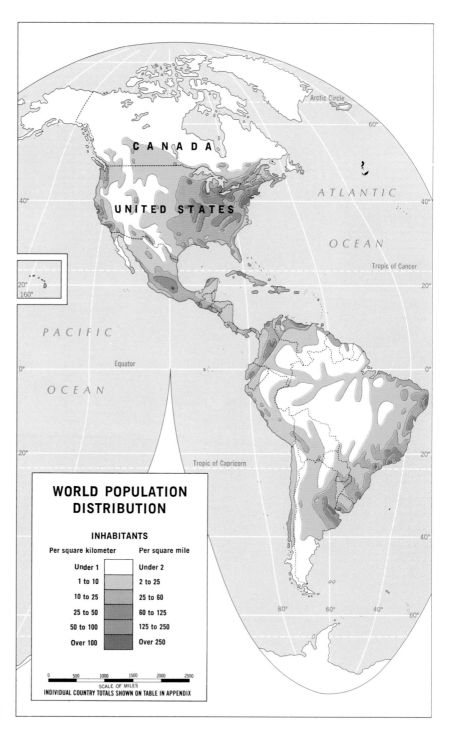

WORLD POPULATION DISTRIBUTION

INHABITANTS

Per square kilometer	Per square mile
Under 1	Under 2
1 to 10	2 to 25
10 to 25	25 to 60
25 to 50	60 to 125
50 to 100	125 to 250
Over 100	Over 250

0   500   1000   1500   2000   2500
SCALE OF MILES
INDIVIDUAL COUNTRY TOTALS SHOWN ON TABLE IN APPENDIX

PLATE 5

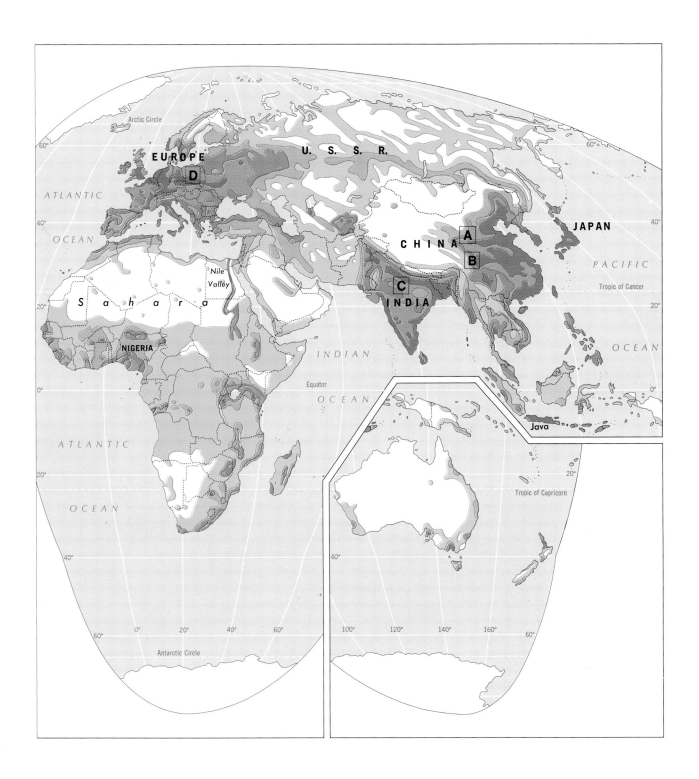

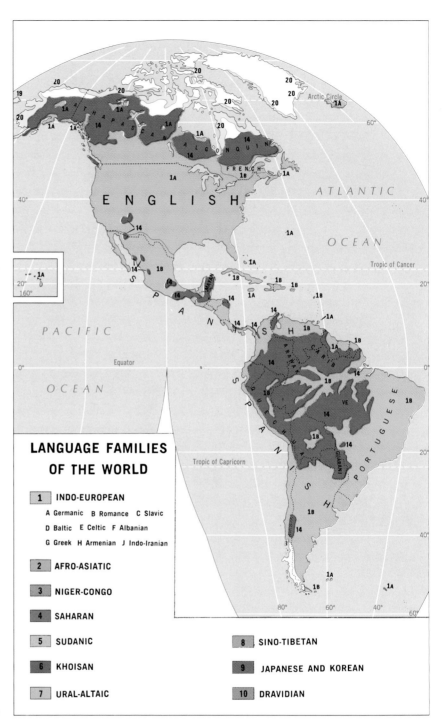

LANGUAGE FAMILIES
OF THE WORLD

1	INDO-EUROPEAN

A Germanic B Romance C Slavic

D Baltic E Celtic F Albanian

G Greek H Armenian J Indo-Iranian

2	AFRO-ASIATIC
3	NIGER-CONGO
4	SAHARAN
5	SUDANIC
6	KHOISAN
7	URAL-ALTAIC

8	SINO-TIBETAN
9	JAPANESE AND KOREAN
10	DRAVIDIAN

PLATE 6

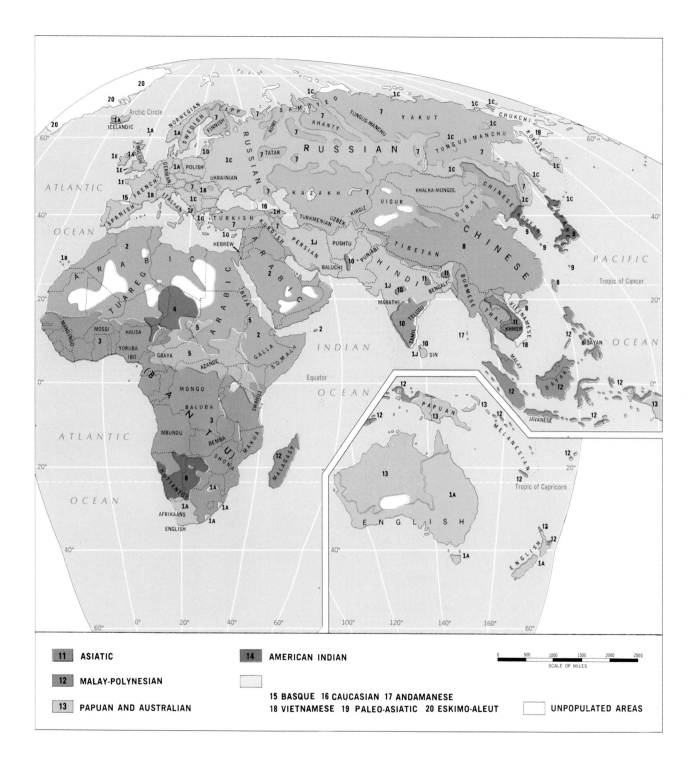

11 ASIATIC	**14** AMERICAN INDIAN	
12 MALAY-POLYNESIAN		
13 PAPUAN AND AUSTRALIAN	**15** BASQUE **16** CAUCASIAN **17** ANDAMANESE	
	18 VIETNAMESE **19** PALEO-ASIATIC **20** ESKIMO-ALEUT	UNPOPULATED AREAS

SCALE OF MILES

0   500   1000   1500   2000   2500

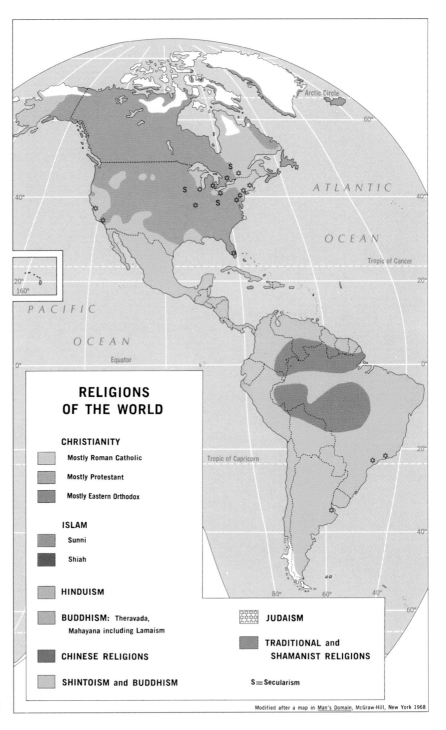

RELIGIONS OF THE WORLD

CHRISTIANITY

Mostly Roman Catholic

Mostly Protestant

Mostly Eastern Orthodox

ISLAM

Sunni

Shiah

HINDUISM

BUDDHISM: Theravada, Mahayana including Lamaism

CHINESE RELIGIONS

SHINTOISM and BUDDHISM

JUDAISM

TRADITIONAL and SHAMANIST RELIGIONS

S = Secularism

Modified after a map in Man's Domain, McGraw-Hill, New York 1968

PLATE 7

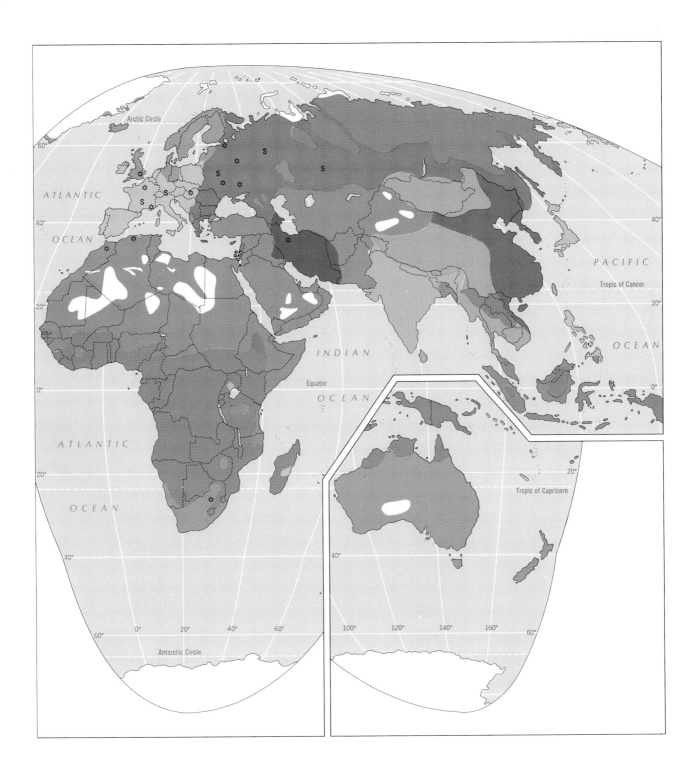

WORLD AGRICULTURE

1 Dairying

2 Fruit, Truck and Specialized Crops

3 Mixed Livestock and Crop Farming

4 Grain Farming

5 Subsistence Crop and Livestock Farming

6 Mediterranean Agriculture

7 Diversified Tropical Agriculture
  -chiefly plantation

8 Intensive Subsistence Farming
  -chiefly rice

9 Intensive Subsistence Farming
  -other crops

10 Rudimental Sedentary Cultivation

11 Shifting Cultivation

12 Livestock Ranching

13 Nomadic and Semi-Nomadic Herding

Nonagricultural Areas

PLATE 8

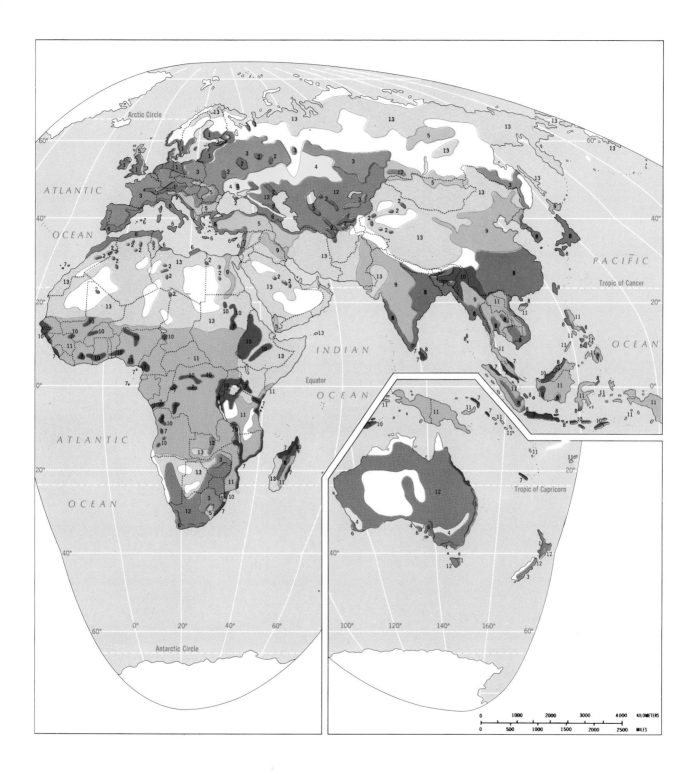

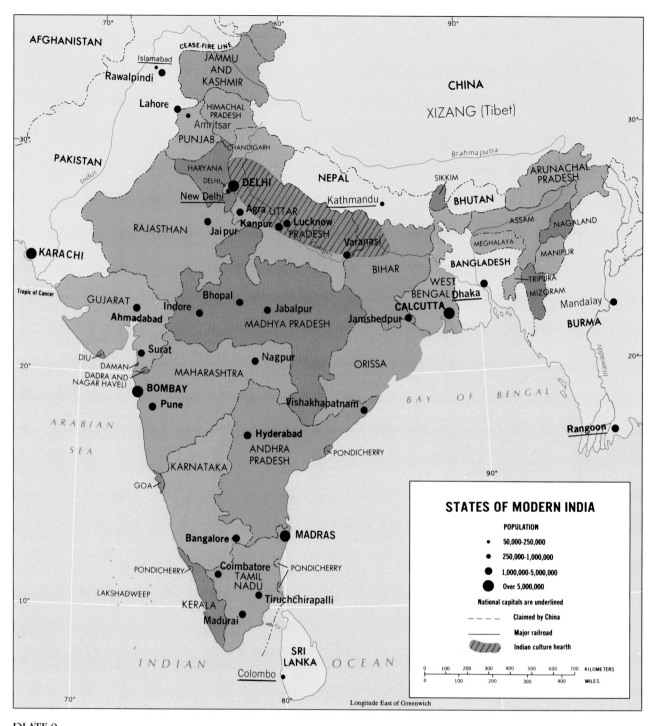

AFGHANISTAN

CEASE-FIRE LINE

Islamabad

Rawalpindi

JAMMU
AND
KASHMIR

CHINA

XIZANG (Tibet)

Lahore

HIMACHAL
PRADESH

Amritsar

PUNJAB

CHANDIGARH

HARYANA

DELHI

DELHI

New Delhi

Brahmaputra

NEPAL

Kathmandu

SIKKIM

BHUTAN

ARUNACHAL
PRADESH

PAKISTAN

Indus

Agra

UTTAR

Kanpur

Lucknow

PRADESH

ASSAM

NAGALAND

RAJASTHAN

Jaipur

Varanasi

MEGHALAYA

MANIPUR

KARACHI

Ganges

BIHAR

BANGLADESH

TRIPURA

MIZORAM

Tropic of Cancer

GUJARAT

Bhopal

Indore

Jabalpur

WEST
BENGAL

Dhaka

Mandalay

CALCUTTA

Ahmadabad

MADHYA PRADESH

Jamshedpur

BURMA

DIU

Surat

DAMAN

DADRA AND
NAGAR HAVELI

MAHARASHTRA

Nagpur

ORISSA

Irrawaddy

BOMBAY

Pune

Vishakhapatnam

BAY   OF   BENGAL

ARABIAN

SEA

Hyderabad

ANDHRA
PRADESH

PONDICHERRY

KARNATAKA

GOA

Rangoon

Bangalore

MADRAS

PONDICHERRY

Coimbatore

PONDICHERRY

TAMIL

NADU

LAKSHADWEEP

KERALA

Tiruchchirapalli

Madurai

INDIAN

OCEAN

SRI
LANKA

Colombo

Longitude East of Greenwich

STATES OF MODERN INDIA

POPULATION

•   50,000-250,000

●   250,000-1,000,000

⬤   1,000,000-5,000,000

⬤   Over 5,000,000

National capitals are underlined

– – –   Claimed by China

———   Major railroad

▨   Indian culture hearth

0   100   200   300   400   500   600   700   KILOMETERS

0   100   200   300   400   MILES

PLATE 9

The
Geography
of Population

On July 7, 1986, the world's major news networks carried a brief but significant report: at some time on that Monday, the earth's human population reached the 5 billion mark as the 5000-millionth person was born somewhere on the globe. The story rated small entries on the inside pages of major newspapers.

Many teachers of population geography remembered the day in 1975, just 11 years earlier, when the global population reached 4 billion. It was headline, front-page news. There were public debates about the future of people and resources on this earth. An organization called ZPG (for zero population growth) organized rallies on college campuses. Public speakers issuing dire warnings drew huge crowds. The impact of growing human numbers on environment and ecology was a hot topic. There were predictions about farmlands turning into desert, about millions facing malnutrition and starvation. If 5 billion in the 1980s, 6 billion in the 1990s . . . then 8 billion by 2010?

Now the fervor has largely faded. The population spiral is racing from 5 billion to 6 billion. But the headlines focus elsewhere: on El Salvador, on the contras, on South Africa. Only when television viewers' senses were touched by searing pictures of Ethiopian refugees, emaciated and helpless, was there a brief revival of the concern and action of a decade earlier. Some compassionate artists organized concerts in the hope that the proceeds would somehow lessen the plight of millions. It amounted to less than the proverbial drop in the bucket. Our attention may have been diverted, but the predictions of the 1970s have not been far off the mark. July 7, 1986 should have been a day of serious reflection on the capacity of our Earth not just to enable billions to survive, but to do so with a reasonable quality and length of life.

As we will see time and again in this chapter, quality and length of life (life expectancy in statistical terms) are among the circumstances influenced by geography. Because we must discuss these and other matters in global terms, it is easy to lose sight of the huge contrasts between regions, within regions—even within countries. Yes, there is hunger, in the form of malnutrition by world standards, in America. Population problems in western India differ considerably from those of eastern India. Diets in southern China are quite different from those of north China. Comfort, balanced nutrition, good health, and other advantages and privileges have much to do with *where* one is born or is able to live in this world. But we must begin with a global view and leave the regional for later.

•　　　　World Population　　　　•

During the late 1980s, the population of the world was growing by about 80 million every year, about 7 million per month, 220,000 per day, or nearly 9000 every hour. Most of the newborn arrive in areas of the world where health-care facilities are inadequate, food is insufficient, diets are deficient, and other conditions are inimical to their well-being or even their survival.

Why have such circumstances not contributed to a sharp decline in the growth of population? If people are faced with starvation in parts of Asia and areas of Africa and if others are malnourished, even in the Western Hemisphere (e.g., in Haiti), why does the population explosion of the twentieth century continue? Will the rate of population growth eventually decline? Can new sources of food be found and developed? These and other questions will be addressed in the following pages.

In the United States and Canada—in the Western world generally—the problems confronting billions of people elsewhere seem remote and unfamiliar. In Europe, the memories of the hunger that afflicted millions during the Second World War are fading rapidly. In North America, there are still substantial numbers of people (especially among the poor and the elderly) whose diets are unsatisfactory. However, in this part of the world, no one need die of hunger, and a day without food or even the absence of a meal is not a normal experience. Certainly population is increasing in North America as well as in less developed countries, but, as will be noted later, the rate of growth is far below that of Africa, Middle and South America, and Asia. The impact of population growth and relocation in the United States and Canada may be felt in deteriorating urban living conditions, worse pollution, larger traffic jams, longer waiting lines—but while we experience some slight discomfort, we do not confront disaster.

Nevertheless, our future is inextricably tied to those other parts of the world where population growth is truly explosive. No country can exist in isolation, no matter how varied and rich its resources. Peoples of Africa and Asia depend heavily on shipments of U.S. grain, just as the United States depends a lot on the import of Middle East oil. The United States does not send its wheat to India only as a humanitarian gesture, it is also a business matter—commercial and political. The specter of world disorder, arising from political pressures that result from widespread hunger and hopelessness, hangs heavily over the comparative comfort of Western nations. In the fall of 1974, after parts of Ethiopia had been afflicted by years of hunger and starvation, the direct result was a political upheaval that terminated more than 40 years of stable (if not progressive) rule under Emperor Haile Selassie. What could not be achieved by attempted coups, religious conflicts, and regional insurrections was accomplished by the desperation of famine. On a larger scale, such a sequence of events could have global implications. Western grain shipments are intended in large measure to diffuse this kind of pressure.

Small wonder, then, that Western nations are in the forefront of the search for means to reduce the rate of growth of the world's population. In the long run, the West has as much to gain from such a slowdown as do the populations of such hungry countries as India and Bangladesh.

There is, however, another side of this issue to contemplate. Hunger is not only and inevitably a matter of human numbers, it is also related to the social and political systems of the countries affected. Before the middle of this century, China was a country of fewer than 350 million people where famines were frequent and starvation was a constant threat. True,

China had been ravaged by war for many years—but even during earlier periods of comparative stability, the specter of starvation always loomed. Today, China has more than 1 billion people and widespread famines no longer occur. Certainly there is local hunger in China and diets generally are not as well balanced as would be desirable. However, the population has reasonable assurance that food will be available on a regular basis. This was accomplished by a sweeping reconstruction of China's rural economy, a reorganization that was part of the Communist program after Mao Zedong took control in 1949. Many aspects of that reorganization were draconian and caused terrible hardships, often unnecessarily. The overall result of China's reconstruction, however, was the defeat of famine—even while the population more than doubled.

What was possible in China is not feasible in many other hunger-afflicted countries. From South Asia to East Africa, governments have neither the power nor the will to make the drastic changes that were forced on postrevolutionary China. The overthrow of Ethiopia's monarchy did not produce a government capable of solving the country's food problems. Physical as well as human factors (prolonged drought, corruption, armed struggles) kept Ethiopia mired in its desperate state. The problem of population and of people's well-being has many sides.

• World Population Distribution •

Before focusing on the problems of the growth process itself, we should become familiar with the distribution of the earth's more than 5 billion people. The human condition (as we noted earlier) is very much a matter of geography. On the other hand, we should not lose sight of the interrelatedness of peoples and regions of the world. Problems in one area can have repercussions in other, distant lands.

Perhaps the most remarkable feature of the world population map (Plate 5) is the persistence of open,

201

even empty, spaces in the most populous lands. Parts of western China remain as empty as the Sahara. In India, where crowding is such a serious problem, you can still travel through miles of sparsely peopled territory. The map depicts population distribution through levels of shading to indicate density per unit area: the darker the colors, the larger is the number of people clustered in an average unit area. The most extensive area of such dark shading lies in East Asia, principally in China, but it extends into Korea and Vietnam and leapfrogs to Japan. More than a quarter of the entire world's population is concentrated in this East Asian cluster, some 1.1 billion in China alone.

Observation of the dimensions of the East Asian population cluster reveals that it adjoins the Pacific Ocean from South Korea to Vietnam and that the number of people per unit area tends to decline from this coastal zone toward the interior. Also visible are several ribbonlike extensions of dense population, penetrating the interior (Plate 5, **A** and **B**). A reference to a physical map of East Asia tells us that these extensions represent populations clustered in the basins and lowlands of China's major rivers. This serves to remind us that the great majority of the people in East Asia are farmers, not city dwellers. True, there are great cities in China, and some of them such as Shanghai an Beijing (Peking) rank among the largest in the world. However, the total population of these and other cities is far outnumbered by the farmers—those who need the river valleys' soils, the life-giving rains, and the moderate temperatures to produce crops of wheat and rice to feed not only themselves, but also those in the cities and towns.

The second major concentration of world population also lies in Asia, and it displays many similarities to that of East Asia. At the heart of this South Asia cluster lies India, but the concentration also extends into Pakistan and Bangladesh as well as onto the island of Sri Lanka. Again, note the riverine and coastal orientation of the most densely inhabited zones and the fingerlike extension of dense population in northern India (Plate 5, **C**). This is one of the great concentrations of people on earth: the plain of the Ganges River.

The South Asia population cluster numbers as many as 998 million people. Our map reflects the influence of physical barriers: the Himalaya Mountains limit the Ganges lowland and the desert takes over west of the Indus River Valley in Pakistan. This is a confined region with a population that continues to grow rapidly. The capacity of the region to support this population under present sociopolitical circumstances has, by all estimates, already been exceeded. As in East Asia, the overwhelming majority of the people here are farmers, but in South Asia, the pressure on the land is even greater. In Bangladesh, some 102 million people are crowded into an area about the size of Iowa. Nearly all of these people are farmers, and even fertile Bangladesh has areas that cannot sustain as many people as others. Over large parts of Bangladesh, the rural population density is between three and five *thousand* people per square mile. To compare: the 1985 population of Iowa was 2.9 million and just 40 percent (by the census) lived on the land rather than in cities and towns. The rural density in Iowa was 30 people per square mile.

Further inspection of Plate 5 reveals that the third-ranking population cluster also lies in Eurasia—and at the opposite end from China. An axis of dense population extends from the British Isles into Soviet Russia and includes large parts of West and East Germany, Poland, and the western U.S.S.R.; it also incorporates the Netherlands and Belgium, parts of France, and northern Italy. This European cluster counts about 700 million inhabitants, which puts it in a class with the South Asia concentration—but there the similarity ends. A comparison of the population and physical maps indicates that in Europe terrain and environment appear to have less to do with population distribution than in the two Asian cases. See, for example, the lengthy extension **D** on Plate 5 that protrudes far into the U.S.S.R. Unlike the Asian extensions, which reflect fertile river valleys, the European population axis relates to the orientation of Europe's coalfields, the power resources that fired the Industrial Revolution. If you look more closely at the physical map, you will note that comparatively dense population occurs even in rather mountainous, rugged country, for example, along the boundary zone between Czechoslovakia and Poland. In Asia, there is much more correspondence between coastal and river lowlands and high population density than there generally is in Europe.

202

Another contrast lies in the number of Europeans who live in cities and towns. Far more than in Asia, the European population cluster is constituted by numerous cities and towns, many of them products of the Industrial Revolution. In West Germany, 94 percent of the people live in such urban places; in the United Kingdom the number is 76 percent; and in France, it is 73 percent. With so many people concentrated in the cities, the rural countryside is more open and sparsely populated than in East and South Asia where about 30 percent of the people reside in cities and towns.

The three world population concentrations discussed (East Asia, South Asia, and Europe) account for over 3.4 billion of the world's approximately 5.2 billion people. Nowhere else on the globe is there a population cluster with dimensions that are even half of these. Look at the dimensions of the landmasses in Plate 5 and consider that the populations of South America, Africa, and Australia combined barely exceed that of India alone. In fact, the next-ranking cluster, comprising the east–central United States and southeastern Canada, is only about one-quarter the size of the smallest of the Eurasian concentrations. As Plate 5 shows, this region does not have the large, contiguous high-density zones of Europe or East and South Asia.

BEYOND EURASIA

The North American population cluster displays European characteristics, and it even outdoes Europe in some respects. Like the European region, much of the population is concentrated in several major cities, whereas the rural areas remain relatively sparsely populated. The major focus of the North America cluster lies in the urban complex along the eastern seaboard from Boston to Washington, which includes New York, Philadelphia, and Baltimore. This great urban agglomeration is called **megalopolis** by urban geographers, who predict that it is only a matter of time before the whole area coalesces into an enormous megacity. However, there are other urban foci in this region: Chicago lies at the heart of one, and Detroit and Cleveland anchor a second. If you study Plate 5 carefully, you will note other prominent North

American cities standing out as small areas of high-density population, including Pittsburgh, St. Louis, and Minneapolis–St. Paul.

Still further examination of Plate 5 leads us to recognize substantial population clusters in Southeast Asia. It is appropriate to describe these as discrete clusters, for the map confirms that they are actually a set of nuclei rather than a contiguous population concentration. Largest among these nuclei is the Indonesian island of Djawa (Java), with more than 100 million inhabitants. Elsewhere in the region, populations cluster in the lowlands of major rivers, for example, the Mekong. Neither these river valleys nor the rural surroundings of the cities have population concentrations comparable to those of either China to the north or India to the West, and under normal circumstances Southeast Asia is able to export rice to its more hungry neighbors. Decades of strife, however, have disrupted the region to such a degree that its productive potential has not been attained.

South America, Africa, and Australia do not sustain population concentrations comparable to those we have considered. Africa's 551 million inhabitants cluster in above-average densities in West Africa (where Nigeria has a population of some 91 million) and in a zone in the east that extends from Ethiopia to South Africa. Only in North Africa is there an agglomeration comparable to the crowded riverine plains of Asia: the Nile Valley and Delta with its over 45 million residents. Importantly, it is the pattern of the Nile agglomeration—not the dimensions—that resembles Asia. As in East and South Asia, the Nile's valley and delta teem with farmers who cultivate every foot of the rich and fertile soil. However, the Nile's gift is a miniature compared to its Asian equivalents. The lowlands of the Ganges, Chang Jiang (Yangtze), and Huang He (Yellow) contain many times the number of inhabitants than the number who manage to eke out a living along the Nile.

The large light-shaded spaces in South America and Australia and the peripheral distribution of the modest populations of these continents suggest that here remains space for the world's huge numbers. Indeed, South America could probably sustain more than its present 271 million if reforms in land ownership and use could take place. At present, although

South America as a whole is not one of the world's hungry regions, there are, nevertheless, areas where poverty and malnutrition occur. Northeast Brazil is especially troubled economically, and certainly South America's capacity will be increasingly tested. In several countries (including Brazil, the largest), population growth is extremely rapid. Australia's environmental limitations hardly qualify that continent as a relief valve for Asia's millions if agriculture is to be the source of support.

This raises an issue that is crucial in any study of population density and the capacity of a country to support its people: the level of technology that country has reached. You will note that Japan, a comparatively small island country (small compared to Australia, at least) has a population in excess of 120 million. Its population density is at least as great as that of parts of China and India, but its farmlands are quite limited, not only by the size of the country, but also by its mountainous character. What makes so large a population cluster in Japan possible—and, indeed, well-fed and prosperous—is Japan's technological prowess, its industrial capacity, and its money-producing exports. Japan imports raw materials from all over the world, converts these materials into finished products in its factories, and exports the products to most parts of the globe. With the income from these exports, Japan can buy the food from world markets that it cannot produce at home. Thus, it is not enough to say that a country cannot support a certain population. We should qualify this by saying that under present conditions of technology and political economy, it can or cannot support a given population. True, Australia could not find a place for tens of millions of Chinese farmers. However, if tens of millions of Japanese came to Australia with their skills and technologies and factories and international connections, Australia would be quite capable of accommodating them—and many more.

204

Skyscrapers tower over substandard housing in Caracas, Venezuela. The world's mushrooming cities cannot accommodate the stream of in-migrants, and urban slum development is a global problem.

DISTRIBUTION AND DENSITY

In our discussion of world population distribution, we have used **population density** figures as the measure. On the face of it, this is a good index to use when comparing world areas. From each country's national census we obtain the total population. The country's area is a matter of record, then all that is required is to calculate the number of people per unit area (per square mile or kilometer) in the world's countries.

However, we already are aware—and Plate 5 reminds us—that the people of any country are not evenly distributed over the national territory. Thus, the result of any calculation may be misleading. The United States, with 3,678,900 sq miles (9,528,400 sq km), had a 1985 population of 239 million (Table 6.1) and its average population density would, therefore, be figured as 66 per sq mile (26 per sq km). This is the **arithmetic density** of the United States. Compare this figure to the enormously high averages for Asian countries such as Bangladesh (1825 per sq mile; 705 per sq km), Japan (840 per sq mile; 325 per sq km), India (600 per sq mile; 232 per sq km), and others. These averages do not, however, take account of the internal clustering of people within these countries. In the case of the United States, the average fails to reflect the emptiness of Alaska, the sparseness of population in much of the West, the concentration of people in the eastern cities. As an even clearer example of the misleading nature of average density figures, consider the case of Egypt, with its 48 million inhabitants. Nearly all these people live in the Nile Valley and Delta. However, Egypt's total area is 386,900 sq miles (about 1 million sq km), so the average density is about 125 per sq mile (48 per sq km). Obviously, the figure is rather meaningless when all those people are crowded onto the Nile's irrigable and cultivable soils and the rest of the country is desert! It has been estimated that 98 percent of Egypt's population occupies just 3 percent of the country's total area.

Can we arrive at a more meaningful index of population density in individual countries? Yes—by relating the total population to the amount of cultivated land in the country rather than to its whole area. In those overpopulated Asian countries, after all, it is the productive land that matters, not the dry areas or the inhospitable mountains. Instead of the arithmetic density, we calculate the **physiologic density**, the number of people per unit area of productive land. In the case of Egypt, while the arithmetic density is 125 per sq mile (48 per sq km) the physiologic density is more than 6200 people per cultivable sq mile (2400 per sq km). In Japan, the physiologic density is even higher, 6466 per sq mile (2400 per sq km). In Europe, the country with the highest physiologic density is the Netherlands, exhibiting the lower figure of 4380 per sq mile (1690 per sq km).

Even these calculations are subject to error, however. In every country there are farmlands of different productivity. Some lands produce high crop yields (more than once a year, in many Asian countries), whereas others are marginal or can sustain only livestock. All these variable levels of production are treated equally in our calculation of physiologic density. Still, the measure is much more useful than the arithmetic density index. Table 6.2 provides an indication of the usefulness of the concept of physiologic density. Note, for example, that the United States and Colombia have almost the same arithmetic density but Columbia has a physiologic density more than four times that of the United States. Egypt's arithmetic density of 125 per sq mile does not look especially high, but its physiologic density tells us what the real situation is in terms of cultivable land and the growing pressure on it. Argentina is the only country on the list with both relatively low arithmetic and physiologic densities, whereas Iran is one of those countries where the arithmetic density looks quite low—near that of the United States—but the physiologic density is rather high. Also note that India's physiologic density, although high, is still moderate compared to those of neighboring Bangladesh and island Japan.

Arithmetic and physiologic density measures are only two indexes used by population geographers to interpret patterns of population distribution. Another measure is the **agricultural population density**. This measure differs from the physiologic density in that the calculations exclude all urban residents, making possible a more nearly accurate estimate of the pressure of people on the rural areas. In a country where a few large cities or towns exist (e.g., in Bangladesh),

205

TABLE 6.1 Population Data for the World's States

STATE	AREA 1000 SQ. MILES	AREA 1000 SQ. KMS	POPULATION (MILLIONS) 1984	POPULATION (MILLIONS) 1988	POPULATION (MILLIONS) 2000	LIFE EXPECTANCY 1986 (YEARS)	ANNUAL INCREASE 1988 (%)	POPULATION DENSITY (NO./SQ MILE) 1988	POPULATION DENSITY (NO./SQ KM) 1988
East Asia	4,394.2	11,380.8	1,132.1	1,165.1	1,301.2	66	1.0	265	102
China	3,691.5	9,560.9	1,044.0	1,071.0	1,190.0	64	1.0	290	112
Hong Kong	0.4	1.0	5.4	5.8	6.7	75	1.0	14,146	5,808
Japan	145.7	377.4	120.0	123.2	128.1	77	0.7	846	326
North Korea	46.5	120.4	19.6	21.4	27.3	65	2.3	460	178
South Korea	38.0	98.4	42.0	44.7	52.0	66	1.6	1,176	454
Mongolia	604.3	1,565.1	1.9	2.0	2.8	62	2.5	3	1
Taiwan	13.9	36.0	19.2	20.2	22.4	73	1.5	1,453	561
Southeast Asia	1,735.4	4,494.6	394.6	425.8	531.5	58	2.2	245	95
Brunei	2.2	5.7	0.2	0.2	0.3	71	2.6	91	35
Burma	261.2	676.5	37.0	39.2	48.2	58	2.0	150	58
Indonesia	741.0	1,919.2	164.3	175.5	219.9	55	2.1	237	91
Kampuchea	69.9	181.0	6.1	6.7	8.5	43	2.3	96	37
Laos	91.4	236.7	3.7	3.9	4.9	44	2.3	43	16
Malaysia	128.4	332.6	15.3	16.6	20.6	67	2.4	129	50
Philippines	115.8	299.9	54.5	61.0	75.5	62	2.5	527	203
Singapore	0.2	0.5	2.5	2.7	2.9	71	1.1	13,471	5,313
Thailand	198.1	513.1	51.7	54.9	65.4	63	2.0	277	107
Vietnam	127.2	329.4	58.3	65.1	85.3	59	2.5	512	198
South Asia	1,710.3	4,429.7	977.4	1,075.6	1,357.9	52	2.4	629	243
Bangladesh	55.6	144.0	99.6	109.7	144.7	48	2.7	1,973	762
Bhutan	18.1	46.9	1.4	1.5	1.9	46	2.0	83	32
India	1,237.1	3,204.1	746.4	821.1	1,017.0	53	2.3	664	256
Maldives	0.1	0.3	0.2	0.2	0.3	52	3.1	1,554	601
Nepal	54.4	140.9	16.6	18.2	24.6	46	2.3	335	129
Pakistan	319.9	828.5	97.3	107.6	148.7	50	2.8	336	130
Sri Lanka	25.1	65.0	15.9	17.3	20.7	68	2.0	689	266
Europe	1,879.6	4,867.6	489.6	495.7	507.5	73	0.3	264	101
Albania	11.1	28.7	2.9	3.2	3.8	71	2.0	288	112
Austria	32.4	83.8	7.6	7.6	7.5	73	0.0	235	91
Belgium	11.8	30.5	9.9	10.0	9.8	73	0.1	847	328
Bulgaria	42.8	110.9	9.0	9.1	9.3	72	0.2	213	82

Czechoslovakia	49.4	127.9	15.5	15.6	16.5	71	0.3	316	122
Denmark	16.6	43.1	5.1	5.0	5.0	75	-0.1	301	116
Finland	130.2	337.0	4.9	5.0	5.0	74	0.4	38	15
France	211.2	547.0	54.9	55.8	57.0	75	0.4	264	102
East Germany	41.8	108.2	16.7	16.7	16.9	72	0.0	400	154
West Germany	96.0	248.6	61.4	60.5	58.6	74	-0.2	630	243
Greece	51.0	131.9	10.0	10.1	10.4	74	0.4	198	77
Hungary	35.9	93.0	10.7	10.3	10.6	70	-0.2	287	111
Iceland	39.8	103.0	0.2	0.2	0.3	77	1.0	5	2
Ireland	27.1	70.3	3.6	3.7	4.1	72	0.9	137	53
Italy	116.3	301.2	56.4	57.4	58.1	74	0.1	494	191
Luxembourg	1.0	2.6	0.4	0.4	0.4	73	0.0	400	154
Netherlands	15.9	41.2	14.4	14.7	15.0	76	0.4	925	357
Norway	125.1	324.2	4.1	4.2	4.2	76	0.2	34	13
Poland	120.7	312.7	36.9	38.2	40.7	71	0.9	316	122
Portugal	34.3	88.8	10.0	10.3	10.8	73	0.5	300	116
Romania	91.7	237.5	22.8	23.0	24.5	71	0.4	251	97
Spain	194.9	504.8	38.4	39.2	41.9	76	0.5	201	78
Sweden	173.7	450.0	8.3	8.4	8.1	77	0.0	48	19
Switzerland	15.9	41.3	6.5	6.5	6.5	76	0.2	409	157
United Kingdom	94.2	244.0	56.0	56.8	57.4	74	0.2	603	233
Yugoslavia	98.8	255.8	23.0	23.5	25.1	70	0.7	238	92
Soviet Union	8,600.4	22,275.0	274.5	285.0	311.0	69	0.9	33	13
North America	7,515.3	19,464.7	261.7	270.6	296.7	75	0.7	36	14
Bahamas	5.4	14.0	0.2	0.2	0.3	69	1.9	37	14
Canada	3,831.0	9,922.3	25.1	26.0	28.4	75	0.8	7	3
United States	3,678.9	9,528.4	236.4	244.4	268.0	75	0.7	66	26
Middle America	1,047.0	2,712.0	132.2	145.0	188.8	66	2.4	139	53
Barbados	0.2	0.5	0.3	0.3	0.3	73	0.9	1,523	588
Belize	8.9	23.1	0.2	0.3	0.3	70	2.5	34	13
Costa Rica	19.7	51.0	2.7	2.8	3.6	73	2.6	142	55
Cuba	44.2	114.5	10.0	10.5	11.6	73	1.1	238	92
Dominican Republic	18.7	48.7	6.4	6.7	8.4	63	2.5	358	138
El Salvador	8.1	21.0	4.8	5.3	7.5	65	2.4	654	252
Guatemala	42.0	108.8	8.0	9.1	13.1	59	3.1	217	84

207

STATE	AREA		POPULATION (MILLIONS)			LIFE EXPECTANCY 1986 (YEARS)	ANNUAL INCREASE 1988 (%)	POPULATION DENSITY	
	1000 SQ. MILES	1000 SQ. KMS	1984	1988	2000			(NO./SQ MILE) 1988	(NO./SQ KM) 1988
Haiti	10.7	27.7	5.8	6.2	8.4	53	2.3	579	224
Honduras*	43.3	112.1	4.4	4.9	6.8	60	3.2	113	44
Jamaica	4.2	10.9	2.3	2.4	2.8	73	1.8	571	220
Mexico	761.6	1,972.5	77.7	85.9	112.8	66	2.6	113	44
Nicaragua	50.2	130.0	2.9	3.5	5.2	60	3.4	70	27
Panama	29.8	77.2	2.1	2.4	2.9	71	2.1	81	31
Puerto Rico	3.4	8.8	3.3	3.4	3.6	74	1.4	999	386
Trinidad and Tobago	2.0	5.2	1.2	1.3	1.5	70	1.8	650	250
South America	6,839.6	17,714.5	264.5	291.8	372.0	65	2.3	43	16
Argentina	1,068.3	2,766.9	30.1	32.2	37.5	70	1.6	30	12
Bolivia	424.2	1,098.7	6.0	6.8	9.2	51	2.8	16	6
Brazil	3,286.5	8,512.0	134.4	150.0	194.7	63	2.3	46	18
Chile	292.1	756.5	11.7	12.7	14.8	70	1.6	43	17
Columbia	439.7	1,138.8	28.2	31.3	38.4	64	2.1	71	27
Ecuador	109.5	283.6	9.1	10.1	13.9	64	2.8	92	36
Guyana	83.0	215.0	0.8	0.9	0.9	68	2.2	11	4
Paraguay	157.0	406.6	3.6	4.3	6.0	65	2.8	27	11
Peru	496.2	1,285.2	19.7	21.2	28.0	59	2.5	43	16
Suriname	63.0	163.2	0.4	0.4	0.5	69	2.0	6	2
Uruguay	68.0	176.1	2.9	3.1	3.4	70	0.9	46	18
Venezuela	352.1	911.9	17.6	18.8	24.7	69	2.7	53	21
Subsaharan Africa	8,394.4	21,741.7	425.6	475.5	682.2	48	2.8	57	22
Angola	481.4	1,246.8	7.8	8.6	12.4	42	2.5	18	7
Benin	43.5	112.7	3.9	4.3	6.4	44	3.0	99	38
Botswana	231.8	600.4	1.0	1.2	1.8	56	3.3	5	2
Burkina Faso	105.9	274.3	6.7	7.5	10.5	45	2.6	71	27
Burundi	10.7	27.7	4.7	5.2	7.1	47	3.0	486	188
Cameroon	183.6	475.5	9.5	10.5	14.8	51	2.7	57	22
Central African Republic	240.5	622.9	2.6	2.9	4.1	43	2.8	12	5
Chad	495.8	1,284.1	5.1	5.4	7.6	43	2.0	11	4

Comoro Islands	0.8	2.1	0.5	0.6	0.7	50	3.1	750	286
Congo	132.0	341.9	1.7	1.9	2.6	47	2.6	14	6
Equatorial Guinea	10.8	28.0	0.3	0.4	0.5	44	2.3	37	14
Ethiopia	472.4	1,223.5	42.0	45.7	65.8	41	2.1	97	37
Gabon	103.3	267.5	1.0	1.2	1.6	49	1.6	12	4
Gambia	4.4	11.4	0.7	0.8	1.1	35	2.0	182	70
Ghana	92.1	238.5	13.8	14.5	20.5	54	3.4	157	61
Guinea	94.9	245.8	5.6	6.5	8.9	40	2.3	68	26
Guinea-Bissau	13.9	36.0	0.8	0.9	1.2	43	1.9	65	25
Ivory Coast	123.8	320.6	9.2	11.1	16.0	51	3.0	90	35
Kenya	225.0	582.8	19.4	22.8	35.8	53	4.2	101	39
Lesotho	11.7	30.3	1.5	1.7	2.3	49	2.5	145	56
Liberia	43.0	111.4	2.2	2.4	3.6	49	3.1	56	22
Madagascar	226.7	587.2	9.6	10.9	15.6	50	2.8	48	19
Malawi	45.7	118.4	6.8	7.8	11.7	45	3.2	171	66
Mali	478.8	1,240.1	7.6	8.3	11.0	42	2.8	17	7
Mauritania	398.0	1,030.8	1.6	2.0	3.0	44	2.9	5	2
Mauritius	0.8	2.1	1.0	1.0	1.1	67	1.5	1,317	509
Moçambique	302.3	783.0	13.4	14.7	21.1	45	2.5	49	19
Namibia	318.3	824.4	1.1	1.2	1.9	48	2.9	4	2
Niger	489.2	1,267.0	6.3	7.1	10.8	43	2.8	15	6
Nigeria	356.7	923.9	98.1	111.7	159.2	49	3.0	313	121
Rwanda	10.2	26.4	5.8	7.0	11.1	47	3.8	686	265
Senegal	76.0	196.8	6.5	7.3	10.5	43	2.9	96	37
Sierra Leone	27.9	72.3	3.7	3.8	4.9	34	1.8	136	53
Somalia	246.2	637.7	6.9	8.2	11.7	41	2.5	33	13
South Africa	471.4	1,220.9	31.7	34.7	44.8	54	2.3	74	28
Tanzania	364.9	945.1	21.2	24.0	35.9	51	3.5	66	25
Togo	21.9	56.7	2.9	3.2	4.8	51	3.1	146	56
Uganda	91.1	235.9	14.3	16.2	24.9	49	3.4	178	69
Zaïre	905.6	2,345.5	31.2	33.1	46.9	50	2.8	37	14
Zambia	290.6	752.7	6.6	7.6	11.1	53	3.3	26	10
Zimbabwe	150.8	390.6	8.7	9.6	14.9	56	3.5	64	25

STATE	AREA		POPULATION (MILLIONS)			LIFE EXPECTANCY 1986 (YEARS)	ANNUAL INCREASE 1988 (%)	POPULATION DENSITY	
	1000 SQ. MILES	1000 SQ. KMS	1984	1988	2000			(NO./SQ MILE) 1988	(NO./SQ KM) 1988
North Africa and Southwest Asia	5,931.0	15,361.4	292.3	325.7	445.4	60	2.8	55	21
Afghanistan	250.0	647.5	14.4	16.1	23.7	37	2.4	64	25
Algeria	919.6	2,381.8	21.4	24.3	34.3	60	3.2	26	10
Bahrain	0.3	0.8	0.4	0.5	0.7	67	2.8	1,642	634
Cyprus	3.6	9.3	0.7	0.7	0.8	74	1.3	194	75
Egypt	386.9	1,002.1	47.1	53.1	71.2	58	2.6	137	53
Iran	636.3	1,648.0	43.8	49.3	68.3	57	2.9	77	30
Iraq	167.9	434.9	15.0	17.1	24.2	62	3.3	102	39
Israel	7.8	20.2	4.0	4.3	5.3	74	1.6	551	213
Jordan	37.7	97.6	3.5	4.0	6.4	67	3.7	106	41
Kuwait	6.9	17.9	1.7	1.9	2.7	70	3.2	275	106
Lebanon	4.0	10.4	2.6	2.8	3.6	65	2.1	700	269
Libya	679.4	1,759.6	3.7	4.2	6.1	58	3.3	6	2
Morocco	275.1	712.5	23.7	24.9	32.1	58	2.6	91	35
Oman	82.0	212.4	1.2	1.4	2.0	52	3.3	17	7
Qatar	4.2	10.9	0.3	0.3	0.5	68	2.9	71	28
Saudi Arabia	830.0	2,149.7	10.8	12.2	17.8	61	3.0	15	6
Sudan	967.5	2,505.8	21.1	24.2	34.2	48	2.9	25	10
Syria	71.5	185.2	10.1	11.3	17.2	64	3.8	158	61
Tunisia	63.2	163.7	7.2	7.6	9.4	61	2.7	120	46
Turkey	300.9	779.3	50.2	54.9	69.7	62	2.5	182	70
United Arab Emirates	32.3	83.7	1.3	1.5	1.9	68	2.3	46	18
North Yemen (San'a)	75.3	195.0	5.9	6.7	9.9	48	3.0	89	34
South Yemen (Aden)	128.6	333.1	2.1	2.4	3.4	48	3.0	19	7
Australia	3,070.1	7,950.9	18.7	19.5	20.8	75	0.8	6	2
Australia	2,966.2	7,682.7	15.5	16.1	17.2	75	0.8	5	2
New Zealand	103.9	268.2	3.2	3.4	3.6	74	0.8	33	13
Pacific	185.8	481.2	4.1	4.4	5.4	55	2.6	24	9
Fiji	7.1	18.4	0.7	0.8	0.9	62	2.4	113	43
Papua New Guinea	178.7	462.8	3.4	3.6	4.5	52	2.6	20	8
Worlda	57,821.3	149,757.2	4,809.7	5,109.1	6,156.5	62	1.7	88	34

aMicrostates included.

TABLE 6.2 Physiologic Density for Selected Countries

COUNTRY	1985 POPULATION (MILLIONS)	AREA (000 SQ. MILES)	ARITHMETIC DENSITY	PHYSIOLOGIC DENSITY
Japan	120.8	143.7	841	6466
Egypt	48.3	386.7	125	6273
Netherlands	14.5	14.4	1077	4380
Bangladesh	101.5	55.6	1825	2900
Colombia	29.4	439.7	67	1336
India	762.2	1269.3	600	1155
Iran	45.1	636.3	71	886
Ethiopia	42.8	471.8	89	809
Nigeria	93.2	356.7	262	791
Argentina	30.6	1068.3	29	590
United States	238.9	3678.9	66	330

Sources: World Population Data Sheet, Population Reference Bureau; Production Yearbook, 1983, FAO; UN population data; World Bank reports.

the agricultural density will be nearly as high as the overall physiologic density. However, where large numbers of the population are crowded into major cities, there will be fewer people on the land and the difference between agricultural and physiologic density will be greater.

Another population index is the *settlement density*, in which the measure represents the amount of area in a country for each city with one hundred thousand people or more. In the United Kingdom, there are about 13,000 sq miles (34,000 sq km) for each city. In the United States, there are 175,000 sq miles (453,000 sq km), and the figure for China is 350,000 sq miles (907,000 sq km).

Before returning to the issues associated with the growth of the world's population, we should remind ourselves of a reality that affects all our calculations and figures: the unreliability of the information that we are compelled to use. When calculating the arithmetic density of a country, we must depend on population counts that may very well be inaccurate (but at least we can be fairly sure of the country's total area). When it comes to figuring the physiologic density, we divide population data of which we are often uncer-

tain and cultivable area data about which we are not sure either. Since population problems are so vital to the world's nations and their future, there has been a major effort in the United Nations to help countries organize their censuses and to increase the effectiveness of data-gathering systems. However, taking a census is an expensive business, and countries that cannot even afford enough food cannot be expected to place a very high priority on census accuracy, which is viewed as a costly luxury. Therefore, of the statistics we have just used and those we will use later, we should realize that many are subject to considerable error.

• Population Growth •

If there is reason to doubt the accuracy of some of the published population statistics, there is no uncertainty about another dimension of world population: its accelerating growth. Never before in human history

• Urban Density •

The problems of overpopulation are felt in many rural areas: in the countrysides of Bangladesh and Burundi, Haiti and Sudan. However, the quality of life for tens of millions of people—recent arrivals as well as long-term residents—has deteriorated in the world's large cities as well.

Mexico City is on its way to becoming the largest urbanized area in the world. In a country with a 1985 population of 80 million, this city has more than 19 million inhabitants, half of them under 18 years of age. Every year more than 400,000 immigrants from Mexico's countryside arrive in the city. Add to this the natural increase, and Mexico City will have more than 30 million residents by the end of the twentieth century.

Mexico City sprawls over more than 400 sq miles (1000 sq kms) of mountain-encircled highland. Local wells have run dry, and water must be pumped in across the mountains. There is no way to dispose of the city's waste except to pump it across the mountains as well. Air pollution is so severe that the city, when approached from the air, seems to lie in a lake of gray-brown smog on many days. Perhaps half the entire population lives in makeshift housing, a vast expanse of ramshackle dwellings lacking most of the basic cooking, laundering, and sanitation facilities.

Yet the immigrants arrive, more than a thousand each day, hoping that their lives will be better in one of the city's thousand formal and informal neighborhoods than they were in the villages left behind. No matter how difficult the new life will be (especially, they know, in the beginning), the arrivals pour their last resources and their highest hopes into the move to the city.

Crowded Mexico City has many urban ills, but there are cities less well off: Lagos in Nigeria (West Africa), with one of the highest urban growth rates in the world but without even a citywide sewer system; Calcutta, India, with far more homeless people than Mexico City.

Shanghai, China's largest metropolis, also exceeds Mexico City in sheer overcrowding, although its population is smaller (about 16 million in 1985) and its growth rate slower. One out of every 8 inner-city residents of Shanghai is homeless, more than half the people do not have toilet or bathing facilities. Pollution-spewing factories stand side by side with crowded dwellings. Piles of industrial wastes fill empty lots; raw sewage seeps into some streets. Rush hour is a daily nightmare, with three times as many riders as there are seats available on buses and trains. People must spend countless hours standing in line for everything from food to medical attention. Health conditions are dismal, not only as a result of the polluted air, but also from the contaminated water and toxin-affected foods. Medical services are inadequate; there is one fully appointed hospital bed for every 1500 residents.

Yet Shanghai, too, attracts people like a magnet. The city grew by leaps and bounds during the 1950s, doubling its population in just *eight* years. The government acted to stop immigration in 1960, but during China's political turmoil of the 1960s and 1970s, home building in China virtually stopped for more than five years. As a result, Shanghai has been trying to cope with the aftermath of these inappropriate government actions, which launched the present episode of runaway, explosive growth. The city has been described as one of the most uncomfortable places in the world to live and work—a designation it shares with other mushroomed urban giants.

have so many people filled the earth's livable space as today nor has world population grown as rapidly as it has during the past 100 years. In 1975, world population reached 4 billion; it reached 4.5 billion in 1981 and passed the 5 billion mark in 1986. It took from the dawn of history to the year 1820 for the earth's population to reach 1 billion. It required a mere *12 years* to add the same number in the 1970s and 1980s. If the growth of population does not slow during the coming decades, the earth will have to support 8 billion

212

people (double the 1975 total) before the first decade of the twenty-first century has ended.

At present, about 131 million babies are born every year and about 53 million people die. This means that we are adding approximately 78 million inhabitants to the world's population every 12 months. Most of this increase occurs in those areas least able to support the new arrivals, and millions among the year's deaths are young children who succumb to starvation or disease.

Can this go on indefinitely? Obviously not. Occasionally, there are signs that explosive population growth will be followed by a marked slowdown in the rate of increase and that world population will actually stabilize during the twenty-first century. This has already happened in some parts of the world, for example, in the United Kingdom, France, and several other developed countries where population growth, after an explosive period of rapid increase, slowed to a trickle. However, the population of France is less than one-twelfth that of India. What is really needed is a decrease in the rate of growth in countries with large populations and high rates of expansion—not only India, but also Indonesia, Bangladesh, Nigeria, Brazil, Mexico, and others.

Sometimes there are signs that a slowdown may be coming. In 1980, the Census Bureau of the United States reported that its research indicated that the world's overall growth rate had declined from approximately 2.1 percent per year, during the years 1965 to 1970, to 1.7 percent during the period from 1975 to 1979. However, when the growth rate was 2.1 percent, the world's population was approaching 4 billion, resulting in 80 million additional inhabitants each year. By the time the rate was down to 1.7 percent per year, the population base was already between 4 and 4.6 billion. Calculate it for yourself: a 1.7 percent increase on a base of 4.6 billion still produces nearly 80 million additions.

Optimism about the future population should be tempered for other reasons. Not only would a slower growth rate still produce a global population of about 6 billion by the end of this century, but the very evidence on which this assumption is based is weak. Census counts in underdeveloped countries have wide margins of error. As already noted, a national census is a matter of considerable organization and substantial cost and many countries simply do not have the means or the resources to carry out a complete census.

It may be reasonable to assume that reported data err toward the low side and that undercounts are more common than overcounts. Furthermore, although there may be recorded declines in the growth rates of populations in certain countries, other areas are showing a reverse trend. During the mid-1980s, Africa's population was growing at 2.9 percent per year—the highest rate in the world—up from 2.4 percent. Middle and South America's populations were growing at 2.3 percent, still far above the world average of 1.7 percent. Figure 6.1 shows the growth rates of individual countries in these regions. Note that the highest growth rates prevail in Africa and Middle and South America (except in the three southernmost countries). In recent years, Kenya has been known as the country with the highest rate of **natural population increase** in the world (4.1 percent). By comparison, the present growth rates of South Asian countries are moderate, and China has a much lower annual growth rate than just a generation ago (1.1 vs. 2.5 percent). Low-growth countries are those of Europe, the U.S.S.R., the United States, Canada, Uruguay, and New Zealand.

DIMENSIONS OF GROWTH

As the figures quoted earlier indicate, the expansion of human population has not proceeded at a uniform rate. It has not been a *linear* process whereby something that grows increases by a uniform amount during a series of equal time periods. If you have $100 and add $10 to it the first year and each successive year, your $100 will become $200 after 10 years, $300 after 20 years, and so on. This is linear growth. However, if you invested your initial $100 at a rate of interest of 10 percent, each increment would be based on the original amount plus previously added interest. After 10 years your $100 would have increased to $259; after 20 years it would have increased to $673. The difference between linear and *exponential* growth is obvious—and the world's population has been growing at exponential rates.

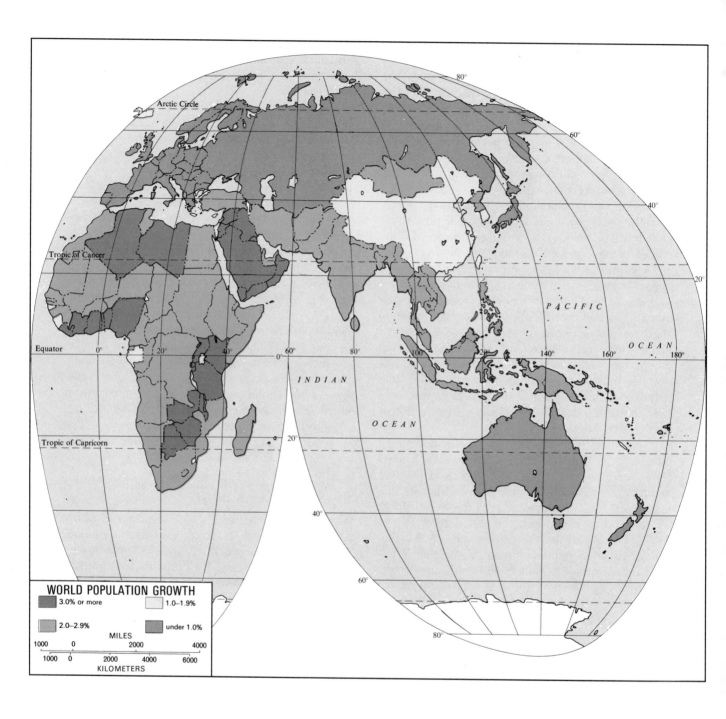

WORLD POPULATION GROWTH

- 3.0% or more
- 2.0–2.9%
- 1.0–1.9%
- under 1.0%

MILES
1000   0   2000   4000

KILOMETERS
1000   0   2000   4000   6000

Arctic Circle

Tropic of Cancer

Equator

Tropic of Capricorn

80°
60°
40°
20°
0°
20°
40°
60°
80°

0°   20°   40°   60°   80°   100°   120°   140°   160°   180°

INDIAN

OCEAN

PACIFIC

OCEAN

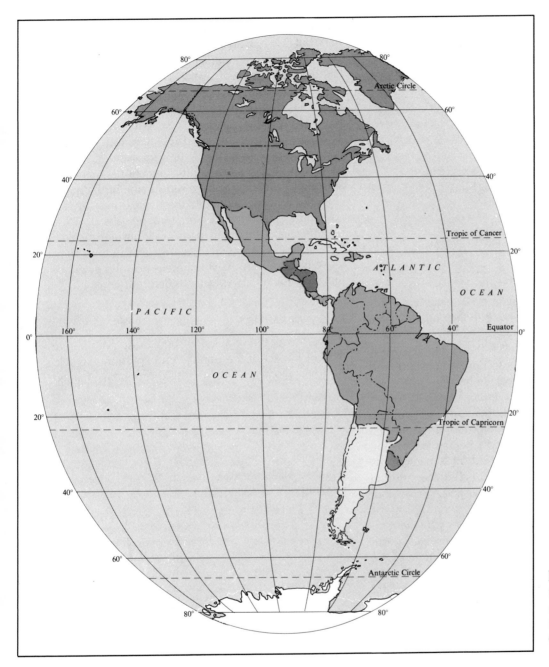

FIGURE 6.1
World population: natural increase (data from United Nations and other sources).

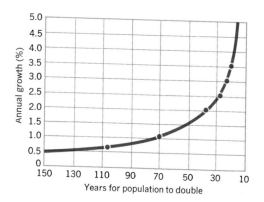

FIGURE 6.2

Population growth rates and doubling time.

humanity has not only been one of increasing numbers, but also one of increasing *rates* of increase.

Among few encouraging swings in recent years has been an apparent overall decline in the rate of growth. A major reduction in the growth rate of China and a smaller, but very significant, decline in India reduced the global rate to 1.7 percent during the late 1970s (which held through 1985), and thus the doubling time is edging slightly upward again. Certainly numerous governments are making efforts to encourage family planning and are subsidizing birth-control clinics. However, although these measures are encouraging, the opposite situation prevails in other countries. In Africa and parts of South and Middle America, rates of natural increase are at all-time highs for some countries. As yet, there is no solid evidence that the population explosion is defused.

EARLY WARNINGS

Concern over the population spiral arose even before the full impact of the population explosion was felt. As long ago as 1798, a British economist named

Another way of looking at exponential growth is by comparing the rate of growth to the **doubling time** involved (Fig. 6.2). Every rate of growth has a doubling time, for example, our $100 invested at 10 percent took just over 7 years to double to $200, and then slightly more than 7 years to become $400, and then still another 7 years to amount to $800. When the growth rate is 10 percent, therefore, the doubling time is just over 7 years. When, during the middle part of this century, the world's human population was increasing at an average rate of 2 percent, its doubling time was 35 years (Fig. 6.2). When, during the mid-1980s, the rate reached 1.7 percent, the doubling time rose to 41 years.

This historical perspective suggests why the term *population explosion* has come into use. It is estimated that the world's population at the time of the birth of Christ was about 250 million. More than 16 centuries passed before this total had doubled to 500 million, the estimated population of 1650. Just 170 years later, in 1820, the population had doubled again, to 1 billion (Fig. 6.3). And barely more than a century after this, in 1930, it reached 2 billion. Now the doubling time was down to 100 years and dropping fast; the population explosion was in full gear. Only 45 years elapsed during the next doubling, to 4 billion (1975). In that decade, the rate of growth was approximately 2 percent per year, and the doubling time (Fig. 6.2) had declined to 35 years. The history of

216

FIGURE 6.3

World population increase since A.D. 1650 and projection beyond 1990.

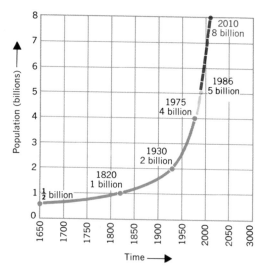

India's population explosion continues. Crowded urban scenes such as this in Varanasi (Benares), on the banks of the Ganges River, mark the cultural landscape of this, the world's second most populous country. Urban living conditions, from Bombay to Calcutta, are below the poverty level for millions.

Thomas Malthus published a long paper, *An Essay on the Principle of Population as It Affects the Future Improvement of Society*, in which he sounded the alarm: population was increasing faster than the means of subsistence. Malthus recognized the nature of exponential growth and pointed out that population increases at what he called a geometric rate. The means of subsistence, he argued, grow only at an arithmetic (linear) rate. Malthus issued revised editions of his essay from 1803 to 1826, and he responded vigorously to a barrage of criticism (for he condemned the social order in England). He suggested that the population growth in Britain might be checked by hunger within fifty years from the first appearance of his warning.

Malthus could not have foreseen the multiple impacts of colonization and migration, and, undoubtedly, he would not have believed that the United Kingdom would have a population of more than 50 million (but Britain does need to import large quantities of food to support that number). Nor was Malthus correct about the linear increase of food production. It, too, has grown exponentially as the acreage being cultivated has expanded, improved strains of

Thomas Malthus, who warned in 1798 that a population crisis was approaching.

sources by taking them from the hands of the few and distributing them among the many. Others have criticized the Malthusian doctrine on the grounds that the earth has proved capable of ever-expanding food production. There are still untapped resources, both on land (e.g., the Amazon and Congo [Zaïre] Basins could be made productive) and in the sea (where globally organized fish harvests could be secured in perpetuity if international planning succeeded). Only technological and political progress needs to be made to achieve these and other advances.

Those who feel the weight of Malthus's concerns (even if not every detail of his argument) are sometimes called the neo-Malthusians, and their number is growing. They point out that human suffering on a scale undreamed of even by Malthus is now occurring, and they argue that it is not enough to simply assert that this planet must inevitably go through a period such as this—even if population stability were to be achieved in the twenty-first century. The time to attack the problem, they say, is now, with vigorous programs of population control and other strategies (of which more will be said later).

seed have been developed, and fertilizers have been applied. However, the rate of increase of food production has not kept up with the rate of population growth, and although Malthus was wrong in terms of timing and detail, the essence of his argument still appears to have merit.

Karl Marx, the nineteenth-century socialist, suggested that the poverty of the masses foreseen by Malthus had less to do with their numbers than with the capitalist system governing them. Far from proposing a reduction in the rate of population growth, Marx wanted to alter the control and use of natural re-

FIGURE 6.4

Population pyramid for Mexico.

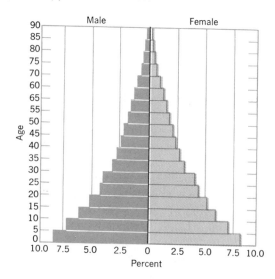

POPULATION STRUCTURE

Maps showing the regional distribution and density of population tell us about the numbers of people in countries or regions, but they cannot reveal two other important aspects of those populations: the number of men and women and their ages. This is very important because a populous country where half the population is very young (say under 15 years of age) has very different problems from a country where a large proportion of the population is elderly. When geographers study populations, therefore, they are interested not only in spatial distribution, but also in **population structure**.

The structure of a population involves the number of people in various age groups, and it is represented by an age–sex pyramid. The age–sex pyramid for Mexico (Fig. 6.4) shows how great the number of children in the lowest three age categories (0–4, 5–9, 10–14) is compared to people in the older groups. In an age–sex pyramid such as that shown in Figure 6.4, males are customarily recorded to the left of the center line, females to the right. Mexico's population structure as reflected by Figure 6.4 reveals a rapid rate of population growth. Mexico has 47 percent of its population below 15 years of age. Families in countries such as Mexico and India are large, with five to six children per couple being the average. Thus, the wider the base of an age–sex pyramid, the faster the population is growing.

When family size decreases, and with it the overall population growth rate, the age–sex pyramid shows lower percentages in the younger age groups, as in the case of the United States (Fig. 6.5). This pyramid, for 1979, reveals the baby boom after the end of the Second World War (the steplike increase below the stable waist of the pyramid). Then, beginning the late 1950s, the increase levels off, and the most recent categories (below 15 years of age, i.e., born since 1965) show a decline. Note how much greater the percentage of older people in the United States is compared to the population structure of Mexico.

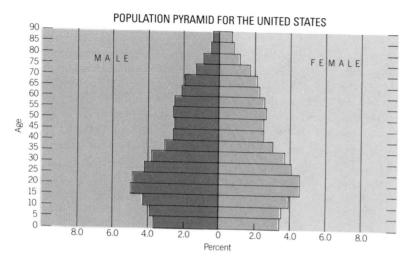

POPULATION PYRAMID FOR THE UNITED STATES

MALE

FEMALE

Age

Percent

FIGURE 6.5
Population pyramid for the United States.

Demographic Cycles

220

The study of population is the field of **demography**, and its spatial component is population geography. The term *demography* appears in such contexts as demographic cycles—to denote population growth processes and stages in populations' expansion—and demographic *transition*—to describe a country's passage through the Industrial Revolution, transforming its growth pattern.

The natural growth of a population is recorded as the difference between the number of births and the number of deaths during a specified period. These two measures, the birthrate and the death rate, are commonly expressed by the number of individuals per thousand. In statistical tables, they are called the **crude birthrate** (the number of live births per year per thousand people in the population) and the **crude death rate** (the number of deaths per year per thousand). For example, in the United States in 1985, the crude birthrate was 16 and the crude death rate 9. If there had been absolutely no immigration or emigration, the rate of increase would have been 7 per thousand, or 0.7 percent. However, because legal and illegal immigration into the United States has been substantial, the estimated rate of increase for 1985 was actually 1.0 percent.

The high birthrates of so many of the world's countries fuel the population explosion; in our lifetime, population change has become a synonym for population growth. However, population change has not always been positive. In world history, there have been times when populations declined greatly and the number of deaths exceeded the number of births. Earlier in this chapter, we described the slow increase of world population until the early nineteenth century; what we did not report was what held the population down. There were times when people succumbed by the hundreds of thousands to epidemics and plagues. Europe between 1348 and 1350 was ravaged by the bubonic plague, when fully a quarter of the entire population died; in the epidemic and its aftermath, many cities and towns were left with fewer than half their inhabitants. It is estimated that the population of England was nearly 4 million when the plague began and just over 2 million when it was over. A student of population change in the second half of the fourteenth century would have been more likely to talk of annual population *decrease* than of growth!

If there was not an epidemic, then famine could reverse a period of population growth. There are records of famines in India and China during the eighteenth and nineteenth centuries in which millions of people perished; Europe was not safe from such disasters either. From Britain to Russia, the ravages of unusual weather periodically caused crops to fail and people to die from hunger. At other times, population gains were largely wiped out by destructive wars, of which Europe and the world have seen many. Thus, the slow and steady increase of world population sometimes shown on graphs as growing from 250 million in A.D. 0 to 500 million in 1650 and 1 billion in 1820 is something of a misrepresentation, for it shows an average of countless ups and downs. Birthrates were high, but death rates were high, too; and there were times when there were many more deaths than live births.

However, things began to change. In Europe (we know much less about Asia, Africa, and the Americas), there was a marked increase in the growth rate during the eighteenth century, and this time there was no major setback to erase the gains made. This was the time of the agrarian revolution, when farming methods improved, yields increased, storage capacities were expanded, and distribution systems were bettered. The Industrial Revolution began to wipe out the old problems in Europe as well: sanitation facilities made the towns and cities safer from epidemics, and modern medical practices began to have widespread effect. Disease prevention through vaccination introduced a new era in public health. Death rates declined markedly. Before about 1750, they probably averaged 35 per 1000 (births averaged under 40), but by 1850 the death rate was about 16 per thousand. Consider what this means in terms of natural growth rate: if in 1750 the birthrate was 39 per thousand and the death rate 35 per thousand, then the rate of natural increase was 4 per thousand or 0.4 percent. In 1850, birthrates were still high, perhaps 36 per thousand, but the death rate was 16 per thousand. The rate of natural increase was now 2.0 percent. The change is especially spectacular in the context of doubling time. In 1750, it was on the order of 150 years; in 1850 it was only 35 years.

What happened to the large number of people that Europe was now generating? Millions left the squalid, crowded industrial cities (and the farms as well) to emigrate to other parts of the world, to North and South America, to Australia, to South Africa, and elsewhere. They were not the first to make this journey. Adventurers, explorers, merchants, and colonists had gone before them. Those early immigrants had decimated the indigenous population in many areas: by conquest, by forced exodus as slaves, and through the introduction of their diseases, against which the local people had no natural immunity. However, when the European colonization started in earnest during the nineteenth century, the Europeans brought with them their newfound methods of sanitation and medical techniques, and the effect was the opposite of what it had previously been. In Africa, in India, and in South America, death rates began to decline as they had in Europe, and populations that had long been caught in cycles of gains and losses commenced new, permanent growth—at increasing rates.

We can speak in only the most general terms about the indigenous populations of the Americas, Africa, Asia, and Australia before this phase of Europeanization. It is speculated that about the time of the first European contact, there were probably fewer than 25 million people in all of North and South America; in Africa, south of the Sahara, there may have been 70 million. (Some recent estimates place these totals somewhat higher.) In China in the mid-seventeenth century, the population may have been less than 200 million; India probably had fewer than 100 million inhabitants. However, there is no doubt about the consequences of European involvement. It reduced the impact of periodic natural checks on the growth of numbers and swept these areas up in the explosive increase prevailing today.

STAGES IN THE CYCLE: THE DEMOGRAPHIC TRANSITION

Demographers who have studied population growth in various parts of the world have argued that the high rates of increase now occurring in much of the underdeveloped world are not necessarily permanent. Today in Europe, for example, the situation is very different from what it was a century ago. In the United

221

The population of a country (of a city or of any other specific region) is a function of three variables: births, deaths, and migration. The population number changes as a result of four conditions: births and in-migration (immigration), which add to the total; deaths and out-migration (emigration), which subtract from it. Births (fertility), deaths (mortality), and migration are called the three **demographic variables**.

To calculate the demographic change affecting a country or region, we use the simple formula

$$TP = OP + B - D + I - E$$

where TP (total population) equals OP (original pop-

ulation) plus B (births) minus D (deaths) plus I (immigration) minus E (emigration). A population's *natural increase* is calculated by using only births and deaths. The demographic change formula explains why calculations based only on births and deaths do not correspond to the figures in Figure 6–1 (World Population Growth). Population change, as recorded on Figure 6–1, also takes emigration and immigration into account. (Obviously, the world figures referred to earlier in this chapter represent the natural increase for the globe as a whole, because our planet does not—yet—experience immigration or emigration!)

Kingdom in 1985, the crude birthrate was 13 and the crude death rate was 12, producing a rate of natural increase of just 0.1 percent. That last figure is reminiscent of the preindustrial period when there also was a small difference between birthrates and death rates, but both were high. Now, both are low. It was in the intervening stage—when birthrates remained high but death rates were lowered rapidly—that Britain's population explosion took place. It is not difficult, then, to discern four prominent stages in the United Kingdom's demographic cycle marked by: (1) high birthrates, high death rates, and a small rate of growth; (2) continuing high birthrates but declining and low death rates and a high rate of growth; (3) declining birthrates but a still-substantial growth rate; and (4) low birthrates, low death rates, and thus a low rate of growth.

This sequence of stages has been observed in the population records of several European countries; on this basis, demographers have defined what they call the *demographic cycle* or demographic transition. Its four stages are

1　*High stationary stage*: high fertility (births), high

mortality (deaths), and variable population but with little long-term growth (Stage 1 in Fig. 6.6).

2　*Early expanding stage*: high fertility and declining mortality (A to B in Stage 2).

3　*Late expanding stage*: declining fertility but—as a result of already-low mortality—continuing significant growth (B to C in Stage 2).

4　*Low stationary stage*: low fertility, low mortality, and a very low rate of growth (Stage 3 in Fig. 6.6).

Since this demographic cycle is indeed a transition, other schemes for recognizing stages in it can be easily devised. Figure 6.6 indicates three major stages, with the middle stage (Stage 2) divided into three phases (A, B, and C). Most demographers, however, prefer the four-stage cycle.

This, then, is what happened in the United Kingdom and much of Europe. Europe, as a continent, currently has a population growth rate of under 1.0 percent after generating its own population explosion at the same time that the United Kingdom's population grew from about 6.5 million in 1750 to 35 million just after 1900. If this is the rule, then we have

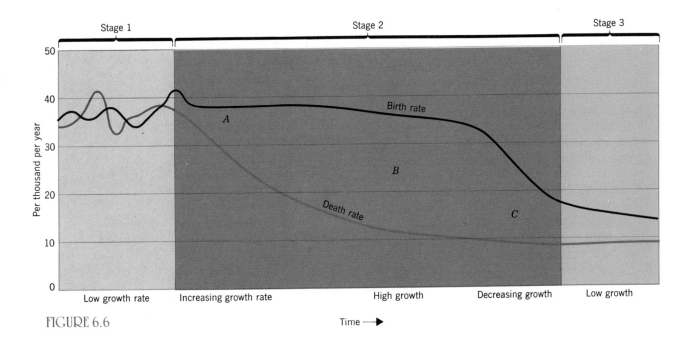

| Stage 1 | Stage 2 | Stage 3 |

Per thousand per year

50

40

Birth rate

A

30

B

20

Death rate

C

10

0

Low growth rate | Increasing growth rate | High growth | Decreasing growth | Low growth

FIGURE 6.6

Time ⟶

reason to be optimistic. It would appear that the population bomb will eventually fizzle out and that, in due course, the European model of low birthrates, low death rates, and a nearly stable population will prevail everywhere.

Or will it? In Europe, the birthrate declined in large part because of the effects of industrialization, urbanization, and general modernization. In much of the developing world, the majority of the people have not been seriously affected by such changes: if the world is to wait for them to occur, the population crisis will intensify. Furthermore, there are quantitative differences between the situation in Europe during the nineteenth century and that prevailing now in certain other parts of the world. When Europe's population revolution began (for such it was), the starting numbers were not great. Britain had between 6 and 7 million residents. Germany had 7 million; France, Belgium, and the Netherlands combined (including some of Europe's most densely peopled lands) had 18 million. Asia's major population clusters were already

much larger when the population revolution commenced there. China may have had over 200 million inhabitants, India over 100 million. Superimpose the growth rates of Stages 2 and 3 on such numbers and the increments take on astronomical proportions.

Therefore, it may be unwise to assume that all countries' demographic cycles will follow the sequence of events that prevailed in the urbanization and industrialization of Europe or to assume that the explosive growth now prevailing in Bangladesh, Mexico, and numerous other countries will simply subside. Optimistic projections suggest that a leveling-off will occur when an ultimate population is reached, but there is little or no hard evidence for this. Occasionally there are hopeful signs in some countries such as Sri Lanka, but these are merely isolated instances, not proof of a permanent, worldwide reversal in population expansion. We repeat: there may be vague, long-range grounds for optimism but, at present, there is no effective machinery to defuse the population bomb.

Migration

Migration (as we noted previously) is the third of the basic demographic variables. In the context of demographic change in individual countries, the statistics that matter are those representing **international (cross-boundary) migration**. But, as population geographers have pointed out, such international migration is but one of several kinds of movement. People move within countries (not affecting national vital statistics), as we do in large numbers in the United States and Canada every year. From rural areas to cities, from Snowbelt to Sunbelt, from east to west Americans move; on the average, an American family moves once every six years. Such movement is recorded in the statistics of states and provinces, but it does not affect national data, the vital statistics of countries as a whole.

Types of Movement

Movement takes on several forms, and these should be quite familiar to us. You may go to classes every weekday, and perhaps to a job as well: you leave your room or apartment, travel to class or work, and return daily, taking approximately the same route each time. This is a form of **cyclic movement**. However, you may also have come from another town, even another state to study at the college or university you now attend. Your arrival for the fall quarter or semester and your return trip after the spring is a different form of movement that involves a lengthy period of residence following your trip. This is called **periodic movement**. After you graduate, you may decide to take a position in a foreign country, say Australia or the United Kingdom. If you decide to move there permanently (as did some who refused to fight in the Vietnam War and took up residence in Canada and Sweden), your move is a form of **migratory movement**.

MIGRATORY MOVEMENT

A permanent relocation, a leaving behind of the old, and the making of a new beginning are all elements of migratory movement. It has numerous causes and many manifestations—so many that it is often impossible to discern the exact reasons underlying people's decisions to seek new abodes. Obviously, there are factors that **push** people away from their homes, and other forces that **pull** them to new, promising locations. Determining or measuring these forces is a difficult proposition. Oppression, discrimination, the threat of war, or natural catastrophe can drive people away. The attraction of better economic opportunity, greater freedom, or security can pull them to new homelands. Always, however, some of the people will move and others will stay behind.

The past five centuries have witnessed human migration on an unprecedented scale, much of it generated by events occurring in Europe. Major modern migration flows include (Fig. 6.7): from Europe to North America (1); from southern Europe to South and Middle America (2); from Britain to Africa and Australia (3); from Africa to the Americas during the period of slavery (4); from India to eastern Africa, Southeast Asia, and Caribbean America (5); from China to Southeast Asia (6); from the eastern United States westward (7); from the western U.S.S.R. eastward (8), and from north-central into northeast China (9). These last three migrations are internal to the United States, the U.S.S.R. and China; our map does *not* show some other significant internal migrations, for example, the south-to-north movement of black Americans during the present century.

Among the greatest human migrations in recent centuries has been the human flow from Europe to the Americas. When we discussed the period of explosive population growth in nineteenth-century Europe, we did not fully account for this process, which kept the total increase far below what it might have been. The great emigration from Europe ([1] and [2] in Fig. 6.7) began slowly. Over the several centuries prior to the 1830s, perhaps 2.75 million Europeans left to settle overseas in the newly acquired sphere of influence. Then the rate of emigration increased sharply, and between 1835 and 1935 perhaps as many as 75 million departed for the New World and other overseas territories. The British went to North America, Australia, New Zealand, and South Africa. From Spain and Portugal, many hundreds of thousands emigrated to Middle and South America. Early European colo-

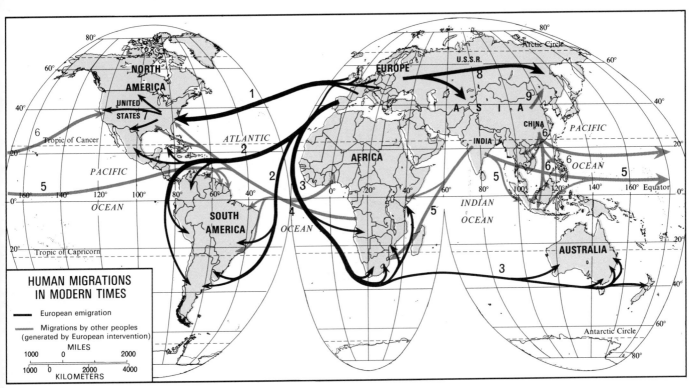

FIGURE 6.7

Major human migrations in modern times.

nial settlements grew to substantial size, even in such places as Angola, Kenya, and Djawa (Java). True, some millions of Europeans eventually returned to their homelands, but the net outflow from Europe was enormous.

This European emigration has had no counterpart in modern world history in terms of size and numbers, but it is not the only major migration flow to have occurred in recent centuries. The Americas were the destination of another mass of immigrants: Africans, transported in bondage, but migrants nevertheless (Fig. 6.8). This migration began during the sixteenth century when Africans were first brought to the Caribbean. In the early decades of the seventeenth century, they arrived in small numbers on the plantations that were developing in coastal eastern North America—so they were among the very first settlers in this country.

It will never be known how many Africans were forced to travel to the Americas; estimates range from 12 to 30 million. As the plantation economy evolved, labor was needed. Prior to the 1830s, the European population in the Americas remained quite small—so, the flow of African slave labor grew. There was a time, before the flow of European immigration intensified, when there were five or six times more blacks in the Americas than whites. The heaviest concentration of Africans in bondage was in the Caribbean and northern and eastern South America. Today the descendants of these people form the overwhelming majority in several Caribbean countries' total population. Actually, despite the substantial black population of the United States at present, the number of enslaved Africans who were taken to this country was quite small compared to the mass that went as forced immigrants to the Caribbean and South America. By 1800, the

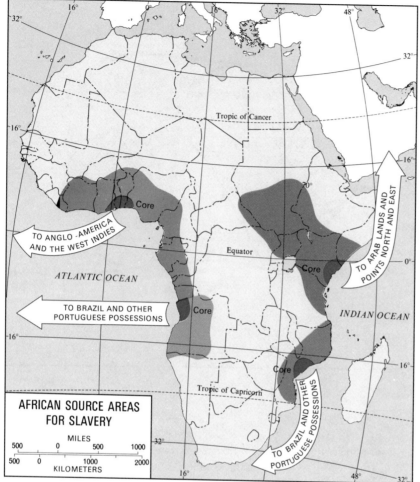

Within the map:

Tropic of Cancer

TO ANGLO-AMERICA
AND THE WEST INDIES

Core

Equator

ATLANTIC OCEAN

TO BRAZIL AND OTHER
PORTUGUESE POSSESSIONS

Core

Core

INDIAN OCEAN

TO ARAB LANDS AND
POINTS NORTH AND EAST

Core

Tropic of Capricorn

TO BRAZIL AND OTHER
PORTUGUESE POSSESSIONS

AFRICAN SOURCE AREAS
FOR SLAVERY

MILES
500   0   500   1000

500   0   1000   2000
KILOMETERS

FIGURE 6.8

Main areas from where African slaves were taken.

black population of the United States was just 1 million.

Voluntary and Forced Migrations

The European and African migrations to the Americas illustrate different types of international migratory movement: *voluntary* and *forced*. As in the case of pull and push factors, the distinction is not always complete, but it does help us understand the underlying causes of major population movements.

In the case of **voluntary migration**, the migrants have options. The decision to move is largely theirs, and they may well have a choice of destinations, as Europeans did when they emigrated to the Americas, Africa, and Australia. Pull factors play a dominant role in such decision making: the notion that life will be materially better with greater opportunity and a higher standard of living in the new homeland. Push factors also play a role because conditions in the source area of the migrants may be such that emigration becomes an ever-better prospect. Hundreds of thousands of Middle American residents, most from Mexico, cross the boundary of the United States, attracted (pulled) by perceived opportunity, but also pushed by a lack thereof in their homelands.

Forced migration takes place when the migrants do not have options, as was the case in the massive movement of African slaves to the Americas. Here it is irrelevant to speak of push or pull factors; the migrants are under the control of other forces. Forced migration still takes place in modern times. During the twentieth century, millions have been forcibly displaced by governments. Among the most recent instances has been the relocation of hundreds of thousands of Ethiopians, ostensibly for practical reasons (to facilitate food provision in the face of mass starvation) but, in fact, for strategic reasons as well, because the country has been embroiled in a civil war. In the 1970s, the government of Uganda suddenly expelled tens of thousands of Asians, loading them on airplanes and sending them to Europe and Asia. Many of these families had lived in Uganda for generations and left all they owned behind. Such cross-boundary forced migrations also change the vital statistics of countries, especially when forced migrations cause even greater flows of *refugees*. International refugee movements (voluntary as well as forced) have changed the demography of many countries, especially in Africa but also in Asia and the Americas.

Refugee Migrations

Migration (as we noted earlier) results from several stimuli. Economic attractions, political pressures, war and conflict, cultural circumstances, and environmental conditions—ranging from prolonged droughts to sudden disasters (floods, earthquakes, volcanic eruptions)—drive people from their homes.

During the twentieth century (especially during the second half of this century), millions of persons have been displaced and become refugees. From Middle America to West and East Africa to Southeast Asia, people have been driven from their homes by war, famine, and political pressures. Often they have simply abandoned their land to join a never-ending column of others walking along some road to no known relief. Life had become impossible at home, and elsewhere it could not be worse. Others, in Vietnam, Haiti, and elsewhere, boarded boats and sailed out to sea in the hope of finding happier shores or, perhaps, to be found and rescued by the vessels of other nations.

Millions of these homeless people have gathered in refugee camps, where international relief agencies and national governments try to help them. Conditions in these camps are often miserable. Food is scarce. Medical services are minimal. Accommodations are basic; many must sleep in the open. The camps are mostly located in those countries least able to handle a flood of hungry immigrants: Sudan, Ethiopia, Zambia, Thailand, Pakistan. The refugee problem has become a global issue, and the stream of refugees grows ever larger.

Refugees are recognized as displaced persons by a special U.N. agency, the High Commission for Refugees. However, they are identified as such only when they cross an international boundary, thus becoming external migrants. Many refugees—perhaps the majority—are not external migrants, but displaced persons within their own country. The hundreds of thousands who clustered in makeshift camps in Sahel countries and Ethiopia, the refugees of Kampuchea, and the Greek Cypriots who were forced to leave their homes by the Turkish invasion of 1974 and its aftermath were all refugees, but they did not cross international borders. One way to classify refugees, therefore, is by grouping them into groups of "recognized" external migrants and "internal" displaced persons. Another terminology for these two categories is *international* and *intranational*, respectively.

An important difference between refugee movement and voluntary migratory movement is the absence of a simultaneous and substantial countercurrent in the case of refugee movement. Refugee out-migration may be followed by a return of all or some of the out-migrants after the conditions that ousted them improve, but such a return migration comes after, not during, their emigration from the threatened zone. The great majority of refugees leave their abode without assurance that they will be able to return at all. Under present data-gathering practices, it is impossible to estimate how many refugees actually do return or have returned, say, in the past 10 years. Still, it is appropriate to differentiate, on the basis of motive and intent, between *permanent* refugees, who leave without the prospect of return, and *temporary* refugees, who move the shortest possible distance away

227

and who would return if the opportunity presented itself. The boat people who left Vietnam by the tens of thousands during the 1970s faced relocation or death at sea; for them, return was no option. Haitian refugees arriving on American shores report that they have nothing to return to and would be imprisoned if they were repatriated. They, too, are permanent refugees in search of a new homeland. Hundreds of thousands of Somali refugees who abandoned their traditional homelands in Ethiopia during the late 1970s, however, would return if political circumstances permitted it.

Refugee
Destinations

Refugee sources and destinations in the 1980s existed primarily in five locations: Middle America, Africa, the Middle East, South Asia, and Southeast Asia. When ranked by percentage of the source population affected, Middle America (including the Caribbean region) leads the world; the United States is Middle America's primary destination (Fig. 6.9). When ranked by absolute total number of refugees, black Africa is the geographic realm most severely affected by refugee problems; in 1985, it was the area where these problems were worsening most rapidly. The Middle East remains an area of persistent refugee problems, and although regional geographers normally place Cyprus in the European sphere, it was a Moslem country—Turkey—that played the principal role in the geopolitical crisis that produced thousands of Cypriot refugees. The geopolitical crisis that began with the Soviet Union's invasion of Afghanistan in 1979 generated one of the world's largest refugee populations: some 3 million Afghans in camps in Pakistan. However, South Asia had other refugee problems as well (as noted later). Probably the most dramatic refugee movements occurred in and near Southeast Asia in the late 1970s and the 1980s. As boat people of Chinese and Vietnamese ancestries poured out of communist Vietnam, some 250,000 Kampucheans were driven into Thailand by the war in their country, which also involved Vietnamese forces. Uncounted hundreds of thousands of Kampucheans were internal refugees of this conflict within Kampuchea itself.

MIGRATION AND THE MOSAIC OF AMERICAN SOCIETY

In July and October 1986, the United States celebrated the centennial of a world-renowned symbol of migration: the Statue of Liberty, beacon to millions who came to this continent. Immigration into the United States first passed the 5-million-per-decade mark in the 1880s, and it reached an all-time high of nearly 9 million during the first decade of the present century. Then immigration began to decline, reaching a low point during the 1930s when only about a half million immigrants reached these shores. Since then, the swing has been upward again: 3 million during the 1960s, 4 million during the 1970s, and a projected 5 million during the 1980s.

These figures reveal the contribution of immigration to the growth of the U.S. population, but they do not reflect the changing origins of the immigrants. During the last half of the eighteenth century, about 90 percent of the migrants came from European-source countries, a majority from Western Europe. There was small immigration from Middle and South America and even smaller inflow from Asia. During the second half of the twentieth century, immigration from Latin America and from Asia has far overshadowed the European influx: forty percent of all immigrants came from Middle and South American countries, more than twice as many as came from Europe. And about one third of the total immigration stream came from Asian countries, mainly from Southeast Asia.

Another situation not reflected by these official data has to do with the flow of illegal immigrants across U.S. borders. The projected figure of 5 million immigrants for the 1980s represents *legal* immigration only. During the mid-1980s, it was estimated that as many as a half million people were entering the United States illegally every year, and possibly many more. Most of these undocumented immigrants came from Mexico, Central America, and the Caribbean. If the estimates are correct, it would mean that more immigrants will have entered the United States during the 1980s than even during the first decade of the century, the period of maximum immigration from Europe.

Such statistics naturally cause concern, but we should keep the numbers in perspective. An entire

228

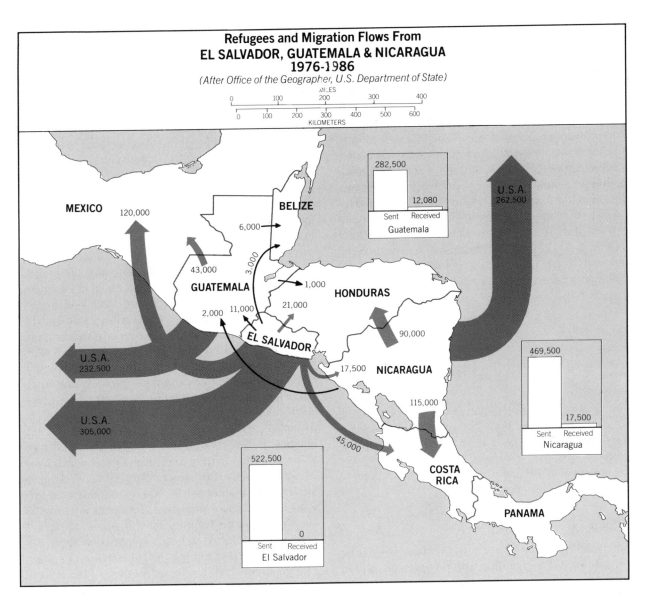

Refugees and Migration Flows From
EL SALVADOR, GUATEMALA & NICARAGUA
1976-1986
(After Office of the Geographer, U.S. Department of State)

FIGURE 6.9

Refugees and migration flows from: El Salvador, Guatemala and Nicaragua.

229

decade of immigration, even if it reaches 9 million, represents just slightly more than 3 percent of the total population. The 1980 census reported that nearly 93 percent of the U.S. population is U.S.-born, and although that figure is declining, it will remain higher than it is for many other countries for years to come. For the country as a whole, there is no tidal wave or flood of migrants.

But the regional geography of immigration presents a different picture. The European sources of the immigration stream of a century ago are etched on the map. Although many immigrants moved westward, hundreds of thousands stayed in the Northeast, and built the cluster of population we see in Plate 5. Now the migrants are coming from the south—to Los Angeles and Southern California, to Texas, and to South

Florida—and they are coming across the Pacific, not the Atlantic. Miami is a city transformed, its Little Havana a more contemporary symbol of the times than the Statue of Liberty. The cities of Texas are developing ever-stronger Mexican imprints. Los Angeles and its surrounding communities constitute a newer form of the American melting pot, where Americans of European descent (Anglos) will, by century's end, no longer constitute a majority. Of course the Vietnamese, Filipinos, and other Southeast Asian immigrants have to some extent been dispersed throughout the United States, and there are large Hispanic populations in northern cities as well as southeastern and southwestern ones. But the cultural mosaic of the country inevitably will reflect both the proximity of the latest sources of immigration and the directions of the migration streams.

· Health and Hunger ·

The geography of population, as a field of study, is closely linked to the study of health and well-being, the field of medical geography. Many of the implications of what we have seen so far in this chapter become reality on maps of food availability, infant mortality, and life expectancy.

GEOGRAPHY OF NUTRITION

Adequate nutrition is not only a matter of eating enough food. The diet must also be balanced, and in large parts of the world that balance is lacking. People need carbohydrates (derived from staples such as rice, corn, wheat, and potatoes), proteins (from meat, poultry, fish, eggs, dairy products), vitamins (from fruits and vegetables as well as other sources), fats, and minerals. The amount of food intake is measured in terms of calories, which are units of energy production in the body. It is impossible to generalize about the number of calories we need without adding many qualifications: people in desk jobs need fewer calories than people engaged in manual labor; young adults need more calories than old persons or children. Peo-

ple of large stature require more calories than smaller people. Men need more calories, on an average, than women. But even with these considerations in mind, it is clear that there are whole countries whose populations are badly underfed. The per person daily calorie intake in the United States, Canada, Australia, and New Zealand exceeds 3200; for such countries as Sri Lanka, India, the Philippines, and Ecuador it is 2000 or fewer. Generally, 2300 calories is taken as the minimum daily consumption level. The U.N.'s FAO gives 2360 calories as the minimum daily requirement, but other agencies place this index slightly lower. Even 2500 is still low, but if we can rely on published figures, about half the people of the world survive on fewer calories than this.

The situation would not be so bad if the people who must survive on too little food could at least have a balanced diet. But they cannot. With few exceptions, the countries where there is a low calorie intake are also the countries where protein is in short supply. This is not surprising when we remember that it takes food to raise the animals that could offset protein deficiencies, food that cannot be given up to animals when people are going hungry. There are places in the world where certain taboos limit people's access to the meat products that might offset their dietary imbalance, but most of all it is shortage of surplus feed that creates the barrier. The result of this is that even persons whose calorie intake is marginally adequate are actually malnourished. There is hidden hunger even in the areas identified as moderate in terms of calorie supply on Figure 6.10. A caution about this map: even published statistical information about calorie intake, especially for the developing countries, can be unreliable. The map should be taken for a general impression only.

It is especially tragic that undernutrition and malnutrition prevail in countries where so large a percentage of the population is young, in the 0–14 age group. A child's brain grows to about 80 percent of its adult size in the first three years of life, and an adequate supply of protein is crucial in this development. If there is not enough protein, brain growth is inhibited and mental capacities are permanently impaired. There can be no subsequent recovery of the loss. Both mental capacity and physical growth are adversely affected by inadequate nutrition, and so the unfortunate youngster born into an environment of deprivation faces a lifelong handicap. In large regions of several countries—the slums of Middle and South American cities, the desert margins of Africa, and extensive rural parts of India—whole populations are so afflicted, their hopes for progress dashed not long after birth.

And so, in many of the world's countries, people cannot expect to live as long as they do in the better-fed regions. Today life expectancy in many European and other Western countries exceeds 70 years (Fig. 6.11). In a number of less fortunate countries, it falls below 50 years and, in some cases, even to 40 years—which is what life expectancy was in Europe two centuries ago.

The life expectancy figures given on Figure 6.11 do not mean, of course, that everyone in those countries lives to that age. The figure is an average that takes account of the children who die young as well as the people who live to old age. The dramatically lower figures for the underdeveloped countries, therefore, reflect primarily the high rate of infant mortality. A person who has passed through the childhood years can expect to survive well beyond the recorded average expectancy. Those low figures for the hungry countries remind us again how hard hit are the children in the underdeveloped world.

HUNGER AND SICKNESS

People who are inadequately fed are susceptible to numerous debilitating diseases. Again the impact on children is devastating. **Kwashiorkor**, a condition rampant in protein-poor tropical and subtropical countries, causes a child to develop a distended belly—a grotesque irony—while the skin discolors and hair loosens and develops a reddish tinge. Eventually liquids begin to collect in swelling limbs, the digestive system fails, appetite is lost. Apathy takes over, and the child dies. These countries' high infant-mortality rates are directly related to the incidence of kwashiorkor, the greatest killer of small children there. Significantly, a child can develop kwashiorkor even when he or she receives enough total calories. Kwashiorkor is a result of malnutrition, not necessarily undernutrition. It tends to develop when a mother stops breast-feeding her child (perhaps because a new baby has been born), and the child is put on a starchy diet.

Recent research indicates that kwashiorkor results not only from protein deficiency, but also from improper storage of grain. When grain is stored for a prolonged period in hot, moist places, a fungus develops on it. This fungus, aflatoxin, is believed to cause kwashiorkor when the grain is consumed by children.

When there are not enough calories *and* a gross lack of protein, a child is likely to develop **marasmus**. The body is thin and bony, the skin shrivels, and the eyes appear huge in the tiny, drawn face.

Kwashiorkor and marasmus are among a host of other threats facing the undernourished. Besides calorie inadequacy and protein deficiency, hungry people may have vitamin deficiencies. Low vitamin A intake is related to diseases affecting the eyes, and these, too, take a heavy toll of children. Beriberi, which affects the nerves, digestive system, and the heart, is related to vitamin B inadequacy. It is frequent in East and South Asia. Insufficiency in vitamin C can produce scurvy; and vitamin D is needed to ward off rickets.

This depressing (and far from complete) list of diseases directly related to hunger does not include the many other infectious diseases that prey upon the ill-protected, malnourished body. Cholera, yellow fever, hookworm, malaria, and numerous other maladies ravage people already weakened by their imbalanced diets, and they take a heavy toll if not of life, then of energy and longevity. And so well-fed tourists and travelers from the comfortable world of plenty see people sleeping on sidewalks, resting on their jobs, working at a snail's pace, and they conclude that those people are not better off because they are lazy, there in Guatemala, in Peru, in Gabon, in Niger, in India, in the Philippines.

FOOD FOR BILLIONS

The world today can actually produce a quantity of food that, if equally distributed among all the earth's inhabitants, would come very close to satisfying the 2300-calorie minimum established earlier. Obviously, something happens to prevent this from taking place. In the first place, a shockingly large amount of the food grown and raised each year is lost through decay

231

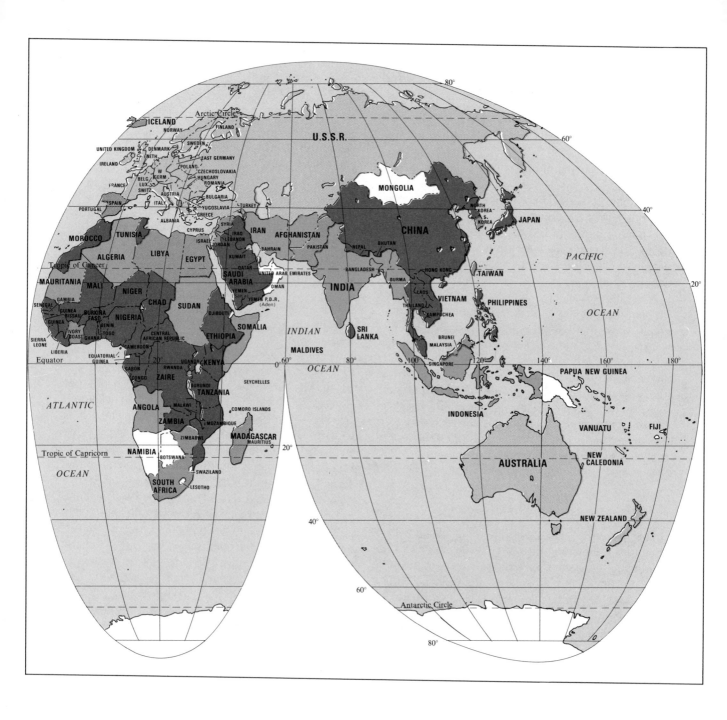

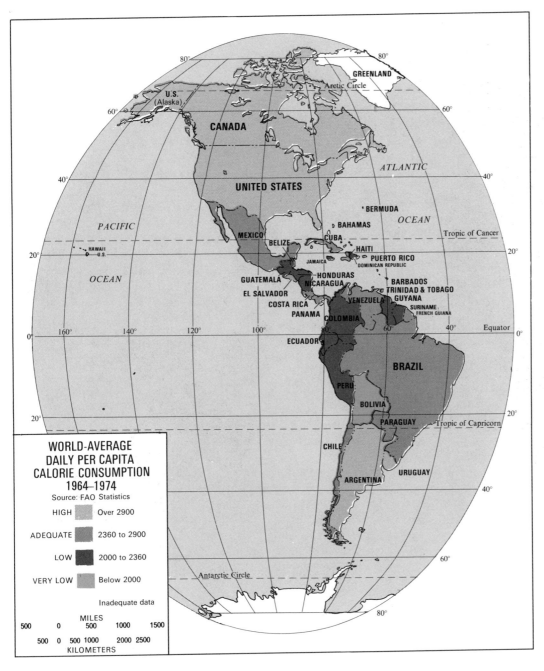

233

FIGURE 6.10
Average daily calorie consumption per person, 1964–1974 (data from United Nations FAO sources).

WORLD-AVERAGE
DAILY PER CAPITA
CALORIE CONSUMPTION
1964–1974
Source: FAO Statistics

HIGH   Over 2900

ADEQUATE   2360 to 2900

LOW   2000 to 2360

VERY LOW   Below 2000

Inadequate data

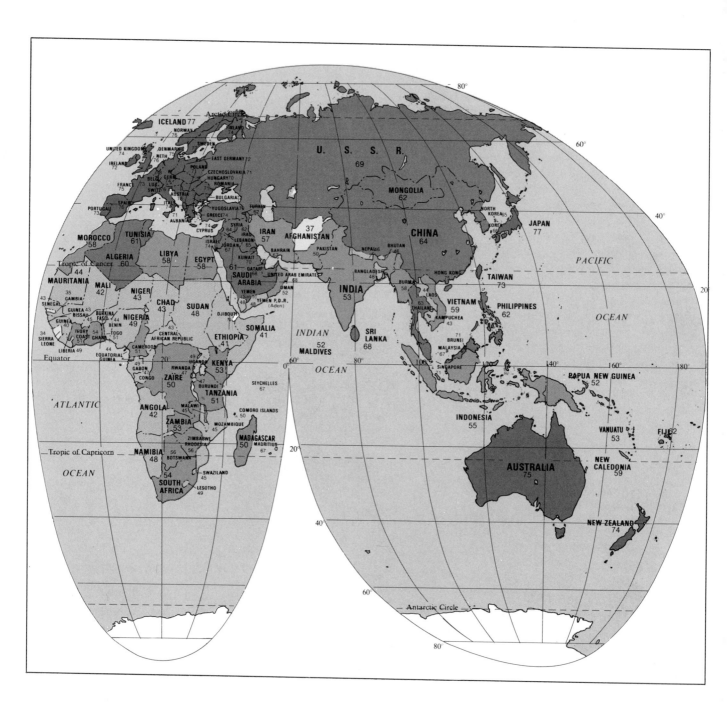

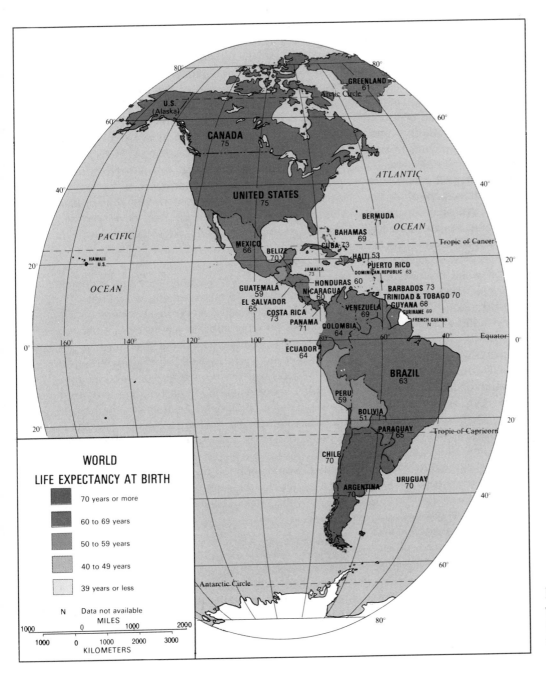

235

FIGURE 6.11
World life expectancy (data from United Nations).

in storage or transportation; through destruction by mice, rats, and other rodents and pests; and through other misfortunes. It has been calculated that these losses amount to between 10 and 20 percent of total world production and that the loss may be as high as 50 percent in some areas. Then there is the matter of distribution. The world's food production is simply not distributed to ward off hunger everywhere. Things have improved a great deal: grain shipments to famine-threatened areas can now stave off disaster when millions would have died in another period of history. But areas still remain in which the supply of food is chronically inadequate or where **seasonal hunger** occurs over and over. What prevails, though, is the fact that even if 2300 calories—minus losses incurred through spoilage and destruction—were available to every person on this globe, it would be inadequate quantitatively and qualitatively. There can be no doubt about it: if all the world's food were allotted evenly to all people, everyone in this world would be malnourished.

The obvious answer to this aspect of the population crisis, then, seems to be the production of more food. But will more food expand the limits to population growth, thereby contributing to an eventual, still greater shortage of food? We are back to Malthus's original warning, but in recent years another dimension has been added. We are destroying streams to create irrigation ditches. We are cutting down forests to clear still more land. We are causing soils to erode and grasslands to be overgrazed. We are exhausting the fishing grounds of the oceans. We must ask ourselves not only whether we can produce enough food to supply added billions of people, but also what damage we will do in the process to our earthly environment, our life-support system.

THE GREEN REVOLUTION

In Chapters 2 through 5, we considered some of the environmental properties of our Earth, and we found that there are limits to its agriculturally productive areas. Forests are still being cut over to make way for new farmlands, and some marginal areas have not yet been brought into production, but in the next several decades, we will not discover any new Ganges plains or Nile deltas. In the 1960s, the optimists among demographers, who would argue that technological progress will overcome population–food crises, scored a point when there was a real breakthrough in the search for new, more productive strains of cereals. Miracle rice, as one such new variety aptly came to be called, permitted the Philippines to end more than 50 years of dependence on rice imports; Sri Lanka's rice production rose by 32 percent in two years. Wheat, too, shared in this **Green Revolution**: Pakistan and India in 3 years increased their wheat harvests by 50 percent. Over a longer period, Mexico, where the wheat experiment first succeeded, tripled its production of this grain. The miracle rice, developed at a research station in the Philippines, has a short maturing time, a high-yield capacity, and a high response to fertilizers.

But even the Green Revolution has not been able to eradicate hunger in our world. It was a reprieve, but it was able to close the gap between population and food needs only partially, despite its dramatic effects on yields. Pessimists pointed out that much of the spectacular increase in yields had to do with exceptionally good weather during the first years of introduction. Others feared the impact of the use of fertilizers on the environment. Economists emphasized that many farmers will not have the capital or credit to participate in the technology of the Green Revolution. And the inexorable march of population growth, they argued, would soon eat up the gains made and produce essentially the same situation as prevailed before, only this time involving many more people.

Failure in Africa

The situation in Africa during the 1980s has underscored the limitations of a Green Revolution. Years of misguided policies and inappropriate priorities, mismanagement, and corruption took an increasing toll on the newly independent African states. Widespread hunger—not seen for many decades—affected unprecedented numbers of Africans and created problems that defied short-term solutions.

What is especially disturbing about the situation in Africa in the 1980s is the contrast with optimistic predictions just 25 years ago when Africa's population was a mere 285 million and self-sufficiency in food appeared within grasp. Now Africa's population is growing at a rate of nearly 3 percent annually and food supply, which would have to double in volume every 23 years just to keep pace, is actually declining. So Africa joins the growing list of regions that depend on food imports from the United States and elsewhere.

The contrast between Africa and Asia is striking. Despite continuing local problems, food production in Asia has increased as much as 20 percent over little more than a decade, whereas in Africa, population grew about twice as rapidly as crop yields did, and it may be said that Africa has barely been touched by the Green Revolution. The reasons for this are diverse. In the first place, the peoples of Asia depend mainly on two crops, rice and wheat, which are grown mainly under irrigation. Africa's variety of crops is much greater (corn, millet, sorghum, wheat, some rice, etc.). The Green Revolution has not yet affected all of these, although a new strain of sorghum has been developed that yields three times as large a crop as the standard variety and that has raised harvests in the Sudan. More problematic is the soil situation. In Asia, most rice and wheat is planted in fertile, deep soil. Africa's soils are not irrigated for the most part, and they are thin, fragile, and easily damaged. Overcropping of such soils can easily occur.

Although it may seem unlikely, there are also labor problems in populous Africa. Africa's population is growing rapidly, but the labor force in the farming areas is not always large enough to handle the farming chores when they must be done to ensure the best yields.

Add to these conditions the poorly developed state of African agriculture generally—from equipment to farming methods to education to transportation—and the slowness of the Green Revolution to have an impact in Africa is understandable.

In Chapter 3, we noted the changes we may be causing to the world's climate as the Industrial Age progresses. The same thing is happening in the quest for more food: the natural environment is threatened. Already, the enormous volume of fertilizers used during the Green Revolution's upsurge are affecting freshwater lakes and streams, having seeped through the soil and into the groundwater. Eutrophication (oxygen deprivation) of many of these water bodies in the developing world has destroyed fish and deprived villagers of much-needed protein. Marginal farmlands are being forced into production and thousands of acres of land have turned into dust bowls when the inevitable years of deficient rainfall came. This contributed also to the amount of particulate matter in the air, perhaps affecting the solar energy balance. Plans (now being executed) to cut severely into the Amazon Basin's forests have raised fears that the oxygen supply they generate and transmit into the atmosphere will be reduced. Rain-making attempts and the diversion of river waters for irrigation purposes all are experiments that threaten change in regional and perhaps world climates, changes that cannot be predicted and, once set in motion, probably not reversed.

Population Policies and Planning

Although international conferences have been convened to consider the population question, collective action in our divided world is not yet possible. A full-scale gathering of the world's states was held in August 1974 in Bucharest, Romania, to seek solutions to the population problem and its attendant issues (nutrition, health, education); but ideological disagreements soon arose, and no far-reaching decisions or actions arose from the conference. Many governments recognize the cost of rapid population growth; few are able to stem the tide. The experiences of one country may or may not be useful to another because so many variables are involved. China was able to reduce its rate of natural increase largely because of the ability of its authoritarian central government to impose its policies on the entire nation. In democratic India, no such central authority exists, and individual states in that country pursue different population policies. In Africa, many countries must now cope with population growth of unprecedented dimensions. Governments realize the consequences, but have neither the power nor the means to control the process. In the pages that follow, some examples of national policies are described.

Japan

The Japanese experience remains a prime example of success in population control. During its nineteenth- and early twentieth-century era of modernization, expansion, and military successes, Japan's leadership encouraged families to have several children; limiting families was actually opposed. However, modernizing Japan also had growing urban centers, which tended to reduce the birthrate somewhat. The combination of circumstances tended to stabilize the rate of growth. With the end of the Second World War, Japan found itself with many hundreds of thousands of refugee nationals who had left the colonies; soldiers came home and rejoined their families, and the American occupation was attended by the introduction of improved medical services and public health. The cumulative effect was an unprecedented increase in the birthrate and a simultaneous drop in death rates; and Japan's rate of growth, which over the decades had averaged about 1.3 percent, suddenly rose to 2.0 percent per year. That represents a doubling time of 35 years, and with Japan's population already about 70 million at that time, it was a crisis situation.

In 1948, the Japanese government took action, and the results show clearly in the population pyramid. The Eugenic Protection Act legalized abortions for "social, medical, and economic reasons." Contraceptives were made available, and family-planning clinics were set up throughout the country. Although contraception and female sterilization (also made available) helped reduce the birthrate, it was the enormous number of abortions that really brought it down. In fact, so many abortions were performed—perhaps 7 to 8 million in a decade—that the Japanese authorities began to worry about the effect on the well-being of the nation and encouraged contraception by means of propaganda and educational programs. In any case, the birthrate, which in 1947 had been over 34 per 1000, was down to 18 per 1000 just one decade after the implementation of the Eugenic Protection Act. It has been reduced to 13 per 1000 (1985); the death rate has declined from 14.2 in 1948 to 7.5 in 1958 and 6.0 percent in 1985. Thus Japan's population, growing at less than 1 percent per year, increases by about 1 million annually, since immigration is very small and contributes little to population growth and emigration, though slightly larger, is similarly insignificant.

India

The problems confronting the formulation and execution of a coordinated population policy are nowhere more evident than in India, one of the non-Western countries that needs such planning most. The first serious effort was begun in the early 1950s with a small investment (1 million Rs) in the promotion of family planning. However, during the 1950s, India's leaders still seemed unaware of the real dimensions of the population explosion then going on; only during the 1960s, when a more recent and comprehensive census became available, did the situation generate a more intensive campaign. During the decade of the 1960s, India spent less than 1 billion Rs on family planning efforts, and during the 1970s spent about 7 billion Rs. Large appropriations have continued during the present decade. Individual states in the Indian federation joined in the program with varying degrees of success. More than fifty thousand family-planning centers were established throughout India.

Yet, the population explosion continued, especially in India's already populous eastern regions. Not all of India's states cooperated equally in the family planning campaign, and there were serious problems in some of the states that tried the hardest. In Maharashtra State, a program was instituted that required sterilization of anyone with three children or more; rioting followed, but not before 3.7 million persons had been sterilized. Other states also engaged in compulsory sterilization programs, but the political costs were heavy. Eventually a total of approximately 22.5 million Indians were sterilized, but this effort could not be sustained.

Everywhere in India's cities, towns, and villages, one can see posters with warnings and exhortations in favor of small families. Various devices were designed to assist women in the poorest and most remote villages to learn and practice the rhythm method, but none worked satisfactorily. Obviously, India cannot wait for countrywide industrialization and urbanization to induce people to have fewer children; the attack must be launched on fertility now and where it is highest. Recent vital statistics from the state of Ker-

Indian women receive birth-control instructions in a family-planning center. The diffusion of information and dissemination of materials pose difficult problems for national and state governments in India. Costs are high, resources limited, results uncertain.

ala, where the rate of population growth was reduced following a state campaign, provide a glimmer of hope that such a decline can, indeed, be accomplished. All of India's efforts notwithstanding, however, the demographic situation in the world's largest democracy remains bleak. The 1985 national growth rate remained at the annual level of 2.2 percent (16.8 million), and India is on the way to becoming the world's most populous country before the end of the first half of the twenty-second century.

China

For nearly 30 years after the Communist government took power, China's demography was as much a mystery to the outside world as China was as a whole. Estimates of China's population (and its rate of natural increase) varied widely. In 1978, the World Bank published its *World Development Report*, which estimated a Chinese population of 826 million. Shortly thereafter, the Chinese government announced that the 1 billion mark had been passed. Guesses about China's population had been wrong by as much as 200 million!

If there were questions about China's numbers, the political and social regime of Mao Zedong left no doubts about its views on family planning. At the 1974 Bucharest conference, the Chinese representative denounced population policies as imperialist designs aimed at sapping the strength of developing countries.

Following Mao's death, China's new leaders expressed very different views. If China was to modernize, they said, the population spiral would have to be brought under control. They wasted no time reversing the Maoist trend. In 1979, the Chinese government launched a policy to induce a married couple to have only one child. This would stabilize China's population at about 1.2 billion by the end of the century.

The policy was applied loosely at first, but when this had less effect than desired, enforcement was severely tightened in 1982. The result was dramatic. In 1970, population increase in China was at a rate of 2.4 percent (as estimated by China's own planners); by 1985, it was down to 1.1 percent. In 1983, when the growth rate had been reduced to 1.2 percent, the United Nations gave China (along with India) its first Family Planning Award.

These statistics are encouraging, but they conceal the hardships faced by families. After 1982, the government made it mandatory for women to use contraceptive devices after they had their one child; if a second child nevertheless was born, one member of the parental couple would have to be sterilized.

Such rules imposed severe hardships, especially on farming families. To do the work of farming, many hands are needed; a large family is a rural tradition in many parts of the world. Therefore, many Chinese families defied the authorities. They kept pregnant women out of sight. They did not register their babies. They prevented inspectors from visiting villages.

At this point, the Chinese government proved its power to rule without impediments from opposition politicians, judges, or human rights advocates. Offenders of the new policies were fired from their jobs, had their farmlands taken away, lost many benefits, and were otherwise disadvantaged in a society suddenly rushing to modernize. In some parts of China, it was much worse: pregnant women known to have one or two children were arrested at work, in the fields, or at home and taken to abortion clinics where their babies were aborted—sometimes after more than six months of pregnancy. The national policies were imposed more harshly in some of China's provinces than in others. In southeast China, according to reports from visitors and Chinese reaching Hong Kong, the campaign was most severe.

Today the evidence indicates that the Chinese central government is not prepared to be flexible anywhere in the country except in those areas where minorities live (and even in such areas, the population control policies have had an impact). Any woman who has one child and becomes pregnant again without official authorization must have an abortion. It is estimated by China's own Ministry of Public Health that nearly 70 million abortions were performed in China between 1979 and 1985. Yet, just a generation ago, abortion was a crime, regarded as murder.

The sterilization campaign has reached many millions as well. In the mid-1980s, more than 20 million Chinese were sterilized *annually* during the height of this campaign, three times as many women as men.

The success of the Chinese population control effort is ensured, not only by government incentives and punishments, but also through the Chinese Communist party, its officials, local chiefs, and members. Through promises of advancements and cash payments for local compliance, party members have become birth-control vigilantes. No village, neighborhood, factory, or collective is without constant scrutiny.

The one-child-only policy has had a particularly sad side effect in a society where sons carry on the family name: widespread female infanticide, the killing of baby girls. The family whose first child is a daughter may abandon or even murder the child and try again later—for a boy. Although the government officially disapproves of this and although families caught engaging in such practices have been prosecuted, there is evidence that hundreds of thousands of cases of female infanticide still go unreported and unpunished in China. Demographers who have studied the ratios of male to female births in China during the 1980s have estimated that the number may exceed three hundred thousand annually. China's own population experts have expressed concern over the developing imbalance; in the China of the future, males will outnumber females quite substantially, with unpredictable social consequences.

Thus, China's relentless drive for zero population growth by the year 2000 has torn the traditions of Chinese society and has brought misery to millions. Chinese government and Communist party officials acknowledge that the policy is severe, but they point to the alternative as a justification: a China in which millions more are born, only to be permanently mired in a cycle of stagnation and poverty. To get ahead, these officials argue, the country cannot allow its gains to be negated by an ever-growing population. The end will justify the means.

The Chinese example underscores several salient

China's central government may have strict population policies, but the provincial governments also have a say in the matter. In May 1986, the People's Congress of the Province of Guangdong proclaimed that certain couples would be free from the one-child rule. Urban couples whose first child was disabled by an accident or *nongenetic* disease, couples who adopted a child after a physician diagnosed sterility, couples who are both only children, and couples with one partner who has worked in an underground mine for more than five years are exempt from the one-child rule.

---

aspects of the population dilemma. It is evidence of the degree of intervention necessary to achieve real change in a comparatively short time. The effectiveness of China's system for imposing government policies on the general population is unmatched in most countries of the world; so it is unreasonable to expect India, for example, to be able to make comparable gains. Most African countries faced with serious population problems have neither China's effective authority nor India's resources to address these issues. Therefore, the population explosion will remain a global concern for many years to come.

•   •
Reading about
Population Geography

This chapter touches on a variety of geographic topics relating to population. Perhaps the best place to start some additional reading is among the standard books on population geography. A recent publication is J. Beaujeu-Garnier's *Geography of Population* (New York: Longman, 1978). Another basic introduction, though less geographic, is John R. Weeks, *Population: An Introduction to Concepts and Issues* (Belmont, Calif.: Wadsworth, 1981). Fundamentals of demography are also well outlined in a volume by R. Tomlinson, *Population Dynamics: Causes and Consequences of World Demographic Change* (2d ed.; New

York: Random House, 1976). J. I. Clarke's introduction, *Population Geography* (Elmsford, N.Y.: Pergamon, 1972), though somewhat dated, remains worth a visit to the library. The same is true of *Population Geography* (New York: McGraw-Hill, 1970) edited by G. J. Demko, H. M. Rose, and G. M. Schnell. Also see D. G. Bennett, *World Population Problems: An Introduction to Population Geography* (Champaign, Ill.: Park Press, 1984); H. R. Jones, *A Population Geography* (New York: Harper & Row, 1981); J. L. Newman and G. E. Matzke, *Population: Patterns, Dynamics, and Prospects* (Englewood Cliffs, N.J.: Prentice-Hall, 1984); G. L. Peters and R. P. Larkin, *Population Geography: Problems, Concepts, and Prospects* (Dubuque, Iowa: Kendall/Hunt, 1983); and R. Woods, *Theoretical Population Geography* (New York: Longman, 1982).

For recent information on population issues and trends, an excellent source is the publication series of the Population Reference Bureau in Washington, D.C. If you are writing a paper on a demographic topic, consult your library's collection of U.N. statistical yearbooks and other publications.

World population distribution has been discussed in a series of books by G. T. Trewartha. The most recent is a volume he edited, *The More Developed Realm: A Geography of Its Population* (New York: Pergamon, 1978). An earlier one is G. T. Trewartha's *Geography of Population: World Patterns* (New York: Wiley, 1972). Another older book that contains much still-useful material is *The Limits to Growth* (New York: Universe Books, 1972), a disturbing treatise by D. H. Meadows and others. A more popular book,

241

written by P. Ehrlich, A. Ehrlich, and J. D. Holdren, *Ecoscience: Population, Resources, Environment* (San Francisco: Freeman), aroused much debate when it was first published; revised and updated in 1977, it is worth an evening's attention. Also see G. L. Tuve's *Energy, Environment, Populations, and Food* (New York: Wiley–Interscience, 1976). Still directly relevant is a three-volume work edited by G. O. Barney, *The Global 2000 Report to the President* (Washington, D.C.: U.S. Gov. Printing Office, 1980; 1981).

Economic-geographic problems are discussed in J. L. Simon, *The Economics of Population Growth* (Princeton, N.J.: Princeton Univ. Press, 1977). On a related topic, see P. A. Morrison, *Population Movements: Their Forms and Functions in Urbanization and Development* (Liège, Belg.: Ordina Editions, 1983). A book with a hopeful title is N. Eberstadt (ed.), *Fertility Decline in Developing Countries* (Chicago: Univ. of Chicago Press, 1980).

On the food question, there is a large amount of material, much of it outside the geographic literature. But a geographer's early contribution has become an oft-quoted classic: G. Borgstrom's *The Hungry Planet* (New York: Macmillan, 1976). Also still worthwhile is L. R. Brown and G. W. Finsterbusch, *Man and His Environment: Food* (New York: Harper & Row, 1972). A more recent contribution is A. T. A. Learmonth, *Patterns of Disease and Hunger: A Study in Medical Geography* (London: David and Charles, 1978). In general, you will have more success in the journal literature than in books for material on the geography of food and hunger.

The geography of health and well-being is discussed in *The Geography of Health and Disease* (Chapel Hill: University of North Carolina Press, 1974), edited by J. Hunter. An interesting collection of readings entitled *Medical Geography* (London: Methuen, 1972) was edited by N. D. McGlashan. A useful book is *The Social Geography of Medicine and Health* (New York: St. Martin's Press, 1983) by J. Eyles and K. J. Woods. The spread of diseases is discussed in W. H. McNeill, *Plagues and Peoples* (New York: Doubleday, 1976). Medical geography is one of those fields of geography that have practical application, see G. F. Pyle, *Applied Medical Geography* (New York: Wiley–Halsted, 1979). A facet of this is the delivery of health care, see G. W. Shannon and G. E. A. Dever, *Health Care Delivery: Spatial Perspectives* (New York: McGraw-Hill, 1974.

Again, some of the most penetrating literature in medical geography lies in the professional journals, for example, "River Blindness in Nangodi, Northern Ghana: A Hypothesis of Cyclical Advance and Retreat" by J. M. Hunter in the *Geographical Review* (Vol. 56, July 1966); and "The Ecology of African Sleeping Sickness" by C. G. Knight, in the *Annals* of the Association of American Geographers (Vol. 61, March 1971). In addition, consult G. M. Howe (ed.), *A World Geography of Human Diseases* (New York: Academic Press, 1977).

On human migration, see two volumes edited by B. du Toit and H. Safa, *Migration and Urbanization* and *Migration and Development* (The Hague: Mouton, 1975). Also see, H. McNeill and R. S. Adams (eds.), *Human Migration: Patterns and Policies* (Bloomington: Univ. of Indiana Press, 1978). Another valuable source is a book edited by L. Kosinski and R. M. Prothero, *People on the Move* (London: Methuen, 1975). Again, there is hardly a volume in the geographic periodical literature that does not contain several articles on some aspect of migration; and for any research paper, the major journals are indispensable. For an overview of some of this literature, see C. C. Roseman, *Changing Migration Patterns within the United States*, a resource paper published by the AAG in 1977.

Also important in connection with migration are R. Simon and C. Brettell (eds.), *International Migration: The Female Experience* (Totowa, N.J.: Rowman and Allanheld, 1986); R. Jones (ed.), *Patterns of Undocumented Migration: Mexico and the United States* (Totowa, N.J.: Rowman and Allanheld, 1984); and P. Ogden, *Migration and Geographical Change* (London/New York: Cambridge Univ. Press, 1984).

Population policies are discussed in S. Johnson, *The Population Problem* (New York: Wiley–Halsted, 1973); and by E. D. Driver in *Essays on Population Policy* (Lexington, Mass.: Heath, 1972). For insight into the impact of policies in an important arena, see H. Yuan Tien, *China's Population Struggle* (Athens: Ohio Univ. Press, 1973).

Malthus's *An Essay on the Principle of Popula-*

tion . . . can be found in any good library in both the original (1798) and the revised (1826) versions. For contrast, look up the U.N.'s volume entitled *The Determinants and Consequences of Population Trends* (New York, 1973). Also see, a volume edited by W. Zelinsky, L. A. Kosinski, and R. M. Prothero, *Geography and a Crowding World: A Symposium on Population Pressures on Social Resources in Developing Lands* (London/New York: Oxford Univ. Press, 1970), dated but still valuable.

Population geography (as we noted) is an interdisciplinary field. Several journals carry population geography articles, including *Demography* (published in Washington, D.C.); the *Population Bulletin* (published by the Population Reference Bureau in Washington, D.C.); *Population and Development Review* (published by the Population Council, New York); and *Population Studies: A Quarterly Journal of Demography* (a publication of the London School of Economics).

CHAPTER SEVEN

Culture
Worlds

ore than 5 billion people inhabit this earth, and they live their lives in thousands of different ways. Their languages differ. Their racial sources vary. Their religious beliefs are not the same. In their systems of education, their architectures, their technologies, and in countless other ways, the societies of the world reveal contrasts and differences. All of the world's societies, large and small, occupy a portion of the earth's surface; thus, all have regional properties. In some special way, all societies have exploited the opportunities of their natural environment and have adapted to its limitations. In the process they have all made their imprint on the landscape: in some cases practically transforming it, in others barely marking it.

Cultural Geography

Geographers study the spatial imprints of cultural expression from several vantage points. Before we consider these approaches, however, we should look more closely at the concept of culture itself. The world **culture** is not always used consistently in the English language, which can lead to some difficulties in establishing its scientific meaning. When we speak of a cultured individual, we tend to mean someone with refined tastes in music and the arts, a highly educated, well-read person who knows and appreciates the ''best'' attributes of his or her society. But as a scientific term, culture refers not only to the music, literature, and arts of a society but also to all the other features of its way of life: prevailing modes of dress, routine living habits, food preferences, the architecture of houses as well as public buildings, the layout of fields and farms, and systems of education, government, and law. Thus *culture* is an all-encompassing term that identifies not only the whole lifestyle of a people, but also their prevailing values and beliefs.

This is not to suggest that anthropologists and other social scientists have no problems with the concept of culture. If you read some of the basic literature in anthropology, you will find that anthropologists

have had as much difficulty with definitions of the culture concept as geographers have had with the regional concept (see p. 248). A culture may be the total way of life of a people—but is it their *actual* way of life (the way the game is played) or the standards by which they give evidence of *wanting* to live, through their statements of beliefs and values (the rules of the game)? There are strong differences of opinion on this; as a result, the various definitions have become quite complicated. Anthropologist E. Adamson Hoebel says in his *Anthropology: The Study of Man* (1982) that culture is ''the integrated system of learned behavior patterns which are characteristic of the members of a society and which are not the result of biological inheritance . . . culture is not genetically predetermined; it is noninstinctive . . . [culture] is wholly the result of social invention and its transmitted and maintained solely through communication and learning.'' This definition raises still another question: How is culture carried over from one generation to the next? Is this entirely a matter of learning, as Hoebel insists, or are certain aspects of a culture, indeed, instinctive and in fact, a matter of genetics? This larger question is the concern of sociobiologists and not cultural geographers, although some of its side issues such as **territoriality** (an allegedly human instinct for territorial possessiveness) and proxemics (individual and collective preferences for nearness or distance in different societies) have spatial and, therefore, geographic dimensions.

But even without these theoretical additions, the culture concept remains difficult to define satisfactorily. In 1952, anthropologists A. L. Kroeber and C. Kluckhohn published a paper that identified no fewer than 160 definitions—all of them different—and from these they distilled their own: ''Culture consists of patterns, explicit and implicit, of and for behavior and transmitted by symbols, constituting the distinctive achievements of human groups, including their embodiments in artifacts . . . the essential core of culture consists of traditional (that is, historically derived and selected) ideas and especially their attached values; culture systems may, on the one hand, be considered products of action, and on the other as conditioning elements of further action.''

246

Music is a powerful force in many cultures. These Moroccan folk musicians entertain a crowd in the town of Taroudant. They move from place to place, collecting coins from their listeners as they perform. Informal performances of this kind take place in Western society as well, from the street corners of New Orleans to the hills of Tennessee.

For our purposes, it is enough to stipulate that culture consists of a people's beliefs (religious and political), institutions (legal and governmental), and technology (skills and equipment). This construction is a good deal broader than that adopted by many modern anthropologists, who now prefer to restrict the concept to the interpretation of human experience and behavior as products of symbolic meaning systems. It is important to remember that definitions of this kind are never final and absolute but rather arbitrary and designed for a particular theoretical purpose. The culture concept is defined to facilitate the explanation of human behavior. Anthropologists today tend to focus on what people know, on codes and values, on the rules of the game. This was not always the case, as the group of definitions shows. Sociologists, political scientists, psychologists, and ecologists have different requirements and would construct contrasting operational definitions. The same is true of cultural geographers. Geographers would be attracted to the Herskovits definition because they have a particular interest in the way the members of a society perceive and exploit their resources, the way they maximize the opportunities and adapt to the limitations of their environment, and the way they organize that part of the earth that is theirs.

This aspect of the geographic approach to the study of culture—the human imprint on the natural environment—is a central theme in human geography. It is no accident that farming is called agri*culture*, because the production and consumption of food are fundamental ingredients of any culture. Farming

248

shapes and fashions the countryside, producing food *and* creating a sometimes-unmistakable landscape. When human communities that had depended on fishing, hunting, and gathering learned to plant crops and harvest them, an agricultural revolution began (see Chapter 8). Probably the first agricultural system to emerge was shifting cultivation, which involved the deliberate and planned planting of several crops in forest clearings. This, in turn, produced the earliest organized control over land use, labor assignments, and food allocation, all elements of culture. Shifting agriculture is still practiced in several areas of the world, and it is discussed in detail later. From these beginnings, there developed highly distinctive forms of sedentary agriculture, which are closely associated with particular cultures. The terraced paddies of East Asia, the huge grain fields of commercial farming on the U.S. Great Plains, the market gardens near European and other Western cities, the village fields of Africa, and the dairy farms of New Zealand all characterize regional cultures, and they represent but a few of the many agricultural patterns on the land. Certain crops (rice, olives, grapes, bananas, cassava) and domestic animals (the camel, llama, yak, elephant) are closely identified with specific cultures and readily evoke images of those cultures.

THEMES IN CULTURAL GEOGRAPHY

To organize the study of so wide a range of phenomena in geographic context, cultural geographers have developed a set of themes in cultural geography. Each of these is briefly discussed in this chapter, but you will see that every one of them could form the basis for an entire course. They are

1 **Cultural landscape**. The imprint of a culture upon the land; the creation of a distinct human landscape on a part of the earth.
2 **Culture hearths**. Certain areas of the world, in ancient and modern times that have been crucibles of cultural growth and strength.
3 **Cultural diffusion**. Ideas, innovations, and inventions have been disseminated and transmitted from their source areas (often culture hearths) to distant lands and peoples.

4   ***Cultural ecology***. The interrelationship between culture and nature, between human society and the natural environment, is a major theme not only in cultural geography, but also in the field of geography as a whole.

5   ***Cultural regions***. A culture, defined on the basis of certain criteria (e.g., its landscapes), extends over a certain definable area of the world.

6   ***Geographic realms***. The human world is divisible into a set of major realms that are multicultural but unified by certain properties and characteristics.

•   ## Cultural Landscape   •

A culture gives character to an area. **Aesthetics** play an important role in all cultures—often a single scene in a photograph or a picture can tell us, in general terms, in what cultural milieu it was made. The architecture, the mode of dress of the people, the means of transportation, perhaps the goods being carried—all reveal a distinctive cultural environment.

The people of any particular culture transform their living space by building structures on it, creating

Cultural landscape in a Turkish setting: the Blue Mosque dominates a section of Istanbul, Turkey. The Bosporus is in the foreground. Eastern European and Middle Eastern architectural styles meet in this pivotal city.

lines of contact and communication, tilling the land, and channeling the water. There are a few exceptions: nomadic peoples may leave a minimum of permanent evidence on the land; some peoples living in desert margins (e.g., the few remaining San clans) and in tropical forest zones (small Pygmy groups) alter their natural environment little. However, most of the time, there is change: asphalt roadways, irrigation canals, terraced hillslopes.

This composite of such features made by humans is conceptualized as the **cultural landscape**, a term that came into general use in geography in the 1920s. The geographer whose name is still most closely identified with this concept is Carl O. Sauer. In 1927, he wrote an article, "Recent Developments in Cultural Geography," in which he produced a deceptively simple definition of this concept. The cultural landscape, he said, constitutes "the forms superimposed on the physical landscape by the activities of man." However, when human activities cause change in the physical or natural landscape itself, does the physical landscape then become a cultural landscape? For example, when a dam is built in the upper course of a river, that dam can affect the whole character of that river downstream, even hundreds of miles away. It can alter the strength of the river's flow and the rate of deposition of sediments in a delta. Does that mean that the river is no longer part of the natural landscape and, therefore, has become a cultural landscape feature? Similar issues are raised by human-induced erosion of untilled soil and by regenerated, formerly cutover forests. Anyone who is interested can trace the debate in the geographic literature. For our purposes, perhaps the best definition is the broadest: the cultural landscape includes all identifiably human-induced changes in the natural landscape that involve the surface as well as the biosphere.

Thus, a cultural landscape consists of buildings and roads and fields and more, but it also has an intangible quality, an "atmosphere," which is often so easy to perceive and yet so difficult to define. The smells and sights and sounds of a traditional African market are unmistakable, but try to record those qualities on maps or in some other way for comparative study! Geographers have long grappled with this problem of recording the less tangible characteristics of the cultural landscape, which are often so signifi-

cant in producing the regional personality.

The more concrete properties of a cultural landscape are a bit easier to observe and record. Take, for example, the urban townscape (a prominent element of the overall cultural landscape) and compare a major U.S. city with, say, a leading Japanese city. Visual representations would quickly reveal the differences, of course, but so would maps of the two urban places. The U.S. central city, with its rectangular layout of the **central business district (CBD)** and its far-flung, sprawling suburbs contrasts sharply with the clustered, space-conserving Japanese city. Again, the subdivision and ownership of American farmland, represented on a map, looks unmistakably different from that of a traditional African rural area, with its irregular, often tiny patches of land surrounding a village. Still, the whole of a cultural landscape can never be represented on a map. The personality of a region involves not only its prevailing spatial organization, but also its visual appearance, its noises and odors, even the pace of life.

Culture Hearths

As long as human communities have existed on this earth, there have been places where people have done well, where they succeeded, where invention and effort were rewarded by an increase in numbers, growing strength, comparative stability, and general progress. And there have been areas where communities did not do well at all. The areas where success and progress prevailed were the places where the first large clusters of human population developed, both because of sustained natural increase and because other people were attracted there. The increasing numbers brought about new ways to exploit locally available resources and also generated power over resources located farther away. Progress was made in farming techniques and consequently in yields. Settlements could expand. Society grew more complex, and there were people who could afford to spend time not in subsisting, but in politics and the arts. The circulation of goods and ideas intensified. Traditions developed, as did ways of life that became the example for other areas far and near. These areas were humanity's early **culture hearths**, the sources of civilization, and from here radiated outward the ideas, innovations, and ideologies that would change the world beyond.

Culture hearths should be viewed in the context of time as well as space. Long before human communities began to depend on cultivated crops or domesticated animals, culture hearths developed—based on the discovery and development of a tool or weapon that made subsistence easier or more efficient. Fishing techniques improved and waterside communities prospered and grew long before the momentous changes of the first agricultural revolution began. Thus the Eskimo, with early and inventive adaptation to a frigid, watery environment, developed a culture hearth, just as the ancient Mesopotamians did. The nomadic Masai and their remarkable cattle-based culture still inhabit the region in which they achieved their culture hearth.

Some culture hearths, therefore, remain comparatively isolated and self-contained, others have an impact far beyond their bounds. When the innovation of agriculture was added to the culture complexes that already existed in the zone of the Fertile Crescent, it soon diffused to areas where it was not yet practiced and affected culture complexes far and wide. In the culture hearth itself, the practice of cultivation led to an explosion of culture, the evolution of an infinitely more elaborate civilization where one innovation followed the other.

Thus it is appropriate to distinguish between culture hearths—many of which have evolved across the earth from the Eskimo Arctic to Maori New Zealand—and the source areas of major early as well as modern civilizations. These latter also began as culture hearths, but their growth and development had wider, sometimes global impact. Early culture hearths (Fig. 7.1) developed in Southwest Asia, North Africa, and South and Southeast Asia, and East Asia in the valleys and basins of the great river systems. The Middle and South American culture hearths evolved thousands of years later, not in river valleys, but in highlands. The West African culture hearth emerged later still and was strongly influenced by innovations from the Nile Valley and from Southwest Asia.

The map of ancient culture hearths (Fig. 7.1) reveals the relationships between these cultural source areas and major rivers or other bodies of water. Mesopotamia, "between the rivers," lay juxtaposed to the Fertile Crescent, one of the places where people first learned to domesticate plants. It was one of the first regions to benefit from the new techniques emanating from the Fertile Crescent; the opportunities for irrigation afforded by the river regimes added new dimensions in planning, cultivation, and harvesting. Substantial settlements developed, and Mesopotamia found itself at the focus of a whole network of routes of trade and movements across Southwest Asia. Stimuli came from the Mediterranean, from Persia (now Iran) and from the north, and innovations were diffused as far away as Egypt and the Indus—and farther still. Not all those inventions were quickly adopted wherever they could have been. The wheel and the chariot, in use in Mesopotamia not long after 6000 B.P., found adoption in Egypt more than two thousand years later.

Ancient Egypt, too, was one of the world's earliest culture hearths. The scene of this culture hearth lay above the delta and below the first cataract, where the Nile Valley lies surrounded by inhospitable country. The region was open to the Mediterranean, but otherwise it was rather inaccessible to overland contact. In contrast to Mesopotamia, which lay open to its surroundings, the Nile Valley was a natural fortress of sorts. There, the ancient Egyptians converted the security of their isolation into progress. The Nile waterway was the area's highway of trade and association, its lifeline. It also sustained its irrigated agriculture and was a great deal more predictable in its cyclic regime than were the Tigris and Euphrates. By the time ancient Egypt finally began to fall victim to outside invaders, from about 3700 B.P. onward, there had emerged a full-scale civilization whose permanence and continuity are reflected by the massive stone monuments its artist-engineers designed and created.

Asian Hearths

Much of the archeological record of the Indus civilization has been buried beneath the present water table in the Indus River Valley, but there is no doubt that here, too, was a quite sophisticated urban culture with

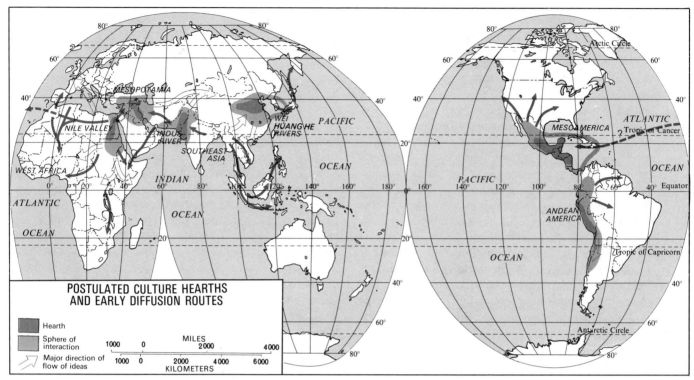

FIGURE 7.1

Ancient culture hearths and probable diffusion routes.

large and well-organized cities in which houses were built of fired brick and consisted of two and sometimes even more levels. There were drainage systems, public baths, and brick-lined wells. Protective fortifications were laid out around the major cities, of which Mohenjo-Daro and Harappa are best known. As in Mesopotamia and the Nile Valley, considerable advances were made in the technology of irrigation, and the culture was based on the productivity of the Indus lowland's irrigated soils. In the pottery and other artistic expression of this literate society lies evidence of contact and trade with the other major lowland civilizations of Southwest Asia and North Africa.

The culture hearth of ancient China was located around the confluence of the Hang (Yellow) and Wei rivers. There, it received stimuli from the Southwest Asian cultures in areas of agriculture, irrigation tech-

niques, metalworking, and possibly even writing. Whether and how these innovations reached northern China still remain matters of conjecture, but the character of the Hang–Wei region as a culture hearth of prime importance is beyond doubt. There is a record of Chinese civilization over five thousand years long, and from 4000 B.P. onward that record is continuous, quite reliable, and proof positive of China's cultural individuality. In fact, this part of China has yielded archeological remains that indicate a human presence many tens of thousands of years ago. Chinese cultural attributes were disseminated across Korea and into Japan as well as southward from the core area into the region that is modern China. China still remains focused on its ancient northern heartland, the only modern country of major stature whose lineage traces back spatially as well as temporally directly to its oldest origins.

Africa

Savanna West Africa received stimuli from the Nile Valley. The savannas of West Africa lie between the desert to the north and the tropical rain forest to the south. For many centuries, the peoples of the desert have traded goods with the peoples of the forests. The forest dwellers needed salt; the desert people could easily provide it. The forest people could produce ivory, gold, spices, hides, wood; from the desert people, they received metal goods, many kinds of cloth, dried fruits, and much else. The peoples of the West African savanna found themselves positioned very advantageously between the trading partners, and their towns grew into market centers. From the Nile Valley, these peoples received some political ideas: about divine kingship, about power inherent in a monopoly over the use of iron, and about military forces exacting tribute and maintaining control. But West Africa was itself also a place of invention and innovation. Some anthropologists say that West Africa along with Southwest Asia, Southeast Asia, and Middle America was one of the four major agricultural complexes of the ancient world, where crops were domesticated and farming techniques invented and refined. And so, in West Africa, a series of states of impressive strength and amazing durability arose that welded diverse peoples into unified entities. Their influence radiated far into the forests, into the desert, and southeastward into black Africa.

The Americas

Tropical Middle America also was the scene of impressive cultural achievements. From central Mexico to Costa Rica lay a region of highly developed indigenous culture with agricultural specialization, urbanization, permanent transport networks, writing, various art forms, religion, and other features. Here lay the great Mayan and Aztec civilizations, in the culture hearth known as **Mesoamerica**. The Mayan state was theocratic, with a complex religious hierarchy; like the ancient Egyptians, the Maya built magnificent palaces, huge temples, and large pyramids (it has been suggested that Egyptian mariners somehow crossed the Atlantic and contributed their knowledge to Mesoamerica). When the Spaniards arrived, much of Mayan civilization was already in ruins, the cities overgrown with tropical vegetation. But on the Mexican plateau, the Aztec civilization thrived. The Aztecs' leading city, Tenochtitlán, was a metropolis for

---

• Acculturation •

When cultures make contact, a process of intercultural borrowing takes place: there is a continuous exchange of traits and elements that results in the modification of the original cultures and the creation of new, blended ones. This process is called **acculturation**. Often, when the cultures have different strengths, the weaker culture will be affected far more than the stronger one.

When the Spanish conquerors overthrew the Aztec state in present-day Mexico, Spanish culture was the strongest, thus the Aztec culture was transformed. Towns and villages, religious practice, and economies were much modified. But Spanish culture also changed. Spanish-Mexican architecture began to reflect some of the motifs of Indian buildings. Aztec-domesticated crops and animals were acquired. The Spanish residents began to wear clothing that revealed the influence of Indian color and cut. Some Indian words entered into Spanish use. Mexican culture became a blend too, and acculturation affected both cultures.

Acculturation differs from *enculturation*, a term that refers to the process in which individuals learn the traditions of a culture and assimilate these; it also differs from **transculturation**, a process all cultures undergo. In transculturation, a culture gains new elements, through invention and borrowing, and loses or changes existing ones.

Mayan structures at Tikal, Guatemala. The theocratic Mayan state produced magnificent architecture, with palaces and pyramids reminiscent of ancient Egypt.

its time, counting perhaps one hundred thousand inhabitants and serving a vast hinterland. The Mayan cities had been primarily ceremonial centers, but the Aztec cities and towns were true central places that interacted with major spheres of influence. In these urban places, Aztec culture recorded remarkable accomplishments in architecture, the arts, and other spheres. But it was in the rural areas that the culture had its most lasting impact. Dozens of crops were domesticated, some of them staples in the diets of many millions of people to this day. And the **chinampa** method of cultivation added uniquely to the productivity of the Aztecs' lands. By this method, practiced along the shallow margins of the area's lakes, layers of aquatic vegetation and lake mud were piled upon each other to create thick, layered mats of alluvial and vegetative material so fertile that several

crops could be grown in them each year. The method is still in use today.

In South America, the mountainous country of the high Andes witnessed the rise of an important culture hearth. The roots of this culture appear to lie in coastal Peru, from where it may have been transmitted into the **altiplanos**, the valleys between the ranges of the Andean chain. These individual **intermontane basins** came to sustain discrete regional cultures. One of these was centered on Cuzco, destined to become the focus of Inca civilization. Gradually other altiplanos were brought under Inca domination, and eventually the Incas built an empire that extended over much of the Andes region (Fig. 7.1). The Incas were superb engineers and extremely effective colonizers. They managed to create a network of communications over their difficult territory by building roads and bridges, they terraced the steepest of slopes and farmed them, and they brought the unifying element of Inca culture wherever they subjugated neighboring peoples. Although they did not match the Aztecs in calendrics, mathematics, or the arts, the Incas did achieve much in fields of architecture (Cuzco had impressive fortresses and public buildings), economics, administration, and agriculture. Cuzco, their ancient headquarters, was a city of perhaps as many as 250,000 people, 200 miles (320 km) inland from the ocean and 11,000 ft (3350 m) above sea level! Even today, with these dimensions, it would be among Andean South America's most substantial urban centers. Certainly the Inca civilization ranks among the great ancient culture hearths.

URBANIZATION IN THE CULTURE HEARTHS

The culture hearths of ancient times were crucibles of civilization. The focal points of these regions were places the world had never known: large clusters of people living permanently in the first cities. It was possible, in largest measure, because there were food surpluses, the long-term availability of food generated by a revolution in agricultural methods (see Chapter 8). Soon the growing towns began to compete against each other, and geography came to play its vital role. The productivity of particular areas had much to do with the development of towns: the proximity of fertile farmlands and the regular and dependable flow of produce favored certain villages over others, and among these villages there were some that evolved into larger settlements. In turn, the ability of the town's residents to organize and to plan was an important determinant of sustained growth. A town that grew to overshadow other settlements developed new requirements, which also involved new risks. The inflow of food staples from a wider area had to be protected. The collection of taxes and tribute from the expanded region of urban dominance must be organized, administered, and maintained. The community itself needed protection against adversaries, which required such collective action as the construction of fortified walls.

Hence, the geographic advantages of certain locations as well as the internal organization of the clus-

tered community played roles in the growth of ancient towns and cities. Not only proximity to productive farmlands, but also the availability of water from surface or underground sources and the defensibility of the settlement's site contributed to the durability of certain towns. On the flanks of the Fertile Crescent, towns in Mesopotamia enjoyed secure food supplies. In the Indus Basin, the first cities were served by carefully constructed and maintained stone-lined wells to tap groundwater supplies. Also significant was the position of towns on ancient routes of travel and trade. Where such routes converged on an urban place, there was contact, interaction, and growth. Less accessible, more isolated places were at a comparative disadvantage.

Urban growth tested the ingenuity of the town's leading citizens. Food must not only be acquired and stored, but also distributed. Essential to such a system of allocation, of course, was a body of decision makers and organizers, people who controlled the lives of others—an **elite**. Such an urban-based elite could afford itself the luxury of leisure and could devote time to religion and philosophy. Out of such pursuits came the concept of writing and record keeping, an essential ingredient in the rise of urbanization. Writing made possible the codification of laws and the confirmation of traditions. It was a crucial element in the development of systematic administration in urbanizing Mesopotamia and in the evolution of its religious-political ideology. The rulers in the cities were both priests and kings, and the harvest the peasants brought to be stored in the urban granaries was a tribute as well as a tax.

Thus ancient cities had several functions. As centers of power, they became political foci, the headquarters of the first statelike entities the world had seen. As religious centers, their authority was augmented by the pressure of the clergy and by temples and shrines; ancient cities in many areas of the world were **theocratic** centers, where rulers were deemed to have divine authority and were, in effect, god-kings. In the Americas, the great structures of Yucatan, Guatemala, and Honduras built by the ancient Maya Indians (including Tikal, Chichén-Itzá, Uxmal, and Mayapán) exemplify such places. As economic centers, they were the chief markets, the bases from

255

which wealthy merchants, land and livestock owners, and traders operated. As educational centers, they included among their residents respected teachers and philosophers. They also had their handicraft industries, which attracted the best craftspeople and inventors. Therefore, ancient cities were the anchors of culture and society, the focal points of power, authority, and change.

The emergence of urban places did not begin simultaneously; after all, the culture hearths themselves did not develop at the same time. The earliest evidence of expanded settlements that may be called towns comes from Mesopotamia about seven thousand years ago, when long-established villages grew larger and took on urban characteristics. The Nile Valley witnessed similar events perhaps eight centuries later and the Indus Valley about 4800 B.P. China's urbanization may have begun simultaneously or perhaps even earlier than that of the Indus Valley. European urbanization, beginning in the insular Mediterranean area, commenced much later, around 2000 B.C. (4000 B.P.), and the great Mayan structures of Middle America were constructed beginning during the last millennium B.C. Urbanization in West Africa, where records are virtually nonexistent, probably dates back about two thousand years.

The earliest towns probably experienced the same growth conditions that later sustained the modern rise of such cities as Paris and London: as the principal centers, crossroads, markets, places of authority, and religious headquarters, they drew the talents, the trade, and the travelers from far around. Where should the beginnings of metallurgy have been based but in these incipient cities? Where would a traveler, a trader, a priest, or a pilgrim rest before continuing a journey? Towns had to have facilities that would not be found in farm villages: buildings to house the visitors, to store the food, to treat raw materials, to worship, to accommodate those charged with the protection and defense of the place. The cities grew—but drew attention at a price. Their contents of food and wealth and their status and regional influence spread their reputations far and wide; soon there were enemies who realized that the takeover of a city would mean power, prestige, and riches. So the early cities were required to build defenses—walls, moats, and the like—and to maintain armed forces. Their troops also came in handy when it was necessary to exact tribute and taxes from the surrounding rural region and to subdue rebellious elements within the city's own walls.

How large were the ancient cities? The indigenous cities of Middle and South America were quite substantial as was noted previously. But for ancient Middle Eastern cities, we only have estimates because it is impossible to conclude from excavated ruins the total dimensions of a city at its height or the number of people that might have occupied each residential unit. However, it seems that, at least by modern standards, those ancient cities were not large. The cities of Mesopotamia and the Nile Valley may have had between ten thousand and fifteen thousand inhabitants after nearly two thousand years of growth and development. That, scholars conclude, is about the maximum size that could have been sustained by then-existing systems of food gathering, distribution, and social organization. So, these urban places were but islands, geographic exceptions in an overwhelmingly rural society. Urbanized *societies* such as we know today did not emerge until several thousand years later.

Cultural Diffusion

The ancient city was not just a cluster of buildings and people, of markets and defenses. It also represented an innovation, a new concept of productive living. Word of these amazing places where goods could be traded for food and where toolmakers made implements for all kinds of uses spread far and wide. People who had visited a city told tales of events that could only happen in such a place. The city became more than a tangible place. It became an image, a vision of far-off power and splendor.

The cities were foci of the culture hearths. They were places of cultural growth and strengthening, invention and innovation, and sources of ideas and stimuli. From these source areas, newly invented techniques, tools, instruments, and ideas about ways of doing things radiated outward, carried by caravan and soldier, by merchant and mariner. Some of the innovations that eventually reached distant peoples were quickly adopted, often to be modified or refined; others fell on barren ground.

The process of dissemination, the spreading of an idea or an innovation from its source area to other cultures, is the process of **cultural diffusion**. Today the great majority of the world's cultures are the products of innumerable ideas and innovations that arrived in an endless, centuries-long stream. Often it is possible to isolate and trace the origin, route, and timing of adoption of a particular innovation, so that the phenomenon of diffusion is a valuable part of the study of cultural geography.

Spatial diffusion occurs through the movement of people, goods, and ideas. Carl O. Sauer—the cultural geographer whose work we encountered in Chapter 1—analyzed the diffusion of agricultural innovations, methods, and crops. The processes of diffusion, he found, are very complicated. This attracted the attention of those geographers who focus on theory and modeling, and diffusion is today much better understood through the pioneering work of another of these theorists, Torsten Hägerstrand.

Hägerstrand, a Swedish geographer, took an elaborate set of assumptions about how diffusion works and developed a model to explain it. He concluded that there are four stages in the process. In the *primary* stage, an innovation or invention appears in its source area and begins to be adopted there. Next comes the *diffusion* stage when the innovation disperses rapidly outward and is adopted in new locales. Thus an invention made in one of those ancient cities would find its way to smaller towns and villages in the surrounding region and take hold there. This is followed by what Hägerstrand called the *condensing* stage during which the innovation spreads into yet-unaffected areas, for example, to farmers in the countryside. And last there is the *saturation* stage when the diffusion process slows down and ends.

EXPANSION AND RELOCATION

Expansion and relocation is a logical sequence of events. But how does the process actually take place?

Further research has produced some interesting results. Spatial diffusion takes place in two ways: **expansion diffusion** and **relocation diffusion**.

In the case of expansion diffusion, an innovation or idea develops in a source location, is adopted there, and spreads outward through a population that is fixed (i.e., a population that is not itself moving or migrating). Later in this chapter, we will discuss the spread of Islam from its source area on the Arabian Peninsula to Egypt and North Africa, to the Middle East and Southwest Asia, to India and Southeast Asia, and to West Africa. This is an instance of expansion diffusion; if we drew a series of maps at 50-year intervals, beginning about A.D. 620, the area of adoption of the Islamic faith would show larger on every map. Certainly there were people moving, traveling from one place to the next carrying Mohammed's message, but the adopting populations were mainly in place, from West African villagers to Indonesian islanders. They received and accepted Islam where they were.

Expansion diffusion takes on two forms. First there is *contagious diffusion*, a word that may remind you of the way a disease spreads—and appropriately so. Here it is the proximity (nearness) of propagator and adopter that is the key. The innovation is likely to disperse throughout the whole population, affecting all (or most) individuals in some way. By actually seeing another person doing something new or by being informed of it, someone is affected. But some ideas are not diffused this way. An innovation that has much relevance in the urban culture of a large city may not have nearly as much of a market in rural areas. In **hierarchical diffusion**, certain people in a society adopt an invention rapidly, but others are left largely unaffected. Something may be especially appropriate for older people or for younger ones—such as break dancing! In a sense, hierarchical diffusion is selective diffusion.

Relocation diffusion occurs when whatever is being diffused enters the area with its carrying agent, most commonly a migrating population. When the Europeans who emigrated from Europe in such large numbers reached American shores, they brought with them countless pieces of cultural baggage that, in turn, they dispersed throughout this continent. The spread of the log cabin during Colonial times in the northeastern United States is often used as an example of this process. This innovation was transplanted from Sweden by seventeenth-century migrants, and it was then adopted by many thousands of other Europeans moving westward into the Appalachians and beyond.

To return to the ancient culture hearths and their innovation-rich cities, we should be careful not to conclude that every idea or invention had just one source. The appearance of a particular technique or device in widely separated areas does not necessarily prove that diffusion did occur. Various cultures in parts of Asia, Africa, and the Americas developed forms of irrigation, learned to domesticate animals and plants, and made other achievements on their own. This is **independent invention**, a process that may be difficult to discern and one that often leads to debate among cultural geographers. Some scholars, for example, have suggested that because the ancient Maya built pyramidlike structures in their cities, there must have been a route by which the pyramid concept diffused from the Nile Valley to Middle America. Others argue that if the ancient Egyptians could invent pyramidal structures, so could the Maya.

STIMULUS AND MATERIAL

Cultural geographers also distinguish between **stimulus** diffusion—the dissemination and spreading of an intangible idea that may, on adoption, be changed almost beyond recognition—and **material diffusion**—the outward spread of a tool or device that can be traced and identified. But whatever is being diffused, there are also factors that work against adoption. One of these is distance. The farther from the source, the less likely is an innovation to be adopted and the weaker are the innovation waves. Related to this aspect is the factor of time. Acceptance of an innovation becomes less likely the longer it takes to reach distant populations. Therefore, **time-distance decay** must be considered in any study of the spread of an idea.

An understanding of the way diffusion processes occur is helpful in the study of many problems in historical cultural geography. Sometimes the appearance of a single item, a fragment of a traditional ceremony, a drawing or carving, a building style, a word in a vocabulary can reveal contacts and linkages and begin to explain spatial variations of cultures. We return to this topic later in the chapter.

Cultural Ecology

The distribution of old culture hearths inevitably leads to questions about the environment under which they thrived. Some ancient cultures lie revealed today by ruins in a desert or in a tropical rain forest. Could the cities and other settlements have prospered under such conditions or have climates changed so much that where there is desert today there may have been ample water supply some thousands of years ago? And in any case, what are the optimum environmental conditions for the evolution and growth of civilizations? Present-day world culture hearths seem to lie in temperate climatic regions, not in the tropics, or in the desert, or in near-polar zones. Does this mean that we can generalize about optimum climatic conditions for the human success story?

As we noted in Chapter 1, Aristotle answered this question affirmatively twenty-three centuries ago when he generalized about the peoples of cold, distant Europe as being "full of spirit . . . but incapable of ruling others" and those of Asia as "intelligent and inventive . . . [but] always in a state of subjection and slavery" (see p. 00). And many geographers after Aristotle have held similar views. How easy it is to view people living in cold climates as "hardy, but not very intelligent," those of the warm tropics as "lazy and passive," and those in the intermediate zones as productive and progressive. Here is how Elsworth Huntington, a twentieth-century geographer, stated it in *Principles of Human Geography*, 5th ed., published in 1940: "The well-known contrast between the energetic people of the most progressive parts of the temperate zone and the inert inhabitants of the tropics and even of intermediate regions, such as Persia, is largely due to climate . . . the people of the cyclonic regions rank so far above those of other parts of the world that they are the natural leaders" (p. 339).

258

From the debate over the impact of the environment on human behavior has emerged a whole new field of geography in which interest focuses *not* on the manner in which environment conditions human behavior, but on the way people perceive their environment and act on the basis of those perceptions. We all have an image of the world around us, and we could argue that our decisions and actions are based on this image rather than on reality. If this is the case, then our concern when we analyze behavior ought not to be with the objective realities of our environment, but with these images we have in our head. What goes into the formation of these images? Undoubtedly an Indian's view of the American Southwest would be quite different from most white Americans' image of that region. A person living in an urban ghetto perceives it quite differently from a person who does not live there. Environmental determinists of the past mostly failed to recognize that the environment that interacts with humans consists not only of landscapes and climates, but also of attributes created by people themselves, all elements of a cultural environment.

Now, when we want to study human activities in the context of environment, the magnitude of the problems involved becomes clearer. Experience, the accessibility of information, psychological state, physiological condition, cultural traditions, social constraints—all these factors come into play. And this is only a short list. The longer it becomes, the less likely we are to engage in the business of sweeping generalizations about climate and behavior!

The doctrine expressed by these statements is **environmentalism** or, more specifically, **environmental determinism**. It holds that human behavior, individually and collectively, is strongly affected by and even controlled, or determined, by the environment that prevails. It suggests that the climate is the critical factor in this determination: for progress and productiveness in culture, politics, and technology, the ideal climate would be, say, that of Western Europe or the northeastern United States. The people of hot, tropical areas or cold, near-polar zones might as well abandon hope. Their habitat decreed that they would never achieve anything even approaching the accomplishments of their midlatitude contemporaries.

And so, for a time, geographers set about their attempts to explain the distribution of present and past centers of culture in terms of the "dictating environment." In the early part of this century that was a major theme in the field: the manner in which the natural environment conditions human activities. Quite soon, though, there were geographers who doubted whether those sweeping generalizations about climate and character, environment and productiveness were really valid. They recognized exceptions to the environmentalists' postulations (e.g., the Maya Empire in Mesoamerica may have arisen in tropical rain forest) and argued that people are capable of much more than merely adapting to the natural environment. As for the supposed efficiency of the climate of Western Europe, this was an interesting idea but not scientifically proved. And surely it was best not to base laws of determinism on inadequate data in the face of apparently contradictory evidence.

These arguments helped guide the search for answers to questions about the relationships between human society and the natural environment in different directions, but the unqualified environmentalist position was still held by some in the past several decades. For a more modern approach, it is interesting to read S. F. Markham's *Climate and the Energy of Nations* (1947), still an environmentalist statement, but with the benefit of Huntington's massive writings and the works of several others behind it. Markham thought that he could detect in the migration of the center of power in the Mediterranean (from Egypt to Greece to Rome and onward) the changing climates

259

of that part of Europe during the most recent several thousand years of glacial retreat. He argued that Egypt was much cooler in its prime than it is now; when the Greeks rose to prominence, Greece's climate, too, was more conducive to the generation of "energetic" people. Rome was so cool that public hot baths were a part of daily life. In all this, Markham saw the northward movement of **isotherms**. Like his predecessors, he tried to determine the optimum temperature for physical and for mental activity, and he was sure that these temperatures prevailed, in turn, in Egypt, Greece, and Rome (and later in Western Europe).

But geographers grew increasingly cautious about such speculative writings, and they began to ask their questions about human society and natural environment in new ways. There still is the old interest in how people are affected by the natural environment, but if generalizations are to be made, they ought to come from detailed, carefully designed research. Everyone agrees that human activity is in certain ways affected by the natural environment, but people are the decision makers and the modifiers, not the complete subordinates the environmentalists believe them to be.

In some ways, this issue remains one of geography's most relevant and interesting themes. Whatever the answer to the arguments of the environmental determinists, there can be no doubt that, even over the comparatively short span of the past ten thousand years, the earth has seen major environmental change, and conditions in the old culture hearths as well as the sites of more recently emerging states have changed. Take those West African states that flourished in the savanna zone for many centuries between about A.D. 500 and 1500. They benefited from a circumstance geographers call regional **complementarity**, in which two adjacent or nearly adjacent regions produce different goods and each needs the other's products. In the case of West Africa, the peoples of the southern, coastal forests needed the products the northern, desert peoples could produce (salt, leather goods, metal items), just as the northerners desired forest products such as foodstuffs, spices, textiles, and ivory. In the savanna zone between the northern desert and the southern forests, there arose a series of well-organized, long-lived states whose leading cities were major market centers where these

goods were exchanged. Timbuktu, for example, was such a central place.

Descriptions of life in that West African region a thousand years ago suggest that environmental conditions were quite different from what the map shows today. The Sahara was narrower, its margins not as dry, its heart more easily crossed. The coastal forests formed a wider belt, the savanna was broad and better watered. Almost certainly the environment here changed drastically, and the change took its toll on the economies of the region's states.

Today such change (in the form of desertification) often is attributed to overpopulation by people and their livestock, especially by overgrazing. But recent research in the savanna zone has produced evidence that there was much iron smelting in the area where the Niger makes its great elbowlike bend. Fuel for that smelting, of course, was the savanna forest. Some archeologists have described the amount of slag (smelting waste) they found as staggering—staggering because of its implications for the forests of the time. Deforestation is nothing new.

When major environmental modification takes place (whether caused by natural process or induced by human societies), the impact on the regional culture can be severe. People may find themselves compelled to move away in large numbers. Or there may be famine, population decline, and cultural weakening. From this broad perspective, therefore, the environmental role in affecting the human condition is beyond question.

Geographers studying ecological matters emphasize that the relationships between human society and natural environment change as technology changes. Climatic change over Japan, for example, would have far less immediate impact than was the case in the West African example just cited. Japan's highly developed economy depends on worldwide sources of raw materials and on global markets for its products. Most (though not all) of the food Japan consumes is bought overseas with money earned from these exports. To Japan, the stability of those foreign links is much more important than are environmental fluctuations at home. The study of cultural ecology thus requires a knowledge not only of environmental factors, but of social conditions as well.

Cultural Regions

The concept of the culture area is another important theme in cultural geography. But before concentrating on culture areas, we should take note of another geographic term: the **region**. Use has been made of this term in previous chapters without any precise definition. And for good reason: because it is as difficult to define region as it is to define culture. Now, however, we are faced with the need to stipulate what the regional concept involves. Like **area**, region is a word commonly used to describe some part of the earth's surface. We all have some idea of what it means: when we talk about conflict in the Middle East or farm problems in the Corn Belt, we have used the regional concept in its broadest sense as a frame of reference (Fig. 7.2). But as geographers, we may have to be more specific than that. Just what is the Middle East? Does it consist of all the countries of the Arab world or only certain ones, and if only certain ones, exactly which ones? Where are the precise limits of this Corn Belt? How large a percentage of farmland must be under corn cultivation for it to qualify as part of this region? These are not frivolous questions. If the regional concept is to be useful, it must have firm foundations.

FIGURE 7.2
A regional definition of the Corn Belt in the United States.

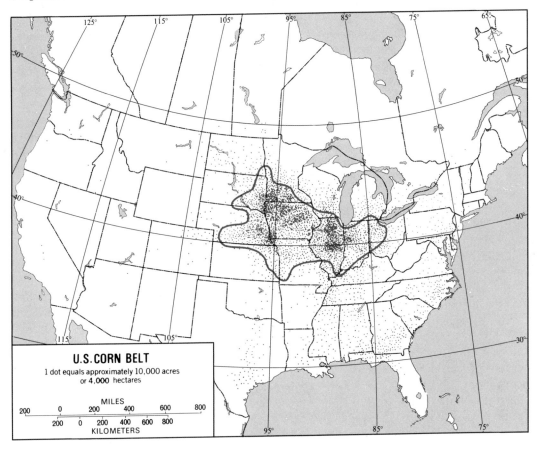

Properties

The first thing we observe about regions is that they have **location**. By our reference to the Corn Belt (or to Indochina or the Amazon Basin or the Chicago area, all regions of a kind) we have in mind a region positioned in a particular location somewhere on this globe. We could actually identify that location by using the grid system (referred to in Chapter 3) to describe it by degrees latitude and longitude. That would be the region's **absolute location**, but such a numerical index would not really have much practical value. Location attains relevance only when it relates to other locations. Thus if we are to recognize the Corn Belt as a region, we would identify it with reference to the climatic zone(s) in which it lies (Plate 3) and the soil regions over which it partly extends (Fig. 4.5). Now we are saying something about the Corn Belt's **relative location**, a much more meaningful criterion.

A second quality of regions is that they have **area**. This seems very obvious, but herein lie some difficult problems. If a region is to have areal extent, its limits must be definable. We would probably all agree that the Rocky Mountains form a physical (physiographic) region in the United States. But taking a map and drawing a line exactly where the Rocky Mountains end and other regions (such as the adjacent Great Plains) begin is not as simple as it might seem. Mountainous regions often peter out in a marginal belt of lower hills before they give way to true plains. Thus a decision has to be made whether those hills lie beyond the limits of the mountain region or are part of it.

A third characteristic of regions is their internal sameness, or **homogeneity**. Each time we mentioned examples of regions, there was something to unify: a high intensity of corn farming in the Corn Belt; forests, heat, humidity, flatness in the Amazon Basin (Fig. 7.3); high relief, slopes, rugged topography in the Rocky Mountains. Professor R. Hartshorne in *Perspective on the Nature of Geography* (1959) defines a region as ''an area of specific location which is in some way distinctive from other areas and which extends as far as that distinction extends'' (p. 93). This distinctiveness may be physiographic, as in several of our examples, but it can also be cultural. A certain type of architecture may prevail over an area, and we recognize a region. In eastern Canada, the French language is spoken by a majority of people in a region centered on Montreal. In Chapter 3 we discussed several climatic regions. In Chapter 6, when we spoke of important population clusters in the world, we thereby regionalized population on the basis of high density.

Thus a region is a part of the earth's surface, some specific locale in which certain predetermined criteria are met. The boundaries of the region are determined by the criteria; they are not something we would go and search for in the field. Take our Corn Belt example again. We would have to establish some criterion (50 percent of farmland under corn, as an instance) and then draw our region's boundary on that basis. Someone else might very well argue that 60 percent is a more appropriate criterion to use. That will change the region's boundaries. Thus the Corn Belt is not an absolute region; it is an artificial construct, a device to use for further discussion. That is true of all regions: they are intellectual devices designed to function as organizing concepts in geography.

Formal and Functional

The kind of region we have so far discussed—in which sameness (homogeneity) is the rule—is called a **formal** region. But regions may be conceptualized on other bases. For example, a large city has a substantial surrounding area for which it supplies goods and services, from which it buys farm products, and with which it interacts in numerous other ways. Its newspapers sell in the smaller towns in this area, the city's manufacturers distribute their products wholesale to subsidiaries there. Any map of road traffic, or of telephone calls, or of radio-station listening will confirm the domination of the large urban center in this, its tributary area, or hinterland. Here again, we have a region, but the region is not characterized by homogeneity, but rather by the city-centered system of interaction that creates it. Appropriately, such a region is referred to as a **functional region**. The region is conceptualized as a system, and it is defined by its interacting parts rather than by its visible homogeneity. And obviously, it can only be recognized through the

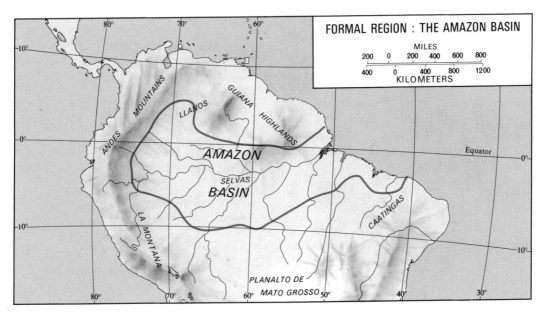

FIGURE 7.3

A definition of a formal (homogeneous) region: the Amazon Basin in South America.

analysis of the many ways the city dominates the activities within it. Where the city's influence wanes lie the boundaries; another city's strength may be taking over. Once again, the functional region is an artificial, a mental, construct, and "finding" it depends on the criteria we set ourselves before we begin our search for it.

What we have just said in rather definite and positive terms about regions is not beyond debate, and no one is a geographer for very long without becoming aware of that. Some geographers have doubts about the usefulness of the whole regional concept; others recognize functional regions but find formal regions unproductive as organizing concepts. Some geographers would even argue that a Corn Belt region does not exist: if we have to imagine its existence through the setting of arbitrary criteria, they argue that there is little or no validity to the end result. The dispute goes on in the scientific journals, but it is difficult to read any substantial part of the geographic literature over the past hundred years or so without concluding that the region has been a central concern and a concept that has had a unifying influence over a

very wide-ranging discipline. Physical as well as human geographers, people interested in systematic work and those interested in countries—all have employed the regional concept in many useful ways.

REGIONS AND CULTURES

In cultural geography, the regional concept has produced the idea of the **culture** region, a portion of the earth's surface where certain cultural traits prevail. The **formal** culture region is most easily defined where culture traits are markedly different from those around it. Localized traditional cultures, for example, Cherokee Indian and Eskimo, possess discrete systems of behavior and are marked by distinctive dwellings, a particular language, a certain economic system, a specific kind of social organization, and unique religious beliefs and practices. Any attempts to delimit the culture region would involve mapping the regions where each of these traits prevails and then creating a composite boundary. This would produce the same problem we encountered when trying to define the Corn Belt as a formal region: certain arbitrary deci-

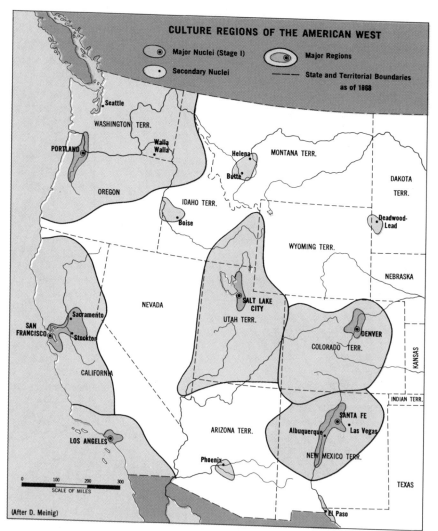

FIGURE 7.4

sions must be made. Culture regions, even localized cases of comparatively small areal extent and with strong identity, tend to merge along transition zones; they do not end along sharply marked boundaries.

The regional concept was effectively used by Professor Donald Meinig in an article entitled "American Wests: Preface to a Geographical Introduction," published in 1972. Using regional criteria, Meinig identified no fewer than six cultural nuclei in the West, each surrounded by a cultural region (Fig. 7.4). Among

these, the Mormon culture region is of special interest because it retains some sharp landscape properties. Although definable as an "American" landscape, the Mormon region is distinct from other Western culture regions. The dominant, impressive, often towering and always meticulously maintained temple rises above a townscape marked by compactness and closeness; houses, normally neatly cared for and comparatively modest, are spaced closely together in the Mormon tradition of communal association. Absent are taverns and bars; quietness is part of the Mormon cultural atmosphere and landscape.

Religion is only one cultural criterion that can be subjected to regional definition and analysis. Languages, dress modes, food preferences, musical expression, and other aspects of culture can be represented regionally, either as individual traits or in some combination. Again, to these formal culture regions we may add functional regions. A dominant urban center such as Jerusalem or Cairo or Rome is the node of a functional culture region, and by mapping how far its cultural reach extends, we can gain insight regarding the importance of such places on the cultural world stage. As always, the region we define will depend on the criteria we select. When we deal with regions in geography, we have theoretical objectives.

•   Geographic Realms   •

The regional concept can be employed at any level of scale, which is one of its advantages—and creates problems, too. If the hinterland of a great city can be defined as a functional culture region and if the membership of a church diocese constitutes a region also, then it will be difficult to establish a hierarchy, a true ranking of regions. Certain regions, obviously, will be *nested* within larger regions.

On a global scale, there are conceptual difficulties as well. We all use such geographic terms as *Latin America* and *Southeast Asia*, employing the regional concept in our everyday conversation. But such vast areas actually consist of more than one region. They are (with few exceptions) multicultural. Different languages, religions, and ways of life coexist within these great units. To mark them as the largest, most comprehensive regional entities, these are called world *geographic realms*. In the past, such global geographic realms were sometimes called culture realms, but most of them (Southeast Asia is a good example) are culturally too complex to be so named. There is more to geography than cultural geography, and there is more to a geographic realm than a culture region.

The best way to become familiar with this world framework of geographic realms is by consulting Figure 7.5. In a way this map is a summary of all the geographic factors covered in this book: the natural world and its landscapes, climates, soils, and plants; and the human world, its cultures and traditions, movements and migrations.

Transition Zones

Two aspects of the map are especially important. The first of these has to do with the lines that separate the geographic realms. These lines look like sharp dividers on the map, but in the real world they normally do *not* represent such sharp contrasts. Only rarely does the cultural landscape change suddenly from one place to the next. More commonly, there is a zone of transition, a belt where cultures make contact and, to some extent, mix. Take, for example, the line that separates Black Africa from the realm called North Africa and Southwest Asia. That line lies mostly across the southern Sahara. Obviously, it represents an *area* where change takes place. Islam, the religion of the north, decreases. Languages change. The styles of houses also change as the square, flat-roofed house of Arab Africa gives way to the round, cone-roofed African house. But in many villages in the transition zone, Islamic and non-Islamic, Arab and African traditions stand side by side.

Some of the zones of transition depicted by the lines on Figure 7.5 are wider than others. For example, the Chinese influence in Southeast Asia is very strong and Southeast Asian imprints exist in southern China. There the line represents a wide zone of change. But where the dividing line lies along high mountains (as in the Himalayas between East Asian [Chinese] and South Asian [Indian] spheres) the transition is more sudden. When you study the world's geographic realms in detail, for example, in a world regional geography course, those differences will become clear.

Regions

The second important aspect of the great geographic realms is that they divide, in turn, into regions. All the great geographic realms contain regions. Europe, for

NORTHERN EUROPE

BRITISH ISLES ①

WESTERN EUROPE

EASTERN EUROPE

MEDITERRANEAN EUROPE

SOVIET WESTERN CORE

UKRAINE

CAUCASUS

CENTRAL ASIA

②

VOLGA URALS

WESTERN SIBERIA

EASTERN SIBERIA

FAR EAST

XINJIANG

MONGOLIA

THE NORTHEAST

⑨

KOREA

TIBETAN (XIZANG) PLATEAU

CHINA PROPER

③A

TAIWAN

PACIFIC

Tropic of Cancer

NON-ARAB NORTH

MIDDLE EAST

THE MAGHREB AND IT'S NEIGHBORS

EGYPT AND THE NILE BASIN

⑥

ARABIAN PENINSULA

PAKISTAN

INDIA ⑧

BANGL

SRI LANKA

INDO-CHINA

PHILIPPINES

OCEAN

⑩

MALAYSIA

⑩A

AFRICAN TRANSITION ZONE

WEST AFRICA

EQUATORIAL AFRICA

EAST AFRICA

⑦

INDONESIA

INDIAN OCEAN

Equator

ATLANTIC OCEAN

SOUTHERN AFRICA

Tropic of Capricorn

1A

CORE AREA

Arctic Circle

Antarctic Circle

MILES

0 1000 2000

0 1000 2000 3000

KILOMETERS

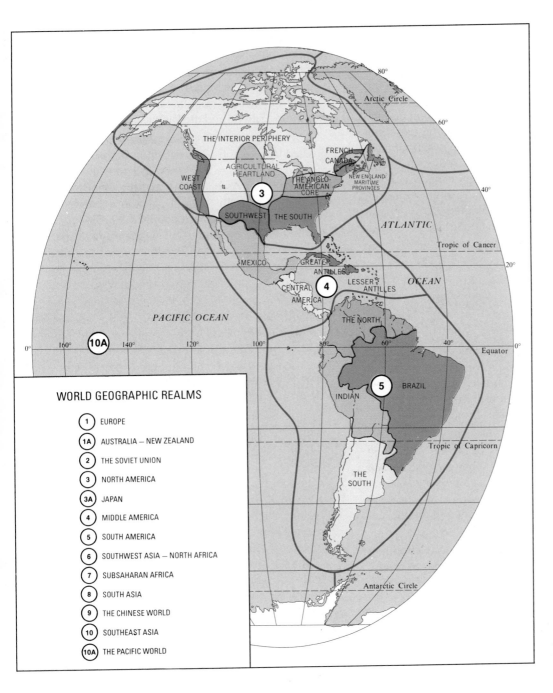

WORLD GEOGRAPHIC REALMS

1. EUROPE
1A. AUSTRALIA — NEW ZEALAND
2. THE SOVIET UNION
3. NORTH AMERICA
3A. JAPAN
4. MIDDLE AMERICA
5. SOUTH AMERICA
6. SOUTHWEST ASIA — NORTH AFRICA
7. SUBSAHARAN AFRICA
8. SOUTH ASIA
9. THE CHINESE WORLD
10. SOUTHEAST ASIA
10A. THE PACIFIC WORLD

267

FIGURE 7.5

example, is a geographic realm. Western Europe, Northern Europe, and the British Isles are regions of the European realm. To make use of our South American example again, Brazil is a region of the South American realm. Brazil is similar enough to all of Latin America to be part of the realm, but different enough to be a region by itself. Brazil speaks Portuguese, not Spanish; its population composition is substantially Afro-American; Indian populations make up a very small part of the total; the Brazilian economy differs considerably from those of other South American countries. These are *regional* differences. When we study the states of the world, an understanding of such regional contrasts is very helpful.

FOURTEEN GEOGRAPHIC REALMS

The map of world geographic realms consists of 14 major units. These are

1   *North America: The United States and Canada.* Marked by productivity, prosperity, a high degree of urbanization, much mobility, and considerable cultural uniformity. Regions based on French-English differences in Canada, ethnic and economic contrasts in the United States.

2   *Middle America.* Consists of all the states and territories in Central America between the United States and Colombia, and all the islands of the Caribbean Sea. Dominated by Mexico. The Central American republics form one region, the Caribbean islands another.

3   *South America.* The South American continent forms a geographic realm. Brazil covers about half the area, has about half the realm's population, and is a region by itself. The countries overlooking the Caribbean Sea form a second region, the countries with major Indian populations (the states of the Andes) a third, and the three southernmost countries are a fourth.

4   *Europe.* Europe for centuries has been one of the world's modern culture hearths. Its regions are the British Isles (small but different enough to be a region), Western Europe, Northern Europe (including the Scandinavian states), Southern Europe (the states of the Mediterranean), and Eastern Europe (where Soviet influence is strong today).

5   *Soviet Union.* Not only its size (largest state in the world), but also its political system and ideology make the Soviet Union a world realm. Its broadest regions are the west (west of the Ural Mountains, where most of the population lives) and the east, including Siberia and the Pacific area.

6   *North Africa and Southwest Asia.* This realm is united by the faith of Islam and its cultural domination. It includes the ''Middle East''; the giant state of Saudi Arabia; Egypt and the other North African states; the oil-rich sheikdoms; and the countries of the northern highlands: Turkey, Iran, and Afghanistan. Some geographers argue that Pakistan also is part of this realm.

7   *Africa.* This geographic realm extends from the southern Sahara to the Cape of Good Hope. It is a realm of many peoples and numerous languages. Regions are West Africa (scene of the ancient culture hearth), East Africa (region of highlands and lakes), Equatorial Africa (centered on the great basin of Zaïre), and Southern Africa (dominated by powerful, white-ruled South Africa at the tip of the continent).

8   *South Asia.* India lies at the heart of this populous realm. Also part of it are Pakistan (one of those transitional states), Bangladesh, Sri Lanka, and Nepal.

9   *China and Its Sphere.* China, the most populous state in the world, dominates this geographic realm, which also includes Mongolia, North and South Korea, and Taiwan. It divides into several regions, including Eastern China, where most of China's 1 billion people live; China's western interior, of which Tibet is a part; and the eastern perimeter, where economic conditions differ (South Korea, Taiwan, Hong Kong).

10 *Japan.* We will discuss Japan as a part of East Asia, but there are good geographic reasons for regarding Japan as a separate realm, small as it is. Japan is one of today's culture hearths.

11 *Southeast Asia.* As the map shows, this realm lies wedged between two giants, India and China. That is why part of it is indeed known as *Indochina*. It has received cultural influences from both, and these are imprinted on the cultural landscape. The most obvious regional subdivi-

sion is between the states of the mainland and those of the islands adjacent.

12 *Australia and New Zealand.* These are European outposts, dominated by Western cultural norms. Indigenous (native) populations were overwhelmed by the European invasion. Only in New Zealand do the original Maori people retain identity. Both states are highly urbanized, prosperous, and productive; Australia in terms of minerals and farm products, New Zealand mainly in farm products.

13 *Pacific Realm.* Fragmented, culturally complex, dominated by vast expanses of water, the Pacific realm nevertheless possesses clearly defined regions in Melanesia, Micronesia, and Polynesia.

14 *Southern Realm.* Antarctica, the frozen continent, lies at the heart of a vast realm of which the Southern Ocean also is a part. This realm, the least populous on earth, covers about one sixth of the globe! It contains no states, but it has economic potential and may become a place of international competition and conflict.

•   Race and Culture   •

When you, as a traveler, enter a country for a visit you will be asked to fill out a questionnaire to be presented, with your passport, to the immigration officer. As often as not, one of the questions on the form will be "What is your race?" And the appropriate answer, of course, is "human." All the people of this world belong to the same species. We all have far more features in common than there are differences between us as individuals or groups.

But the answer "human" on the questionnaire is not the one that is wanted. In fact, it may get you into trouble. The term **race** has come to refer to an undeniable reality of our human existence. People differ physically from each other, and those differences have regional expression. Even after centuries of movement and migration, mixing and intermarriage, peoples with distinct physical traits remain clustered in particular areas of the world even in our mobile world.

Thus, we use the term *race* in quite a different way: we speak of the European, African, and Polynesian races of humanity. It may not be quite correct to do so (indeed, some anthropologists have proposed that the whole concept of race be abandoned), but it is an inescapable fact of life.

Race, then, is first and foremost a biological concept; it refers to people's physical features. A racial group such as those just mentioned (European, African) is recognized because it has a distinctive combination of such physical traits, the product of a particular genetic inheritance. This inheritance has been determined by many centuries of isolation and inbreeding during which a certain dominant set of genes, a gene pool, evolved for each racial group. Man presumably evolved from a common stock, but after man radiated outward from the source area (perhaps Southwest Asia, North and East Africa) spatial and social isolation began to play its role in generating discrete gene pools and racial groups in Asia, Australia, Africa, Europe, and the Americas. Humanity was differentiated into what used to be called the white, black, red, yellow, and brown races. As the concept of race was refined, other terms emerged: Caucasoid, Negroid, Amerindian, Mongoloid, Australian. Anthropologists have recently been using a nine-unit racial classification of humanity.

1 *European.*

2 *Asian,* peoples of China, Japan, inner Asia, Southeast Asia, and Indonesia.

3 *Indian,* peoples of the Indian subcontinent.

4 *African,* peoples of Africa south of the Sahara.

5 *American,* the indigenous, Indian populations of the Americas.

6 *Australian,* the original peoples of Australia.

7 *Melanesian.*

8 *Polynesian.*

9 *Micronesian.*

RACIAL REGIONS

Ours is an age of movement and migration, contact and interaction. Yet associations between physical type and regional focus persist. The spatial representation of these associations (Fig. 7.6) is a significant

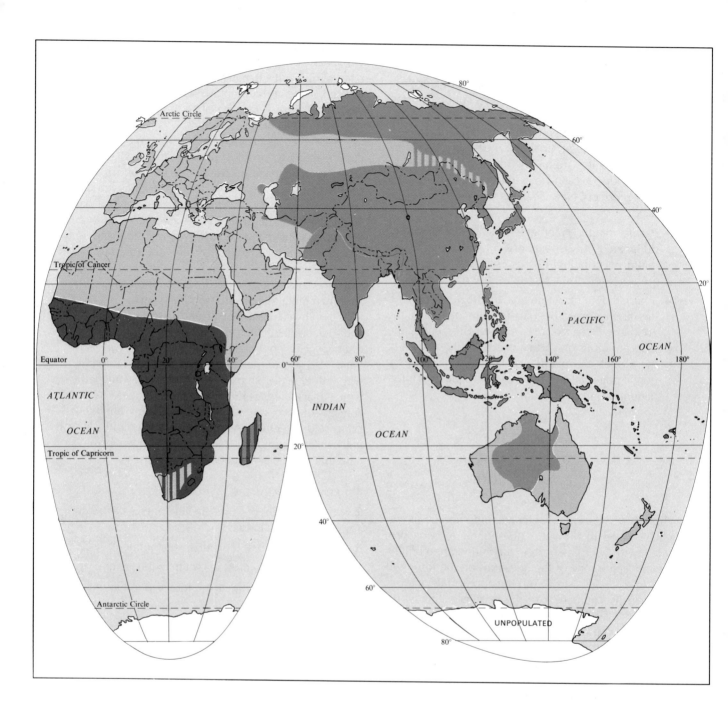

Arctic Circle

80°

60°

40°

Tropic of Cancer

20°

PACIFIC

OCEAN

Equator   0°   20°   40°   60°   80°   100°   120°   140°   160°   180°

ATLANTIC

OCEAN

INDIAN

OCEAN

Tropic of Capricorn

20°

40°

60°

Antarctic Circle

UNPOPULATED

80°

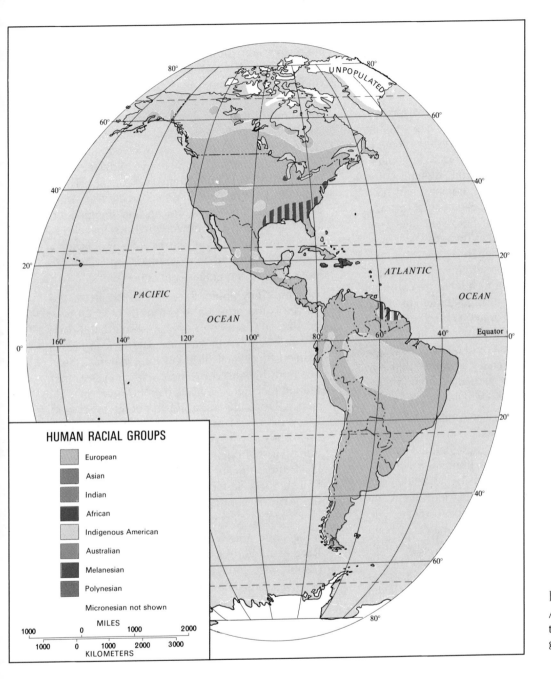

80°

UNPOPULATED

60°

80°

60°

40°

40°

PACIFIC

OCEAN

ATLANTIC

20°

20°

OCEAN

PACIFIC

20°

OCEAN

Equator

0°

160°

140°

120°

100°

80°

60°

40°

0°

20°

20°

40°

HUMAN RACIAL GROUPS

European

Asian

Indian

African

Indigenous American

Australian

Melanesian

Polynesian

Micronesian not shown

60°

MILES

80°

1000   0   1000   2000

1000   0   1000   2000   3000
KILOMETERS

271

FIGURE 7.6
A regional classifica-
tion of human racial
groups.

ingredient of the map of world geographic realms discussed earlier.

1 **European Geographic Race.** Although not the largest human racial group numerically, the European race is the world's most widely dispersed. As Figure 7.6 shows, this racial group inhabits not only its source area, the western zone of Eurasia, but also North Africa from Mauritania to Somalia and Southwest Asia from Afghanistan to Turkey. This means that Arabs, Iranians, and Turks all are members of the European geographic race. Great mobility marked the European race even before its momentous expansion to the Americas, Africa, and Australia began. For many thousands of years, these movements stirred European populations from interior Asia to the North Sea, from Arabia to Morocco. As a result, although there are substantial variations of traits (skin color ranges from very light to quite dark, for instance), there are few sharp breaks that could be mapped. European racial types are marked by rather thin hair that tends to turn gray with age, high-bridged noses, and considerable body hair.

2 **Indian Geographic Race.** The Indian geographic race occupies the geographic realm generally referred to as South Asia, including India, Pakistan, Bangladesh, Sri Lanka, and extreme northwestern Burma (Fig. 7.6). This is a physiographically well-defined realm, with effective barriers against interaction along much of its margin, especially to the north and east. Internally it has also been marked for many centuries by strong cultural barriers based on the caste system of Hindu society. Most Indians speak languages that are related to European languages and, in fact, belong to the Indo-European language family. In racial appearance, too, there are notable similarities between Indians and Europeans. The most significant contrast is the much darker skin color of Indians, but the face, high-bridged nose, head hair and beard, and the pattern of graying and balding with age all resemble European features.

3 **Asian Geographic Race.** The Asian (also *Asiatic*) geographic race inhabits the opposite (eastern) zone of Eurasia, including China and Japan, Korea and Mongolia, all of Southeast Asia except extreme western Burma, and, remarkably, part of Madagascar off the southeast African coast. This is the largest geographic race, numbering as many as 1.6 billion people. Anthropologists are not unanimous that this huge cluster constitutes a single racial type, but it is essentially not more varied than the European race. One common trait is the epicanthic fold, or inner-eye fold; faces tend to be broad, with comparatively flat noses. These qualities give the Asian race a characteristic countenance; generally dark, coarse, and straight hair; and mainly brown eyes. Although Asians tend to be regarded as being of short stature and having a dark skin color, inhabitants of Xizang (Tibet) are quite tall and Mongolians are rather light skinned.

4 **African Geographic Race.** Africa south of the Sahara is the principal realm of the African geographic race, although Africans also inhabit parts of North and South America. The African race is marked by several distinguishing features. Head and body hair are curly, more so than in any other human racial group. Body hair, however, is quite sparse. Prognathism (protrusion of the lower jaw) is common, as is eversion (outward turning) of the lips. The nose if often low-bridged, short, and quite flat. Although there is a considerable range in stature, Africans are relatively tall and long-limbed. Dark skin, of course, is another feature of the African geographic race although, again, much variation occurs.

5 **Indigenous American (Amerindian) Geographic Race.** As Figure 7.6 shows, the American geographic race today prevails in only a few comparatively small regions of the Americas where once (still in rather limited numbers) they were the sole inhabitants. The Amerindians may have entered the Americas during the interglaciation before the most recent advance of the Pleistocene ice, perhaps thirty to forty thousand years ago, but some immigrants may have come across Pacific waters to South America. In any case, they remained isolated in this realm for thousands of years, sparsely peopling most of it and clustering in a few significant areas (especially the Andes and the Mexican plateau). Although the Amerin-

272

dian groups exhibited much local variation, the epicanthic fold occurs fairly commonly, especially in women. Eye color tends to be dark, the head hair is coarse and straight, and the body hair is sparse. Stature varies considerably, but is frequently stocky. The skin color is medium, but tans darkly.

6 **Australian Geographic Race.** When the European invaders arrived in Australia to settle, just two centuries ago, the Australian indigenous population probably numbered barely three hundred thousand, consisting of numerous small groups living in almost complete isolation. This led, of course, to considerable local variation in physical features, but it is nevertheless appropriate to group the aboriginal Australians together as a single geographic race. Although skin color is characteristically dark (again, tanning contributes to this), among children hair is sometimes reddish or even blond, a feature that occasionally persists into adulthood. The face is often marked by pronounced protrusion, not just of the lower jaw, but of both upper and lower jaws. Brow ridges are also pronounced, so that the eyes appear deep-seated; noses are flat, lips are thick, but not, as among Africans, everted. Teeth often are unusually large, a characteristic feature.

7 **Melanesian (Papuan) Geographic Race.** The Melanesians inhabit New Guinea and islands to the east (shown only partially on the map projection used in Fig. 7.6). These people have experienced long-term isolation, as have the indigenous Australians, and they are among the world's most dark-skinned races. In other ways, however, they differ from the Africans (whom they resemble superficially). Although their hair is long and woolly and their stature is fairly tall, they have high-bridged, rather broad noses, heavier brow ridges, and full, but not everted lips. The Melanesians (*mela* means black) were formerly called the oceanic Negro peoples in recognition not only of their darkness, but also of their prognathism.

8 **Polynesian Geographic Race.** The Polynesian culture region extends from New Zealand in the south to the Hawaiian and Midway Islands in the north and from the islands of Fiji in the west to Easter Island in the east. The Fijian islands themselves lie in a transition zone between Melanesia and Polynesia. This vast oceanic region, with its scattered islands, was perhaps the last to be occupied by human communities. The Polynesians are known for their excellent physique, their strong build, medium-colored skin, dark hair and eyes, and comparatively tall stature. Their adaptation to their fragmented habitat and their capacity to establish effective circulation systems in an oceanic zone as large as Asia are matters of special cultural-geographic interest.

9 **Micronesian Geographic Race.** The Micronesian race inhabits the Pacific islands to the west of Polynesia and north of Melanesia. It is believed to have resulted from the mixture of Southeast Asians and Melanesians and is characterized by dark-skinned, brown-eyed individuals whose hair is black and quite frizzy. Although *micro* means small, Micronesians are not especially small of stature. They are shorter than the Polynesians, but their average height is similar to that of the Melanesians.

These nine geographic races are defined on the bases of proximity, physical appearance, and blood chemistry. They do not account for *local* races, such as the Khoisan-speaking Bushmen and Hottentots of southern Africa or the Afro-Americans of North America. It would be possible to identify dozens of exceptions to the fundamental framework represented in Figure 7.6 but none would diminish its basic validity.

• Language and Culture •

Language is a vital element of culture. It is our means of communicating, and in literate societies, the use of written words increases greatly the efficiency with which culture is transmitted from generation to generation. This is not to suggest that peoples who do not write (preliterate societies) cannot transfer their culture from one generation to another; but writing is an

273

enormous asset and, indeed, a necessity in complex, advanced cultures.

What is a language? The term is defined in many different ways. Look in any dictionary. For our purposes, the question of definition is not important as long as we realize that a language is a language no matter how large or small the vocabulary or how large or small the number of people using the language. As we discuss language in the spatial context of culture, we should also remind ourselves now and then that languages are always changing. That reminder should hardly be necessary, considering how our own use has changed over the past few years.

In the context of cultural geography, we are obviously interested in the way languages are distributed across the world, how these distribution patterns came about, and how they are changing. This, in turn, raises the question of classification. Scholars continue to differ on what is a discrete language and what is a dialect; depending on the definitions we decide to accept, there are between twenty-five hundred and thirty-five hundred languages in use in the world today (plus their dialects!), over a thousand of them in Africa alone. Clearly we cannot begin to study the relative location of around three thousand items across the world without first bringing some order to the mosaic.

In the classification of languages, terms are used that are also employed in biology, and for the same reasons: certain languages are related to each other, some are not. Languages grouped in a **family** are thought to have a shared origin (such as Indo-European, Niger-Congo in Plate 6). In the **subfamily**, their commonality is even more definite (Germanic, Romanic in Plate 6). These are divided into language **groups** that consist of **individual** languages (the scale of Plate 6 does not permit this level of detail).

As Plate 6 reveals, the Indo-European language family dominates not only in Europe, but also in much of Asia (the Soviet Union and India, among other countries), North and South America, Australia, and in parts of southern Africa. Indo-European languages are spoken by about half the world's peoples, and English is the most widely used Indo-European language today. Linguists theorize that a lost language they call Proto-Indo-European existed somewhere in east–central Europe and that the present languages of the Indo-European family evolved from this common heritage. In the process, as vocabularies grew and changed and peoples dispersed and migrated, differentiation took place. Latin arose out of this early period, to be disseminated over much of Europe during the rise of the Roman Empire; later Latin died out and was supplanted by Italian, French, and the other Romance languages.

DIFFUSION

The map of world language families tells us much about human migrations. Indo-European languages are spoken as far east as India (Plate 6); Hindi is one of the languages in this family. This reflects the western source of the peopling of modern India and the route taken by the ancient migrants through Southwest Asia and Iran. The oldest inhabitants of India, who speak Dravidian languages, were driven southward into the tip of the peninsula by the newcomers from the northwest. The linguistic map is part of the proof.

Other migration streams carried Indo-European languages to the Americas, Australia, South Africa, and other regions overseas. In the process, English became the most widely used Indo-European language, but not without modifications. In *Word Geography of the Eastern United States* (1949), H. Kurath describes the development and distribution of dialects of American English from three source areas on the East Coast. The New England (*Northern*) dialect spread across the northern United States, covering the Great Lakes and upper Midwest states. The Middle Atlantic (*Midland*) dialect diffused from its source area in southern New Jersey, Delaware, and Maryland into the heart of the eastern United States. The *Southern* dialect eventually extended across the approximate area of the Gulf and Atlantic coastal plain. Even today these dialect regions continue to exist, recognizable by the accents of representative speakers and sometimes contrasting vocabularies.

In the United States as elsewhere, English was influenced by the arrival of immigrants who did not speak English. In the northern U.S., Germans, Scandinavians, Italians, Slavs, and Jews all complicated the ethnic and linguistic mosaic. These clusters long retained their traditional languages (and other attributes of their cultures as well). And in the slow process of assimilation into evolving American culture, they contributed to the modification of regional English dialects. In the South, the forms of English spoken by the black population—with a strong element of pidgin—had a major impact on the evolution of the regional southern English dialect.

THE WORLD LANGUAGES

The map of world language families cannot tell us how many persons speak the languages mapped, although a comparison between this map and that of world population (Plate 5) does provide some general insight. Table 7.1 reports the approximate number of speakers of the major world languages. It reveals that although the Indo-European languages, taken together, are spoken by more of the world's peoples than any other language family, Chinese is the single largest language in terms of numbers of speakers;

English ranks second. The figures in Table 7.1 should be viewed as only approximations for several reasons. English, for example, is not only spoken by 250 million North Americans, 60 million Britons and Irish, nearly 20 million Australians and New Zealanders, and by millions more in countries with smaller populations, it is also used as a second language and as a *lingua franca* by hundreds of millions of people in India, Africa, and elsewhere—a situation that is not reflected by the table showing "native" speakers. French, too, is more widely used than its 75 million first-language speakers suggest. Furthermore, figures in Table 7.1 are based in some instances on population data that are not reliable. The regional languages of India (Indo-European as well as Dravidian) are among the world's largest, but exact data on the number of speakers are unobtainable.

It is noteworthy that Table 7.1 does not list any languages spoken in Africa south of the Sahara as major world languages. African languages are the topic of a later section of this chapter, but Plate 6

TABLE 7.1   Estimated Speakers of World Languages, 1986

LANGUAGE FAMILY	LANGUAGE	SPEAKERS (MILLIONS)
Indo-European	English	355
	Spanish	250
	Hindi	235
	Russian	215
	Bengali	160
	Portuguese	140
	German	100
	Punjabi	80
	French	75
	Italian	60
Sino-Tibetan	Chinese	1000
	Thai	45
	Burmese	30
Japanese and Korean	Japanese	120
	Korean	60
Afro-Asiatic	Arabic	150
Dravidian	Telugu	65
	Tamil	60
Malay-Polynesian	Indonesian	110

provides one of the reasons for the absence of African languages: the extreme fragmentation of the African language map. Africa south of the Sahara, as a geographic realm, still has a relatively modest population (about 450 million in the late 1980s), but about 1000 languages remain in use there. Not including the Afro-Asiatic languages of North Africa nor the Indo-European languages spoken by whites on the continent, Black Africa's languages must still be grouped into four families (3, 4, 5, and 6 in Plate 6). In terms of the number of speakers, Hausa is estimated to be the largest African language, with perhaps as many as 35 million. Hundreds of African languages have fewer than 1 million users.

Other language families not included under the headings of Table 7.1 can be seen (Plate 6) to constitute dwindling, often marginally located, or isolated groups. Austro-Asiatic languages (11), spoken in interior locales of eastern India and in Kampuchea (Khmer) and Laos, are thought to be survivors of ancient languages spoken in this realm before modern cultural development (and numerous external invasions) took place. Some scholars place Vietnamese in this family, others do not. The Papuan and indigenous Australian languages (13), although numerous and quite diverse, are spoken by fewer than 10 million people today. The languages of American Indians (14) remain strong only in areas of Middle America, the high Andes, and northern Canada. Languages of the Eskimo-Aleut family (20) continue to survive on the Arctic margins of Greenland, North America, and eastern Asia.

If you spend some time looking carefully at the map of world languages, some interesting questions arise. Consider, for example, the large island of Madagascar, off the East African coast. The predominant languages spoken on Madagascar are not of an African language family, but belong to the Malay-Polynesian family, the languages of Indonesia and its neighbors. How did this happen? Surely Madagascar's closeness to Africa would suggest a very different situation. Actually, the map reveals a piece of ancient history still not well understood. Long ago, seafarers from the islands of Southeast Asia crossed the Indian Ocean. They may, in fact, have reached the East African coast first, and then sailed on to Madagascar.

There they established settlements. Africans had not yet sailed across the strait between continent and island, so there was no threat to the Indonesian-Malayan presence. The settlements grew and prospered, and large states evolved on a Southeast Asian model. Later, Africans began to come to Madagascar, but by that time the cultural landscape had been forged. Take an atlas map and compare the names of Madagascar's places to those across the water in Africa. The language map reveals a fascinating piece of historical geography.

MULTILINGUALISM

Language can unite; it can also divide. Countries in which only one language is spoken are fortunate—and rather few. Such *monolingual* countries include Japan in Asia, Uruguay and Venezuela in South America, Poland in Europe, and Lesotho in Africa. However, even in these countries there are small numbers of people using other tongues: more than a half million Koreans in Japan, for example. In the modern world, there is no truly monolingual country left. English-speaking Australia has more than 180,000 speakers of aboriginal languages. Dominantly Portuguese-speaking Brazil has nearly 1.5 million Indian-language speakers.

Countries in which more than one language is in use are called *multilingual* countries. In some of these countries, the linguistic fragmentation reflects strong cultural pluralism and the existence of divisive forces. This is true in former colonial areas where peoples of diverse tongues were thrown together by the force of foreign interests, as happened in Africa and Asia. This also occurred in the Americas: Indian languages are spoken by more than half the people in substantial areas of Guatemala and Mexico, although these countries tend to be viewed as Spanish-speaking. Countries never colonized in comparable ways, but peopled from different cultural sources, may also display multilingualism. Canada's Quebec Province is the heart of that country's French-speaking sector, and Canada's multilingualism mirrors a cultural division of considerable intensity. In Belgium, a fairly sharp line divides the northern Flemish-speaking half of the country from the Walloon (French-speaking) south.

The mosaic of languages is complex and diverse. People of neighboring countries often do not speak a mutually intelligible language. The owner of a freighter whose business it is to carry cargo up and down the coast from one country to the next faces difficult problems of communication.

Over time, however, languages have developed in response to these very problems. A durable trade route would lead to long interaction in which people borrowed words from various languages and created their own **lingua franca**—literally, Frankish language.

Why Frankish language? Because the earliest of these trade languages was probably the one that developed after the Crusades when the Mediterranean Sea was opened to commerce. The traders from the ports of southern France—the *Franks*—followed the routes of the Crusaders, revitalized the ports of the eastern Mediterranean, and introduced their own tongue. This was a mixture of French, Italian, Greek, Spanish, and local Arabic. If you wanted to do business, you had to speak the Frankish language—the *lingua franca*.

Thus, a *lingua franca* is a language used in trade, commerce, and general communication among peoples who mostly speak other languages of their own. In coastal West Africa, a *lingua franca* developed that is a mixture of English and West African languages, and you can hear it spoken from Freetown, Sierra Leone, to Port Harcourt, Nigeria, and beyond. This language is appropriately known as Wes Kos, and naturally it stands apart from the Guinea subfamily of languages. A more famous *lingua franca* is Swahili, spoken in Kenya, Tanzania, Uganda, and even in eastern Zaïre and northern Moçambique in modified forms. Swahili is essentially a Bantu language, but much modified by the tongue brought to East Africa by Arabic traders. Although various forms of Swahili are spoken in eastern Africa, it serves effectively as a means of communication in a strongly multilingual area. Still another African *lingua franca* is Hausa. This language is spoken across much of West Africa's Sahel zone, from eastern Senegal to Nigeria, where it serves in commerce and general contact.

In the Caribbean area, forms of pidgin based on English have come to serve as *linguae francae*; their use extends even to non-English-speaking territories, such as Dutch-influenced Suriname. Elsewhere, *linguae francae* facilitate communication among hundreds of millions of people. Chinese serves as such for a region far beyond China itself in eastern Asia, and bazaar Malay functions as a *lingua franca* in much of Southeast Asia. Melanesian pidgin serves in the western Pacific's numerous archipelagoes, and it resembles Caribbean pidgin, being based chiefly on English.

These languages of trade do not fit neatly in the system of classification shown in Plate 6, but in the ancient as well as the modern world, they perform important functions.

Multilingualism takes several forms. In cases such as Canada and Belgium, it has regional expression—that is, the different languages prevail in separate areas, so that one can speak of French-speaking Canada as a region and Flemish-speaking Belgium as a distinct part of that country. Canada and Belgium are *bilingual* countries, that is, two languages dominate in each. However, many other countries use three or more languages, and the potential for division and misunderstanding is even greater. We have already noted India's 15 major (and many minor) languages and the hundreds of languages in use in the four dozen countries of Black Africa.

The problems of multilingualism affect different countries in different ways. Despite nearly 60 years of Russian influence, the Soviet Union is still marked by a complex linguistic pluralism. Lithuanian and Latvian are still used in the west, Armenian and Caucasian in the south, and various Ural-Altaic languages (7) in Moslem central Asia. Russian may be *known* by as many as 210 million Soviet citizens and is used when appropriate, but 15 million Uzbeks prefer to speak their own language, as do 7 million Kazakhs, 4 million Georgians, and many millions of other Soviet peoples.

In Nigeria, Africa's most populous country, three major languages dominate, but Nigeria is not trilingual: more than a dozen other languages are spoken by more than a million people each. The three strongest languages are Hausa, the old *lingua franca* of the interior—long used by traders along the Sahara's edge—and spoken in Nigeria by perhaps 25 million; Yoruba, leading language of the southwest, region of great cities and many villages (18 million); and Ibo, the major language of the southeast (16 million). Unlike the Soviet Union, there is no single, dominant domestic language that can be used as a national tongue.

TOPONYMS

One of the most interesting areas of study in linguistic cultural geography involves **toponyms**, or place names. This research owes a great deal to the work of George R. Stewart, whose book *Names on the Land: A Historical Account of Place-Naming in the United States* (1958) remains a classic work. Place names can reveal much about the contents of a culture area, even when time has erased other evidence. A cluster of German place names in Pennsylvania, or French place names in Louisiana, or Dutch place names in Michigan reveals not only national origins there, but may also provide insights into language and dialect. Further, the reconstruction of past cultural landscapes is aided by the analysis of the place names themselves. Many place names consist of two parts: a specific (or given) and a **generic** (or classifying). For example, the name Battle Creek (Michigan) consists of a reference to an event (specific) and a landscape feature (generic). Many such names mark the landscape: Johns-town, Pitts-burgh, Nash-ville, Chapel Hill, Little Rock, and so on. As it happens, certain generic names can be associated with each of the three Eastern culture cores (p. 000), so their diffusion can to some extent be traced across the continent.

Any foreign visitor to the United States is impressed by the rectangular layout of many cities and the use of compass directions for showing the way ("Four blocks east, then three blocks north"). This aspect of U.S. culture appears in map as well labels. Many towns are named *East* Lansing, *West* Chester, and so on. But this is most strongly a northern phenomenon, and the westward dispersal of New England culture traits can be traced in part by this kind of place naming.

The Stewart classification of place names has produced 10 categories, including descriptive names (Rocky Mountains), associative names (Mill River), incident names (Battle Creek), possessive names (Castro Valley), commemorative names (Saint Louis), commendatory names (Pleasant Valley), folk-etymology names (Owlpen), manufactured names (Tesnus [sunset spelled backward]), mistake names (involving historic errors in identification or translation), and so-called shift names (relocated names, double names for the same feature). Each of these categories contains cultural-geographic opportunities. For example, the capital of the Soviet Union has an associative place name. Moscow is actually spelled Moskva in Russian, but *kva* is Finnish for water. A check on other toponyms in the Moscow region confirms the ancient presence of Finnish tribes in what is now the Russian-Soviet heartland. Again, the southern tip of South America, Cape Horn, can be categorized as a mistake name. The Dutch named this area Cape Hoorn, after a Dutch town. The English interpreted this as Cape Horn, which the Spanish, in turn, translated into *Cabo Hornos*—meaning cape of ovens!

Obviously, the whole area of language in cultural geography affords countless opportunities for research and study. Language is a critical element in any attempt at the reconstruction of past cultures; the transmittal of oral literature and its interpretation opens up whole new possibilities. The professional storyteller in an African village is not just a picturesque figure of incidental interest. He holds in his tales the history and psyche of the people of whom he speaks. The study of dialects and the spatial character of modifications in words can tell us much about peoples' movements, their external contacts or isolation, former distribution, and more. And then there is the relevant issue brought up by so many in arguments about the value in learning a foreign language.

Language can reveal much about the way peoples view reality, their own culture, as well as other cultures. In their structure and vocabulary and in their capacity (or inability) to express certain concepts and ideas, languages reflect something of the way people think and perceive their world. There are African languages that have no word, no term, for the concept "god." Others (in Asia) really have no system for the reporting of chronological events, no time scale. It would be difficult for us to understand these peoples' perception of the world about them, preoccupied as our culture is with the supernatural and dating and timing. Learning the language would be a first, timid step. But the rewards could be enormous.

Religion and Culture

In many parts of the world, especially in non-western areas, religion is so vital a part of culture that it practically *constitutes* culture. Thus it is not surprising that we would have difficulty in defining exactly what a religion is. The phenomenon manifests itself in so many different ways—in the worship of the souls of ancestors living in natural objects such as mountains, animals, or trees; in the belief that a certain living person possesses particular capacities granted him or her by a supernatural power; and in the belief in a deity or deities, as in the great religions. In some societies, notably those of the Western, industrialized, urbanized, commercial world, religion has become a rather subordinate, ephemeral matter in the life of many people. But in Asia and parts of Africa, religious doctrine may exert tight control over behavior—during the daytime through ritual and practice and at night in prescribing the orientation of the sleeping body.

If we cannot define religion precisely, we can at least observe some of the properties of this element of human culture. There are, of course, sets of doctrines and beliefs that relate to the god or gods central to faiths. In each faith, there will also be a number of more-or-less complex rituals through which these beliefs are given expression. Such rituals may attend significant events in peoples' lives: birth and death, attainment of adulthood, marriage. They are also expressed at regular intervals in a routine manner, as is done on Sundays in most of the Western world. Prayer at mealtime, or at sunrise and sundown, or at night when retiring, or in the morning when arising commonly attends religious ritual. Such ritual is likely to involve the use of the religion's literature, if a literature exists; we are most familiar with the Bible and the Koran.

Religions—especially the major, world-scale faiths such as Christianity and Islam—have produced vast and complex organizational structures. These bureaucracies have a hierarchy of officers and command a great deal of wealth; rank and status are confirmed by sometimes elaborate and impressive costumes and by a certain authority over people's lives. These religious officers maintain and police an approved set of standards of behavior and protect the code of ethics peculiar to each religion.

From this brief statement of the properties of religion, it is evident how strongly an entire culture may be dominated by the precepts of the prevailing faith. Some scholars who study religion have argued that people's dependence on some supernatural power over which they have no control tends to lead to widespread apathy, a sense of impotence to bring about change. But religion has also inspired others to compete and to achieve. In any case, the idea that a "good" life has rewards and that "bad" behavior risks punishment must surely have enormous cumulative effect on cultures. Modes of dress, the wearing of religious articles, the kinds of food and drink people should and should not consume, even the location and structure of houses may be determined by the rules of religion. Even in secular (nonreligious) societies of Europe, the United States, and the Soviet Union, the continuing effects of religious morality are felt in many ways: in the calendar and weekly work cycle, holidays, place names, prominent architectural landmarks, political systems, and public morality. The slogan "IN GOD WE TRUST" appears on all the coins and paper money printed in the United States. Imagine, then, the role of religion in an almost totally Roman Catholic society such as Spain or in an Islamic state such as Pakistan. Perhaps to an unsurpassed degree, religion sustains and perpetuates culture in Is-

279

lamic societies. It may be true that organized religion is losing adherents in secular states of the modern world, but the imprints of religion on culture seem indelible. Religious doctrine ordained that much of India is vegetarian (and overrun by holy animals), that the eating of pork is taboo in Moslem countries (eliminating the profitable pig from domestic economies), that eating meat on Fridays was forbidden for Roman Catholics.

ORIGIN AND DISTRIBUTION

Human geographers are naturally interested in the locational characteristics of the major religions, their source areas, dispersals, distributions, and present patterns of adherence. The spatial distribution of the major religions is depicted in Plate 7, which necessarily shows the dominant religions in world regions, not the intricate mosaic that would be revealed by larger-scale maps. For example, India (except for the northwest) is shown as a Hindu region, but other religious faiths also continue to survive there (Islam remains strong in several parts of the country). Plate 7, should, therefore, be viewed as a generalization of a much more intricate set of distributions. With this caveat as background, the map does reveal the dominance of the Christian religions, the wide dispersal of Islam, the coincidence of Hinduism with one of the world's major population concentrations, and the survival of Buddhism.

Christian Religions

The principal religions of the world are numerically represented in Table 7.2. As the table reveals, the religions of Christianity presently have the largest number of adherents and are the most widely dispersed around the world. More than 1 billion Christians include about 340 million in Europe and Russia, a quarter of a billion in North America, over 200 million in South America, nearly 160 million in Africa, and 110 million in Asia. Thus, Christians account for 40 percent of all the world's members of major religions. Islam, with fewer than 600 million faithful, is a distant second, about as large as Roman Catholicism alone.

Like other major religions, Christianity is a di-

vided faith. Roman Catholicism constitutes the largest constituency of Christianity. Plate 7 reveals the strength of Roman Catholicism in Europe, areas of North America, and Middle and South America. The Protestant churches prevail in Northern Europe, in much of North America, in Australia and New Zealand, and also in South Africa. The Eastern Orthodox churches, the major faiths of the old Russia, are now diminishing in the atheist Soviet Union, but they still count as many as 64 million adherents in Europe and the Soviet Union, in Africa (where a major cluster survives in Ethiopia), and in North America.

Faiths of Islam

The faiths of Islam (also a divided religion) dominate northern Africa and Southwest Asia, extending into the Soviet Union and China, and they include outlying clusters in Indonesia, Bangladesh, and southern Mindanao in the Philippines. Islam is heavily represented along the East African coast, survives in Albania, has an outlier at South Africa's Cape of Good Hope, and has adherents in the United States. Islam has close to 600 million faithful, more than half of them outside the culture realm often called the Islamic World. A substantial Islamic minority still resides in India.

As Table 7.2 shows, the Moslem faiths are strongest in Southwest and Southeast Asia. Two thirds of its nearly 600 million adherents are located in Asia and another 160 million in Africa, mainly north of the Sahara. The salient division within Islam is between Sunni Moslems (who constitute the great majority, as reflected by Plate 7) and the Shi'ah (or Shi'ite) cluster, which is concentrated in Iran. A comparison between Fig. Plate 7 and Fig. Plate 5 will prove that the largest Moslem country does not lie in the comparatively sparsely populated Southwest Asian–Northern African source area, but in Southeast Asia: Indonesia is the world's largest Islamic state, with about 140 million Moslems.

Hinduism

In terms of number of adherents, Hinduism ranks after Islam as a world religion, but there are significant differences among the Hindu, Christian, and Islamic faiths. The Hindu religion is without a comparable

TABLE 7.2 Estimated Membership of Major World Religions[a] (in millions)

RELIGION	NORTH AMERICA	SOUTH AMERICA	EUROPE AND U.S.S.R.	ASIA	AFRICA	PACIFIC AUSTRALIA AND NEW ZEALAND	WORLD
Christianity	253	201	339	110	157	19	1079
Roman Catholic	139	190	179	59	62	5	634
Protestant	109	10	113	45	82	13	372
Eastern Orthodox	6	<1	45	3	9	<1	64
Islam	2	<1	21	399	159		581
Hinduism	<1	<1	<1	473	1	<1	475
Buddhism	<1	<1	<1	257			258
Chinese Religions	<1		<1	187			187
Shintoism	1		1	34			36
Judaism	8	1	5	4	<1		18

[a]Blank space indicates fewer than 0.1 million adherents. Data from various sources, including U.N. records and offical church bulletins. Published statistics vary widely, and these figures should be viewed as rough approximations.

ecclesiastical organization, lacking the kind of bureaucracy that is familiar to Christians and Moslems. Certainly there are holy men, but they represent literally thousands of gods. Thus, unlike Christianity or Islam, Hinduism is polytheistic, fragmented by numerous cults, and without a prescriptive book, such as the Bible or the Koran. Again, unlike Christianity or Islam, Hinduism remains concentrated in a single geographic realm, the region of its source (it is regarded as the world's oldest organized religion). The vast majority of the 475 million Hindus live in India, although the faith also extends into Bangladesh, Burma, Sri Lanka, and Nepal.

Buddhism

Buddhism, another religion that had its source in India, is now a minority faith in that country, but still remains strong in Southeast Asia, China, and Japan. Buddhism's various faiths are estimated to have about 260 million adherents; again, this is a strongly regional religion, as Plate 7 shows. Shintoism, the Japanese ethnic religion closely related to Buddhism, has some 35 million adherents, a number that is declining under the pressure of Japan's modernization.

Chinese Religions

The Chinese religions, too, have elements of Buddhism mixed with local Chinese belief systems. As we will note later, the traditional Chinese religions never involved concepts of supernatural omnipotence. Confucianism was mainly a philosophy of (earthly) life, and Taoism held that human happiness lies in one's proper relationship with nature. Chinese Buddhism was a pragmatic version of what the Buddha had originally preached, and it related well to Chinese religious styles. The faiths survive in China today, but we do not know in what strength. The data given in Table 7.1 should be considered as rough estimates only. They suggest that Confuciansim and Taoism today have fewer than 200 million adherents.

Judaism

Our map shows Judaism to be distributed through parts of the Middle East and North Africa, the Soviet Union and Europe, and parts of North and South America. Judaism is one of the world's great religions, but apart from the state of Israel, it is now scattered and dispersed across much of the world. Today Judaism has about 18 million adherents.

Traditional Faiths

Finally, Plate 7 shows large areas in Africa and several other parts of the world as "Traditional and Shamanist." Shamanism occurs in various forms in many parts of the world. It is a community faith in which people follow their **shaman**, who is a religious leader, teacher, healer, visionary, but in the ancient Chinese tradition, a man of *this* world not of another. Such a shaman appeared to various peoples in many different parts of the world—Africa, Indian America, Southeast Asia, and East Asia where shamanism may first have taken form. Scholars of religion have noted that there have been similar effects of such an appearance on the cultures of peoples scattered far and wide across the world, and they employ the term *shamanism* to identify such faiths. We might guess that if these shamanist religions had developed elaborate bureaucracies—which they did not—and had sent representatives to international congresses (as do the Christian faiths), then they would have negotiated away their differences and created still another world religion. However, unlike Christianity or Islam, the shamanist faiths are small in scale and comparatively isolated.

Shamanism, then, is traditional religion, an intimate part of a local culture and society. However, not all traditional religions are shamanist. Traditional African religions involve beliefs in a god as creator and indivisible provider, in divinities both superhuman and human, in spirits, and in a life hereafter. Christianity and Islam made inroads into traditional religions, but as the map indicates, they have failed to convert the African peoples, except in particular areas. Where Plate 7 shows traditional religions do exist, they remain in the majority today.

The major religions originated in a remarkably small area of the world. Judaism and Christianity began in what is today the Israel–Jordan area. Islam arose through the teachings of Mohammed, a resident of Mecca in western Arabia. The Hindu religion, which has no central figure and is a complex, multifaceted faith, originated in the Indus region of what is today Pakistan, long before Christianity or Islam. Buddhism emerged from the teachings of Prince Siddhartha, a man who renounced his claims to power and wealth in his kingdom, located in northeast India,

to seek salvation and enlightenment in religious meditation.

These source areas coincide quite strongly with the culture hearths shown in Figure 7.1, and there can be no doubt that while developments in other spheres were occurring in these regions—urbanization, irrigated agriculture, political growth, increasingly complex social orders, and legal systems—religious systems also became more sophisticated. Like the technological innovations, the faiths diffused far and wide.

Geographic Expressions of Faith

Earlier in this chapter, we identified six leading themes in cultural geography: cultural landscapes, culture hearths, cultural diffusion, cultural ecology, cultural regions, and geographic realms. The cultural geography of religion provides good examples of several of these themes. The spatial relationship between ancient culture hearths and the rise of major faiths already has been mentioned. In addition, the great religions' dispersal exemplifies diffusion processes. Cultural landscapes are strongly influenced by religion. And the regional geography of religion often is well defined. In the discussion that follows, the focus is on the geographic expressions of religion.

GLOBAL CHRISTIANITY

Not only does Christianity have more adherents than any other faith, but Christian religions also are the most widely dispersed on earth. Of its three great branches (Roman Catholic, Protestant, Eastern Orthodox), the two largest are strongly represented in nearly all the world's major geographic realms, and the third still has strength in its Eurasian core area.

Diffusion

The dissemination of Christianity occurred as a combination of expansion and relocation diffusion. The

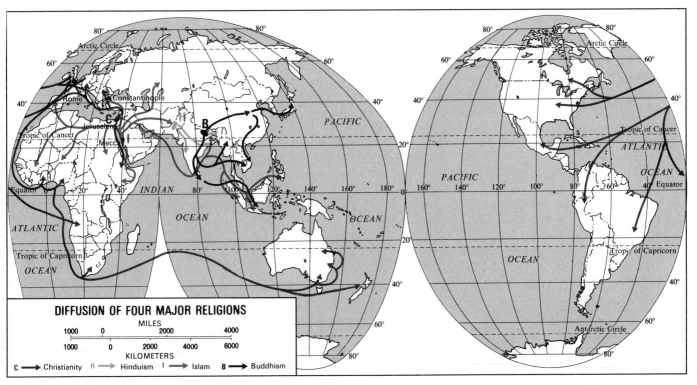

FIGURE 7.7

Major diffusion routes of the world's leading religions.

worldwide dispersal of Christianity was accomplished by the era of colonial acquisition on which Europe embarked in the sixteenth century. Spain invaded Middle and South America, bringing the Catholic faith to those areas. Protestant refugees, tired of conflict and oppression and in search of new hope and freedom, came in large numbers to North America. A patchwork of missionary efforts produced mixed conversions in much of Black Africa; Catholicism made inroads in Zaïre (Congo), Angola, and Moçambique. A very small percentage of the people in formerly British India were converted to Christianity; Catholicism scored heavily in the Philippines during the period of Spanish control.

Today Christianity is the most widespread and largest religion, and although the number of its adherents may be declining in some places, it is still gaining adherents in many areas. The faith has always been characterized by the aggressive and persistent proselytism of its proponents, and Christian missionaries created an almost worldwide network of conversion during the colonial period (Fig. 7.7).

Cultural Landscape

The cultural landscapes of Christianity's various branches reflect the changes that the faith has undergone over the past centuries. In medieval Europe, the cathedral, church, or monastery was the focus of life. In the towns, other buildings clustered around the tallest—the tower, steeple, and spire of the church, the beacon that could be seen (and whose bells could be heard) for miles in the surrounding countryside. In the square or plaza in front of the church, the crowds would gather for ceremonies and festivals in which the faith played its role, whether the event was primarily religious or not. Good harvests, military victories, public announcements, and all else was done in the shadow of the symbol of religious authority.

The Reformation, the rise of secularism, and the decline of organized religion's power are all reflected in the cultural landscape as well. The towns in Europe in which the cathedral still rises above the townscape are reminders of the region's Roman Catholic-domi-

Religion has affected architecture in all cultures. From Indian villages to European cities, religious structures mark and often dominate the scene. One of the leading European examples is the great Notre Dame cathedral in Paris, whose elaborate interior is shown here.

nated history. Protestants tend to regard a house of worship differently; it need not be especially large, imposing, or ornate. In Protestant regions, therefore, churches tend to blend into the local architecture and may be identifiable only by a sign that announces the name of the coming Sunday's speaker and the title of the sermon. In most large cities, cathedral and church now stand in the shadow of another kind of structure: the multistory office building, the skyscraper, symbol of the power of commerce and money. Churches are likely to be built outside the CBD, where land costs less. Except for some large-scale projects (such as the great Washington Cathedral), most modern religious structures are not as impressive and elaborate as those of earlier times. Other kinds of symbolism are needed now to sustain the church in an age of televised sermons and drive-in places of worship.

We should note that certain denominations, especially those that may be regarded as the more conservative, have more durable cultural landscapes in which older traditions of the authority and the influence of the church remain visible. In the United States, the best example is undoubtedly the Mormon culture region; to a lesser extent the Northeast, with its history of Catholic power and particular combinations of denominations, is another.

The cultural landscape also carries the imprint of death. It is appropriate to relate this topic to the cultural landscape of Christianity because no faith (other than perhaps those of China) uses land so liberally for the disposition of the departed. Hindus, Buddhists, and Shintoists cremate the dead, and it is noteworthy that this practice prevails in regions where living space and farmland are at a premium. However, Christian faiths bury their dead (as does Judaism), often with elaborate rituals and in spacious cemeteries that are carefully tended and resemble parks in their layout. Class expresses itself even in death; some graves are marked by a simple tombstone, others are elaborate, templelike marble structures. Whatever the nature of the monuments to the dead, however, the more impressive aspect of this culture trait is the amount of space that is devoted to graveyards and cemeteries, even in severely crowded urban areas where land prices have risen enormously—a reflection of the tradition of the power of the church. This also reveals the reality that cemeteries and funeral-related establishments represent a significant economic enterprise in Western cultures, further evidence of the secularization of formerly religious functions and responsibilities.

Regions

Earlier in this chapter we identified the Mormon culture region as an element in American cultural geography. Other religious regions in the United States can also be defined: the strongly Catholic New England area; the Baptist-dominated South; the strongly Lutheran upper Midwest; and the Spanish-Catholic Southwest. In the mobile, culturally interdigitated United States, these regions are not sharply defined, but regionalism can be very rigid. Northern Ireland, for example, is compartmentalized into majority-Catholic and majority-Protestant regions; although all the people are Christians, the divisions are the dominant fact of life.

The culture regions of Christian denominations in the United States are better known and understood than similar regions in other geographic realms. Data on religious affiliations are more nearly accurate and available for the United States, Canada, and Europe, but similar information for the Communist-controlled countries is not provided, and the religious geography of Black Africa and the world of Islam is also less accessible. It is obvious that a great deal of research remains to be done in this interesting field.

ISLAMIC TRANSFORMATION

The faith of the Moslems is the youngest of the major religions, having grown from the teachings of Mohammed (Muhammad), born in A.D. 571. According to Moslem belief, Mohammed received the truth directly from Allah in a series of revelations that began when the prophet was about 42 years of age. During these revelations, Mohammed involuntarily spoke the verses of the Koran (Qur'an), the Moslems' holy book. Mohammed, a student of religion even before this event, admired the monotheism of Christianity and Judaism; he believed that Allah had already manifested himself through other prophets (including

Jesus). But he, Mohammed, was the real and ultimate prophet.

Mohammed became a towering figure in Arabia during the seventh century A.D. Born and soon orphaned, he grew up under the tutelage of his grandfather, then leader of Mecca (Makkah); after his grandfather's death, an uncle continued his upbringing. After his visions, Mohammed at first had doubts that he could have been chosen to be a prophet, but once convinced by continued revelations, he committed his life to the fulfillment of the divine commands. In those days, the Arab world was in religious and social disarray, with various gods and goddesses admired by peoples whose political adjustments were at best feudal. Soon Mohammed's opponents sensed his strength and purpose, and they began to combat his efforts. The prophet was forced to flee Mecca for the safer haven of Medina (al Madinah); from this new base, he continued his work.

The precepts of Islam constituted, in many ways, a revision and embellishment of Judaic and Christian beliefs and traditions. There is but one god, who occasionally reveals himself to prophets; Islam acknowledges that Jesus was such a prophet. What is earthly and worldly is profane; only Allah is pure. Allah's will is absolute; he is omnipotent and omniscient. All humans live in a world created for their use, but only to await a final judgment day.

Islam brought to the Arab world not only a unifying religious faith, but also a whole new set of values, a new way of life, a new individual and collective dignity. Apart from dictating observance of the Five Pillars of Islam (repeated expressions of the basic creed, frequent prayer, a month of daytime fasting, almsgiving, and at least one pilgrimage to Mecca), the faith prescribed and proscribed in other spheres of life as well. Alcohol, smoking, and gambling were forbidden. Polygamy was tolerated, although the virtues of monogamy were acknowledged. Mosques made their appearance in Arab settlements, not only for Friday prayer, but also to serve as social gathering places to bring communities closer together. Mecca became the spiritual center for a divided, far-flung people for whom a joint focus was something new.

The stimulus given by Mohammed, spiritual as well as political, was such that the Arab world was

285

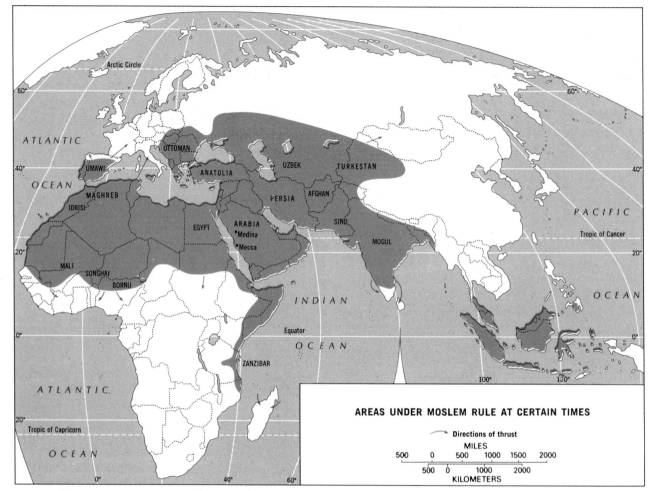

FIGURE 7.8

Areas that, since the rise of Islam, have been strongly influenced by the faith; Moslem domination occurred at some time in history.

mobilized overnight. The prophet died in 632, but his faith and fame continued to spread like wildfire. Arab armies formed, they invaded and conquered, and Islam was carried throughout North Africa. By the early ninth century A.D., the Moslem world included emirates extending from Egypt to Morocco, a caliphate occupying most of Spain and Portugal, and a unified realm encompassing Arabia, the Middle East, Iran, and most of what is today Pakistan (Fig. 7.8). Moslem influences had penetrated France, attacked Italy, and

invaded what is today Soviet Central Asia as far as the Aral Sea. Ultimately, the Arab empire extended from Morocco to India and from Turkey to Ethiopia. The original capital was at Medina in Arabia, but in response to these strategic successes, it was moved, first to Damascus and then to Baghdad. In the fields of architecture, mathematics, and science, the Arabs far overshadowed their European contemporaries, and they established institutions of higher learning in many cities, including Baghdad, Cairo, and Toledo

(Spain). The faith had spawned a culture, and it is still at the heart of that culture today.

Diffusion

The spread of Islam from its Arabian source area is a classic example of expansion diffusion, and its subsequent dispersal to Malaysia, Indonesia, South Africa, and the New World resulted from relocation diffusion. Unlike Hinduism, similarly diffused through the relocation process, Islam attracted converts wherever it took hold, and new core areas soon proved to be effective source areas for further dispersal. As Plate 7 shows, Islam's regions include not only North Africa and Southwest Asia, but also Bangladesh, Malaysia (to be a Malay *is* to be a Moslem), and Indonesia. As Table 7.2 indicates, although Islam's adherents are concentrated mostly in Asia (nearly 400 million) and Africa (about 160 million), there are also 21 million Moslems in Europe and the U.S.S.R. (mostly in that country's Asian republics), and perhaps as many as 2 million in the Americas. Islam is experiencing a modern resurgence, and its expansion appears likely to continue.

Cultural Landscape

Islamic cities, towns, and villages are dominated by elaborate, ornate, sometimes magnificently designed and executed mosques, whose balconied minarets rise above the townscape. As in Catholicism of old, the mosque often constitutes the town's most imposing, most carefully maintained building: the focus of life. From the towering minarets the faithful are called to prayer, filling the streets and paths as they converge on the holiest place, there to bend to the ground in disciplined worship.

At the height of Islam's expansion into eastern North Africa and southern Europe, the Moslem architects combined their skills with the Roman structures of an earlier age; from this coalescence came some of the world's greatest architectural masterpieces, in-

Islam spread from its source area in Arabia to Southern Europe, Southeast Asia, and West and East Africa (among other distant areas). The townscapes of East African cities such as Dar es Salaam, Mombasa, and Nairobi carry the Islamic imprint: shown here is one of Mombasa's numerous mosques.

cluding the Alhambra Palace in Granada and the Great Mosque of Córdoba. During the eleventh century, the practice of glazing the tiles of domes and roofs developed, and to the beautiful arcades and arched courtyards were added the exquisite beauty of glasslike, perfectly symmetrical cupolas. Moslem architecture represents the unifying power of Islamic monotheism, the perfection, vastness, indeed, endlessness of the spirit of Allah.

In architecture, Islam achieved its greatest artistic expression, its most distinctive visible element. Even in the smallest town, the community contributes to build and maintain its mosque in tribute to Allah. As such, it symbolizes the power of the faith and its role in the community; its primacy in the cultural landscape is a confirmation of the reality that in the Moslem world, religion and culture are one.

Regions

As the map shows, Islam is no more a unified faith than is Christianity. The religion's principal line of fragmentation formed early on, in fact, almost immediately after the prophet's death, and it was caused by a conflict over his succession. A complicated struggle arose, punctuated by murder, warfare and eventually by doctrinal disagreements. The orthodox *Sunni* Moslems ultimately prevailed, but the *Shi'ite* Moslems, who were masters at political manipulation, survived as small minorities throughout the Moslem world. Then, early during the sixteenth century, an Iranian (Persian) dynasty made Shi'ism (or Shi'ah) the only legal faith of that Moslem empire—an empire that extended into present-day southern Azerbaydzhan, eastern Iraq, and western Afghanistan and Pakistan. This gave the Shi'ite sect unprecedented strength and created the foundations of its modern-day culture region centered on the state of Iran (Plate 7). Approximately 10 percent of all Moslems (nearly 60 million persons) adhere to the beliefs of Shi'ah.

In a general way, the differences between the Sunnis and the Shi'ites may be viewed as a matter of practicality and earthly knowledge as opposed to idealism and the supernatural. Sunni Moslems believe in the effectiveness of family and community in the solution of life's problems; Shi'ites follow the "infallible" *imam*, the sole source of true knowledge. Whereas Sunni Moslems are comparatively reserved, Shi'ite Moslems are passionate and emotional (as recent events in Iran and Lebanon have demonstrated). The death of their early leader's son, Husayn, the child of Ali, is celebrated with intense processions during which the marchers beat themselves with chains and cut themselves with sharp metal instruments.

Aggressive Shi'ah has influenced Sunni Islam in several ways. The passion motive has been diffused eastward into Pakistan, Afghanistan, and India, where Sunni masses engage in similar rituals. The admiration and, indeed, veneration of the Shi'ite's early leader, Ali, has diffused throughout Sunni Islam and is reflected in the respect shown to his family's descendants—the *sayyids* of East Africa and the *sharifs* of North Africa—by all Moslems. The revolutionary fervor of Iran during the late 1970s and the 1980s stirred all of Islam, although it also produced violent conflict along the cultural boundary between Iran and its Sunni-dominated neighbor, Iraq.

ANCIENT HINDUISM

Chronologically, the Hindu religion is the oldest of the major religions and one of the oldest extant religions in the world. It emerged without a prophet or a book of scriptures, and without evolving a bureaucratic structure comparable to those of the Christian religions. Hinduism appears to have begun in the region of the Indus Valley, perhaps as long as four thousand years ago. The fundamental doctrine of the faith is the **karma**, which involves the concept of the transferability of the soul. All beings are in a hierarchy; all have souls. The ideal is to move upward in the hierarchy and then to escape from the eternal cycle through union with the **Brahman**, at the top of the ladder. A soul moves upward or downward according to one's behavior in the present life. Good deeds and adherence to the faith lead to a higher level in the next life. Bad behavior leads to demotion. All souls, those of animals as well as humans, participate in this process and the principle of *reincarnation* is a cornerstone of the faith: mistreat animals in this life, and chances are that you will *be* that animal in a future life.

Hinduism's doctrines are closely bound to Indian society's **caste system**, for castes themselves are steps on the universal ladder. However, the caste system locks people in social classes, reducing mobility—an undesirable feature manifested especially in the lowest of the castes, the *untouchables*. Until a generation ago, the untouchables could not enter temples, were excluded from certain schools, and were restricted to performing the most unpleasant tasks. The coming of other religions to India, the effects of modernization during the colonial period, and especially the work of Mahatma Gandhi, India's famous leader, loosened the social barriers of the caste system and bettered the lot of 90 million untouchables.

Hinduism, born on the Indian subcontinent's western flank, spread eastward across India and soon, before the advent of Christianity, into Southeast Asia. It would penetrate by first attaching itself to prevailing traditional faiths and then slowly supplanting them. Later, when Islam and Christianity appeared and were vigorously promulgated in Hindu areas, Hindu thinkers sought to assimilate certain of the new teachings into their own religion. For example, elements of the Sermon on the Mount now form part of Hindu preaching and Christian beliefs led in some measure to the softening of caste barriers. In other instances, the confrontation between Hinduism and other faiths led to the formulation of a compromise religion. Thus, the monotheism of Islam stimulated the rise of the *Sikhs*, who disapproved of the worship of all sorts of idols and who disliked the caste system but, nevertheless, retained concepts of reincarnation and the karma.

As Fig. 7.7 shows, the Hindu faith evolved in what is today Pakistan, reached its fullest development in India, and spread, to some extent, into Southeast Asia. Hinduism, however, has not been widely dispersed. In Southeast Asia, Islam overtook Hinduism, and most of the results of its early diffusion have been erased. In overwhelmingly Moslem Indonesia, the island of Bali remains a Hindu outpost, but Djawa (Java) retains only architectural remnants of its Hindu age.

Diffusion

Hinduism has not spread by expansion diffusion in modern times and has remained, essentially, a cultural

Hindu traditions still prevail in far-flung places, witnessing the faith's early and wide diffusion. Life on Bali, in Moslem Indonesia, remains dominated by modified Hindu beliefs. Shown here is one of many Hindu temples that diversify Bali's cultural landscape. The characteristic columns and steps beckon the faithful.

289

religion of South Asia. During the colonial period, however, many hundreds of thousands of Indians were transported to other areas of the world, including East and South Africa, the Caribbean, northern South America, and Pacific islands (notably Fiji). This relocation diffusion did establish Hindu outposts in those far-flung locations; although the Hindu clusters themselves survive, they did not form the foci of growing Hindu regions, and few non-Indians were converted to the faith.

Cultural Landscape

Hinduism is more than a faith: it is a way of life. Meals are religious rites; prohibitions and commands multiply as the ladder of caste is ascended. Pilgrimages follow prescribed routes, and rituals are attended by millions. Festivals and feasts are frequent, colorful, and noisy. Hindu doctrines include the belief that the erection of a temple, modest or elaborate, bestows merit on the builder and leads to heavenly reward. As a result, the Hindu cultural landscape—urban as well as rural—is characterized by countless shrines, ranging from small village temples to structures so large and elaborate that they are virtually holy cities. The location of shrines is important because there should be minimal disruption of the natural landscape; the temple should be in a "comfortable" position (say under a large, shady tree) and near water wherever possible, because many gods will not venture far from water and because water has a holy function in Hinduism. A village temple should face the village from a prominent position and offerings must be made frequently. Small offerings of fruit and flowers lie before the sanctuary of the deity honored by the shrine.

Thus the cultural landscape of Hinduism is the cultural landscape of India, the cultural region. Temples and shrines, holy animals by the tens of millions, distinctively garbed holy men, and the sights and sounds of endless processions and rituals all contribute to an atmosphere without parallel; the faith is a visual as well as an emotional experience.

BUDDHA'S LEGACY

Buddhism appeared in India during the sixth century B.C. as a reaction to the less desirable features of Hinduism. It was by no means the only protest of its kind (*Jainism* was another), but it was the strongest and most effective. As previously mentioned, the faith was founded by Prince Siddhartha, known to his follows as Gautama, heir to a wealthy kingdom in what is now Nepal. He was profoundly shaken by the misery he saw about him, which contrasted so sharply against the splendor and wealth that had been his own experience. The Buddha (enlightened one) was perhaps the first prominent Indian religious leader to speak out against Hinduism's caste system. Salvation,

he preached, could be attained by anyone, no matter what his or her caste. Enlightenment would come to a person through knowledge, especially self-knowledge: the elimination of covetousness, craving, and desire; the principle of complete honesty, and the determination not to hurt another person or animal. Thus, part of the tradition of the karma still prevailed.

Following the death of the Buddha in 489 B.C. at the age of 80, the faith grew rather slowly until, during the middle of the third century B.C., the Emperor Aśoka became a convert. Aśoka was leader of a large and powerful state that covered India from the Punjab to Bengal and from the Himalayan foothills to Mysore. Aśoka not only set out to rule his country in accordance with the teachings of the Buddha, he also sent missionaries to the outside world to carry the Buddha's teachings to distant peoples. Buddhism now spread as far south as Sri Lanka, and later west toward the Mediterranean, north into Tibet, and east into China, Korea, Japan, Vietnam, and Indonesia, all over a span of some 10 centuries (Fig. 7.7). However, while Buddhism spread to distant lands, it began to decline in its region of origin. During Aśoka's rule, there may have been more Buddhists in India than Hindu adherents. But after that period of success, the strength of Hinduism again began to assert itself. Today Buddhism is practically extinct in India, although it still thrives in Sri Lanka, in Southeast Asia, in Nepal and Tibet, and in Korea. Along with other faiths, it also survives in Japan.

Buddhism is fragmented into numerous branches, of which the leading are two sectarian groups, Mahayana Buddhism and Theravada Buddhism. Theravada Buddhism is the monastic faith of the source, Gautama's teachings, and it survives in Sri Lanka, Burma, Thailand, Laos, and Kampuchea. It holds that salvation is a personal matter, achieved by worldly good behavior and religious activities, including periods of service as a monk or nun. Mahayana Buddhism, practiced mainly in Vietnam, Korea, Japan, and for centuries in China, holds that superhuman, holy sources of merit do exist and that salvation can be aided by appeals to these sources. The Buddha is regarded as a divine savior. Mahayana Buddhists do not serve for periods as monks, as the Theravadans do, but spend much time in personal meditation and wor-

Although Buddhism is in decline in its source area, it still thrives elsewhere. In Sri Lanka (Ceylon) the Temple of the Tooth, in the highland town of Kandy, attracts many thousands of Buddhist adherents in a continuous stream. Within its elaborate walls is a magnificent gold shrine reputed to contain one of the Buddha's teeth. This is one of Buddhism's holiest places.

ship. Other branches of Buddhism include the Lamaism of Xizang (Tibet), which combines monastic Buddhism with the worship of local demons and deities, and Zen Buddhism, the contemplative form prevalent in Japan.

Buddhism is experiencing a period of revival that started two centuries ago and has recently intensified. It has become a universal religion, unlike Hinduism, and has diffused to many areas of the world. However, the faith has suffered severely in its modern hearth in Southeast Asia. Militant Communist regimes have attacked the faith and destroyed its institutions in Kampuchea, Laos, and (to a considerable extent) in Vietnam as well. In Thailand, also, Buddhism has been under severe pressure, afflicted by rising political tensions from which the influential faith could not escape. Yet the appeal of Buddhism's principles ensured its continued diffusion, notably in the Western world.

Cultural Landscape

When the Buddha received enlightenment, he sat under a large tree, the Bo tree at Bodh Gaya in India, now a place of pilgrimage for Buddhists of all branches and sects. (The Bo tree now growing on the site is believed to be a descendant of the original tree). The Bo tree has a thick, banyanlike trunk and a wide canopy of leafy branches. Because of its association with the Buddha, the tree is protected, revered, and the object of pilgrimages wherever it is believed that Buddha may have taught beneath its branches. It has also been diffused as far as China and Japan as a symbol of the faith, and it marks the townscape of numerous villages and towns.

Buddhism's architecture includes some magnificent achievements, including the famed structures at Borobudur in central Djawa (Java). The shrines of Buddhism include bell-shaped structures that protect

While the Buddha's teachings were attracting converts in India, a religious revolution of quite another sort was taking place in China. The great Chinese philosopher K'ung-Fu-tzu (Confucius), born in 551 B.C., countered Chinese cults that included beliefs in heaven and the existence of the soul, ancestor worship, sacrificial rites, and shamanism. He wrote and preached that the real meaning of life lay in the present, not in some future abstract existence, and that service to one's neighbors, friends, and acquaintances should supersede any service to spirits and ghosts. During the same period, an older contemporary of Confucius, Lao-tzu, published a book on *The Way and Its Virtue* (*Tao-te-ching*), a prescription for life that held humanity and nature are one and that people should learn to live in harmony with nature as but an insignificant element in the great universal order.

Both Confucianism and Taoism had great impact on Chinese life. Confucius was appalled at the suffering of ordinary people in his country at the hands of feudal lords, and he urged the poor to assert themselves. He was no prophet who dealt in promises of heaven and threats of hell; Confucius denied the divine ancestry of China's aristocratic rulers, educated the landless and the weak, disliked supernatural mysticism, and argued that human virtues and abilities, not heritage, should determine a person's position and responsibilities in society.

Notwithstanding his earthly philosophies, Con-

fucius took on the mantle of a spiritual leader after his death in 479 B.C. His teachings diffused widely throughout East and Southeast Asia; temples were built in his honor all over China. From his writings and sayings emerged the Confucian Classics, a set of 13 texts that became the focus of education in China for two thousand years. In government, law, literature, religion, morality, and in every conceivable way the Confucian Classics were the Chinese civilization's guide. Elements of Buddhism, introduced into China during the Han dynasty, also formed part of the society's belief system. Buddhism's reverence for the aged, the departed, and nature in general made it easily adaptable to Chinese philosophies.

Over the centuries Confucianism (with its Taoist and Buddhist ingredients) became China's state ethic, but time did modify Confucius' ideals. Worship of and obedience to the emperor became a part of Confucianism, for example.

During the twentieth century, political upheavals in China led to reactions against Confucian philosophies, first during the Republican period after 1912 and more seriously under the Communist regimes after 1949. But Confucianism has been China's beacon for a very long time. It will be difficult to eradicate two millennia of cultural conditioning in a few decades: the dying spirit of Confucius will haunt physical and mental landscapes in China for years to come.

relic mounds, temples that enshrine an image of the Buddha in his familiar cross-legged pose, and large monasteries that tower over the local townscape. The *pagoda* is perhaps Buddhism's most familiar structure, its shape derived from the relic (often funeral) mounds of old, every fragment of its construction a meaningful representation of Buddhist philosophy.

ROOTS OF JUDAISM

The oldest major religion west of India, Judaism grew

from the belief system of the Jews, one of several Semitic, nomadic tribes that traversed Southwest Asia about 2000 B.C. The Jews led an existence filled with upheavals. Moses led them from Egypt, where they suffered repression, to Canaan, where an internal conflict arose and the nation split into two, Israel and Judah. Israel was subsequently wiped out by enemies, but Judah survived longer, only to be conquered by the Babylonians. Regrouping to rebuild Jerusalem, the Jews fell victim to a sequence of alien powers and saw their holy city destroyed again in A.D. 70. The Jews were driven away and scattered all over the region and eventually into Europe and much of the rest of the world. For many centuries—indeed until late in the eighteenth century—they were persecuted, denied citizenship, driven into ghettos, and massacred. As the twentieth century has proved, the world has not been a safe place for Jewish people even in modern times.

In the face of such constant threats to their existence, it has been their faith that has sustained the Jews, the certainty that the Messiah would come to deliver them from their enemies. The roots of Jewish religious tradition lie in the teachings of Abraham, who united his people and during whose time the faith began to take shape. Among the religions adhered to by the ancient Semitic tribes, that of the Jews was unique in that it involved the worship of only one god—who, the Jews believed, had selected them to bear witness to his existence and works. But their religion also showed the many contacts with other beliefs and other peoples. From **Zoroastrianism**, which arose in Persia during the sixth century B.C., Judaism acquired its concepts of paradise and hell, angels and devils, judgment day, and resurrection.

Modern times have seen a division of Judaism into several branches. During the nineteenth century, when Jews began to enjoy greater freedom, a Reform movement developed that was aimed at adjusting the faith and its practices to current time. But there were many who feared that this would cause a loss of identity and cohesion, and the Orthodox movement sought to retain the old precepts as effectively as possible. Between these two extremes is a branch that is less strictly orthodox but not as liberal as the reformers: the Conservative movement. These developments are responses to the acculturation processes that have affected Jewish societies ever since their worldwide dispersal began.

The objective of a homeland for Jewish people, an idea that gained currency during the nineteenth century, produced the ideology of **Zionism**. The Zionist ideals were rooted in a determination that Jews should not be absorbed in, and assimilated by, other societies. The goal of a Jewish state became a reality in 1948 when Israel was created under U.N. auspices on the shores of the eastern Mediterranean.

As Table 7.2 shows, the Jewish faith presently has fewer than 20 million adherents, but the distribution of Jews proves that Judaism is indeed a global religion. Eight million reside in North America, about 5 million in Europe (including the U.S.S.R.), and the total for Asia, over 4 million, includes the Jewish population of Israel itself.

Secularism and Fundamentalism

Religions in the modern world exhibit contrasting trends. Today many hundreds of millions of people do not practice any religion. In fact, church membership figures as reported in Table 7.2 do not accurately reflect the number of active practitioners of a faith. In the modern world, secularism—indifference to, or rejection of, religious ideas—is on the rise.

The growth of secularism may be attributed to several factors and conditions. Some national governments actively discourage church membership and activity, as is the case in a number of major Communist countries. Greater personal freedom of choice and democracy permitted people not only to choose with whom to worship, but also whether to worship at all. The dwindling power of the church in modernizing societies has been combined with the failure of many faiths to change with the times. Organized religion's role during colonial times, and in major wars, also has led to disaffection and rejection.

But this rise of secularism is paralleled by a resurgence of fundamentalism in other places. Where modern and traditional societies make contact, the traditional faith and its adherents may feel threatened. This may lead to a rejection of the new and a reconfirmation of the virtues of the old. In recent years, several Islamic countries have experienced this, notably Iran, where the shah engaged in a program of modernization and Westernization during the 1960s and 1970s. This effort ran counter to the traditional Shi'ite religion and its leaders. Ultimately a religiously based revolution ousted the shah, and Iran reverted to fundamentalist rule.

Religion continues to dominate or strongly affect the lives of many hundreds of millions of people in this world; in various ways, this influence is imprinted on the cultural landscape, from tiny Buddhist villages to major Islamic cities. The cultural geography of religion is a large part of human geography in general.

293

A culture's landscape is its geographic expression, its regional imprint, its durable manifestation of history, aesthetics, objectives, and priorities.

The cultural landscape, we have learned, consists of an assemblage of visible attributes, a composite of images and impressions that define the regional culture. However, the landscape of culture is in the mind, as well as in the eye, and intangible qualities also contribute to its formation, qualities that would not appear on a photograph nor on an ordinary map. Enter an African or an Arab town and you may be greeted by sounds of music and smells of cooking, the shouts of traders and the bustle of people—all of which will remain essential ingredients of your perception of these cultures.

Cultural geographers differentiate between the tangible, visible elements of the cultural landscape and the more abstract by differentiating between expressions of *material* and *nonmaterial* culture. On the face of it, this seems an easy distinction among, say, buildings and roads and cultivated land, on the one hand, and music, aromas, and the sounds of language, on the other. In fact, it is not so simple. There is no absolute, clear-cut division between material and nonmaterial culture. Take, for example, music as an aspect of nonmaterial culture. Certainly music and musical expression are nonmaterial elements of culture. However, only unaccompanied song would truly represent nonmaterial culture; other music is made with instruments and represents material culture. Thus, the cultural geography of musical styles and forms and their diffusion would, indeed, be a study in nonmaterial culture. The study of the development, use, diffusion, and distribution of musical instruments, however, would venture into material culture.

Nonmaterial culture usually is taken to include, in addition to music, theater and dance, plays and literature, art (painting), food habits and preferences, law and legal systems, language, and religion.

Material culture, the more important contributor to the cultural landscape, includes architectural forms and styles and the clothing and other adornments worn routinely by the people. Architecture represents a culture system in several ways. We have already taken note of the importance of religious structures; governmental architecture also contributes importantly to a townscape. Commercial architecture (such as high-rise office buildings in American downtowns) and domestic architecture characterizing people's dwellings also are key elements in the overall cultural landscape.

The personality and character of a region are as surely defined by the dress of its inhabitants as by the properties of regional architectural styles. Modern or traditional, garments such as the kilt, kimono, and turban are closely associated with particular locales. Headgear, footwear, hair fashions, jewelry, even tatooing and cosmetics typify regional cultures—all contributing to the overall impression a region creates.

PERCEPTUAL REGIONS

Concepts of culture and region (we have noted) are intellectual constructs to help us understand the nature and distribution of phenomena in human geography. We have studied a wide range of these phenomena, ranging from the tangible to the abstract. All of us have in our minds a picture of a ''Chinese,'' ''African,'' and ''Latin'' cultural landscape—among many others—based on what we have seen and learned about these geographic realms and regions.

Thus, we all carry impressions and images of regions. These perceptions are based on our collection or mental ''bank'' of knowledge about such regions, and material as well as nonmaterial culture is part of this. The landscape and other qualities of the natural environment also contribute to the overall perception of a region, and sometimes this is the dominant image. The somewhat unfortunate term *Sunbelt*, for example, is a regional term that has become popular in recent years. It summarizes the chief attraction of a perceived region in the southern United States, extending from Florida to California, for persons retiring to a more comfortable climate and for others seeking to escape the rigors of colder climates.

Just where is this Sunbelt, this region of consensus? Anyone who attempts to draw lines on maps to delimit what is part of the Sunbelt and what is not faces a difficult task. The same is true for even more

294

Imposing and representative of power and authority are these government buildings, part of the Capitol complex in Washington, D.C. The architectural style is borrowed from the ancient Greeks to convey durability and permanence as well as cultural continuity.

familiar and older regional names such as the *South* or the *Midwest* or the *Northeast*. It would not be correct to identify such regions as nonmaterial culture regions because elements of material culture visible in the landscape certainly help constitute them. However, nonmaterial elements—language and dialect, music and art—also contribute to our impressions. Such regions, therefore, are called *perceptual* regions, regions perceived to exist. They are also known as *popular* regions and as *vernacular* regions.

Perceptual regions are not bounded by visible or clearly defined boundaries. Drive southward from, say, Pittsburgh or Detroit, and it will be impossible to discern a specific place where you enter the South. Instead, you will note features in the cultural landscape that you perceive to be associated with the South until, at some point, they come to dominate the area to such a degree that you will say to yourself, "I

am really in the South now." This may result from the music on a radio station you turn to, an item on a menu at a roadside restaurant (grits e.g.,), the sound of Southern accents, or a succession of Baptist churches in a town you drive through. These impressions now become part of your overall perception of the South as a region.

Perceptual regions can be studied at more than one scale. When world geographic realms are discussed (in another geography course), we note that these are fundamentally perceptual spatial units; the identification of Southeast Asia or Black Africa as discrete realms of the world results from much the same perception process as that producing an image of the South in the United States. Again, perceptual regions can be studied on a larger scale, and thus in greater detail. In an interesting article, Professor Terry Jordan analyzed "Perceptual Regions in Texas" (1978)

The Astrodome in Houston represents a relatively new trend in urban architecture: the enormous enclosed stadium within which the environment can be controlled. The dimensions of these structures are such that they form a dominant element in the urban landscape.

and gave spatial identity to commonly used names such as *Panhandle, Gulf Coast, Permian Basin*, and *Metroplex*. Other states of the United States have similar perceptual regions, and defining them is a challenging exercise. It is one thing to refer loosely to a region perceived to exist; it is quite another to put it on the map!

The perceptual regions of the United States and southern Canada were given concrete spatial form in a study by Wilbur Zelinsky. In an article entitled ''North America's Vernacular Regions'' (1980), Zelinsky delimited 12 major perceptual regions on a series of maps: the South, Midwest, New England, Southwest, the East, the West, the Pacific, the Atlantic, the North, the Gulf, the Northeast, and the Northwest. Figure 7.9 summarizes these regions (it deletes two small and comparatively unfamiliar names), and, of necessity, it shows overlaps between certain units. For example, the more general term *the West* obviously incorporates more specific regions such as the Pacific Region and part of the Northwest.

This problem of defining and delimiting perceptual regions can be approached in several ways. One way involves an elaborate system of interviews in which people residing within, as well as outside of, the postulated region are asked to respond to questions about their home and cultural environment. Zelinsky, in his 1980 study, used a different technique: he analyzed the contents of the telephone directories of 276 metropolitan areas in the United States and Canada, noting the frequencies with which businesses and other enterprises list themselves with regional or locational terms (e.g., ''*Southern Printing Company*''). The resulting maps, in several instances, show quite a close similarity between regions so defined and those of culture regions in North America as delimited by geographers, indicating that perceptions about regions are sometimes quite accurate.

In a sense, the perceptual region as a concept in cultural geography summarizes folk and popular culture, material and nonmaterial culture. It is an idea that can be explored at virtually all levels of detail. It

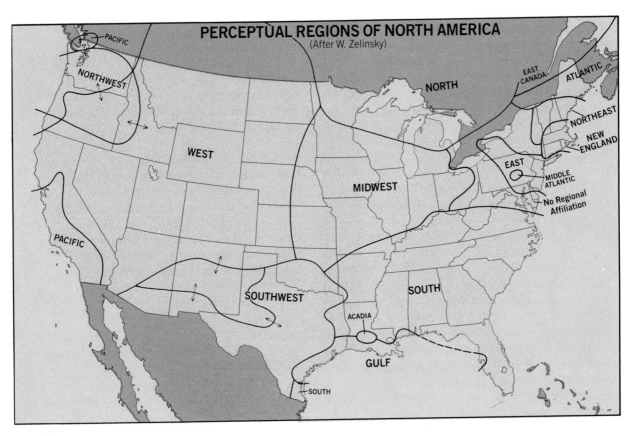

PERCEPTUAL REGIONS OF NORTH AMERICA
(After W. Zelinsky)

FIGURE 7.9

involves landscapes and environments, natural as well as social. In certain ways, it integrates everything we have studied so far in this book. At a smaller scale than the U.S. example, the great world geographic realms also constitute perceptual regions, a useful framework to keep in mind when reading the chapters that follow.

Reading about Cultural Geography

Chapter 7 begins with a discussion of the concept of culture. For more complete background on this issue,

the best sources are the several introductory anthropology texts you will find in the library and bookstore. A prominent anthropology text is E. A. Hoebel, *Anthropology: The Study of Man* (New York: McGraw-Hill, 1982), from which the quotation used in this chapter is taken (p. 246). The article by A. L. Kroeber and C. Kluckhohn that distilled 160 definitions of the culture concept was "Culture: a Critical Review of Concepts and Definitions," *Peabody Museum Papers* (Vol. 47, No. 1, 1952). Several of the definitions of culture on page 248 are derived from F. M. Kessing, *Cultural Anthropology: The Science of Custom* (New York: Holt, 1966).

Another good source in this context is D. Kaplan and R. A. Manners, *Culture Theory* (Englewood Cliffs N.J.: Prentice-Hall, 1972). C. Geertz's *Interpretation of Cultures: Selected Essays* (New York: Basic Books,

1973) also contains many ideas about the meaning and application of the culture concept.

Cultural geographers also have addressed this topic. In Chapter 1, reference was made to the seminal article on cultural landscape by C. O. Sauer. The most useful approach probably is to investigate several of the introductory volumes such as *Introducing Cultural Geography* by J. E. Spencer and W. L. Thomas (2d ed.; New York: Wiley, 1978). Also see *The Human Mosaic: A Thematic Introduction to Cultural Geography* by T. G. Jordan and L. Rowntree (New York: Harper & Row, 1986). Also consider *Human Geography in World Society* by J. F. Kolars and J. D. Nystuen (McGraw-Hill, 1974) and *Introduction to Cultural Geography* by S. N. Dicken and F. R. Pitts (Waltham, Mass.: Ginn, 1970), which is still worth reading. Note that no strong difference seems to exist between *cultural* and *human* as the terms are used in the titles of these books. J. E. Spencer and W. L. Thomas discuss the different meanings of the two, describing *human* geography as having a broader scope and a more contemporary orientation than *cultural* geography (p. 17).

For the source of the environmental debate in cultural geography, it is necessary to go back to the older literature first. For an indication of prevailing ideas during the first half of the century, an easily accessible book is E. Huntington's *Principles of Human Geography* (5th ed.; New York: Wiley, 1940). Modern geographic approaches to environmental questions are quite different, and an excellent research source is the volume edited by I. R. Manners and M. W. Mikesell, *Perspectives on Environment* (Washington, D.C.: AAG 1974).

On the regional concept, see J. R. McDonald's *Geography of Regions* (Dubuque, Iowa: W. C. Brown, 1972). Also see R. E. Dickinson, *Regional Concept* (London: Routledge & Kegan Paul, 1976). Of course, the geographic journals contain many articles dealing with the regional idea in geography, in theory and in practice. You may find it useful to look again at G. J. Martin and P. E. James' *All Possible Worlds: A History of Geographical Ideas* (New York: Wiley, 1981) for some insight into the evolution of the regional concept in geography. D. Meinig's "American Wests: Preface to a Geographical Introduction" appeared in the *Annals* of the Association of American Geographers (Vol. 62, June 1972).

On questions of race, see F. Hulse, *The Human Species* (New York: Random House, 1975), and several of the anthropology texts mentioned earlier. Also see R. Freedman and B. Berelson, "The Human Population," in *Scientific American* (Vol. 231, September 1974).

On the geography of language, there is an excellent chapter in *The Human Mosaic* by Jordan and Rowntree, listed above. Still quoted is P. L. Wagner's article "Remarks on the Geography of Language" in the *Geographical Review* (Vol. 48, 1958). Several topics in cultural geography, including language and religion, are discussed by W. Zelinsky in *The Cultural Geography of the United States* (Englewood Cliffs, N.J.: Prentice-Hall, 1973). On toponyms, begin with G. R. Stewart, *Names on the Land: A Historical Account of Place-Naming in the United States* (Boston: Houghton Mifflin, 1958). Also see Stewart's *Names on the Globe* (New York: Oxford Univ. London Press, 1975). For something more specific, there is J. Rydjord, *Kansas Place-Names* (Norman: Univ. of Oklahoma Press, 1972). On African languages, see J. H. Greenberg, *The Languages of Africa* (Bloomington: Indiana Univ. Press, 1963). When you look up the language chapters in various anthropology texts, you will find that they tend to focus on origins and structure, less on distribution and diffusion. Regional geographies often have useful sections on language patterns.

Religious origins and dispersals are discussed in D. E. Sopher's *The Geography of Religions* (Englewood Cliffs, N.J.: Prentice-Hall, 1967). I. R. al Faruqi and D. E. Sopher produced a *Historical Atlas of the Religions of the World* (New York: Macmillan, 1974). Also see J. R. Shortridge's article, "Patterns of Religion in the United States," in the *Geographical Review* (Vol. 66, October 1976).

The perception of regions is the topic of "Perceptual Regions in Texas," by T. Jordan in the *Geographical Review* (Vol. 68, October 1978). Also see W. Zelinsky, "North America's Vernacular Regions," *Annals* of the Association of American Geographers (Vol. 70, March 1980).

An interesting and productive source for ideas

and materials in the cultural geography of North America is an atlas edited by J. Rooney et al., *This Remarkable Continent: An Atlas of United States and Canadian Societies and Culture* (College Station: Texas A & M Univ. Press, 1982).

Cultural geography is served by a specialized journal, the *Journal of Cultural Geography*, published by Bowling Green State University.

299

Geography
of Farming

E|ven in developed, urbanized societies, where most of the people live in cities and towns, farmers have transformed much more of the land than the urbanites. Travel by road out of a large city or by air from a suburban airport and soon the built-up scene will be left behind. Then, for hour after hour, there is the mosaic of agriculture: the fields and furrows, fences and hedges, contoured slopes and irrigation ditches. The cultural landscape of farming can have strong identity, from the meticulously maintained terraces of Southeast Asia's paddies to the etched squares of the American Township-and-Range system, from manicured tea plantations to machine-sown wheatfields. All of this is the farmer's legacy on the land.

But the topic of this chapter is not the cultural geography of farming. Rather, we turn now to economic geography, the geography of sustenance. The field of economic geography is concerned with the various ways in which people earn a living and how the goods and services they produce in order to earn that income are spatially expressed and organized. Economic geographers tend to group occupations broadly into four major aggregates of productive activities that combine together to form the total *spatial economy*. *Primary* activities involve the extractive sector in which workers and the natural environment come into direct contact—especially **agriculture** and mining. *Secondary* economic activity comprises the *manufacturing* sector, in which raw materials are transformed into finished industrial products. *Tertiary* activities entail the *services* sector, including a wide range of livelihoods, from retailing to finance to education to routine office-based jobs. The fast-growing *quaternary* sector in technologically advanced societies involves the collection, processing, and manipulation of *information*.

This chapter treats primary economic activities, particularly agriculture—the deliberate tending of crops and livestock in order to produce food (and fiber). Although farming employs less than 3 percent of the labor force in the United States today, in most other countries agriculture is, by far, the leading employment sector. Thus the overwhelming majority of the world's peoples—industrial and technological progress notwithstanding—still farm the soil for a liv-

ing. However, as with other aspects of human geography, agricultural practices and systems vary sharply from region to region. Farming in the tropical rain forest of Africa is something very different from growing rice in Asia—and Asia's paddyfields look nothing like the vast wheat farms of the North American Great Plains. This chapter will provide an overview of these contrasts, focusing on both traditional and modern commercial agriculture.

• Ancient Livelihoods •

The processes whereby food is produced, distributed, and consumed form a fundamental part of every culture. The way land is allocated to individuals or families and the manner in which it is used, the functions of livestock, food preferences and taboos are all elements of culture. Food consumption is often bound up with religion: religious dogma prohibits adherents of Islam from eating pork and Hindu believers from consuming meat. Various forms of partial or total abstinence occur among human cultures, including periodic fasts. Such prescriptions, like the religions that generated them, tend to be old and persistent, and change comes slowly, even when change would seem to be beneficial. The Islamic prohibition against pork deprives millions of people of a useful source of nutrients. India's huge population of holy cattle puts a drain on already inadequate food supplies.

Most of the food we consume comes, directly or indirectly, from the soil. It may be that an age of synthetic foods is approaching and that the technological age will bring a declining dependence on the soil; but for the time being, practically everything we consume is the product of a process or a sequence of processes that begins in the earth itself. Farming has become the basis of existence all over the world.

BEFORE FARMING

Viewed in the perspective of all human history, farming actually is a rather recent innovation; its beginnings date back a mere twelve thousand years, perhaps less. Even today some societies survive much as they did before agriculture was developed, by hunting and by gathering what nature has to offer, and sometimes by fishing. Peoples who subsist this way have been pushed into difficult environments by more powerful competitors (as southern Africa's San were by the Bantu and the white Europeans), and their survival seems to involve the weathering of one crisis after another. Drought is the worst enemy. It withers the vegetation, kills and drives off the wildlife, cuts off natural water supplies such as springs. Still the San, the Australian indigenous peoples, the inhabitants of interior New Guinea, and several other communities in the Americas, Africa, and Asia manage to sustain themselves in the face of great odds. They do so by knowing and exploiting their environment exhaustively. Every seed, root, fruit, berry, and beetle is sought out and consumed. Hunting is done with poisoned spears, bows and arrows, clubs, and sticks. When the people do not depend on a particular water hole for their own water, they may poison it and follow the animals that have drunk there. But without agriculture and without the practice of storing food in preparation for future periods of shortage (very few hunting-gathering peoples do this), life is difficult. It was easier where the land was more productive, but most of the remaining hunting-gathering peoples have been driven from better-endowed areas into dry, cold, isolated, and less hospitable environments.

We should, therefore, not be misled into the assumption that the human groups that survive today as hunters and gatherers are entirely representative of the early human communities that lived by the same means. Although Europe's plainlands opened up following the most recent glacial retreat, our distant ancestors hunted the mammoth and other plentiful wildlife. They set elaborate traps and cooperated in driving wildlife to areas where it would be vulnerable. Communities then were much larger than the present-day San clans. Very early on there were peoples who subsisted on hunting, gathering, and fishing who had learned to specialize to some extent in some area of production. The oak forests of parts of North America provided an abundant harvest of nuts, sometimes enough to last more than a full year, so Indian com-

munities collected and stored this food source. Other Indians near the Pacific Ocean coast became adept at salmon fishing. The buffalo herds of interior North America also provided sustenance for centuries before being virtually wiped out. In more northerly regions, people followed the migrations of the caribou herds. The Eskimo and the Ainu (now confined to northern Japan) developed specialized fishing techniques.

Place

Thus we can deduce to some extent the means by which our preagricultural ancestors survived. Undoubtedly certain of the hunting-gathering communities found themselves in more favorable locations than others. For example, it may be that forest margins provided such advantages. There, people could gather food in the forest when hunting yielded poor results, then return to hunting when the opportunities improved again. Possibly those communities could become semisedentary and stay in one place for a length of time, perhaps to create a more-or-less permanent settlement. That is one of the contributions the eventual development of agriculture made: it permitted people to settle permanently in one location, with the assurance that food would be available in seasons to come.

Early Technology

The capacity of early human communities to sustain themselves was enhanced by their knowledge of the terrain and its exploitable resources as well as by their ability to improve tools, weapons, and other equipment. Such technological advances came slowly, but some of them had important effects, each time expanding the resource base. The first tools used in hunting were simple clubs, tree limbs that were handle-thin at one end and thick and heavy at the other. These were used not only to strike trapped or pursued animals, but also were thrown at hunted wildlife. The use of bone and stone and the development of spears made hunting far more effective. The fashioning of stone into hand axes and, later, handle axes was a crucial innovation, which enabled hunters to skin

303

their prey and cut the meat; it was now also possible to cut down trees and to create even more efficient shelters and tools. The controlled use of fire was also an early and critical achievement of human communities. Fires caused by natural conditions (lightning, spontaneous combustion of surface-heated coal) provided the first opportunities for control. Excavations of ancient settlement sites suggest that attempts were made to keep a fire burning continuously once it was captured. Later, it was learned that fire could be generated by manual means, by rapid hand rotation of a wood stick in a small hole, surrounded by dry tinder. Fire became a focus for the settlement, the camp fire a symbol of the community. It was a means of making otherwise unpalatable foods digestible. It was used in hunting as a way to drive animals into traps or over cliffs. This greatly enhanced ancient communities' capacity to modify the natural landscape.

In the meantime, tools and equipment were developed as well. Perhaps the first tool of transportation ever devised was a strong stick carried by two men over which hung the limp body of a freshly killed gazelle. Simple baskets were fashioned to hold gathered berries, nuts, and roots. Various forms of racks, packing frames, and sleds were developed to facilitate the transport of logs, stones, firewood, and other heavy goods. Fishing became a more important means of survival for communities situated along rivers or on shorelines, and primitive rafts and canoes soon made their appearance (controlled use of fire made the dugout canoe possible). Even before the momentous developments involving animal and plant domestication, rudimentary forms of metalworking had emerged, although true *metallurgy*, the technique of separating metals from their ores, came later. There is evidence that fragments of native copper, nuggets of gold, and pieces of iron from meteorites were hammered into arrowheads and other shapes by craftspeople in ancient communities. Equipment made of stone also went far beyond simple knives and axes. Stone pots and pounders, grinders, and simple mills were developed to treat seeds, grains, and other gathered edibles. Meat was roasted, other foods were cooked; dietary patterns and preferences began to develop. Therefore, long before the revolutionary changes in animal and plant **domestication** occurred, pre-

agricultural human communities and settlements were characterized by considerable complexity not only in dwelling forms, but also in tools, utensils, and weapons; food preferences and taboos; and related cultural traits.

FISHING

It is quite likely that our distant ancestors learned to fish, and added dried fish to their diets during the warming period that accompanied the melting of the last of the Pleistocene glaciers. Perhaps fifteen to twenty thousand years ago, these glaciers melted, sea levels began to rise, and coastal flatlands were inundated as the seawater encroached on what we today call the continental shelves. We may surmise that until this time coastal waters over most of the earth had been cold and rough and that shorelines were marked by steeply dropping relief. Thus, coastal areas were not the most hospitable parts of the human habitat and marine life was not nearly as plentiful as it was to become.

When glacial melting began and water levels rose, the continental shelves became shallow seas, full of coastal lagoons and patches of standing water. The sun warmed these thin layers of water very quickly, and soon an abundance of marine fauna flourished. Coastal regions became warmer, more habitable places to establish settlements, and numerous communities moved to the water's edge. There, people were able to harvest all kinds of shellfish, and they learned to cut small patches of standing water off from the open sea, thereby trapping fish. Equipment was invented to aid in catching fish: harpoons, with which larger fish could be speared, and baskets, suspended in streams where fish were known to run.

In several areas of the world, human communities were able to achieve a degree of permanence by combining hunting and fishing with some gathering and by making use of the migration cycles of fish and animal life established during the changing climatic regimes. Indian peoples along the North American Pacific coast, Eskimo peoples on Arctic shores, the Ainu of Japan and coastal East Asia, and communities in coastal Western Europe caught salmon as they swam up rivers and negotiated rapids and falls. Huge accumulations of fish bones have been found at prehistoric sites near such locations. When the salmon runs ended, people stalked deer during their annual spring and fall movements, trapping them where they would habitually cross rivers in narrow valleys that they would traverse each season or in some other favorable place. The summer salmon runs and the wildlife migrations of fall and spring would leave only the winter to endanger the permanence of the settlement. It had been learned that dried meat could remain edible for months and, undoubtedly, the coldness of winter provided natural refrigeration that retarded spoilage. However, the early fishermen and hunters had their bad years too; sometimes the months of winter brought hunger and death. Such winters forced the riverside dwellers to abandon their settlements in pursuit of the distant herds.

Techniques and Tools

The development of fishing as a means of total or partial subsistence was attended by the invention of a wide range of tools and equipment. Among the earliest innovations for catching fish and other aquatic life was a simple stone trap used in tidal channels. Stones would be removed during the incoming tide, permitting fish to enter an inlet; they would then be replaced at high tide. When low tide drained the closed-off pool, water could seep out through the stones, but the fish would be trapped. Crescent-shaped stone traps on a tidal flat had the same effect, but traps were eventually refined by the creation of basketlike, wicker devices that could be used in stream channels, and there were nets of rough twine that could be stretched across an inlet or placed off a shoreline.

The fishing spear was to fishing what the arrow was to hunting, and various kinds were invented, ranging from simple pointed sticks to more refined harpoonlike spears. The use of hooks and bait led to the invention of hooks made from wood, bone, horn, and seashells. Most important, however, the resources of rivers and seas induced people to construct the means to pursue them, and from the first simple rafts there developed more elaborate canoes and sailing boats. Their role in the eventual worldwide diffusion of humankind hardly needs emphasis.

Agricultural Beginnings

It is likely that human groups began to domesticate plants simultaneously in several different parts of the world. Scholars are still not certain about the sequence of events, but independent invention in areas as widely separated as the Middle East, Central and South America, and Southeast Asia seems plausible. The process itself, which occurred between ten and twenty thousand years ago, was not necessarily a clear-cut "agricultural revolution," as is sometimes claimed. It is more likely that people slowly came to realize that it was possible to do more than simply gather wild crops. Perhaps rotten fruit, thrown away on refuse heaps near settlements dropped seeds that began to grow, the plants eventually bearing new crops. Possibly this suggested that roots and plants from the forest could be transplanted to a site near the settlement; roots and tubers could be stuck in the ground and they would "take," thus adding to the food obtained by gathering. Some scholars have suggested that the first cultivation probably occurred in association with some other means of survival, and it has been theorized that the fishing villages of ancient times, which afforded a measure of permanence of settlement, may have been the places where the earliest forms of agriculture were developed.

It is one thing to plant a root or tuber and leave it to grow, but seeding is quite another thing. Again, we may assume that seeding and harvesting were learned after the planting of root crops. Seeding involves the gathering of grains from plants, the preparation of a piece of land (and the invention of appropriate tools), then the sowing, followed perhaps by some weeding and general tending to the crop.

AGRICULTURAL HEARTHS

Where could agriculture have developed, more than ten thousand years ago? Again, we are compelled to

305

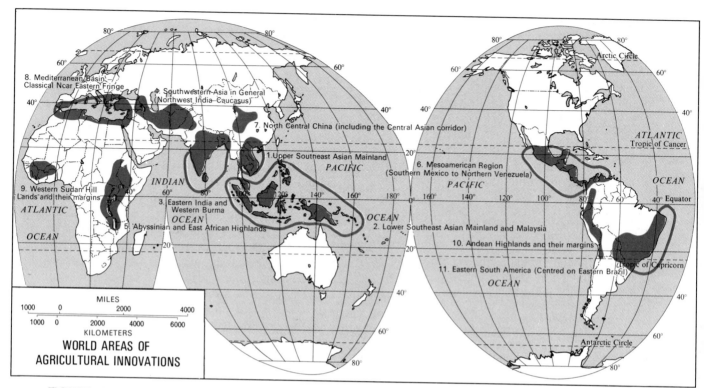

FIGURE 8.1

World areas where agricultural innovations are believed to have occurred.

speculate to some extent. In a classic study entitled *Agricultural Origins and Dispersals* (1952), Carl O. Sauer suggested that Southeast Asia may have been that agricultural hearth, the place where the first farmers lived (Fig. 8.1). He points out that the conditions of natural vegetation, climate, and soils must have been favorable for such a development and that the knowledge about cultivation gained there diffused across China, India, and even Africa. Other scholars have suggested the region of agricultural invention that existed in Central and South America, where the first cultivation of roots and cuttings was also followed by the more sophisticated practice of seeding. The Middle East, both in North Africa and in Southwest Asia, was another leading region of plant domestication. Here lies the famous Fertile Crescent, perhaps the single most important area of agricultural innovation. China's source area (surrounding the confluence of the Huang He and Wei rivers) also may have been a site of independent agricultural invention. Furthermore, although it took a long time for scholars to recognize it, Africa, too, shared in the innovation process. Significant areas of plant domestication lay in West Africa and in the East African Great Lakes region (Fig. 8.1).

The source regions of crop plant domestication, mapped in Figure 8.1, reveal a surprisingly global distribution. Cultural geographers Joseph E. Spencer and William L. Thomas, in their book *Cultural Geography* (1969), emphasized that particular local groupings of plants constituted the basic ingredients of each regional development complex. In the Mesoamerican region, for example, maize (corn), squashes, and several kinds of beans became the primary base for the region's agricultural development. In Southeast Asia, taro, yams, and bananas were the main plants. In

Southwest Asia the domestication process centered on wheat, barley, and some other less significant grains.

From these areas of agricultural invention and early development, the new techniques of cultivation diffused to other parts of the world. It is thought that millet, a small-seed grain, was introduced to India from West Africa, and sorghum, another grain crop, from West Africa to China. The watermelon spread from West Africa, first to nearby regions but eventually all over the world. Corn (maize) spread from Middle America into North America. Later, after the Portuguese brought it across the Atlantic, it became a staple in much of Africa. The banana came from Southeast Asia, as did a variety of yams. The process of dispersal unfolded slowly for many thousands of years, but the worldwide communications network established with the expansion of Europe during the past five hundred years greatly accelerated it.

Animal Domestication

While our distant ancestors learned to plant crops, they also began to keep animals as livestock. Once again, we are uncertain about the nature and timing of the transition from hunting to animal domestication; hunting had been a mainstay of human communities for many tens of thousands of years. Perhaps animal domestication came about when communities grew larger and better organized and, as a result, became more sedentary. Animals became a part of the local scene, not to be hunted, but to be kept as pets or for some other (e.g., ceremonial) purpose. Just as plant domestication still produces unanswered questions, the origins of animal domestication are a matter for speculation, but it is likely that animals attached themselves to human settlements as scavengers and even for protection and thus contributed to the idea that they might be tamed and kept. Any visitor to an African wildlife reserve can observe that when night falls, a permanent camp will be approached by certain species (gazelle, zebra, monkeys) that spend the night near and sometimes even within the camp's confines. With daybreak, the animals wander off, but the dangers of the next nightfall bring them back. Similar behavior probably brought animals to the settlements of the ancient forest farmers. Hunters might bring back the young offspring of an animal killed in the field and raise it. Such events probably contributed to the concept of animal domestication.

Just when this happened is also a matter for guesswork. Some scholars believe that animal domestication came earlier than that of the first plants, but others place animal domestication as late as eight thousand years ago—well after the practice of farming had started. In any case, the goat, pig, and sheep became part of a rapidly growing array of domestic animals, and in captivity, they changed considerably from the wild state. Archeological research indicates that when such animals as wild cattle were penned in a corral and their offspring also domesticated, they developed different varieties from those that remained in the wild. Protection from predators led to the survival of strains that would have been eliminated in the wild, and inbreeding entrenched those strains in captivity. Our domestic version of the pig, the cow, and the horse differ considerably from those first kept by our ancestors.

How did the ancient communities select their livestock, and for what purposes were livestock kept? It is thought that wild cattle may have been domesticated first for religious purposes, perhaps because the shape of their horns looked like the moon's crescent. Apparently, cattle were strongly associated with religious ritual from the earliest times and (as we have noted) there remain societies today where cattle continue to hold a special position as holy animals. However, those religious functions may also have led to their use as draft animals and as suppliers of milk. If cattle could pull sledlike platforms used in religious ceremonies, they could also pull plows. If cattle whose calves were taken away continued to produce milk and needed to be milked to provide relief, cattle could be kept for that specific purpose.

As in plant domestication, it is possible to identify certain regions where the domestication of particular animals occurred. In Southeast Asia, the presence of several kinds of pigs led to their domestication along with the water buffalo, chickens, and several other bird species (ducks, geese). In South Asia, cattle were domesticated and came to occupy an important place in the regional culture. Later, the domestication of the Indian elephant was accomplished, although the use of the elephant as a ceremonial animal, as a beast of burden, and as a weapon of war never did involve successful breeding in captivity (and some scholars argue that the elephant never really became a domesticated animal). In Southwest Asia and adjacent areas of Northeast Africa, domesticated animals included the goat, sheep, and camel. In the expanses of inner Asia, the yak, horse, species of goats and sheep, and reindeer were among domesticated animals. In the Mesoamerican region (including the Andes from Peru northward and Middle America to the latitude of central Mexico), the llama and alpaca were domesticated animals along with a species of pig and the turkey.

Although regional associations such as these can be made, you should regard them with caution. When animal domestication began, there were numerous species of a large variety of fauna, and these were domesticated simultaneously. The water buffalo, for example, was probably domesticated in Southeast and South Asia during the same period. Camels may have been domesticated in inner Asia as well as Southwest Asia. The pig was domesticated in numerous areas. Different species of cattle were domesticated in regions other than South Asia, where the zebu emerged from early domestications. Dogs and cats attached themselves to human settlements very early (they may have been the first animals to be domesticated) and in widely separated regions. Only a few animals have a comparatively specific source, including the llama and the alpaca, the yak, the turkey, and the reindeer.

Again, as in the case of crops, the dispersal of domesticated animals—first by regional expansion, later by worldwide diffusion—blurred the original spatial patterns of domestication. Chickens are now part of virtually every rural village scene, from Indonesia to Ecuador. Donkeys (probably first domesticated in Southwest Asia) now serve as beasts of burden the world over. Goats and sheep, cattle and horses, dogs

and cats are globally distributed. Even the elephant made its appearance not only in China, but also in ancient Europe as part of Hannibal's Carthaginian forces in combat against Rome.

Efforts at animal domestication still continue today (as we noted in Chapter 5). Among Africa's huge herds of antelope and other wildlife, there are species that may be capable of domestication as livestock, for example, the large eland, a potential source of meat in a region of imbalanced diets. Several rural experiment stations in Africa's savannalands are working to find ways to control and breed the region's wildlife. There has been some success with a species of eland, but less with various species of gazelles. Africa's powerful buffalo have not proved susceptible to domestication, however. Indeed, only about 40 species of higher animals have been domesticated worldwide.

Thus, the process of animal domestication, set in motion more than eight thousand (and perhaps as long as twelve thousand) years ago, still continues. Communities that were able to combine the cultivation of plants and the domestication of animals greatly lessened their dependence on single or limited food resources, and this achievement was a critical one in the evolution of human civilization.

Shifting Cultivation

The ancient farmers learned to plant crops and keep animals, but they knew little about conservation—how to contain disease when their animals died, how to keep the soil productive, how to fertilize, how to irrigate. Probably they were forced to abandon farming areas in subtropical regions after the soils became infertile and crops stopped growing. The farmers would move on to another parcel of land, clear the forest, and try again. This practice of **shifting cultivation**, like hunting and gathering, still goes on. In tropical areas, where the redness of the soil signifies heavy leaching but luxuriant natural vegetation thrives, a plot of cleared soil will carry a good crop at least one time, perhaps two or three. But then the area is best

308

left alone to regenerate its natural vegetation and to replenish the nutrients lost during cultivation. Several years later the plot may yield a good harvest once again.

Shifting cultivation is a way of life for many more people than hunting and gathering. Between 150 and 200 million still sustain themselves this way in Africa, Middle America, tropical South America, and in places in Southeast Asia and Melanesia. At one time, this was the chief form of agriculture in the inhabited world, just as hunting and gathering were previously prevailing modes of existence. It goes by various names: **milpa** agriculture, patch agriculture, and so on. As a system of cultivation, it has changed little over thousands of years of practice.

The controlled use of fire played a major role in the development of shifting agriculture as a technique of farming. Trees were cut down and all existing vegetation burned off; the resulting layer of ash contributed to the soil's residual fertility. The crops planted in these cleared patches were those found growing in their native regions: tubers in the humid, warm tropical areas; grains in the more humid subtropics; vegetables and fruits of various kinds in cooler zones. Shifting cultivation gave ancient farmers their earliest opportunities to experiment with various plants, to learn the effect of weeding and crop care, to cope with environmental vagaries, and to discern the fading fertility of soil under sustained farming.

The process of shifting agriculture involves a kind of natural rotation system in which areas of forest are used without being permanently destroyed. It does not require a nomadic existence by the farmers because usually there is a central village and parcels of land in several directions that are worked successively. When the village grows too large and the distances to workable areas become too great, a part of the village's population may move to establish a new settlement some distance away. This implies, of course, that population densities in areas of shifting agriculture cannot be very high: there has to be room. But high population densities were rare in ancient times; today, shifting agriculture prevails in areas where population densities are far lower than, say, in regions such as the Nile Delta or the Ganges Valley.

Shifting agriculture appears destructive, wasteful, and disorganized to people who are accustomed to more intensive forms of farming. There are not the neat rows of plants, carefully turned soil, precisely laid-out fields. But, in fact, shifting agriculture conserves both forest and soil; its harvest yields are substantial, given the environmental limitations; and it requires better organization than uninitiated observers might imagine. Patch agriculture is a complex adjustment to a fairly delicately balanced set of environmental conditions, and many an outsider has learned this lesson expensively. European farmers in Africa sometimes cleared the land—seemingly underutilized and so freely available—plowed the soil, and planted their crops. Soon the plants would die, often after only one apparently promising season, and no amount of fertilizer seemingly could restore them. And, of course, there was no way to quickly get the original vegetation back. So the land lay abandoned, barren, unusable to shifting cultivator or white farmer. The methods of the shifting cultivators have many merits, ancient as they may be.

We tend to think of agricultural geography in terms of cash cropping (i.e., farming for sale and profit), plantations, ranches, mechanization, irrigation, the movement of farm products, marketing, exports and imports, and so on. When we associate certain crops with particular countries, these are usually cash crops: Brazilian coffee, Colombian tobacco, Egyptian cotton, Australian wool, and Argentinian beef, for example. But the fact is that a great number of the world's farmers are not involved in all this. Hundreds of millions of farmers use their plots of land in the first instance to grow enough food to survive, sometimes with only marginal success. Their chief objective is survival, not profit. The shifting cultivators are subsistence farmers, as are the nomadic pastoralists who follow their life-sustaining herds of livestock on their cyclic migratory routes. But the subsistence farmers of many lands (Fig. 8.2) cannot migrate, nor is there the space to practice shifting agriculture. They are con-

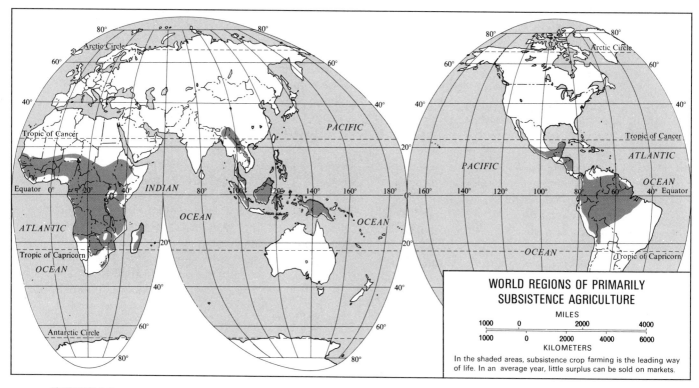

FIGURE 8.2

World distribution of regions of subsistence agriculture.

In the map inset:

WORLD REGIONS OF PRIMARILY SUBSISTENCE AGRICULTURE

MILES

KILOMETERS

In the shaded areas, subsistence crop farming is the leading way of life. In an average year, little surplus can be sold on markets.

fined to a small plot of soil from which they must wrest the means to survive, year after year, good year and poor year. And very likely they do not even own the soil they till.

Scholars customarily divide agricultural societies into categories such as *subsistence, intermediate*, and *developed* or, using different terms to express the same idea, *primitive, traditional*, and *modern*. Actually, the world's many thousands of societies lie along a continuum from the most rudimentary to the most complex, the digging stick giving way to the ox-pulled plow that, in turn, yields to the power-driven tractor. Thus such categorizing is for purposes of simplification and discussion, not to establish absolute divisions. Indeed, the term **subsistence** when it relates to farming is itself not beyond debate. Sometimes it is used in the strictest sense of the word—that is, to refer to farmers who grow food only to sustain themselves

and their families or communities and who, in effect, do not enter into the cash economy of the country at all. Those remaining societies where shifting agriculture is practiced—in remote areas of South and Middle America, Africa, and South and Southeast Asia—would qualify under this definition. On the other hand, farm families living at the subsistence level, but who sometimes sell a small quantity of produce (perhaps to pay their taxes) would not. Yet the term ought surely to apply to societies where small-plot farmers may be able sometimes to sell a few pounds of grain on the market, but where poor years threaten hunger; where poverty, indebtedness, stagnation, and sometimes tenancy are ways of life. Indian peoples in the Amazon Basin, sedentary farmers of Africa's difficult savannas, villagers in much of India, peasants in Indonesia all share subsistence not only as a way of life, but also as a state of mind. Experience has taught the farmer and his father and grandfather before him that moments of comparative plenty will be paid for by times of scarcity, that there is no point in expecting fertilizer to become cheaper and improved seed strains to solve age-old problems.

It is tempting to try to think of ways the subsistence agriculturalist might be helped to escape the cycle in which he and his family find themselves, when higher productivity, healthier children, better nutrition, longer life spans might be the rewards. European colonial powers put such thinking into practice in a variety of ways: by demanding taxes that compelled the farmer to raise some funds, by genuine assistance in the form of land-consolidation schemes, soil surveys, and by initiating compulsory cropping schemes (the last sometimes leading to local famines as the remaining food crops could not support the people). Newly independent governments in the underdeveloped world have initiated cooperatives, subsidized large-scale agricultural projects, encouraged better farming methods. But to encourage farming for profit in areas that have since time immemorial seen the subsistence way of life involves numerous risks. For example, land held communally by true subsistence-agriculture societies is parceled out to individuals when cash cropping becomes a major element in life. Social organization, too, ceases to be communal, and the system that ensured an equitable distribution of surpluses breaks down. The distribution of wealth becomes stratified, with poor and nearly destitute people at the bottom and rich landlords at the top. It is thought by scholars that the innovation of irrigation may have contributed to the first differentiation of subsistence society, but there is no doubt that the greatest number of farmers in intermediate societies find themselves in their present situation as a result of the changes brought to their part of the world by expanding, imperializing Europe.

This Indian farm family in Peru's high mountains ekes out a precarious living by planting its small plot of land with potatoes, the local staple. Drought and cold can destroy the crop, leaving these people with an uncertain future; malnutrition is a common occurrence. The specter of crop failure looms constantly. Opportunities for improvement (such as fertilizing the soil) are precluded by debts.

An Agricultural Revolution

In our changing world, we sometimes tend to look on the Industrial Revolution as the beginning of a new era, the sole stimulus for development and modernization. In doing so, we lose sight of another revolution, which began even earlier and had enormous impact on Europe and on other parts of the world: the modern agricultural revolution. This was perhaps a less dramatic development than the revolutionary changes that affected Europe's industries and cities, but it had far-reaching consequences nonetheless. It began slowly, and at first took hold in a few widely scattered places. Unlike the Industrial Revolution, its origins and diffusion cannot be readily traced. However, after centuries of comparative stagnation and a lack of innovation, farming in seventeenth- and eighteenth-century Europe underwent significant changes. The tools and equipment of agriculture were modified to do their job better. Methods of soil preparation, fertilization, crop care, and harvesting improved. The general organization of agriculture, food storage, and distribution were made more efficient. Productivity increased to meet rising demands. Europe's cities were growing and their growth produced new problems for the existing food supply systems. By the time the Industrial Revolution gathered momentum, progress in agriculture made possible the clustering of even larger urban populations than before. Had an agricultural revolution not been in progress, the course of the Industrial Revolution would undoubtedly have been very different.

The Industrial Revolution itself contributed importantly to sustain the agricultural revolution. The harnessing of power made mechanization possible, and tractors and other machines took over the work done for so long by animals and human hands. The cultural landscape of commercial agriculture in some regions changed as much as the urban landscape of industrializing cities did: fields of wheat and other grains, sown and harvested by machine cloaked entire countrysides. Even where tradition continued to prevail—in the vineyards of France's Bordeaux and the ricefields of East Asia—the impact of modernization was felt, if not in harvesting methods then in research

leading to improved fertilizers and more productive crop strains. The agricultural revolution still continues, sometimes punctuated by such dramatic breakthroughs as the so-called Green Revolution of the 1960s.

Farming, as we are aware, is by no means possible everywhere on this earth. Vast deserts, steep mountain slopes, frigid polar zones, and other environmental obstacles prevent agriculture over much of our globe. Where farming *is* possible, the land and the soil are not put to the same use everywhere. The huge cattle ranches of Texas represent a very different sort of land use than the dairy farms of Wisconsin—or the paddies of Taiwan. Within a few miles of the subsistence lifestyles of forested Middle America lie rich plantations whose products are shipped by sea to North American markets. We can observe the differences around our own cities: travel by car or train from Chicago or Cincinnati or St. Louis into the countryside and you can see the land use change. Close to the city the soil is used most intensively, perhaps for vegetable gardens of crops that can be sent quickly to the nearby markets. Farther away, the fields are larger and time becomes a lesser factor. The cornfields of Iowa are dotted by grain elevators for storage and shipment, and some of the grain may be consumed not in Chicago or St. Louis, but in India or Bangladesh. Still more distant from the Midwestern cities, we enter the pasturelands of the Great Plains, almost an opposite extreme from those vegetable gardens: the land is allocated by thousands of square miles rather than by acre.

What factors and forces have combined to produce the spatial distribution of farming systems existing today? This is a very complicated question to which no complete answer may be possible. It involves not only the effect of different conditions of climate and soil, variations in farming methods and technology, market distributions, and transportation costs, but it also has to do with the domination of the world by an economic structure that favors the developed Western world: the United States and Canada, Europe, and the Soviet Union. Decisions made by colonial powers in Europe led to the establishment of plantations, from Middle America to Malaya, the products of which were grown not for local markets, but for consumers in Europe; U.S. companies similarly founded huge plantations in the Americas. The European and Western impact transformed the map of world agriculture just as the pattern of European farming was changed by the agricultural revolution. The end of colonial rule did not simply signal the termination of the agricultural practices and systems that had been imposed on the formerly colonial areas. Even food-poor underdeveloped countries must continue to grow commercial crops for export on some of their best soils where their own food should be harvested. Agricultural systems and patterns long entrenched are not quickly or easily transformed.

FARMING ON THE MAP

During the nineteenth century (we noted in Chapter 1), several European geographers began to study, systematically, the economic and political geography of their communities and regions. The work of some of these geographers was so productive that it is still being discussed and debated today. The first scholar to study and identify the factors governing the development of commercial agricultural patterns on the ground and on the map was the economist-turned-geographer, Johann Heinrich von Thünen (1783–1850). Europe during von Thünen's time was changing quite dramatically. Population growth was unprecedented, towns and cities were expanding, and the demand for food was increasing proportionately. Von Thünen himself was a farmer. In 1810, he acquired a fairly large estate not far from the town of Rostock in the northeastern German state of Mecklenburg. As he farmed his lands, von Thünen became interested in a subject that still concerns economic geographers today: the effects of distance and transportation costs on the location of productive activity. For four decades, he kept meticulous records of all the transactions of his estate; while doing so, he began to publish books on the spatial structure of agriculture. His series, under the title *Der Isolierte Staat (The Isolated State)*, in many ways constitutes the foundation of location theory, and von Thünen's works are still being discussed and debated today.

Von Thünen called his work *The Isolated State* because he wanted to establish, for purposes of analysis, a self-contained country devoid of outside influences that would disturb the internal workings of the spatial economy. Thus, he created a sort of regional laboratory or **model**, within which he could identify the factors that influence the spatial distribution of farms around a central city. In order to do this, he made a number of limiting assumptions. First, he stipulated that the soil quality and climate would be uniform throughout the region. Second, there would be no river valleys or mountains to interrupt a completely flat farming surface. Third, there would be a single, centrally positioned city in the Isolated State, which was surrounded by an empty wilderness. Fourth, von Thünen postulated that the farmers in the Isolated State would haul their own products to market (no transport companies here) and that they would do so by oxcart, directly overland, straight to the central city. This, as you can imagine, is the same as assuming a system of radially converging roads of equal and constant quality, and, with such a system, transport costs would be directly proportional to distance.

Von Thünen combined these assumptions with what he had learned from the actual data collected while running his estate, and he now asked himself what the optimal spatial arrangement of agricultural activities would be in his Isolated State. He concluded that farm products would be grown in a series of circular zones, radiating concentrically outward from the central city (Fig. 8.3). Nearest to the city, those crops would be grown that yielded the highest returns and were the most perishable, for example, vegetables; dairying would also be carried on in this innermost zone. Farther away, you would find more durable goods such as potatoes and grains. Eventually, because transport costs to the city increased with distance, there would come a line beyond which it would be uneconomical to produce crops. There the wilderness would begin.

Von Thünen's model, then, incorporated five zones surrounding the market center (Fig. 8.3). The first and innermost belt would be a zone of intensive agriculture and dairying. The second zone, von Thünen said, was an area of forest used for timber and firewood. Next, there would be a third belt of increasingly extensive field crops. A fourth zone was

313

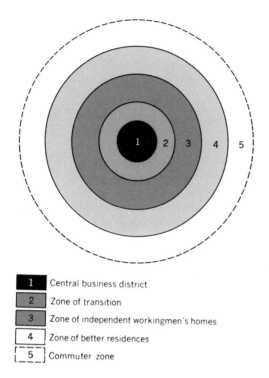

1   Central business district

2   Zone of transition

3   Zone of independent workingmen's homes

4   Zone of better residences

5   Commuter zone

FIGURE 8.3

The rings of Von Thünen's "Isolated State."

occupied by ranching and animal products. The outermost zone was the unproductive wilderness, across which no outside trade occurred—thereby making the Isolated State a self-sufficient spatial economy.

If the location of the second zone, the forest, surprises you, remember that the forest was still of great importance during von Thünen's time as a source of building materials and fuel. All that was about to change with the onslaught of the Industrial Revolution, but there are lots of towns and cities left in the world that are still essentially preindustrial.

Still, the real world does not present situations comparable to those postulated by von Thünen. Transport routes serve certain areas better than others. Physical barriers impede the most direct communications. Rivers and mountains disrupt the surface. External economic influences invade every area. Each of these (and many more) complications serve to distort and modify von Thünen's theoretical conclusions—

and yet, von Thünen had a point. We can see modern cities still ringed by recognizable agricultural zones, even though modern transport networks, suburbanization, and numerous other outlying developments have disrupted older patterns so much that those zones are sometimes difficult to identify. Some decades ago you might have found economic geographers willing to discern Thünian patterns in the farmlands around, say, Chicago; these days they might talk about remnants of such patterns (see Fig. 8.4). However, a concentric zonation still prevails around many smaller centers in America and in many places in the rest of the world.

Let us carry von Thünen's concept of concentricity some steps farther. In his Isolated State, the consumers in that central city, by virtue of their purchasing power and their willingness to pay certain prices for particular products, could determine to a great extent the spatial structuring of the farming economy. Although the real-world situation would be greatly changed as many of von Thünen's assumptions are swept away, his Isolated State model, nevertheless, points to a significant relationship between consumers and producers. By the size and quality of its demands for certain products, an urban market strongly influences the behavior and decisions of farmers in its hinterland. The city is more than a market: it also is a center of political power and a source of economic influence. Farmers who have taxes to pay derive their incomes from the sale of their products on urban markets, but it is not the farmers who decide how high those taxes shall be or what prices their products will command. When regions of the world are struck by food shortages and famines, note that the areas worst affected are not the cities, but the rural areas—where, one would imagine, such problems could be staved off by local production. This reflects the power of cities the world over, and it is one of the reasons for the tidal wave of rural-to-urban migration that in the 1980s characterizes countries from Brazil to India. Urban centers dominate rural areas in many ways, and even though the farmers produce the sustenance of city people, it is the city folk who, in large measure, control the lives of farmers. Thus, von Thünen's agricultural zones represent decisions made by farmers about the use of their

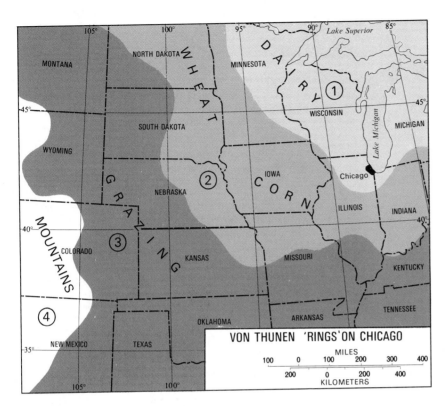

FIGURE 8.4
Von Thünen's model applied to the U.S. Midwest.

land, but those decisions were impelled by require-ments and demands made by urban inhabitants.

If a single urban center can generate certain pat-terns of land use, a group of cities or a large urbanized region has an even wider impact. In the United States, we can observe nearly an entire continent oriented toward the population core centered on the great coalescence of northeastern cities—the megalopolis stretching from Boston through New York, Phila-delphia, and Baltimore to Washington, D.C. The "rings" are not smooth and regular as von Thünen saw them in his hypothetical Isolated State, of course. Climatic variations, soil differences, terrain irreg-ularities, transportation networks, and such local in-terruptions as irrigation projects are but a few of the distorting factors. Nonetheless, we can record an ap-propriate sequence of concentric zonation: fruits and vegetables concentrate along the East Coast, with out-liers in Florida and southern California, which possess year-round growing seasons; dairying is clustered in the northeastern interior, exhibiting a belt extending west from New England through the Great Lakes states; mixed farming and feed-grain production are located in the adjacent zone of the Corn Belt; wheat cultivation is found beyond the Corn Belt, concen-trated in two regions of the Great Plains that are similarly distant from the northeastern seaboard; and cattle and sheep grazing is found even farther away from Megalopolis, in the remote trans–Rocky Moun-tain west. Although the original scale of the Isolated State has been greatly expanded, the fact remains that the logic of von Thünen's fundamental geographic theory is sustained by what we still observe in the field.

An Agricultural Revolution

• *Viticulture and Viniculture* •

One of the most interesting forms of agriculture is the growing of grapes (viticulture) for the purpose of making wine (viniculture). This is an ancient pursuit, vineyards having stood in Mesopotamia and pharaoic Egypt. Later, winegrowing diffused to Crete and Greece, into the Roman Empire, and (through the work of Roman winegrowers) into France, Germany, and even England. After Columbus's journeys to America, the first vines were planted in Mexico, from where the industry spread into California. The colonists carried viticulture to southern South America, to South Africa's Cape, to Australia, and to New Zealand.

When grapes are cultivated for winemaking purposes, a complicated process unfolds. It is not just a matter of bringing in a large harvest of ripe fruit, as in the case of other crops. A large harvest, in fact, may mean inferior grapes and poor wine. Viticulturists must learn to prune their vines to keep the harvest limited. And the timing of the harvest is crucial. Too early: the grapes may be too acid (and the wine sharp and unpleasant). Too late: there may be too much sugar, and a flabby, unappealing wine.

The vine is a hardy plant, but the grape is a fragile fruit. Summer warmth (and very limited rainfall) and sunshine are needed; winters must be cool but not too severe. Slopes must be turned to the sun just right; high winds, hail, and other hazards can doom a vintage. Humidity, causing rot, is a constant threat in many vineyards. Viticulture and physical geography are closely associated.

Italy, base of the Roman Empire of old, remains the world's largest producer of wine. France, where the Romans introduced viniculture, is the second-largest producer, and French premium wines continue to command the highest prices. Spain ranks third, although the Soviet Union, which does not publish reliable statistics, may produce as much wine as Spain in an average year, or even more. Argentina, largest producer in the Southern Hemisphere, follows.

Certain wines are closely identified with their regional sources: champagnes (France), sherries (Spain), ports (Portugal), Chiantis (Tuscany, Italy) among them. A wine can be a telling representative of the culture of a place, its climate, soil, traditions, preferences. It all starts in the vineyard, in the hands of a specialized farmer.

World Patterns

If Thünian patterns of agricultural land use can be discerned around individual cities, urban regions, and even on a whole continent, can world agriculture be interpreted in this context? We know that Argentinian beef is consumed in U.S. homes as are Brazilian coffee, Colombian tobacco, and sugar from Caribbean countries. Europeans import tea from Asia, cotton from Africa, wheat from Australia. Cuban sugar goes to the Soviet Union. While von Thünen was studying his Isolated State as a closed system totally unaffected by outside influences, the Europe of von Thünen's time was in the process of mobilizing much

of the world to its own sustenance and comfort. Europe's cluster of colonializing powers functioned much as von Thünen's central city did, as a market for agricultural products from all around but with an added dimension: Europe manufactured and sold in its colonies the finished products made from imported raw materials. Thus, the cotton grown in Egypt, the Sudan, India, and other countries colonized by Europe was cheaply bought, imported to European factories, made into clothes, which were then exported and sold, often in the very colonies where the cotton had been grown in the first place.

Obviously we should not view world agriculture in the kind of isolation assigned by von Thünen to his experimental model. The evolution of a worldwide transport network, of ever growing capacity and efficiency, continually changed the competitive position of various activities. The beef industry of Argentina, for example, found a world market when the invention of refrigerator ships made long-distance transportation possible for what was previously a highly perishable commodity. European colonial powers did not simply permit farmers in their dependencies to decide on the crops they would grow for export to Europe: they frequently imposed the cultivation of specific crops on traditional farmers. American economic power made itself felt in Middle and South America, creating an empire of plantations and orienting production to U.S. markets. Thus the forces tending to distort any world Thünian zonation are many.

And yet we can observe in this world pattern of agriculture, represented in Plate 8 and in Figures 8.5 to 8.10, elements of the same sort of order von Thünen recognized in his Isolated State. We have already noted that the Industrial Revolution had a major impact on agriculture, and when we attempt to describe the relationships between the urbanized core areas of the world and agricultural patterns, we should view the urban centers as urban-industrial cores, not merely as markets. To these urban-industrial cores, in Europe, in North America, and in the Soviet Union, flow agricultural (and industrial) raw materials and resources from virtually all parts of the inhabited world. Superimposed on the regional zones and on the continental-scale zonation we saw in North America, therefore, is a world system that also has elements of concentric zonation.

It is important to recognize, as we examine world agriculture in the following pages, that the pattern we observe has come about only in part through the decision-making processes envisaged by von Thünen. In the Third World, much of the farming (other than for local subsistence) is a leftover from colonial times, but it cannot simply be abandoned, because it continues to provide a source of revenues that are badly needed—even if the conditions of sale to the urban-industrial world are not often favorable. In the Caribbean region, whole national economies depend on sugar exports (the sugar having been introduced by the European invaders centuries ago). Selling the harvest at the highest possible price is an annual concern

for these countries, but they are not in a very good position to dictate: sugar is produced by many countries in various parts of the world as well as by farmers in the technologically developed countries themselves. Thus, it is the importing countries that fix tariffs and quotas, not the exporters. In the ideological conflict with Cuba, the United States cut off its imports of Cuban sugar. Although the Cuban export trade was cushioned by alternative buyers in Canada and the Soviet sphere, it was a staggering blow. The wealthy, industrialized importing countries can threaten the very survival of the economies of the producers—much as the farmers in von Thünen's Isolated State were at the mercy of decisions made by the buyers in the central-city marketplace.

There are signs that the producing countries are seeking to unite, to present a common front to the rich, importing nations (as the OPEC countries did in the energy arena). But what was possible in the petroleum field will not prove as easy in other areas. Countries that share world market problems in one particular commodity do not face the same difficulties in another, so the importing countries are in a position to divide and rule. Also, a general withholding of the harvest not only endangers economies, it stimulates domestic production in the importing countries. Although sugarcane accounts for more than 70 percent of the commercial world sugar crop each year, farmers in the United States, Europe, and the Soviet Union also produce sugar from sugar beets. Already in Europe and the Soviet Union, these beets produce 25 percent of the annual world sugar harvest. Collective action by countries producing sugarcane could serve to raise that percentage.

COLONIAL IMPRINTS: COTTON, RUBBER

Farming in many former colonial countries, then, was stimulated and promoted by the colonial powers, representing imperial interests. Cotton and rubber, two industrial crops, are good examples. Today cotton is grown in the United States, in Northeast China, and in Soviet Central Asia: a fourth large producer, India, owes its modern cotton fields to colonial Britain, which vastly expanded traditional cultivation. But cotton cultivation was promoted on a smaller scale in

317

Cotton is one of the cash crops grown in many parts of the former colonial world, primarily (but not exclusively) for export to the developed countries. Japan and the European Economic Community are major markets. Cotton farmers in the United States also compete in these lucrative markets. The cotton mill shown here, near Aurangabad, India, is based on fields in the surrounding area developed during the British period.

numerous other countries: in Egypt's Nile Delta, in the Punjab region shared by Pakistan and India, in the Sudan, in Uganda, in Moçambique, in Mexico and Brazil. Cotton cultivation expanded greatly during the nineteenth century when the Industrial Revolution produced machines for cotton ginning, spinning, and weaving that multiplied productive capacity, brought prices down, and put cotton goods in reach of mass buyers. As they did with sugar, the colonial powers laid out large-scale cotton plantations, sometimes under irrigation (e.g. the famed Gezira Scheme, in the triangle between the White Nile and the Blue Nile in the Sudan). The colonial producers would receive low prices for their cotton; the European industries prospered as cheap raw materials were converted into large quantities of items for sale at home and abroad. Today many of the former colonial countries have established their own factories producing goods for the home markets, and synthetics such as nylon and

rayon are giving cotton industries increasing competition. Still, the developed countries have not stopped buying cotton, and for some developing countries cotton sales remain important in the external economy. Japan, the United Kingdom, and Western European countries continue to import cotton fiber, but the developing countries have a formidable competitor for those markets in the United States, whose cotton exports also go there.

The case of rubber is rather more complicated. Initially, rubber was a substance not cultivated but gathered: collected from rubber-producing trees that stood (among many other tree species) in equatorial rain forests, mainly the forests of the Amazon Basin in South America. Those were the days around 1900, when the town of Manaus on the Amazon River experienced an incredible rubber boom; a similar, if less spectacular, period of prosperity was experienced by the rubber companies exploiting the Congo (Zaïre)

Basin in Africa.

The boom in wild rubber was short-lived, however. Ways were sought to create rubber-tree plantations, where every tree, not just some among many, would produce rubber, where the trees could be given attention, and where collecting the rubber would be more efficient and easier. Seedlings of Brazilian rubber trees were planted elsewhere, and they did especially well in Southeast Asia. Within two decades after the Amazonian rubber boom, nearly 90 percent of the world's rubber production came from new plantations in colonial territories in Malaya, the (then) Netherlands East Indies (Indonesia) and neighboring countries.

For many decades, as more uses for rubber were found, demand grew continuously. The advent of the automobile was an enormous boost for the industry, and even today most of the rubber produced is used in the manufacture of vehicle tires. The plantations thrived. Then the Second World War broke out, and the United States, a major consumer of rubber, was cut off from its sources of supply in Southeast Asia. This stimulated the production of synthetic rubber—a good example of substitution for a resource in short supply (Chapter 5). Before long, more synthetic rubber was made than natural rubber. After the war, that trend continued. In 1984, when total rubber production (synthetic and natural) was just below 13 million tons, nearly 9 million of it was synthetic. The United States, the world's largest synthetic-rubber producer, accounted for 25 percent of the total. Almost all (90 percent) of the 4 million tons of natural rubber came from countries in Southeast and South Asia: three countries, Malaysia (36 percent), Indonesia (28 percent), and Thailand accounting for the bulk. The colonial pattern remains entrenched.

If there was a great rubber boom in the Amazon Basin, and rubber plantations increased in Liberia, why did Southeast Asia gain and keep the lead? This had less to do with the natural environment than with the availability of labor and with political circumstances. The colonial powers knew that Southeast Asia combined conditions of tropical environment with labor availability and political stability that neither Amazonian South America nor Equatorial and West Africa could match. The Liberian rubber plantations did develop well (but have suffered in the post-

independence period); and lately efforts have been made to introduce rubber plantations along the Amazon River in northern Brazil. In the mean time, however, Southeast Asia continues to hold its natural-rubber production lead.

Luxury Crops

Similar considerations—a combination of suitable environment and available labor—led the European colonial powers to establish huge plantations for the cultivation of luxury crops such as tea, cacao, and coffee (Fig. 8.5). Coffee was first domesticated in northeast Africa (in the region of present-day Ethiopia), but today it also thrives in Middle and South America, where 70 percent of the world's annual production is harvested; the United States buys more than half of all the coffee sold on the world markets annually, Western Europe imports most of the remainder. Compared to coffee, tea is consumed in lesser quantities by weight and in greater amounts in the areas where it is grown: India, China, Sri Lanka, and Japan. Whereas coffee is the beverage cultivated and consumed dominantly in the Americas, tea is the Eurasian equivalent: it goes from the Asian producing areas to the United Kingdom and Europe. Tea is a rather recent addition to Western diets. It was grown in China perhaps two thousand years ago, but it became popular in Europe only during the nineteenth century. The colonial powers—chiefly the British—established enormous tea plantations in Asia, and thus began the full-scale flow to the European markets, which continues today.

319

World Agriculture: Farming Systems

Mindful of Thünian and colonial and technological factors, let us now examine the world map of agriculture (Plate 8) more closely. Of course, the map at this scale cannot represent individual crops or detailed patterns (some of which are shown on regional maps

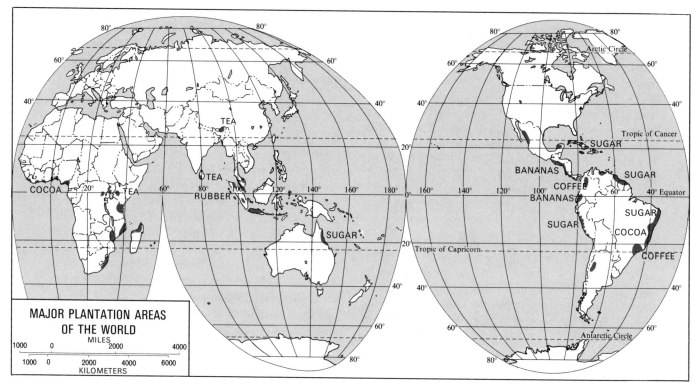

FIGURE 8.5

Plantation areas of the world.

later in this chapter). The world-scale map, however, does reveal the clustering of significant dairying regions (1) near the industrial heartland of Western Europe and the large cities of eastern North America. Other visible dairying zones lie in the northwestern United States, New Zealand, and Australia, all areas that export their dairy products to distant markets. At this scale, numerous local dairying zones near major cities cannot be shown. Fruit, truck, and specialized crops (2) exhibit a comparatively small areal extent, and again only a limited number of these can be mapped here including the Gulf–Atlantic Coastal Plain in the eastern United States, one of the larger of these regions. Regions of Mediterranean agriculture (6) reflect the global distribution of the particular environmental conditions under which this kind of farming takes place (see regions *Csa* and *Csb* on the map of global climates, Plate 3).

Mixed livestock and crop farming (3) cover large areas in eastern North America and most of the North European Plain from Spain northeastward into the heart of the Soviet Union. In the Southern Hemisphere, mixed farming regions are smaller; they include areas in southern Brazil, Uruguay, northeastern Argentina, and south-central Chile. This form of agriculture also occurs in South Africa, Zimbabwe, and on the Canterbury Plain of New Zealand's South Island.

In Plate 8, regions of grain farming are divided into those chiefly producing wheat and those where rice is the main crop. Wheat is grown commercially on extensive fields in the central United States and Canada (4), the Ukraine and eastward in the Soviet Union, Argentina's Pampa, and southern Australia; it is a subsistence crop in western India (9) and northern China. Rice is a subsistence crop on small paddyfields in eastern India (8), Japan, southern China, and Southeast Asia.

Plate 8 also underscores the regional extent of farming primarily under shifting cultivation (11). These include the Amazon region of South America, areas of Central America, much of tropical and equatorial Africa, interior and island Southeast Asia, and Pacific islands, including New Guinea. It should be remembered that any world map such as this is highly generalized, and not all the available land shown is, in fact, subject to shifting cultivation; large areas of equatorial rain forest remain untouched. However, the map of world climatic regions (Plate 3) provides some indication of the reasons behind the spatial pattern shown in Plate 8: these are low-latitude regions of equatorial and tropical climates with precipitation regimes (and soil properties) that do not favor permanent, sedentary cultivation.

Livestock ranching (12) also extends over large regions in both the Northern and Southern Hemispheres, but this single category conceals local variations. As will be noted in detail later, regions of large-scale commercial livestock ranching occur in the Great Plains of North America, in and near Argentina's Pampa, in Australia and New Zealand, on South Africa's *veld*, and in the interior Soviet Union. The livestock industries of Venezuela, Colombia, Bolivia, Paraguay, Mexico, and Zimbabwe are smaller.

Plate 8 underscores the earth's environmental limitations as well. Vast regions are nonagricultural (unnumbered, in white) or suitable only for nomadic and seminomadic herding (13). Add to these regions those under shifting cultivation, and in combination they cover more than two thirds of the world's land area.

A final observation relating to Plate 8 brings us back to our original theme. Plantation agriculture introduced by European colonial powers (7) still continues from Indonesia to the Caribbean, including the more productive areas of Africa as well. Often these lie as well-tended, carefully manicured estates amid fields of the poorest traditional subsistence agriculture, evidence of the capacity of the colonialists to exploit land and labor to their advantage.

ORIENTATION TO THE DEVELOPED WORLD

The wealth and power of Europe and the Western World could reorganize economies half a world away, and land and labor from Southeast Asia to South America were placed in the service of European and North American consumers (Fig. 8.5.) Europe's plantations in the colonial world transformed huge areas from subsistence to cash cropping; they caused large reaches of land to be taken from their indigenous users (a process termed **land alienation**); they generated enforced migrations involving millions of people for labor purposes; and they produced enormous wealth for some of their owners (companies as well as individuals). Some of the plantations established in South America cover more than 400 sq miles (1000 sq km). Although many plantations are smaller, the average plantation is much larger than the average farm in the United States. And even though the colonial era is nearly at an end, the plantation still survives. The character of plantations is changing, but the institution remains.

Plantations are owned by private individuals or by corporations (such as the United Fruit Company, now known as United Brands, which has large banana plantations in Middle American countries). In recent years, plantations formerly owned by whites have been bought or taken over by the governments of newly independent states. But those governments have had to acknowledge the continuing relationships with the markets of the richer nations and have sometimes set up corporations to run the plantations, simply substituting for the white owners, as has been done in Kenya. In a few cases, the plantations, once taken over, have been divided among peasants. This was a great experiment in Mexico, part of the revolution that began in 1910 and stipulated by the constitution of 1917. As of the latter year, landowners with more than 750 acres (300 ha) had their lands taken over for redistribution to Indian farming families. In other cases, large estates were expropriated and incorporated in collectives and communes, as in the Soviet Union and China. But all these are changes of the twentieth century. Originally the plantation was an individual or a corporate enterprise.

Whether a United Brands plantation, a Venezuelan hacienda, or a Liberian rubber estate, today's remaining plantations still are predominantly in the hands of people of European extraction. Large estates were plentiful in Europe during the days of colonial expansion (von Thünen's time!), and the possession of

321

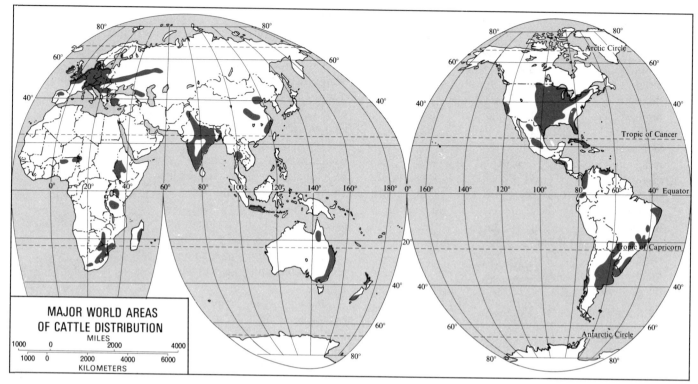

FIGURE 8.6

Distribution of the world's largest cattle herds.

land was a matter of great prestige. Slaves, indentured laborers, tenants, or hired labor did the work (tenancy still survives), and millions of acres around the world that lie under sugar, cotton, coffee, tea, bananas, and tobacco were first laid out as plantations. The map showing the distribution of present and past plantations (Fig. 8.5) suggests a Thünian zone of plantations ringing the world far from the Western markets that demand their products.

DISTANT LIVESTOCK

Nothing evokes a world Thünian scheme quite as much as does the world distribution of livestock whose products are destined for the markets of the developed countries (Figs. 8.6 and 8.7). In a zone lying beyond the plantations, in the southernmost countries of the Southern Hemisphere, ranchers raise great herds of cattle and sheep whose products, although also sold on local markets in Argentina, South Africa, Australia, and New Zealand, are sent thousands of miles to be sold in the cities of Europe, the Soviet Union, and North America.

Cattle Herds

The map (Fig. 8.6) might give the misleading impression that India is part of this process. India does have millions of cattle, but the religious prohibition on eating meat prevents the use of this big herd for consumption—local or external. Thus it is important, in considering Figure 8.6, to differentiate between production and export of livestock products. In terms of beef and veal production (veal is the meat of a calf normally not more than 12 weeks old), the United

States leads all countries, with about one third of the world total, virtually all of it consumed at home. When it comes to exports, Argentina is in the lead, with about one third (by weight) of all the meat traded internationally each year. Argentina's cattle herd is concentrated in the Pampa, where climatic conditions are favorable and where alfalfa can be grown as a fodder crop much as corn is grown in the United States to fatten livestock. Argentina's beef goes primarily to the United Kingdom and several European countries; some comes to U.S. markets. After Argentina, Australia ranks as the next most productive beef-exporting country, followed by New Zealand. It is important to remember the high cost of beef, in terms of grain consumed: Europe's markets can command such use of the soil thousands of miles away, when those same areas could produce enormous tonnages of grain to support the hungry populations of less-distant, poorer countries.

Sheep

Sheep are raised not only for wool, but also for their meat, mutton. Sheep are able to live off land that could not sustain cattle, and subsistence farmers raise them not only for their fiber and meat, but also for milk (and the cheese made from it), and even for the skin, used in the production of leather garments and other items. The world distribution of sheep (Fig. 8.7) should, as in the case of cattle, be seen in this context. The dense sheep population in the United Kingdom, Eastern Europe, and Southwest and South Asia consists of herds raised for local markets; the South American, South African, Australian, and New Zealand sheep herds are maintained largely for external (European) markets. The valuable export item, of course, is their wool: textile makers in the United Kingdom, Western Europe, Japan, the United States, and the Soviet Union use wool imported from Southern Hemisphere countries.

FIGURE 8.7
World distribution of sheep and hogs.

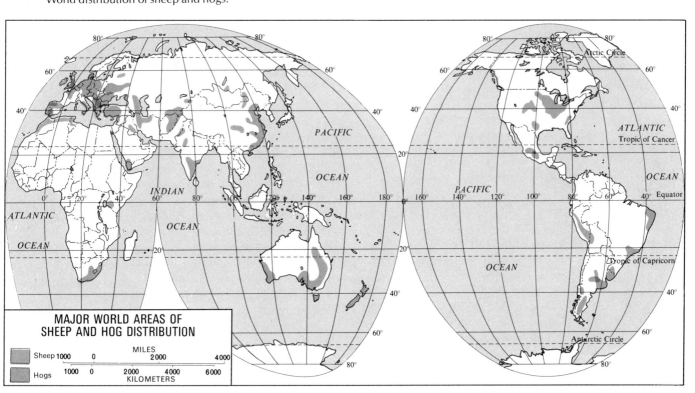

MAJOR WORLD AREAS OF
SHEEP AND HOG DISTRIBUTION

Sheep

Hogs

Wool is a major source of revenue for distant New Zealand, and the lower slopes of the country's mountainous topography afford excellent pastures. Here Merinos are gathered for clipping at Glentanner Station in Canterbury. This flock numbers about 9000 sheep, and in the summer months they graze all along the slopes of the Ben Ohau Range in the background—right up to the ice and snow.

Other Livestock

Goats, hogs, and poultry are among livestock that contribute importantly to the diets of people in many parts of the world. Hogs, the source of pork, bacon, and lard, are a most useful alternative where meat sources are hard to come by. Hogs (or pigs, or swine) can survive in widely different environments. In the United States and Europe, they are raised commercially and fed corn- and potato-derived feed (Fig. 8.7). But in many other parts of the world they are simply scavengers, surviving on leftovers and garbage. Thousands of Chinese villages would be incomplete without their hog population (China is probably the world's leading hog producer), and hogs abound also in Brazil, Mexico, and other South and Middle Ameri-

can countries. Hogs could serve even more effectively to diversify diets if religious taboos (chiefly Moslem) did not exclude them from large areas of the world. There are no hogs in any great number in North Africa, Southwest Asia, or Southeast Asia, and very few in India and Africa. Almost anomalously, the major production (other than China's) is in the already meat-rich Western countries that trade among themselves in ham, pork, bacon, and lard.

If hogs can survive under difficult circumstances, goats can thrive on even poorer land. They manage to climb the steepest slopes after the smallest tuft of vegetation, and they live in large numbers in even the most arid of populated countrysides. From Africa to Southern Europe to the Middle East to Southern Asia,

324

Chapter 8   Geography of Farming

the goat provides milk, meat, skin, even hair to its owners. Goats require little care and multiply rapidly; one negative aspect of their large numbers is that in their persistent search for food they contribute to soil erosion and land denudation.

Poultry, perhaps the oldest of all domesticated livestock, also provides its owners with a set of valuable commodities in return for little attention or land. Chickens live as scavengers, but can also be fed systematically. In terms of weight, they generate the best return per pound of feed expended. They produce meat and eggs, crucial in diets where proteins are badly lacking. In the technologically advanced countries, the poultry industry is as thoroughly organized as that of cattle or hogs, and the local markets consume enormous quantities of poultry products. Breeding has created specialized strains, some noted for their eggs, others for their meat. The poultry industry has not been immune from efforts to establish plantationlike production systems in colonial countries: a prominent (and disastrous) experiment was aimed at the creation of a huge poultry scheme in Gambia, West Africa.

When we first viewed the model developed by von Thünen, we noted that high-priced, perishable items such as dairy products were produced nearest to the central market. And still today, in cities around the world, milk, butter, cheese, and similar items are produced in the immediate environs. On our map of world agriculture (Plate 8), a Dairy Belt is discernible in North America, showing a marked orientation and proximity to the population core of the East. Europe has a similar concentration. But when we check an atlas map showing the world distribution of the dairy industry, it seems at first to contradict what we have just reported. Far away from the European and North American markets, New Zealand and Australia generate dairy products and export them in quantity. Again, it is a matter of the primacy of the European market: Europe produces a large volume of milk and other dairy items, but its consumers demand and can afford more. And just as refrigerator ships made long-distance beef exports possible, refrigeration eliminated time as a factor in the export of cheese, butter, and other dairy products from Southern Hemisphere countries.

This should not suggest that the consumption of dairy products is a monopoly of the wealthy midlatitude counties. The dairy industry is nowhere as well developed as it is there, but in other parts of the world, people also consume milk when they can obtain it, and dairy belts have sprung up around major urban centers from Santiago to Seoul. Milk is a prominent item in Masai diets; in India, where there are so many cattle, milk is about the only substantial benefit from that overpopulation. Milk is a vital dietary ingredient, especially for young children, and hundreds of millions of people do not consume nearly enough of it because it is too expensive to obtain. Goat's milk serves as a substitute in some areas, and condensed and powdered milk stand on the shelves of village stores from Bolivia to Burma. But the ample supply of dairy products that is commonplace in the wealthy Western countries is an exception and a privilege in this other world.

CRUCIAL GRAINS

People in the wealthy countries as well as the poor, in the well-fed nations as well as the hungry are sustained chiefly by the major grain crops, the *cereals*. Along with two root crops, the potato and the cassava, the three leading cereals (rice, wheat, and corn) account for more than three quarters of the food supply produced on earth each year. Grain crops of lesser, local importance such as sorghum, millet, barley, oats, and rye (among others); fruits (the banana is a staple in a few areas); vegetables; meats; and dairy products make up the remainder.

Rice

Probably as many as half the people of the world depend on rice for their daily sustenance. The map showing present areas of substantial rice cultivation (Fig. 8.8) indicates that rice continues to be grown predominantly in and near the region in which it was first domesticated. Compared to the areas where wheat and corn are cultivated (Figs. 8.9 and 8.10) the extent of the rice-producing areas seems rather limited. But the map should be viewed in the light of

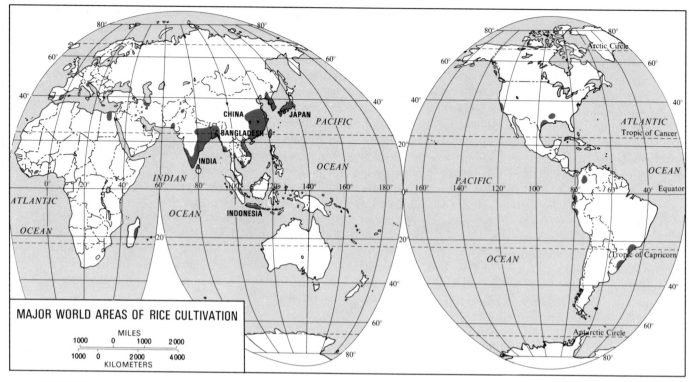

MAJOR WORLD AREAS OF RICE CULTIVATION

MILES
1000   0   1000   2000

1000   0   2000   4000
KILOMETERS

FIGURE 8.8

World distribution of rice-growing areas.

Growing rice in Asia's paddies requires large numbers of workers to plant, raise, and harvest the crop. After the fields have been turned by water-buffalo-pulled plow and the seeds planted, the shoots are transplanted to other paddies. These farmers are at work in the paddies of Sri Lanka.

per acre yields, which for rice are much higher than for wheat—nearly twice as high, in fact. Fast-maturing rice strains have been developed, and now two crops can be harvested on the same acre of land during a single growing season, with time left over for some vegetable growing. There are even places where three rice crops are harvested in a single year.

Rice is grown overwhelmingly as a local subsistence crop. Other grains, especially wheat, enter long-distance international trade in much greater volume. As Figure 8.8 indicates, rice is the staple for much of South, Southeast, and East Asia—especially in eastern India, Bangladesh, Indochina, Thailand, the Philippines, Indonesia, southern China, and Japan. China's agricultural regions, as shown on Figure 8.8 and Plate 8, display a transition from various kinds of rice (grown on upland fields and lowland paddyfields) in southern China to wheat and other grains farther to the north. In India, the maps show wheat and millet

replacing rice toward the country's west. In general, rice needs warmer and moister conditions than wheat, and it is useful to compare Figure 8.8 to Plate 3, the map of world climatic regions. It is also interesting to compare Plate 5 to Figure 8.8 (the world's population distribution). This suggests just how large a percentage of the world's human population is concentrated in, and near, rice-producing areas and how crucial rice is as a staple food. Nearly 90 percent of all rice grown is produced in East, Southeast, and South Asia, and a great majority of people there (about 75 percent) are in some way involved with its cultivation. Such regional association of production and consumption also occurs with other major grains, but it is strongest in the case of rice. The producing countries also trade with each other in rice, but there are no large-volume exports of rice to Europe or North America. Instead, wheat moves in large quantities toward rice-producing, but food-deficient, regions.

FIGURE 8.9
World distribution of wheat-growing areas.

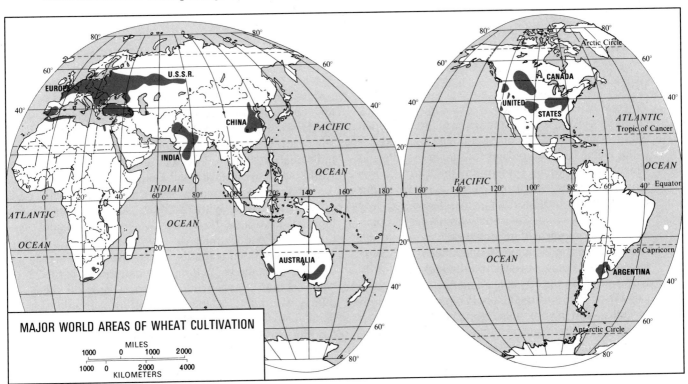

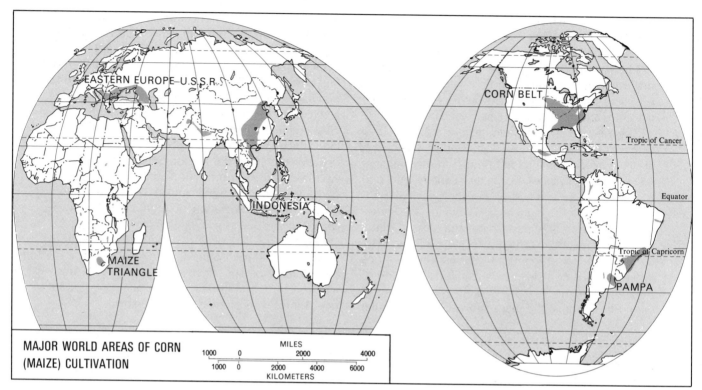

FIGURE 8.10
World distribution of commercial corn- (maize-) growing areas.

Wheat

Whereas rice is the subsistence crop of the developing countries of Asia, wheat is the staple of the developed world. This generalization should be qualified because wheat is cultivated and consumed in India and China as well as in Europe and the U.S.S.R., the United States, Argentina, and Australia. However, our map of world wheat-producing areas (Fig. 8.9) clearly reflects this cereal's association with the richer countries. First domesticated in Southwest Asia, wheat still grows on the shores of the Mediterranean, from Syria to Morocco. Today, however, the bulk of the world's wheat stands in the Soviet Ukraine, the North American Great Plains, Argentina's Pampa, southeastern Australia—and in western India and northern China. It is a spatial distribution very different from that of rice.

Whereas subsistence rice is the crop of both small

paddyfields and meticulous attention by its cultivators, commercial wheat is the crop of huge fields and mechanized production. True, wheat is grown in India and China as rice is, but on small plots and with much individual attention. However, in the U.S.S.R. (the world's leading producer), the United States, Canada, Australia, and Argentina, wheat is raised on an infinitely larger scale, with machines doing the seeding and harvesting. Thus, here we have a situation that distorts our Thünian view of the world, for the wealthy, developed countries are not the importers of wheat (except the U.S.S.R. in recent decades), but its chief exporters. As providers of sustenance for the hungry world, the wealthy countries are thus able to perpetuate their considerable control over the fortunes of poorer nations. Whereas rice is largely consumed in the country where it is produced—only about 5 percent of all the rice produced annually enters international trade—wheat moves in giant cargo ships across oceans to distant consumers. Tonnages vary with harvests in successive years, but between 15 and 20 percent of the world's annual wheat production is usually exported to hungry buyers.

Corn

If rice is Asia's grain crop, corn (or maize, as it is called in other parts of the world) is the cereal of the Americas. First domesticated by American Indians and not known to other peoples until the sixteenth century, corn today grows in practically all parts of the world and is a subsistence crop in Africa and Asia as well as in America. But the world's leading area of corn production still lies in America: the U.S. Corn Belt (Fig. 8.10).

Corn is a versatile crop capable of growing in moist as well as comparatively dry climes, in valleys and on high mountain slopes—and its growing season is less than three months. It has become a primary subsistence crop in Africa, where yields per acre are naturally far lower than those of the favored U.S. Corn Belt. In the United States, however, most of the corn produced is not consumed directly but used to feed livestock. Only about 10 percent of the U.S. corn production (which amounts to more than half of that of the entire world) is made into cornbread, breakfast cereals, cornmeal, margarine, and related products. A small quantity is consumed as a vegetable. Corn is also exported in comparatively small quantity to

Wheat cultivation in the United States is an operation of large dimensions, mechanized at every level. These huge grain elevators dominate the scene at Hutchinson, Kansas. The Farmers' Cooperative Elevator in the foreground of the photograph, one of the largest in the world, has a capacity of more than seventeen million bushels of grain.

Western Europe and Japan; of course, there is also some local, regional trade in certain areas. But very little corn enters international trade, nor is corn nearly as significant in the combat of world hunger as wheat or rice.

Other Crops

Although not a cereal, we should take note of the growing role of soybeans in the provision of much-needed vegetable protein in hungry countries. The soybean plant shares with corn its ability to grow under quite unfavorable conditions. It is able to withstand drought because of its deep root system, which also helps improve the soil. First domesticated in East Asia, the soybean now grows in most areas of the world, and the United States has become the leading producer, surpassing even China. In Asia, the soybean serves as a staple as corn does in Africa; it is consumed fresh, fermented into beverages, ground into meal, and dried for storage. In the United States, most of the soybean harvest is either ground into meal and used for animal feed or the oil is squeezed out of the beans and used for food (the soybean oil also has industrial uses). The versatility of the soybean and its nutritional qualities are reflected by a growing consumption in the United States as food. You may have seen meat substitutes in supermarkets—products made to look and taste like meat but much cheaper; their base is likely to be the soybean. Not surprisingly, international trade in soybeans and soybean products is also growing substantially. The United States is now the world's leading exporter as well as producer, selling more than two thirds of all the oil, three quarters of the meal, and about 90 percent of all the beans bought on international markets.

Although numerous other cereal and bean crops are grown in various areas of the world, none compares in importance to rice, wheat, corn, or soybeans. Barley, for example, has a distribution pattern not unlike that of corn and is used both as food and animal feed as well as in the production of liquors and beer. Millet and sorghum are hardy grains that sustain subsistence farmers in local areas from Africa to China. Rye is an important bread grain in Europe; oats ranks behind corn as a feed crop. In some areas of the world (notably West Africa), the peanut is a staple.

Neither are the root crops a match for the cereals in the sustenance of the world's billions. The potato (the white potato, to distinguish it from other varieties) was grown by American Indian peoples before the European invasion, and the Europeans took it back to Europe—where it became a staple from Ireland to Russia, sustaining the great seventeenth- and eighteenth-century population explosion. The potato provides starches, proteins, iron, and some vitamins; it is used both as a food staple and as animal feed. In terms of nutrient value, the potato is the only crop that can match rice, but although it is spreading in many areas the great bulk of the world's potato crop is still produced and consumed in Europe and the Soviet Union. Still less significant in the world perspective, but locally the basis for survival, are such root crops as the yam and the cassava (also called manioc)—these plants can grow with comparatively little attention and under conditions of heat and humidity that preclude the maturing of a grain crop.

IN RETROSPECT

If it appears strange that, in a world as hungry as ours, enormous tonnages of corn and potatoes can be fed to animals, huge tracts of land can be set aside for unessential luxury crops, and large numbers of people labor on the soil to produce frills rather than food, we should remind ourselves one final time of the Thünian model and its implications. What von Thünen saw locally (and what we noted on regional and even continental scales) now has worldwide expression. The range and variety of products familiar to us on the shelves of supermarkets is an anomaly in this world of simple, ill-balanced, often inadequate diets. A global network of farm products is oriented to that one fifth of the world's population that is highly urbanized, powerful, and wealthy, and that lives in the United States and Canada, Europe, the U.S.S.R., and Japan. The farmers in distant lands may be even less able to make decisions on the use of their land than those in von Thünen's Isolated State, for it is the developed world that decides what will be bought at what price, and it is the wealthy countries that have the choices,

330

not the poor. From the proceeds gained through the cultivation of tobacco, coffee, and sugar, the hungry countries must buy staples for survival—staples often grown by the farmers of the wealthy world. The colonial age may have come to an end, but the age of dependence has not.

Reading about Agricultural Geography

The geography of farming is agricultural geography, and this is the rubric under which most books in this area appear. One of the most interesting is actually a translation from the German by H. F. Gregor (one of this country's leading specialists in the field) of a book by B. Andreae, *Farming Development and Space: A World Agricultural Geography* (New York: de Gruyter, 1981). Theory and practice are effectively intertwined in the discussions of crops, climates, and cultures. A book with historical and non-Western dimensions is T. P. Bayliss-Smith, *The Ecology of Agricultural Systems* (London/New York: Cambridge Univ. Press, 1982).

Also see W. Manshard, *Tropical Agriculture: A Geographical Introduction and Appraisal* (New York: Longman, 1974), still a wide-ranging, insightful work. A more basic book is W. B. Morgan and R. J. C. Munton, *Agricultural Geography* (New York: St. Martin's Press, 1972). Another major text is L. Symons's *Agricultural Geography* (Boulder, Colo.: Westview, 1979). And H. F. Gregor's *Geography of Agriculture: Themes in Research* (Englewood Cliffs, N.J.: Prentice-Hall, 1970) still has much to offer.

Agricultural geography forms part of the wider field of economic geography, and the books cited at the end of Chapter 10 often contain very useful sections on this topic. Thus we could cite P. Bartelmus's book entitled *Environment and Development* (Winchester, Mass.: Allen & Unwin, 1986) in both places; remember it when development problems are again discussed.

As the dates of publication of many of the books cited here suggest, the flow of printed work in this specialized area is smaller than in many other fields of geography. This is true also in the journal literature, where agriculture receives rather less attention than urban and other economic topics.

No background reading in agricultural geography would be complete without a look at C. O. Sauer's *Agricultural Origins and Dispersals* (New York: American Geographical Society, 1952). A second edition, issued by the Massachusetts Institute of Technology in 1969, contains a new introduction and three additional articles by Sauer.

A volume by A. N. Duckham and G. B. Masefield, *Farming Systems of the World* (New York: Praeger, 1970), contains a wealth of information and opinion on world agriculture. Also see A. H. Bunting (ed.), *Change in Agriculture* (New York: Praeger, 1970). An older standard work still in the library and still useful is H. F. Gregor's *Environment and Economic Life* (New York: Van Nostrand, 1963). Another older and still-relevant book was edited by C. K. Eicher and L. W. Witt: *Agriculture in Economic Development* (New York: McGraw-Hill, 1964). Also productive is J. R. Tarrant's *Agricultural Geography* (New York: Wiley–Halsted, 1974).

On regional agriculture, see *Economies and Societies in Latin America* (New York: Wiley, 1973) by P. R. Odell and D. A. Preston, a book that ranges widely over that region's livelihoods, including, of course, farming. Also see the first part of J. P. Garlick and R. W. J. Keay (eds.), *Human Ecology in the Tropics* (New York: Wiley–Halsted, 1977) and *Russian Agriculture* by L. Symons (Wiley–Halsted, 1973). See as well a volume by P. P. Courtenay, *Plantation Agriculture* (London: Bell, 1965).

Concern has been rising over the problems of dry and marginal lands. A good place to start reading on this is the desertification issue of *Focus*, published by the American Geographical Society in New York in 1977. Also consider Jen-Hu Chang, *Climate and Agriculture; an Ecological Survey* (Chicago: Aldine, 1968) as well as a volume edited by H. E. Dregne, *Arid Lands in Transition*, publication No. 90 of the American Association for the Advancement of Science, Washington, D.C. 1970. Also look up the special

331

issue. "The Human Face of Desertification," of *Economic Geography* (Vol. 53, October 1977).

On questions relating to the spatial structure of agriculture, the best place to look remains the journal literature. When you examine the major geographic journals, you will find articles dealing with agricultural topics, ranging from discussions of farming in particular areas or regions to analyses of specific crops under certain environments. *Economic Geography*, published by Clark University (Worcester, Mass.), is a good source, although some of the articles are quite technical. When you do this research, you will note that von Thünen's ideas still permeate the literature. See, for example, "The Pertinence of the Macro-Thünian Analysis" by A. Kellerman, in *Economic Geography* (Vol. 53, July 1977); and "The Spatial Expansion of Commercial Agriculture in the Nineteenth Century: Von Thünen Interpretation" by J. R. Peet, also in *Economic Geography* (Vol. 45, July 1969). Von Thünen's own work, edited by P. Hall, was published as *Von Thünen's Isolated State* in a translation by C. M. Wartenberg (Elmsford, N.Y.: Pergamon, 1966).

332

CHAPTER NINE

The Geography of Urban Areas

The great cities of today are the culture hearths of modern society. Here concentrate the might of nations, their technology and industry, their governments and decision makers. In the cities lie the sources of innovation, the centers of power and organization, the institutions of state. The cities are the crucibles of culture, the places where traditions are forged.

Cities, of course, are giant clusters of humanity. People have clustered in urban communities for thousands of years—indeed, ever since specialization began. When community members could confine their activities to specific functions (e.g., toolmaking, teaching, or food marketing), it was advantageous to perform those functions in a central place. By today's standards, the earliest central places were mere villages, clusters of a few thousand people perhaps. But in ancient times, these were the nodes of change and progress, places where certain people had the free time to meditate, contemplate, teach, and vie for influence and power.

For thousands of years, this clustering process continued; over time, some very large cities emerged. Some of those ancient cities—from Rome at the heart

The ancient city of Mohenjo Daro was the largest of several urban centers that prospered in the Indus Valley (in present-day Pakistan) about 2500 to 1500 B.C. Extensive stonework can be seen in this photograph; the round structures are the remains of permanent wells that served the occupants of these buildings.

of the Empire to Tenochtitlán in Aztec Middle America—measured their populations in hundreds of thousands. But the societies for which they formed the urban foci never became truly urbanized societies. The *proportion* of people living in cities remained quite low, perhaps between 5 and 10 percent of the total. For the world as a whole, it is estimated that as recently as 1800, only about 3 percent of the people lived in urban areas.

The Industrial Revolution changed the geography of urbanization. The rise of large-scale manufacturing stimulated urban growth of unprecedented dimensions, and during the nineteenth century, many cities in Western Europe and North America mushroomed. Soon other cities (not primarily industrial centers) also began to grow more rapidly. The urban spiral had been set in motion, and it continues to this day. Villages are growing into towns, and towns are being absorbed into larger urban complexes called *metropolitan* areas. In several parts of the world these metropolises are coalescing into vast **conurbations**. In the eastern United States, this process is creating a grand geographic expression of the urbanization process, a supercity extending from Boston along the seaboard to Washington. This has been called **megalopolis** (mega = great, polis = city), the ultimate product of decades of urban growth in this region. Megalopolis, however, is not the only manifestation of this process in the modern world.

•　　　　World Urbanization　　　　•

Over the past three decades, the population clustered in cities world wide has more than doubled to a total of just over 2 billion, thereby making urbanites of more than 4 out of every 10 people on earth. The significance of this latest increase in global urbanization can be seen in Figure 9.1. Throughout this century, urban population has been increasing at an accelerating rate, particularly since 1950; in fact, if

337

FIGURE 9.1

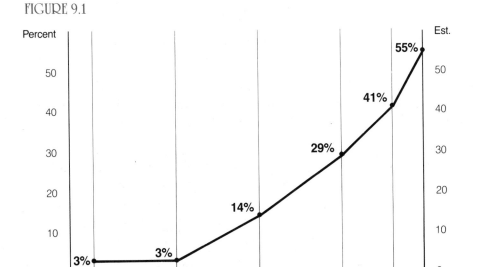

PERCENT OF WORLD POPULATION RESIDING IN SETTLEMENTS OF 5000 OR MORE, 1800–2000

projections hold, the urbanization pace of the second half of the twentieth century will just about double all world urban growth that occurred before 1950.

The 1988 map of world urban areas (Fig. 9.2) reveals the distribution of metropolises containing populations of more than 1 million. It underscores the concentration of large metropolitan agglomerations in eastern North America, Western Europe, and East Asia—notably Japan.

North America

In North America, the anchors of several emerging megalopolitan regions are shown, including the Boston–Washington, Chicago–Detroit–Pittsburgh, San Francisco–Los Angeles–San Diego, and Montreal–Toronto–Windsor conurbations. Yet another, in peninsular Florida, is still marked only as a set of discrete comparatively small cities, but recent population growth in the state has been rapid; a regional megalopolis is developing quickly, centered on Miami–Fort Lauderdale–West Palm Beach, growing northward along the Atlantic coast toward Jacksonville and westward across the burgeoning Orlando and Tampa areas of central Florida.

Europe

In Europe, England's conurbations are approaching megalopolitan unification: London and its immediate environs lie at the center of an expanding population cluster of 20 million, and the industrial cities of South Wales and the Midlands are now but a short distance away. On the European mainland, a major urban complex is emerging in western West Germany, in the Ruhr–Rhine zone that includes such cities as Düsseldorf, Essen, and Cologne. In the Netherlands, planning for a triangular megalopolis (Amsterdam–Rotterdam–the Hague) is a matter of leading national priority; an attempt is being made to guide this megalopolis, already being called *Randstad* (ring city), into a twenty-first-century megacity showcase, complete with parks, spacious housing, good communications, excellent public transportation, and optimally distributed social services. Belgium, too, is experiencing the development of a coalescing urban complex, with Brussels and Antwerp the twin foci.

Other major urban agglomerations in Europe include the region centered on Paris, the rapidly developing Po Plain of northern Italy, and the Central European complex that extends from East Germany's Saxony to Poland's Silesia. Elsewhere, there are major individual cities such as Moscow, Leningrad, and Madrid, that are not yet true multicity urban regions. Of course, we should keep the scale of developments in Europe in perspective, and Figure 9.2 is worth another look in this regard. The whole urbanized area, from Britain's Midlands to West Germany's Ruhr–Rhine region, extends over an area not much larger than North America's megalopolis. Yet Europe's political fragmentation and cultural diversity has led us to identify discrete urban units by country within vast multination urban core.

Japan

Large-scale megalopolitan development outside North America and Western Europe is occurring only in Japan. The Tokyo–Yokohama and Osaka–Kobe–Kyoto conurbations are of enormous dimensions, and they are growing toward each other along Honshu Island's Pacific coast.

A GLOBAL PHENOMENON

This is not to suggest that the urbanizing trend that has generated the Bosnywash megalopolis and Randstad, Holland is not being experienced in other areas of the world. Johannesburg in South Africa lies at the center of a smaller-scale megalopolis that includes several medium-sized cities of the Witwatersrand region. However, in general, conurbanization outside Europe, the United States, and Japan has not reached comparable proportions. Even where a high percentage of the population is urbanized, the numbers are not yet large enough to create huge urban complexes: Australia has several large cities, but no true conurbation. In Argentina, Buenos Aires is a city of world proportions, but its nearest substantial neighbor, Rosario, lies 200 miles (320 km) away. The area between São Paulo and Rio de Janeiro in Brazil also does not yet exhibit the urban development to support the near-future emergence of a coalescing supercity. This mirrors the situation in the underdeveloped countries (UDCs),

• "Law" of the Primate City •

Great cities have always reflected regional cultures. In 1939, geographer Mark Jefferson published an article entitled "The Law of the Primate City" in which he described the dominant city much as Sjoberg did later—as a place that reflects the essence of the culture at whose focus it lies. The "law" states that "a country's leading city is always disproportionately large and exceptionally expressive of national capacity and feeling." It refers, obviously, to cities of the twentieth century that represent national cultures of modern times, but the notion can be extended to earlier periods too. As Sjoberg stated, cities of preindustrial, feudal, and preliterate societies were also products as well as reflections of their cultures.

Although Jefferson's notion is rather imprecise, it is supported by numerous cases, past and present. Kyoto as the **primate city** reflective of old Japan, Tokyo of the new; Paris as the personification of France; and London as the repository of the culture and history of a nation and empire—all are so deeply etched in the urban landscape. These are primate cities indeed! In Europe, such cities as Athens, Lisbon, Prague, and Amsterdam may no longer in every case be disproportionately large, but they

remain exceptionally expressive of the cultures they represent. Beijing (Peking) in China, Lahore in Pakistan, Ibadan in Nigeria, and Mexico City are also examples of primate cities—in some cases no longer the largest regionally, but quintessentially representative of national capacity and feeling. Jefferson's generalization, like Sjoberg's, might be adapted to conform to stages in a culture's growth. Thus, Beijing is the primate city of the old China, Shanghai (now China's largest) of the new; Rio de Janeiro reflects Brazil's historic evolution, whereas São Paulo represents the vigor of its contemporary society.

Today the primate city exists most prominently in: (1) countries with dominantly agriculture-based economies such as Bangladesh, Indonesia, and Ethiopia; (2) countries with a recent history of colonial subjugation, including Kenya, Zimbabwe, and Senegal; and (3) less-developed countries such as Sri Lanka, Liberia, and Nicaragua. In many of these cases, only part of Jefferson's thesis is fulfilled: the cities are disproportionately large, but they do not necessarily express national capacity and feeling, having a foreign-influenced past.

where large cities tend to stand alone as islands in a rural-dominated scene.

Those islands in the underdeveloped world, however, are becoming increasingly crowded in the late 1980s. The intensifying urban trend prevails in Africa, Asia, and Middle and South America: collectively, cities in these three regions grew faster in the 1970s than in the 1960s—and the most recent estimates suggest an even more rapid growth rate in the late 1980s. It is also worth keeping in mind that although the overall percentage of urban population is lower in the underdeveloped countries, the much greater *absolute* number of people there makes even a 1 percent cityward shift a major movement. In India, for example, 1 percent of the national population totals nearly 8 million, more than all the people who

reside in the Delhi–New Delhi conurbation (and India's urban population percentage has advanced from 18 to 26 percent between the 1960s and the late 1980s).

As many Third World urban islands are becoming engulfed by a tidal wave of in-migration, a number of underdeveloped countries' metropolises have achieved some of the world's highest totals. Whereas such DC (developed-country) urban agglomerations as Tokyo, New York, and London have long ranked in the 10-million-plus category, they are now joined by Mexico City (19+ million), São Paulo (17 million), Rio de Janeiro (13 million), Buenos Aires (11 million), Shanghai (16 million), Chongqing (14 million), Beijing (13 million), Calcutta (11 million), Bombay (10 million), and Seoul (10 million).

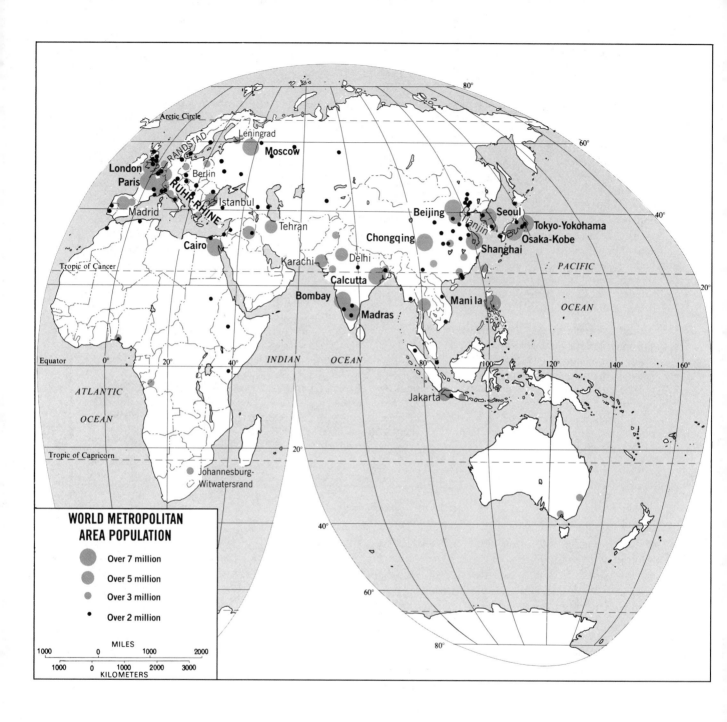

**WORLD METROPOLITAN
AREA POPULATION**

- ● Over 7 million
- ● Over 5 million
- ● Over 3 million
- • Over 2 million

London
Paris
RANDSTAD
Madrid
RUHR-RHINE
Berlin
Leningrad
Moscow
Istanbul
Tehran
Cairo
Karachi
Delhi
Bombay
Calcutta
Madras
Beijing
Chongqing
Tianjin
Shanghai
Seoul
Tokyo-Yokohama
Osaka-Kobe
Manila
Jakarta

Johannesburg-
Witwatersrand

Arctic Circle
80°
60°
40°
Tropic of Cancer
20°
Equator
0°   20°   40°   80°   100°   120°   140°   160°
PACIFIC
OCEAN
ATLANTIC
OCEAN
INDIAN   OCEAN
Tropic of Capricorn
20°
40°
60°
80°

MILES
1000   0   1000   2000
1000   0   1000   2000   3000
KILOMETERS

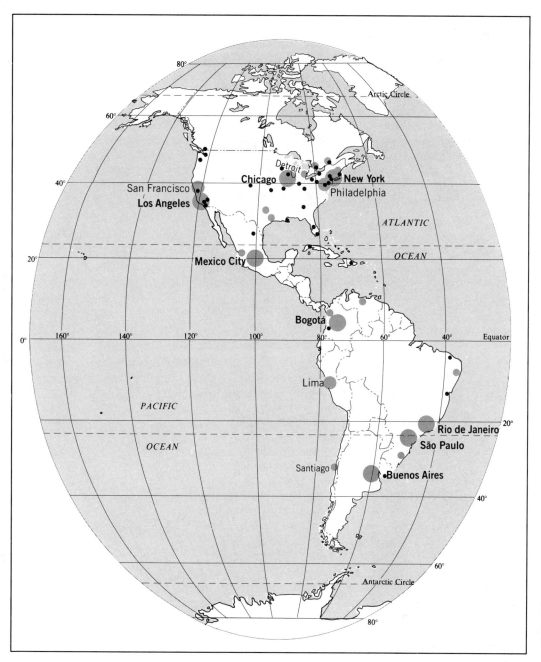

FIGURE 9.2
Major urban centers of
the world.

Two views of ancient Zimbabwe. Several of the cone-shaped towers that formed a part of this structure (called the *temple*) have been destroyed; the one shown here is the best remaining example. The narrow passage reveals the precision of the stonework. No mortar was used. Many questions concerning Zimbabwe remain unanswered.

As we saw in Chapter 6, people migrate to urban areas with expectations of a better life, which all too often fail to materialize. Particularly in the underdeveloped realms, but also in the industrial cities of developed regions, people are often crowded together in overpopulated apartment buildings, in tenements, and always in the teeming slums. New arrivals from other cities, from nearby smaller towns, and from the rural countryside add to the already substantial natural increase rate of the resident population; no housing expansion could possibly keep up with this massive inflow. Huge areas of new shantytowns develop almost overnight—mostly without the barest amenities. However, in-migration is not deterred, and multiplying millions of people will spend their entire lives in urban housing of wretched quality.

Miserable living conditions for urban immigrants notwithstanding, the world's cities continue to beckon. Cities are humanity's arenas of interaction, opportunity, achievement, innovation, and progress. They have always been the nucleus of power, foci of culture, and centers for the arts and sciences. Philosophers, composers, architects, writers, scientists, artists, and religious leaders choose cities as their home base, contributing to the excitement and ferment that great cities generate. The world's greatest orchestras, museums, theaters, architectural achievements, monuments, engineering feats, and other outstanding human accomplishments are concentrated in large metropolises. Many of the best universities are located in and around major cities. Skyscrapers, subways, elevators, domed stadiums, and countless other innovations have emerged in the city. Nowhere is the variety of available goods and services greater than in metropolitan centers. Life in the core city may be problematic for millions of urbanites, but there is always the sense of being where the action is. Thus, the spiral set into motion several thousand years ago has not ended.

Focus on the City

The study of urban places in human geography encompasses a broad range of approaches and interests.

We have already noted the application of the cultural-landscape concept to urban areas, because the townscape in many ways reflects and represents the culture that generates it. Earlier in this century, Carl Sauer proposed a view of urban places as manifestations of the relationship between human societies and natural landscapes. Culture was the "factor"; the natural landscape was the "medium"; urban plan and structure were the "forms." This raised questions about the resulting cityscape. How did the actual, physical place where a city grew affect its development? A city might have begun as a settlement on an island, where protection was provided by nature later, that same island might lose its protective advantage and be the source of severe space limitations. How did the relative location of a city—its regional position vis-à-vis other settlements—influence its growth and development? Some cities emerged as focal points for large prosperous regions, with little competition from other nearby urban centers. Certain other cities found themselves in less favored locations, perhaps marginal to productive regions or overshadowed by more successful competitors. Questions such as these have led to detailed case studies and enhanced geographers' understanding of the evolution of urban places.

Another important contribution came from regional geographers, who were interested in the interrelationships between cities and their surrounding areas. Every city and town has such a region, for which it forms the focus and where its influence is paramount. Farmers in that region sell their products on the city's markets; customers come to shop and do business in the city's commercial areas; the city's newspapers are read throughout this zone. In many other ways, the city's dominance and its numerous interconnections can be recorded and mapped. The term **hinterland** came into use to identify such trade areas, and much was learned about the ways in which cities interact with these tributary regions. This approach was also quite useful in a historical geographic context. An understanding of the relationships between preindustrial cities and their hinterlands in one area of the world was helpful in discerning similar spatial linkages elsewhere.

Urban geographers are also curious about the spacing of cities, towns, and smaller settlement clusters. In general, large cities tend to lie farther apart

343

Ibadan, Nigeria's second city, with about two million people, has long been a state capital. Lagos, the federal capital, is the country's largest urban center; Ibadan lies about 160 kilometers (100 miles) inland. Like many other African cities, Ibadan has developed a modern sector adjoining its traditional areas. In this photograph, multistory skyscrapers overlook the sprawling market and residential area in the background.

than smaller ones, towns lie closer together still, and villages are separated by even shorter distances. What forces influence the development of such arrangements? Is it possible to establish an idealized model of regional structuring, based on a set of simplifying assumptions, to predict where urban places of various sizes would be spatially positioned? Then, relating the model to a complex real-world region and accounting for the inevitable distortions, could predictions be made regarding the growth of certain places and the decline of others? This, again, is the domain of location theory, still another approach to studying the spatial properties of urban agglomerations.

A fourth geographic view of the city concentrates on the internal patterning of the metropolis itself. In a sense, this is an application of regional geography to urbanized areas in order to identify the spatial structure of cities and suburbs and the manner in which various intrametropolitan subregions (clusters of com-

mercial, residential, industrial, and other activities) are arranged. It is one thing to speak generally of a city's downtown or its CBD, but it is quite another to precisely delimit such microscale regions on maps. When geographers began to produce detailed studies of spatial form, certain common patterns emerged that suggested the operation of growth processes that should also be analyzed. This research contributed strongly to our comprehension of the forces at work in shaping the evolution of the modern metropolis.

DEFINITIONS

The study of cities and other urban places is complicated by problems of definition and terminology. When it comes to global comparisons, we should remember that census taking in underdeveloped countries still involves much guesswork and estimations; but even in the United States with its modern methods of data gathering, there are difficulties. Much of a city's growth over a certain period may occur outside its municipal boundaries—adjacent to, but technically not within the city. But the census takers have no alternative: they must continue to use the city's boundaries to determine the population of the city proper. A city's growth may, therefore, be recorded as only a few thousands of people when, in fact, tens of thousands entered the urban area but settled in suburbs outside the actual city limits.

Furthermore, few cities consist of only one municipality. San Francisco is only one such municipality among a contiguous grouping that also includes (among others) Daly City, South San Francisco, San Mateo, and, across the Bay, Berkeley, Oakland, and Alameda. The city of Miami lies surrounded by municipalities such as Coral Gables, South Miami, West Miami, Hialeah, North Miami Beach, and a number of other contiguous but discrete places. Indeed, Miami's urban area (or **conurbation**) sprawls from Dade County northward into Broward County. Thus, the concept of the **urbanized area** is now used to count all the people living not only within all the cities that make up an urban region, but also in all adjacent built-up areas. In the case of New York City, the difference between city proper and urbanized area is shown by the (latest available) 1984 census figures: city, 7.16 million; urbanized area: 17.81 million.

Another unit of measurement employed to help define urban populations is the metropolitan statistical area (MSA), which is county based and includes the central city or a set of coalescing cities counting at least fifty thousand inhabitants. But counties must satisfy certain criteria to be included in MSAs; when an urban agglomeration fills two counties and just edges into a third, that third county may not yet qualify. So, the MSA population statistic is often less useful for comparative purposes than the urbanized area.

Comparisons between cities, therefore, must be based on standardized frameworks. There is little point in comparing conditions in San Francisco and Miami, for example, without employing data that cover all of the Bay area and all of the Dade–Broward County urbanized areas. Miami is merely a sector of the South Florida urban agglomeration, and San Francisco lies at the heart of a much larger conurbation.

At the other end of the urban continuum, a different problem arises. The terms *urban* and *rural* have different connotations in different countries. In Canada, settlements with more than one thousand people are classified as urban. In the United States, incorporated settlements numbering over twenty-five hundred inhabitants have long been taken to be urban and smaller ones as rural villages. In India, places with more than five thousand residents are classified as urban, but in Japan the lower limit is thirty thousand! Nor are the differences exclusively a matter of size. In the United Kingdom, for example, *urban* is a question of status, not just dimension; communities achieve urban status through application and government approval. In Italy, *urban* is considered to be a matter of employment. More than 50 percent of the economically productive population must be engaged in nonagricultural pursuits for a settlement to be classified as an urban center. All this complicates worldwide comparisons. When one country reports its level of urbanization as, say, 43 percent of the total population and another reports 72 percent, it is necessary to check the criteria used in each or the comparison is meaningless.

345

• Cities: Functions and Classes •

When someone asks you a question about your hometown, you are likely to identify it by name, and then add that it's a university town (or a port city, or a mining town, or a farm village). In doing so, you try to summarize the nature of your hometown: to define what kind of an urban place it is. You are, in effect, classifying it with other places of the same type, disregarding size. There are mining towns large and small, industrial centers ranging from a few thousand to several million inhabitants. What these places share is not the same rank in the hierarchy we just discussed, but the same assemblage of dominant functions. So, this is one of the bases geographers use to classify urban places.

At first, classifying cities seems quite easy. You could jot down a list rather quickly: there are resort towns, retirement centers, ports, fishing towns, and so forth. These (usually small) urban places are essentially *unifunctional*, that is, their economic base is dominated by a single activity. But when we look deeper into the matter, we find that port cities also have stores (and thus retail functions), factories (and thus industrial functions), universities, and other facilities. This means that many port cities must be multifunctional, that is, they possess more than one sphere of economic activity, among which the transportation function (of the port) is the chief one. But what about a city that does not have such a dominant function as a port city has or as you might find in a resort town or in a mining town? There are many cities—such as Denver, Kansas City, and Memphis—where no single function stands out to typify the place. On what basis can such cities be included in a generally applicable system of classification?

One approach lies in the activities of the labor force in cities. Where do people find employment? The census gathers such information: it determines the place of employment of the labor force in major cities and reports the number of workers engaged in each economic activity. In 1943, C. D. Harris published an article in the *Geographical Review*, "A Functional Classification of Cities in the United States," in which he used these data to develop a system according to which all U.S. cities could be classified. Thus he recognized two types of manufacturing cities: the M^1 type—in which the total employment in manufacturing is at least 74 percent of the total employment in retailing, wholesaling, and manufacturing combined—and the M type—in which that figure is reduced to 60 percent (Fig. 9.3). In addition to these two classes of manufacturing cities, Harris recognized eight other classes of cities: (1) retail centers, in which employment in retailing is at least half the total number of people employed in retailing, wholesaling, and manufacturing combined and (2) wholesale cities in which employment in wholesaling is unusually strong. Others in the classification are (3) transportation centers, in which transportation and communication industries employ at least 11 percent of the labor force; (4) diversified cities, where the census data yield no clear-cut, paramount function; (5) mining towns, where more than 15 percent of the workers are employed by mining industries; (6) university cities, whose chief industry in constituted by the campus of a large educational institution; (7) resort-retirement towns, where the majority of the gainfully employed are in the service of recreational or retirement-oriented establishments; and (8) political centers, including not only Washington, D.C., but also a number of state capitals.

It is not surprising that many geographers have tried to modify the Harris classification. This can easily be done by adding new classes of cities or by suggesting changes in the quantitative boundaries used by Harris to establish his system—which were admittedly somewhat arbitrary. But despite these reconsiderations, the pattern shown in Figure 9.3 stayed much the same and is still quite current. Manufacturing cities of both classes (M^1 and M) are heavily concentrated in the northeastern United States and in the North American heartland generally. Retailing centers, on the other hand, tend to lie outside this manufacturing zone and, with wholesale cities, are concentrated in a wide belt that extends from north to south across the country from North Dakota to Texas. Diversified cities show a stronger concentration in the eastern half of the country; transportation centers lie both along the coast and along major rivers and in places where railroads converge. University towns, as the map shows, are especially prevalent in the midsection of the country where many small towns are dominated by large universities. Resort-retirement towns, predictably, concentrate in the southern, warmer sections of the United States. Finally, mining towns are associated (of course) with mineral deposits—mainly the coalfields.

Since the Harris classification was proposed, a number of other systems have been suggested, all designed to improve on that effort. This has not proved an easy task: the Harris article of 1943 has become something of a classic in the literature of urban geography. In 1955, H. J. Nelson published an article entitled "A Service Classification of American Cities," in which he made an effort to overcome a

346

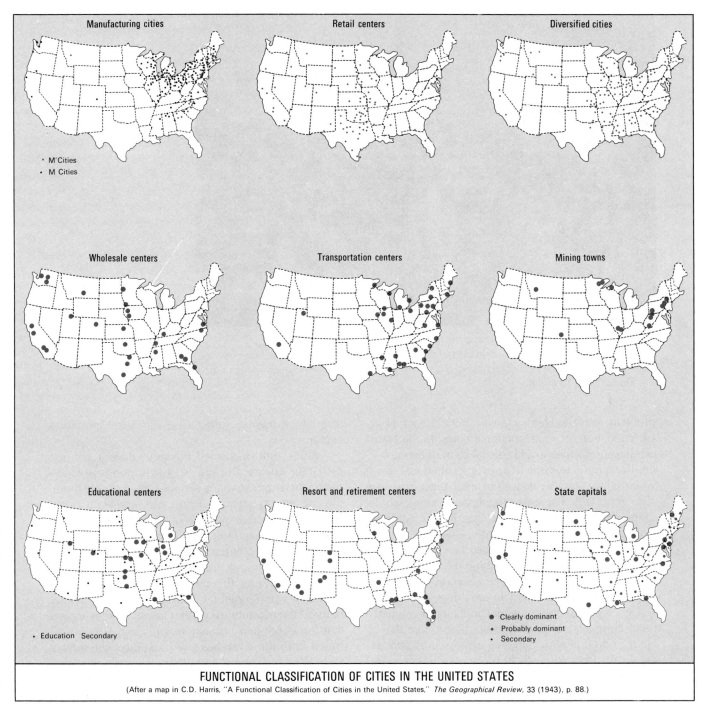

FUNCTIONAL CLASSIFICATION OF CITIES IN THE UNITED STATES
(After a map in C.D. Harris, "A Functional Classification of Cities in the United States," *The Geographical Review,* 33 (1943), p. 88.)

FIGURE 9.3

The Harris system of city classification in the United States.

A dramatic view over the conurbation at the heart of which lies Tokyo—the largest agglomeration of its kind in the world today. Little is left of the natural landscape in this vast urban region; a cultural landscape has been superimposed. In 1988, the population of this area was approaching 30 million.

348

problem with the Harris classification: its lack of accommodation for multifunctional cities. In the Harris classification, cities would specialize in only one economic sphere (except in the diversified category) when it was very evident that other functions were also important. For example, there is more wholesaling in New York or Chicago than in many of those cities classified as wholesale cities—but you will not find New York or Chicago mapped as wholesale cities in Figure 9.3. That is because other functions of these two great cities are similarly so important that the only feasible class to put them in was diversified cities. Nelson, trying to overcome that difficulty, developed a more complex statistical formula based, like that of Harris, on employment figures expressed as percentages of the labor force. Nelson's calculations centered on an index that revealed how far cities exceeded (or failed to reach) the average derived from all available data for every city tested. This gave a more objective measure of specialization than Harris had achieved: Chicago, for example, could score high as a whole-

sale city, a manufacturing city, *and* a transportation center.

Cities with diversified functions have a self-perpetuating quality. Because so many goods are produced there and so many services are available there, the city attracts other producers and service providers who, in turn, add to the draw of the place. As long ago as 1909, the economic geographer A. Weber drew attention to this phenomenon, which has much to do with the continuing prosperity of older cities (e.g., Paris) that, during the Industrial Revolution, found themselves positioned far from the raw materials that boosted the economies of other towns. Weber argued that the continued primacy of Paris and London resulted from the advantages of proximity—the advantages to manufacturers and other technical operations in locating close together. The availability of specialized equipment, a technologically skilled labor force, and a large local market are among these advantages. This process of urban growth has been called **agglomeration**; it can be measured and quantified as one expression of what geographers describe as the *multiplier effect* of large cities.

This interest in identifying the categories or classes of cities and discerning their economic-geographic bases is not just an academic exercise. As times change, so do the functions of many cities. Cities once dominated by *basic* economic activities (the production of goods and services for customers outside the city) have seen their economies change toward *nonbasic* activities (services for the residents of the city itself). The governments of such cities will respond by seeking to attract new basic activities, for example, by making low-priced land available for the development of industrial parks for new, modern manufacturing plants. When one economic sector in a diversified city fails, for example an aging steel plant or a declining port, the city's administration must respond. An immediate question will be: How nearby is the competition? A map showing the relative location of urban functional specializations will be a key element in the decisions to follow.

Location and the City

The rules by which some cities thrived and others failed over time were the rules of production, commerce, finance—and geography. The permanent advantages of some cities, the assets of relative location that had long sustained them, now assumed even greater importance. Human geographers studying cities in this context refer to two locational properties: the city's **site** (the internal physical and cultural attributes) and its **situation** (the external geographic aspects that involve its wider regional position). All urban places, therefore, are simultaneously imbued with both site and situational qualities. Let us examine each set of characteristics in greater detail.

SITE

The site of a town or city can be a major factor in its development. The term refers first to the actual, physical qualities of the place the city occupies: whether it lies in a confining valley, on a section of a coastal plain, on the edge of a plateau, or perhaps on an island. In Europe, a site's defensibility was often an important factor in the development of cities. A good example is Venice, a small coastal settlement near the Po Delta at the head of the Adriatic Sea that grew into a city of wealth and splendor. The city could have continued to develop on the mainland, but there was little natural protection to be found on that site. So, the rich merchants built their warehouses, their mansions, and their monuments to the accumulated wealth on a group of islands offshore, thereby deter-

The Île de la Cité in the Seine River, Paris, France. The crowded island was Paris' source (see text); the Notre Dame cathedral, symbol of the city's later regional primacy (page 351) can be seen in the upper right.

ring attackers from land. Today the site of Venice is causing some unforeseen problems: it is sinking and many of its old historic buildings are endangered by rising water.

Another example of the role of site in the emergence of a major city is Paris. The capital of France was founded on an island in the middle of the Seine River; again, a place where security and defense were enhanced—and also a place where the Seine could be crossed and the cross traffic controlled. Exactly when settlement on this *Île de la Cité* began is not known, but it probably occurred in pre-Roman times. Paris functioned as a Roman outpost for several centuries; later, the Roman regional administration there was replaced by other authority—always making use of the defensible qualities of the island city. Eventually, the rather small island proved too restrictive for Paris's growth, and the city sprawled onto the nearby riverbanks. However, authority remained centered on the island—thereby wedding cultural-landscape site advantages to those of the natural environment—and potential adversaries knew that there was no point in challenging the outskirts if there was no prospect of conquering the *Île de la Cité*.

The role of the *Île de la Cité* is evident in Figure 9.4. The island still lies central to the whole Paris region, and the town's first defensive wall—built by Philip-Augustus in the twelfth century—encloses the area centered on it. As the photograph shows, the *Île de la Cité* is not large, but it still contains some of Paris' primary landmarks, including the great Notre Dame Cathedral. Thus, it remains a vital functional region of the city.

Many cities owe their origins and early stimulus to advantages of site. However, a site can also become less beneficial over time, inhibiting urban development or requiring massive adjustment and possibly even causing a competitive disadvantage. In Kenya, for example, the city of Mombasa, East Africa's largest port, was founded on an island. The original port developed on the eastern side of the island where sailing vessels could most conveniently enter comparatively shallow Mombasa Harbor. Its insular position effectively protected Mombasa from adversaries on land as well as sea. European colonists built a large fort adjacent to the port, and the Arab-African town grew alongside. However, when motor-driven ships began to reach Mombasa during the nineteenth century, the port was too shallow to accommodate them. It was necessary to begin construction of an entirely new facility on the other side of the island, where the Kilindini Channel was deeper. The new port was located several miles from the old town, and Mombasa's urban growth was totally redirected. Ultimately, the growing metropolitan population (now more than five hundred thousand) was forced to spread onto the mainland, and one of Mombasa's leading modern problems has been its inadequate connection to these mainland suburbs. Many workers must make a time-consuming daily round trip across one of the two traffic-choked bridges or aboard an overcrowded ferry.

Thus, an originally advantageous site may in time present problems. Cities located on slopes between hills and the sea may need to reclaim land from the sea in order to accommodate their growth pressures, as was done in Cape Town and San Francisco. Cities built in mountain valleys may have benefited from coolness and pleasant weather, but today many suffer from severe pollution problems. Cities built where a water supply was once ample, now overtax those same resources. The defensibility of certain sites led to their original selection for clustered settlements; that defensibility is meaningless now, but the escarpments, swamps, and other impediments to enemies of old are likely to form obstacles to urban expansion today. Urban sites affect city growth in different ways at different times in history.

SITUATION

Even more important than a city's site is its situation. By situation is meant the position or relative location of the city with reference to surrounding areas of productive capacity, the extent of its hinterland, the distribution of competing towns, its *accessibility*—in short, the greater regional framework in which the city finds itself. Our previous example, Paris, combines advantages of site with a most fortuitous situation. The city lies near the center of a large and prosperous farming region, and as a growing market and distribution point, its focality over time increased continuously (see inset, Fig. 9.4). The Seine River, a

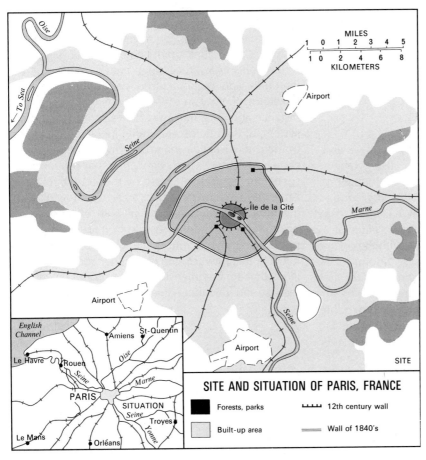

MILES
1 0 1 2 3 4 5
1 0 2 4 6 8
KILOMETERS

Airport

Oise

To Sea

Seine

Île de la Cité

Marne

Seine

Airport

Airport

SITE

SITE AND SITUATION OF PARIS, FRANCE

■ Forests, parks   ┴┴┴┴ 12th century wall

▢ Built-up area   ═══ Wall of 1840's

English Channel

Le Havre Rouen

Amiens St-Quentin

Oise

Seine

Marne

PARIS

SITUATION

Seine

Troyes

Yonne

Le Mans

Orléans

FIGURE 9.4

Paris: site and situation.

navigable waterway, is joined near Paris by several tributaries, and all of them are navigable, which readily connects Paris to various parts of the Paris Basin and to many areas beyond. To the west, the Seine River runs into the English Channel, and although today's oceangoing ships cannot reach Paris, the smaller vessels of the past could—however, even today only one transshipment at Le Havre or Rouen is necessary. Paris, thus, had superior connections to territories overseas when the time came, but in the city's longevity and steady growth, its internal connections were perhaps even more important.

The rivers that join the Seine (the Marne, Oise, and Yonne) from the northeast, east, and southeast as well as the canals that extend them even farther connect Paris to the regions of the Loire, the Rhône–Saône, the Lorraine industrial area, and the industrial complex on the French–Belgian border (Fig. 9.4, inset). Furthermore, Paris always lay at the focus of land transport routes. As the authority based on the *Ile de la Cité* gained greater political influence in surrounding regions, Paris found itself near the very center of one of Europe's largest and most enduring nation-states. Subsequent governments did everything possible to improve the city's connections with the country generally. Under Napoleon's regime, for example, a whole network of roads was laid out, which was specifically designed to intensify the interaction between Paris and the rest of the country. Thus, Paris' situation proved time and again to be its most valuable asset, its guarantee of continued primacy.

When Situation Changes

The situational qualities of a city can improve and change for the better, but they can also worsen. Changes in transport systems, resource production, agricultural productivity, or other regional transformations can put a city at a disadvantage. We can see such effects in the formerly busy commercial center of a town that has suddenly been bypassed with the opening of a new expressway. Restaurants, gasoline stations, and other commercial establishments feel the effects as connectivity patterns are changed and traffic declines. That same superhighway may improve the situation qualities of other towns that it does connect.

Historically, the Industrial Revolution had similarly varied impacts on the cities and towns of Europe. New resource needs arose; some urban centers found themselves positioned very near coalfields, their situational assets enhanced. Other towns were more remote from the foci of industrial activity, and they soon lagged. The need for intensified circulation and the construction of railroad networks, paved roads, and additional waterways completely modified the competitive circumstances of cities in regions affected by such economic-geographic change. Europe's experience is already familiar to us from earlier chapters; it is now time to turn our attention to urban development in the United States.

In the United States, the Industrial Revolution occurred almost a century after it did in Europe, but when it finally did cross the North Atlantic in the 1870s, it took hold so successfully and advanced so robustly that only 50 years later America was surpassing Europe as the world's mightiest industrial power. Thus, the far-reaching economic, demographic, and societal changes experienced in Europe's industrializing countries were greatly accelerated in the United States, fueled further by the arrival of more than 25 million European immigrants who were overwhelmingly concentrated in the major manufacturing centers. The impact of industrial urbanization oc-

curred simultaneously at two levels of generalization. At the national level, or macroscale, a network or system of new cities rapidly emerged that specialized in the collection, processing, and distribution of raw materials and manufactured goods and that were linked together by an ever-more efficient web of long-distance and local transport routes. Within that urban system, at the microscale, individual cities prospered in their new roles as manufacturing centers, generating a wholly new internal structure that still forms the spatial framework of most of the central cities of America's large metropolitan areas. We now examine the burgeoning urban trend at both of these levels, beginning with the macroscale of *interurban* geography: the study of individual cities as points in a network that interact with each other and serve hinterlands that are commensurate in territorial extent with the size and functional diversity of each city.

• The Network of Cities •

The rise of the national urban system in the late nineteenth century was based on the traditional external role of cities: providing goods and services for their hinterlands in return for raw materials. This function was present since Colonial times, with preindustrial North American cities in the early nineteenth century (quite small in size and spatial extent) located at the most accessible points on the existing transportation network so that movement costs could be minimized. Since handicrafts and commercial activities were already concentrated in cities, the emerging industrialization movement naturally gravitated toward them. These urban centers, of course, also contained concentrations of labor and capital, provided a large market for finished goods, and possessed favorable transport and communication connections. Furthermore, they could readily absorb the horde of newcomers who would cluster by the thousands around new factories built within rail corridors in, and just outside, these compact cities. Their constantly growing incomes, in turn, permitted the newly industrialized cities to invest in a bigger local infrastructure of private and public services as well as housing, thereby converting each round of industrial expansion into a new stage of urban development. Moreover, this whole process unfolded so quickly that planning was impossible; almost literally, America awoke one morning near the turn of the twentieth century to discover it had built a number of large cities. This rise of the national urban system, unintended though it may have been, was a necessary byproduct of industrialization, without which rapid U.S. economic development could not have taken place. A far-flung hierarchy of cities and towns now blanketed the whole continent and came to serve its local populations with the conveniences of modern life.

EVOLUTION OF THE AMERICAN URBAN SYSTEM

Even though it first emerged during the Industrial Revolution period from 1870 to 1910, the U.S. urban system was in the process of formation for several decades preceding the Civil War. The evolutionary framework of the system from 1790 to 1970 is best summarized within the four-stage model developed by John Borchert, which he based on key changes in transportation technology and industrial energy. The preindustrial *Sail-Wagon Epoch* (1790–1830) was the first stage, marked by slow and primitive overland and waterway movements. The leading cities were such northeastern ports as Boston, New York, and Philadelphia—none had yet emerged as the primate city— which were at least as heavily oriented to the European overseas trade as they were to their still rather inaccessible western hinterlands (although the Erie Canal was opened as this epoch came to a close). Next came the *Iron Horse Epoch* (1830–70), dominated by the arrival and diffusion of the steam-powered railroad, which steadily expanded its network from east to west until the first transcontinental line was completed as the epoch ended. Accordingly, a nationwide transport system had been forged, coal-mining centers boomed (to keep locomotives running), and—aided by the easier and cheaper movement of raw materials—small-scale urban manufacturing began to spread outward from its New England hearth, where the factory-system innovation had been transplanted from Britain in the early 1800s. The national urban system started to take shape as New York

advanced to become the primate city by 1850, and the next level in the hierarchy was increasingly occupied by such booming new industrial centers as Pittsburgh, Detroit, and Chicago. This economic-urban development process crystallized during the third stage—the *Steel-Rail Epoch* (1870–1920)—which coincided with the American Industrial Revolution. Among the massive forces now shaping the growth and full establishment of the national metropolitan system were the rise and swift dominance of the all-important steel industry along the Chicago–Detroit–Pittsburgh axis (as well as its coal and iron ore supply areas in the northern Appalachians and the Lake Superior district, respectively); the increasing scale of manufacturing that necessitated greater agglomeration in the most favored raw material and market locations for industry; and the steel-related improvements of the railroads—much more durable tracks of steel (which replaced iron), more powerful steam locomotives, heavier and larger (also refrigerated) freightcars—which permitted significantly higher speeds, longer hauls of bulk commodities, and the more effective linking of hitherto distant rail nodes. The *Auto-Air-Amenity Epoch* (1920–70) comprised the final stage of American industrial urbanization and maturation of the national urban hierarchy. The key innovation was the gasoline-powered internal combustion engine, which underwrote ever-greater automobile- and truck-based regional and metropolitan dispersal. Furthermore, as technological advances in manufacturing spawned the increasing automation of blue-collar jobs, the U.S. labor force steadily shifted toward a new emphasis on white-collar personal and professional services to manage the industrial economy: a productive activity that responded less to traditional cost- and distance-based location forces and more strongly to the amenities (pleasant environments) available in suburbia.

SPATIAL STRUCTURE OF THE URBAN SYSTEM

In our previous discussion of settlements, and particularly the U.S. urban system, we have used the terms *hamlet, village, town,* and *city* in such a way as to imply a hierarchy, a ranking of population concentrations according to certain criteria we did not specifically define. The term *hierarchy* originally had a religious connotation (*hierarch* means religious leader) to denote the organization of a ruling body of clergy into stratified ranks. Today we use it to identify an order or gradation of phenomena, each grade being superior to the one below it and subordinate to the one above. Thus a hierarchy is analogous to the steps of a ladder. The question is: Can criteria really be employed to confirm the hierarchy implied by such terms as *town* and *city* or is the continuity too tight to grant these words any meaningful utility?

Obviously, population size alone will get us nowhere. If we plot all the settlements in the United States on a piece of graph paper, ranked from the largest to the smallest, there would be a near-continuous string of sizes, with real breaks showing up only in the largest categories. It would not be possible to discern a gap between "villages" and "towns." However, we could look at those settlements in another way—not in terms of size, but in terms of the functions they perform and make available to their respective surrounding service areas.

Thus, a **hamlet**, usually a settlement of less than one hundred people, contains the smallest number of services, probably less than 10. There might be a general store, a gas station—perhaps a coffee shop. However, you would not find a post office, a church, or a grocery—for those more specialized services you would have to go to a *village*. In a village in, say, the Wisconsin dairy region, you could expect perhaps 60 to 70 establishments, including a couple of gas stations, a restaurant, one or two bars, a farm elevator, a doctor, a grocery or two. Obviously, such a village would draw customers from a much wider area than the hamlet; in fact, people from the nearby hamlets have to come to the village for many services. Again, a *town* of about twenty-five hundred inhabitants has a still greater variety of functions and services. Here we find a greater degree of specialization, and people would come for a particular kind of doctor or dentist, a bank, and stores selling such goods as hardware, furniture, and appliances. Certain of the town's services are not available in either villages or hamlets; so, towns have wider trade areas. Thus, rather than counting heads, we might define a town as a place where a certain assemblage of goods and services is available, with a hinterland that includes the service areas of surrounding hamlets and villages.

This approach can be extended to include cities, metropolises, and still larger conurbations. The advantage of this classification system is that we do not rank settlements simply according to their population sizes but rather as a reflection of their strength as places of trade and commerce. This approach to the determination of a hierarchy of urban places takes into account something that mere population numbers do not reveal—the interaction between the urban center and its trade area. We might also look at the whole problem from the opposite perspective: instead of measuring the availability of goods and services in the urban centers themselves, we could identify the surrounding service areas and thereby determine the *economic reach* of each settlement. A ranking can then be based on this economic reach, which generates a measure of **centrality**.

Centrality is a situational property that is crucial to the development and persistence of urban places and their concomitant service areas. To these hinterlands, towns and cities function as *central places*; even a village is a central place to its small tributary area. How, then, do all these service areas relate to each other? Do they overlap? Do towns of approximately the same size lie about the same distance away from each other? What rules govern the arrangement of urban places on the landscape? These are critical questions if we are to understand the structuring of urban hierarchies. Geographers have discovered many answers, and this inquiry constitutes some of the most significant theoretical work in urban geography to date. Let us take a closer look.

CENTRAL PLACE THEORY

In Chapter 1 we identified several geographers who produced early, significant theoretical work. One of these geographers was Walter Christaller. In 1933 Christaller published a book entitled *The Central Places of Southern Germany*. In this very important volume, which was not translated into English until the 1950s, Christaller laid the groundwork for **central place theory**. Addressing himself to questions such as those we have just raised, he attempted to develop a model that would show how and where central places in the hierarchy (hamlets, villages, towns, and cities) would be functionally and spatially distributed with

This small South Dakota town has a comparatively low number of services, but it still functions as a central place to a substantial area. Many more small towns such as this exist than larger cities—a phenomenon Christaller sought to explain through central-place theory.

respect to one another. In his effort to discover the laws that govern this distribution, Christaller began with a set of simplifying assumptions. The surface of the ideal, laboratorylike region would be flat and without physical barriers. Soils would have equal, unvarying fertility. He also assumed an even distribution of population and purchasing power and a uniform transport network that permitted direct travel from each settlement to the other. Finally, Christaller assumed that a constant maximum distance, or *range*, for the sale of any good or service produced in a town would prevail in all directions from that urban center.

Christaller's idea was to calculate the nature of the central place system that would develop under such idealized circumstances and then compare that model to real-world situations, explaining variations and exceptions. Some places, he realized, would be more central than others. The central functions of larger towns would cover regions within which sev-

eral smaller places with lesser central functions and service areas are nested. What was needed, Christaller reasoned, was a means to calculate the degree of centrality of various places. In order to do this, Christaller identified *central goods and services* as those provided only at a central place. These would be the goods and services a central place would make available to its consumers in a surrounding region—as opposed to services that might be available anywhere (without local focus) and that are unlike those produced for distant, even foreign, markets and, therefore, of no relevance to the local consumers. Next came the question of the *range of sale* of such central goods and services: the distance people would be willing to travel to acquire them. The limit would lie halfway between one central place and the next place where the same product was sold at the same price—because under the assumptions Christaller used, a person would not be expected to travel 11 miles (17 ½

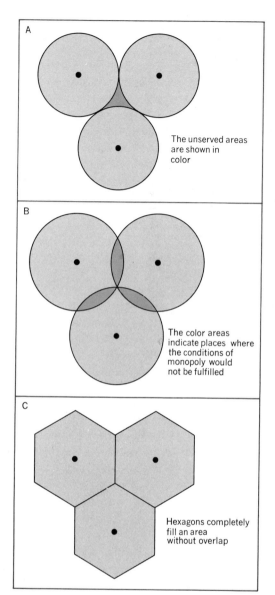

A

The unserved areas are shown in color

B

The color areas indicate places where the conditions of monopoly would not be fulfilled

C

Hexagons completely fill an area without overlap

FIGURE 9.5

Hexagonal and circular service areas around central places.

km) to one place to buy an item if it were possible to go only 9 miles (14½ km) to purchase it at another.

Each central place, therefore, has a surrounding *complementary region* within which the town has a monopoly on the sale of certain goods because it alone can provide such goods at a given price and within a certain range of travel. From what we have

just said, it would seem that such complementary regions would be circular in shape. However, when we construct the model on that basis, problems arise: either the circles adjoin and leave unserved areas, or they overlap, and when they do there is no longer a condition of monopoly. These two problems, and their resolution—a system of perfectly fitted hexagonal regions—is sketched in Figure 9.5.

Now Christaller proceeded to assemble his model for the spatial arrangement of central place systems. To illustrate the result, let us take a specific case. Assume that a certain central place provides 100 different kinds of goods and services. We rank these in order of importance from 1 to 100. Thus, the product ranked number 1 will draw consumers from the widest area and exhibit the largest range and furthest limit, whereas that ranked number 100 has the least importance and the smallest service area. If we now placed five central places, each producing these 100 central goods and services (Christaller called these **G** places), 60 miles away from each other, the following would happen. The largest hexagonal region (outlined in red on Fig. 9.6) surrounding the centrally positioned city marks the upper limit of the range of the product ranked number 1. The product ranked number 2 has an outer limit marked by a circle just inside the hexagon; number 3 (not shown) has a limit again slightly smaller than that of number 2, and so forth. Now, as we continue to draw successively smaller circles for each of the 100 products, we begin to leave an area outside of, say, circle 20 where there would be no service (in theory, of course). By the time we draw the circle for the item ranked number 30, this unserviced area is large enough to accommodate its own hexagon and central place; so, a new network of places and hexagons (the black-lined set surround the towns at the corners of the city-oriented light red hexagon in Figure 9.6) is born. Christaller called these secondary centers **B** places. Now both **G** places (our cities) and **B** places (our towns) provide goods ranked 31 and lower, and as we proceed and reach circles for products ranked somewhere between 55 and 60, we find that the situation we just described repeats itself, and once more there is room for a system of central places of still lower order, in this case, the red-lined hexagonal system centered on the villages, called **K** places by Christaller. In Figure 9.6,

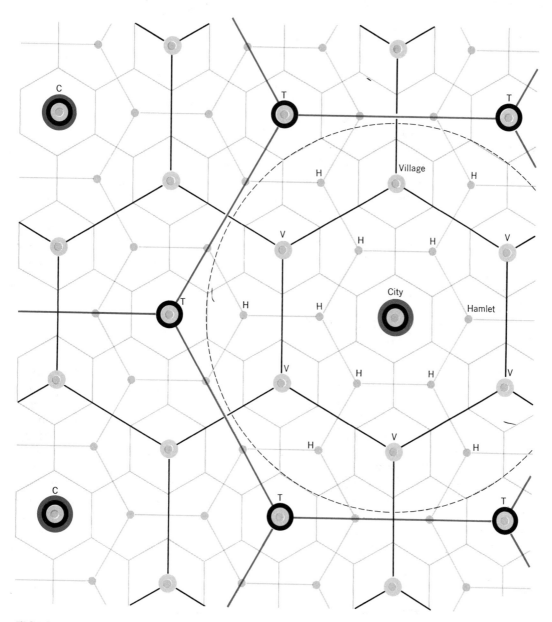

FIGURE 9.6

Christaller's model of a central-place hierarchy. The cites (C) represent Christaller's G places; the towns (T) are what he called B places; the villages (V) are his K places.

the ultimate projection of the model also includes hamlets, so that a fourth set of complementary regions (the gray hexagonal lattice) emerges.

The Real World

Any casual inspection of an actual landscape, however, demonstrates the purely theoretical nature of Christaller's proposed hexagonal distribution of central places. Some regions appear to approximate this pattern better than others, but there are several critical explanations for empirical (real-world) departures from theory. In part, towns are not evenly distributed because resources (environmental and human) are not evenly distributed. Also, social scientists have realized that many people do not behave in such a way that their costs are minimized. Finally, as Christaller himself observed, urban places are not all the same size. Most regions, therefore, contain a varied array of small, medium, and large urban places in complex interrelation with each other.

From the beginning, Christaller was aware that many factors in the real world would operate to distort his model. His book brought on a storm of controversy, and the debate still continues even after numerous modifications of the system. Immediately, there were geographers who saw hexagonal systems everywhere; others saw none at all. Soon there were attempts to alter the Christaller model and to relate it to various areas of the world. Christaller himself joined the debate; in 1950, he published a new article entitled "The Foundations of Spatial Organization in Europe." In it, he insisted that he had been correct all along, writing, "When we connect the metropolitan areas with each other through lines, and draw such a network of systems of the map of Europe, it indeed becomes eminently clear how the metropolitan areas everywhere lie in hexagonal arrangements."

Christaller received support from the research of geographers who applied his ideas empirically to regions in Europe, America, and elsewhere. In China, the North China Plain south of Beijing has the sort of uninterrupted flatness that comes close to Christaller's assumption of territorial homogeneity. Over centuries of human settlement, China's rural population has become distributed into nearly 1 million farm villages, many of them on the North China Plain. These villages each have populations of several hundred, and they are surrounded by their own farm fields. Groups of these villages (averaging about 18) lie in the hinterlands of large market towns. Each village lies within walking distance (about 4 miles [6–7 km]) of the closest market town. For centuries, the market towns (Christaller's **K** places) were the crucial places where peasant and merchant met and goods and ideas were exchanged. Here the peasant would sell the farm's produce, buy goods and services, and talk to people from other villages. The activity space (territorial range) of the Chinese peasant has always been very limited, but the market town was part of the network of still larger market centers—including, at the top of the urban hierarchy, China's biggest cities. In the market towns, the peasants were exposed to the Chinese world beyond their own villages, so that these places served as more than markets for the exchange of goods—they also functioned to integrate and control the Chinese nation.

The Chinese hierarchical network of central places in particular areas strongly resembles elements of the Christaller model, as G. William Skinner described in a series of articles entitled "Marketing and Social Structure in Rural China" (1964–65), based on empirical studies in a region in Sichuan Province similar to the North China Plain. However, even on the North China Plain there are distortions; for example, villages tend to cluster on higher ground and away from streams in order to reduce the flood danger. Still, the pattern of the Chinese urban system confirms that Christaller's concepts, based as they were on observations in a totally different geographic realm, have universal application.

Christaller may have been overly optimistic about the validity of his model, but he set the pattern for future research on central place structures. Such research has focused on the nature and impact of city size on the model (since a large city dominates smaller urban places, even for the same product or service), the effects of physical barriers and cultural differences, the cumulative agglomerative effect of time, the failure of people to make rational decisions (some people *will* travel those 11 miles instead of 9), and other issues. After all, the hexagons need not be perfectly regular; they could be distorted, having some shorter and some longer sides, and maybe the distorted model comes closer to reality than the regular one. Changes in transportation, consumers' ability to travel ever-longer distances, advertising and image making for places and products, and changing demands of the marketplace have loosened the bonds between central places and their immediate hinterlands. Earlier we referred to the impact that a freeway bypass can have on the service center whose main street formerly carried all that traffic; the town declined. However, competition can also stifle a small settlement and render it virtually dormant. On the other hand, a small town may find itself increasingly in the orbit of a nearby large city, and may become a new satellite—resulting in an opposite kind of change.

Obviously, then, Christaller's model should be viewed as a beginning—but it is an important one. If you look at the periodical literature in geography, you will see a number of articles published since the 1950s in which geographers have tried to unravel the complexities that Christaller, through his assumptions, circumvented. In the process, our understanding of the forces that influence the distribution of urban places and their growth and decline is strengthened.

• The Internal Spatial Structure • of Cities

Cities are not simply disorganized, random accumulations of buildings and people. Instead, they have functional structure—they are spatially organized to perform their functions as places of commerce, production, education, residence, and much more. Just as Christaller raised questions about the spacing of cities and towns, we could prepare a model of the internal layout of the "ideal" city. How and where are the various sectors of the city positioned with respect to each other? If there are forces that govern the distribution of central places on the landscape, surely there are forces that affect the way cities are internally

organized? It is not difficult to think of one of these forces: the price of land. This tends to be highest in the central city, and then declines irregularly outward. So you would not look for a spacious residential area near the heart of a city nor for the downtown area on a city's periphery!

Just by using the terms, *residential area* and, *downtown*, we reveal our awareness of the existence of a regional structure in cities. Yes, cities have regions. When you refer to downtown, or to the suburbs, or the municipal zoo, you are in fact referring to urban regions where certain functions are predominantly performed (retailing, residing, recreation, respectively, in the cases just mentioned). All these urban regions, or zones, of course, lie adjacent to one another and together make up the total city. But how are they arranged? Is there any regularity, any recurrent pattern to the alignment of the various zones of the city that perhaps reflect certain prevailing processes of urban growth? In other words, do the city's regions constitute the elements of an urban structure that can be recognized to exist in every urban center, perhaps with modifications related to such features as the city's particular site, its size, functional class, and so forth?

CONCENTRIC ZONES

One way to go about solving this problem is to study the layout of a large number of cities, to compare the resulting maps, and to determine recurrent features. In the most general terms, it would soon become clear that cities have central zones, consisting mainly of the central business district, and outer zones, where sprawling suburbs and shopping centers lie. Between the central and outer zones lies the middle zone, an ill-defined, often rather mixed and disorganized area, where change is frequently evident—for example, in the deterioration of housing and the development of slums.

This suggests that metropolises have a certain concentricity in their spatial arrangement; this impression was first formalized by sociologist Ernest Burgess in the 1920s when he studied the layout of Chicago. Although downtown Chicago lies near Lake Michigan, and the whole urban area is, therefore, semicir-

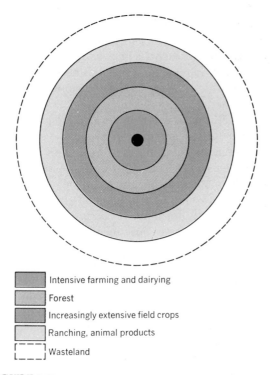

Intensive farming and dairying

Forest

Increasingly extensive field crops

Ranching, animal products

Wasteland

FIGURE 9.7

The Burgess concentric model of urban spatial structure.

cular in shape, Burgess, nonetheless, suggested that his *concentric zone model* applied here as well as in most American cities. Burgess recognized five concentric urban zones in his model rather than three, counting the central business district as the innermost zone (region 1, Fig. 9.7). This, the heart of the city, contains shops, offices, banks, government buildings, and hotels; the larger the city, the more likely is the CBD to be divided into several subdistricts of specialization such as a theater district, a financial district, and a government center. The CBD is a zone marked by high land values and by maximization of floor space through the construction of skyscrapers. Transport lines converge on this central zone, disgorging thousands of workers into streets that bustle with traffic during peak hours but may be nearly deserted by evening after the business day ends.

Surrounding the CBD, Burgess pointed out, lay a transitional area, a zone of residential deterioration (later the scene of postwar urban renewal programs) also marked by the encroachment of business and light manufacturing. This is region 2 (Fig. 9.7), the inner-city zone of persistent urban blight today, of tenements and slums, inadequate services, and deepening poverty and dislocation. Still farther out from the CBD lay what Burgess called the "zone of workingmen's homes" (region 3), a ring of closely built, but adequate, residences of the urban blue-collar labor force. The next zone consisted of middle-class residences (4), outer-central city/inner suburban areas, characterized by greater affluence and spaciousness. At this distance from the CBD, local business districts (early shopping centers) made their appearance. Finally, we reach what was in Burgess's time the "urban fringe," the commuters' zone (5) that consisted of communities that were, in effect, dormitories for the CBD, where most of the economically active residents went to work. Here lie some of the urban area's highest-priced residential areas. Here, too, urban and rural areas intermingled along the leading edge of the advancing city, and small rural towns were often engulfed by the outward pulsations of advancing urbanization.

Burgess also suggested that the zones he identified are not static, but are mobile and encroach on each other. As a city grows, he argued, the whole system expands or "pops" outward, so that CBD functions invade the adjacent portion of region 2 and the characteristics of zone 2 begin to affect the inner edge of zone 3, and so forth. Zone 5, therefore, will lie ever farther from the center of the city as the inner zones pop out. This certainly is something we can see occurring in many cities—but the Burgess model also had its shortcomings when put to the test against reality. It did not account adequately for heavier industries within cities, which formed not a ring around the city center, but several concentrations in certain transport-related areas of the metropolis. Moreover, a number of urban facilities cannot move outward, as the Burgess popout concept would require: railroad yards, port complexes, factory clusters, and other massive capital-investment facilities tend to remain entrenched. In fact, as Chicago itself has demonstrated, this is true even of certain residential areas, especially when they possess strong ethnic linkages. On the other hand, the experiences with airports tended to confirm Burgess' idea. Chicago's Midway Airport was an edge-of-town field when he conducted his research, but it grew into one of the nation's busiest by 1940. Then urban sprawl overtook Midway, thus more distant O'Hare Airport was built in the 1950s. Next time you fly into Chicago, note what has happened to all that erstwhile open space around O'Hare! Some decisions will soon have to be made as the suburban frontier continues to advance: should there be a new airport still farther away, an expansion of existing facilities, or perhaps an airport out in Lake Michigan on an artificial island?

SECTORS

A more specific objection to the Burgess concept casts doubt on its central theme: its concentric framework. Do cities really have concentric-zone structures? When you travel into a city via a major roadway, you will note that certain functions tend to cluster along that major artery. Rarely do you see expensive homes along truck-congested, four-lane highways, rather there are factories, auto dealerships, and the like. This means that arterial routes interrupt the Burgess concentricity scheme—so much so that it is possible to argue that the major structural features of a central city are sectoral or wedge-shaped rather than concentric. This is precisely what economist Homer Hoyt concluded in his 1939 study of urban layout, *The Structure and Growth of Residential Neighborhoods in American Cities*. He proposed the **sector model** of urban structure, based on data specifically related to residential land use. Hoyt showed that much could be learned from an analysis of one aspect of urban life: the rent that people in residential areas pay. Following Burgess's idea, we would conclude that rent consistently increases with distance from the CBD: the quality of housing is lowest in zone 2 and highest in zones 4 and 5. Hoyt, however, argued that a low-rent area could extend all the way from the center of a city to its outer edge. Similarly, high-rent sectors could be recognized (in cities he studied) that were flanked by intermediate-rent areas. For example, a high-rent sector might extend along a very fast, convenient, uncluttered route into the city center (rapid transit or highway), but where communications were somewhat less

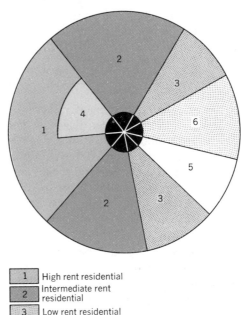

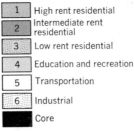

1	High rent residential
2	Intermediate rent residential
3	Low rent residential
4	Education and recreation
5	Transportation
6	Industrial
■	Core

FIGURE 9.8

The sector model of urban spatial structure derived by Hoyt.

361

efficient, this would be reflected by the local dominance of intermediate-quality residential areas.

When we view the overall spatial structure of any central city we know, we can probably discern elements of a pie-shaped arrangement of urban activities (Fig. 9.8). Industrial plants may be concentrated along one or several radial arteries that connect the city to raw material supplies and hinterland markets. A low-rent residential sector may adjoin this industrial complex; a high-quality residential sector may fan outward from the central city in another part of town. Major intercity transport lines may be bunched so closely together that a transportation sector can be recognized. A major university and several schools can also form a discrete sector, as our map suggests. Furthermore, as is so often the case when conflicting interpretations exist, it is often possible to discern elements of both concentricity and sectorial development in cities we know well.

SEPARATE NUCLEI

In 1945, urban geographers Chauncy Harris and Edward Ullman, having concluded that both the concentric and sector models left too many unresolved problems, built a model in which the metropolis was seen as consisting of a number of discrete functional areas that each formed and revolved around their own separate nuclei. Once again, the idea arose from something we can see every day as we live in or visit cities: the CBD is losing its dominant position as the undisputed nucleus of the metropolitan city, and it is getting competition from other, vigorously growing business centers that are, in effect, secondary nuclei. Thus, the **multiple nuclei model** suggested that an urban area may grow and develop not around just one functional focus, but several. These multiple foci include not only the CBD itself and subsidiary business centers, but any other land use activity—a university, a port, or a government complex—that can stimulate urban growth and attract other related functions. In their diagram (Fig. 9.9), Harris and Ullman showed schematically what such a city would look like; note that there is none of the regularity of the Burgess and Hoyt models. A city that comes to mind immediately in this context is Los Angeles, but nucleated development of this sort is occurring in most of the world's larger cities today.

362

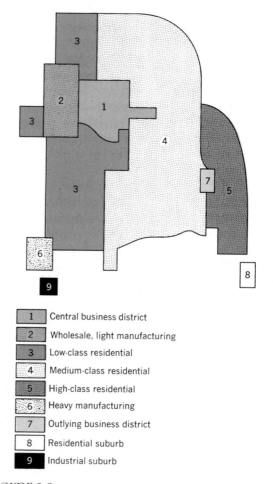

1	Central business district
2	Wholesale, light manufacturing
3	Low-class residential
4	Medium-class residential
5	High-class residential
6	Heavy manufacturing
7	Outlying business district
8	Residential suburb
9	Industrial suburb

FIGURE 9.9

The Harris–Ullman multiple nuclei model of urban spatial structure.

URBAN REALMS

The multiple nuclei approach, however, fails to account for all the spatial-structural complexities of the contemporary American metropolis; although it accurately reflects the decentralization and nucleation of certain urban functions, much has changed since 1945. In the early postwar period, as James Vance documented in his study of the San Francisco Bay Area, rapid population dispersal to the outer suburbs not only created distant nuclei, but also reduced the volume and level of interaction between the central city and these emerging suburban cities. This strengthened the self-sufficiency of the new *outer cities* of the suburban ring, where locational advantages produced an ever-greater range of retailing and employment activity. By the 1970s, outer cities became increasingly independent of the CBD to which these former suburbs had once been closely tied, and they began to duplicate—and even overtake—certain high-order functions of the central city. The metropolis of the 1980s, therefore, consists of a set of *urban realms*, each of separate and distinct economic, social, and political significance and strength. In his later book, *This Scene of Man* (1977), Vance developed the urban realms model further, suggesting that the internal structure of each of the suburban realms as well as the central city is influenced by four factors: the terrain; the dimensions of the metropolis of which all are parts; the range, volume, and type of economic activity in each realm; and the transport networks that create accessibility and interconnect the realms.

An Application

A good idea of the utility of the urban realms model (schematically illustrated in Fig. 9.10) can be seen in its application to metropolitan Los Angeles, where the realm structure is readily apparent. Five discrete urban realms have emerged around the central city (Fig. 9.11), creating a suburban ring that extends as far as 50 miles (80 km) from the CBD. Clockwise from the

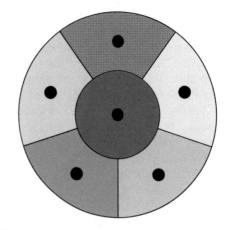

FIGURE 9.10
The urban realms model.

363

FIGURE 9.11

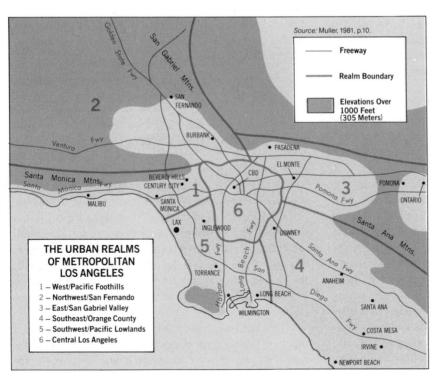

THE URBAN REALMS
OF METROPOLITAN
LOS ANGELES

1 – West/Pacific Foothills
2 – Northwest/San Fernando
3 – East/San Gabriel Valley
4 – Southeast/Orange County
5 – Southwest/Pacific Lowlands
6 – Central Los Angeles

Source: Muller, 1981, p.10.

— Freeway
— Realm Boundary
— Elevations Over 1000 Feet (305 Meters)

west these are: (1) the *West Realm*, typified by Santa Monica, Beverly Hills, and the coast-facing ribbon extending toward Santa Barbara along the Santa Monica Freeway; (2) the *Northwest Realm*, essentially the San Fernando Valley between the Santa Monica and the San Gabriel Mountains; (3) the *East Realm*, which follows the San Gabriel Valley eastward to the desert margin; (4) the *Southeast Realm*, Orange County, one of America's fastest growing urban areas, centered on Costa Mesa and Anaheim; and (5) the *Southwest Realm*, straddling the San Diego Freeway from near Inglewood in the north to Long Beach in the south, and dominated economically by Los Angeles International Airport (LAX) and aerospace activities. Approximately in the middle lies the sixth urban realm of this vast metropolitan complex, *Central Los Angeles*, at the hub of the freeway network, but serving more as a crossroads than as a regional core. The CBD itself does contain a cluster of high-rise commercial buildings and a group of cultural and sports facilities, but it is overshadowed in many categories of economic power by just one of the five surrounding realms, Orange County.

The Los Angeles application could be replicated in any other large American metropolis today. The phenomenon of outer-city development is reflected even on an ordinary road-atlas map in the rings of beltways or circumferential freeways that encircle the central city, often connecting suburban areas more effectively with each other than with the inner city. In effect, it was this expressway system that destroyed the regionwide centrality advantage of the core city's CBD, making most places on the urban freeway network just as accessible to the rest of the metropolis as only downtown had been before the 1970s. Industrial and commercial employers quickly realized that most of the advantages of being located in the CBD were now eliminated. Several companies in the office-based services sector chose to remain and even enlarge their presence downtown, but many others, together with myriad firms in every other sector, chose to respond to the new geographic reality by voting with their feet—or rubber tires—and headed for outlying sites. As early as 1973, America's suburbs surpassed the central cities in total employment; by the mid-1980s, even some major metropolises in the Sun-

belt were experiencing the suburbanization of a critical mass of jobs (greater than 50 percent of the urban-area total).

As the outer city grew rapidly and its functional independence heightened, new suburban downtowns were spawned after 1970 to serve their new local economies, the leading concentrations located near key freeway intersections. These multipurpose activity nodes mainly developed around big regional shopping centers with prestigious images. They attracted scores of industrial parks, office companies, hotels, restaurants, entertainment facilities, and even major-league sports stadiums. As these new downtowns of the outer city flourish in the 1980s, they are attracting tens of thousands of nearby suburbanites to organize their lives around them—offering workplaces, shopping, leisure activities, and all of the other elements of a complete urban environment—thereby not only loosening remaining ties to the central city, but to other portions of suburbia as well.

Thus, the urban realms' model constitutes the latest step forward in the geographic interpretation of urban structure. It clearly demonstrates that today's outer cities are not satellites of the central city—they have become mighty, coequal partners in the shaping of the decidedly *polycentric* metropolis of the late twentieth century.

• The Evolving American City •

In many respects the city is a microcosm of society. In the city are revealed a culture's achievements, aspirations, successes, and failures. The city is a museum, a manufacturer, a market, a cross-section of national attributes. A crucible of social and economic currents, the city draws the privileged and the disadvantaged, the skilled and the untrained. It is, as noted earlier, a crossroads of culture—a place of ferment, interaction, change, and energy.

The contemporary metropolis, however, is also a place of harsh and deepening contrasts. The tenements of the inner city and the shining new com-

364

plexes of the outer suburban city are worlds apart. Whereas some people dine in exclusive restaurants, others cannot afford a balanced meal. Successful businesspeople travel to the swank offices of suburban downtowns and new CBD towers, riding in air-conditioned luxury automobiles while other workers by the thousands crowd onto the gritty vehicles of deteriorating public transporation systems. New arrivals come with hope, often without adequate resources; others leave, in disappointment and failure or attracted by better opportunities elsewhere. Always there are the examples of those who have succeeded in the city: the executive who has risen through the ranks, the worker whose long service resulted in promotions to the top, the family that was able to leave the wrong side of the tracks. Perhaps more than anything else, the city still symbolizes opportunity.

But the position of the central city within today's polycentric metropolis of realms is an eroding one. No longer the dominant metropolitanwide center for urban goods and services, the CBD is being reduced to serving the less affluent residents of the innermost realm and those working there. As core-area manufacturing employment has declined precipitously during the last two decades, many large cities adapted successfully by promoting a shift toward the growing service industries. Beyond the CBD zones, the vast inner city remains the problem-ridden domain of low- and moderate-income people, most forced to reside in ghettos. Financially ailing big-city governments are unable to fund adequate schools, crime-prevention programs, public housing, and sufficient social services, and the downward spiral, including abandonment in the old industrial cities, continues unabated in the late 1980s.

THE TROUBLED INNER CITY

The intrametropolitan deconcentration process that has been underway for several decades has been accompanied by worsening problems in the inner central city. In the older industrial cities, the inner city has become a landscape of inadequate housing, substandard living, and widespread decay. New York City typifies the situation: here nearly 3 million persons (plus another million or so illegal aliens who are not officially tabulated in the census), mainly representing minorities, are crowded into apartment buildings, averaging five stories, that were built as walkups 75 to 100 years ago. Most of these buildings are now simply too old, worn out, and unsanitary; many suffer from rodent infestations. Still, these apartments are overfilled with people who must live with this discomfort and who cannot escape the vicious cycle that put them there.

Yet there is something of value that remains in many of the aging, stressed neighborhoods of New York and other Manufacturing Belt cities. In the squalid poverty of ghetto and slum life, there persists a sense of neighborhood, social structure, and continuity. Urban renewal in the form of anonymous high-rises and the relocation of the people may so disrupt the sense of community that the liabilities ultimately outweigh the assets. Drug abuse, crime, vandalism, and other social pathologies afflict newer areas of the inner city as well as the old, and frustration builds to unleash the kind of furor that devastated large areas during nationwide urban rioting in the 1960s. The burgeoning outer cities of the suburban ring may reveal what ultramodern urbanization can be, but the residential areas of the inner city languish in the wake of metropolitan decentralization.

PROSPECTS FOR DOWNTOWN

Spend a night in the downtown area of an American city and compare your experience with a similar one in Europe—in Paris, Amsterdam, or Vienna. In those European cities, many streets are busy late at night, lights are on in windows above shopping streets, in apartment buildings. In the American CBD, the streets are quiet and dark. Except for some limited activity in theater districts and in such tourist areas such as New Orleans's French Quarter and San Francisco's waterfront, the streets are dormant, the buildings dark. Why the contrast? One great difference lies in the continuing residential function of the downtown area in European cities. The commercial skyscraper has made its appearance, and certain European cities are taking on American aspects architecturally, but where the Second World War did not destroy the inner city, old patterns of mixed retailing and residential living still prevail. Thus, downtown stays alive, and its continuity is reflected by old establishments, always dependent on local, neighborhood patronage that still survives. What is happening in the American city may not, therefore, by symptomatic of Western cities in general, but a regional phenomenon born of the special circumstances of American society. By building exclusive, expensive apartment towers near the city center, attempts have been made for over a decade to lure better-off residents back into the downtown area. However, the rising prices of property and spiralling tax assessments have driven the bulk of those less affluent people who might have remained—the heterogeneous population base the heart of the city needs—away. So, the CBD zone lies surrounded by the sickest sector of the metropolis, whose diseases spill over into the unsafe deserted concrete canyons.

These realities notwithstanding, the downtown area does contain many of the central city's crucial assets. If many establishments have joined the move to the outer cities, certain others have not. Great museums, research libraries, world-renowned orchestras, leading universities, attractive recreational facilities, and other amenities still grace the centers of American cities. The crowds that fill the sidewalks each business day and the automobiles and other vehicles that create traffic jams give evidence of the continued vitality of this, the original heart of the industrial metropolis. The concept of the skyline index tests the vertical development of the downtown area, not the generally low-rise suburban scene. If the downtown areas of St. Louis, Detroit, Cleveland, and other northern and northeastern cities are troubled by the events of the postwar period, there are other (mostly Sunbelt) downtowns—Houston, San Francisco, San Diego—that display many signs of well-being.

Although the federal government has lately refused to offer meaningful financial aid, the plight of the central cities has been generating private-sector action since the early 1970s—in the form of downtown commercial revitalization. However, the opportunities for CBD expansion at this stage of urban American history are definitely limited and, in too many cities, for each shining new skyscraper that goes up several old commercial buildings are simul-

taneously abandoned. Even in many newer Sunbelt cities such as Denver, Houston, and Los Angeles, whole forests of new CBD office towers contain suburban commuters whose only contact with the city below is the short drive between the freeway exit and their building's parking garage. Widespread attempts at residential revitalization in and around the CBD also occurred in many cities during the 1970s, but the numbers of people involved were far less significant than was first believed. Since most of this reinvestment was undertaken by people already residing in the central city, a hoped-for return-to-the-city movement by supposedly disaffected suburbanites clearly did not take place. Moreover, the limited lasting downtown-area neighborhood redevelopment that did materialize involved *gentrification*—the upgrading of residential areas by new, higher-income settlers; but in order to succeed as we noted earlier, this usually requires the displacement of established lower-income residents, an emotional issue that has sparked many bitter conflicts and cast a pall over continued gentrification efforts in the 1980s.

THE OUTER SURURBAN CITY

The attraction of country life with city amenities, reinforced by the discomforts of living in the heart of many central cities, has for many decades propelled those who could afford it to move to the suburbs and more distant urban fringes. In postwar times, the automobile made mass commuting possible from suburban residences to downtown workplaces, and the kind of suburbanization familiar to North Americans and other Westerners is a characteristic of urbanization in the most mobile of developed societies.

Suburbanization holds special interest for human geographers because it involves the transformation of large areas of land from rural to urban uses, affects large numbers of people who can afford to express their spatial preferences, and rapidly creates distinct urban regions complete with industrial, commercial, and educational as well as residential components. In a sense, the suburbs reveal their occupants' idealized living patterns more accurately than any other urban zone, because their layout can be planned in response to choice and demand; elsewhere in the me-

366

• A Skyline Index •

Cultural geographers often view the city in a different manner than urban geographers. The cityscape, the special atmosphere of ethnic neighborhoods, images and perceptions of the metropolis are cultural geographic topics far removed from the statistical analyses favored by many urban geographers. However, cultural geographers also have ways to classify aspects of cities, and an especially interesting one was devised by Larry Ford (*Journal of Geography*, 1976). This classification was based on a *skyline index score*—a measurement based on the number of buildings exceeding 300 ft (90 m) in height within the city center. Although we associate New York City and Chicago with impressive skylines, there are smaller cities that rank relatively low in population, but that exhibit high rankings according to this index. After New York and Chicago, for example, comes Dallas, ranked number 7 by population in the United States in 1980. Houston's

skyline index score places it as number 4 (5 by population), followed by San Francisco (13), Atlanta (30), Pittsburgh (29), Tulsa (62), New Orleans (21), and Albany (43). Some of the largest cities in the United States have a low rank according to the index: Los Angeles, the second largest U.S. city, ranks 19, Baltimore ranks 23, and Cleveland 29.

The skyline index score reflects a visual impression, but it also has a quantitative base. The downtown skyscraper has long been a symbol of modern urban America, evincing the dominance of office-based commerce and finance in recent decades. It is an appropriate criterion for a cultural-urban classification, although, of course, it is subject to rapid outdating. Several of the Sunbelt cities, from Miami and Tampa to Los Angeles and San Diego, are experiencing building booms that could alter their ranking over a short term.

tropolis, there are too many constraints imposed by preexisting land use arrangements. In *Suburban Growth* (1974), editor J. H. Johnson suggests that in the suburbs, "life and landscape are in much closer adjustment" than in the older parts of the urban area, thus providing geographers with a direct expression of the behavior of contemporary urban society and, therein, some important clues about the nature of the urban future.

A prominent manifestation of suburbanization relates to internal accessibility and movement. Earlier stages of development saw the suburbs emerge at the urban frontier, thereby extending the leading edges of the urbanized area. The great majority of the suburban workers originally commuted to work in, or near, the central city; but today (as we have seen) those inner-oriented suburbs of the early postwar period have evolved into self-sufficient entities, outer cities in their own right and with their own economic base and cultural activity. Thus, movement patterns are becoming redirected, with internal urban-realm travel ever

more important. For work trips especially, the deconcentration of employment has produced a reorientation, so that the dominant intrametropolitan commuting flow is now suburb to suburb; although work travel into the central city is still sizable, *reverse commuting* (city to suburb) has more than doubled in the United States since the early 1970s.

The overall importance of suburban life in the United States is underscored by the results of the 1980 census, which indicated that no less than 45 percent of the entire U.S. population resided in the suburbs (up from 37 percent in 1970); the remaining 55 percent of the American people were divided between the central cities (30 percent) and non-metropolitan or rural areas (25 percent). Of the population living in metropolitan areas, 60 percent resided in the suburbs, which in 1980 counted 102 million inhabitants, whereas the central cities totaled 68 million. Another indication of the importance of suburbia comes from regional growth-rate figures: during the 1970s, the country's suburban population grew by

18.2 percent, more even than rural areas (+ 15.1 percent); the central cities grew by only 0.1 percent (+80,000 on a base of 67.9 million). Thus, the suburbs have unmistakably become the essence of the late-twentieth-century American city.

The attractiveness of these outer cities (as noted previously) includes a combination of urban advantages with suburban spaciousness. To be able to escape from the pressures of the central city for the leafy tranquility and other amenities of the country, not only on weekends but on a daily basis, became an American ideal. The suburbs made that possible, and along the metropolitan periphery a whole new expression of urban life became reality. To be sure, certain suburbs had their shortcomings (as many sociologists pointed out), but many of the same inequities of social-class divisions and residential segregation by income and race had long characterized the population mosaic of the central city too. However, the outer cities also had freshness, vigor, newness, productivity, and originality: they developed free from the central city straitjacket, producing novel cultural landscapes; and they have increasingly functioned as social and economic pacesetters in the 1970s and 1980s. True, these have included the lawnmower-barbecue-snowshovel townscapes decried by some social scientists, but they have also generated ultramodern industrial and office parks with unprecedented attention to workers' comforts and efficiency and to harmonious adjustment to environment. The spacious superregional shopping malls at the heart of many suburban downtowns, architecturally innovative office complexes, and other attributes of the outer city had an impact on the old central cities, where long-overdue redevelopment began—with hopes of luring businesses and residents back to a rejuvenated urban core. However, these efforts have not slowed the centrifugal drift, and the mass move to the outer city continues unabated.

The outward movement of people and establishments that has transformed suburbs and urban fringes has also had a devastating effect on millions of workers who are employed in the inner zones of the central cities. Not everyone has an equal opportunity to join the great outward march. An employee whose firm moves from the central city to a distant suburb

(and whose inner-city apartment may have been a short bus ride away) may find housing in the outer city to be much less affordable and often unavailable. Furthermore, modern technology continues to eliminate jobs: when a central-city plant closes and relocates 20 miles away in a suburban industrial park, it is the skilled white-collar worker who benefits. The poorer, unskilled worker cannot follow, cannot afford to live in the suburbs even if his or her job survived the transfer. The situation is not always this stark, of course, but decentralization and the emergence of the new intrametropolitan structure has had a powerfully negative impact on job opportunities for workers residing in the inner cities. It is a major element in our growing urban dilemma, which suffers from other worsening problems as well.

PERSISTENT URBAN PROBLEMS

The difficulties of the CBD and adjacent central-city areas (discussed in the previous section) represent but part of the enormous dilemma facing urban governments as they struggle to halt deterioration of the landscape and the living and working conditions it represents. Four prominent problems beset the American city: the nature of urban administration, declining revenues, stressed environments, and congested internal circulation.

As we have noted, urbanized regions such as the San Francisco Bay Area, Greater New York, metropolitan South Florida, and other sprawling urban complexes consist of not one but a myriad of local municipalities. All too often, such political fragmentation leads to a serious lack of coordinated governmental action; there are conflicting interests, expensive duplications of services, and long-term entrenched divisions. Thus, planning on an integrated overall basis, with long-range objectives to benefit the entire metropolis, is almost impossible because there is insufficient cooperation within multigovernmental urban regions.

The deconcentration of people and economic activity exacerbates an intensifying problem. When the 1980 census figures were announced, several big-city mayors, representing their administrations, objected to the recorded declines and some cities even sued the census bureau to force it to test population totals' validity (leaders of the largest cities were especially bitter because of the growing number of uncounted illegal aliens who were settling there). They did so because they realize that declining numbers would mean declining federal and state support, and any undercount would be very costly. Already the deconcentration process has significantly reduced the inflow of tax funds to the central city, eroding the services that government can make available, slowing urban renewal and housing improvement, and contributing to the breakdown of social order. Many inner-city schools, for example, are so beleaguered by vandalism and violent disorder that their educational functions are impaired—the cities simply do not have the resources to combat the situation. Minority populations, clustered in, and near, the inner city, stand to lose the most—and do.

The problems of the city are not just reflected by broken school windows, abandoned tenements, or the fear of crime and violence in the streets. Sprawling metropolitan areas also have a pervasive impact on the natural environment: on the climate, which is modified by pollutants spewed into the air by automobiles and factories; on waterways fouled by municipal, industrial, and even recreational wastes; on open spaces such as waterfronts, parks, and beaches soiled by the use of thousands. To keep the city livable requires constant protection of its natural environment and open spaces, but powerful pressures work in the opposite direction. The record of failure is etched on any big-city zoning map that has become a patchwork of microgeographic inconsistencies.

The problem of deteriorating intraurban circulation has severely negative implications for the central city. To function adequately, to be an acceptable and reasonable place to live and work, a city must have an efficient internal movement system. The city historically provided advantages to its various functions by permitting them to cluster closely together. However, that advantage erodes sharply when communication and transportation deteriorate. Traffic jams on major arteries have become commonplace. Bus and rapid-transit systems have only partially relieved the situation. Some rapid-transit systems such as the (San Francisco) Bay Area Rapid Transit and particularly Miami's

Metrorail were so poorly planned that they made little difference; others, notably in Washington and Atlanta, show greater promise, but they must be kept well maintained (at great expense) or face the horrendous deterioration that now plagues New York City's subways. The enormous and continuing increase in the number of automobiles in the United States has produced unprecedented urban traffic congestion, and no amount of freeway building (as Houston and Los Angeles pathetically demonstrate) seems capable of ameliorating this condition.

A Global Phenomenon

In this chapter, we have concentrated our attention on North American and European cities, the urban places with which we are most familiar. But urbanization is a global phenomenon, and not all the processes we observed in America occur in all the world's cities. There are, for example, cities that remain—the Industrial Age notwithstanding—essentially preindustrial by modern standards. If we tried to find evidence of the forces behind the structural models described earlier, we would fail. Urban regions in such places do exist, of course, but they do not create the townscapes we recognize instantly as *downtown* or *suburban*. The focus of such an urban center is likely to be a religious structure, perhaps a central mosque; the most important public place is likely to be a great square or plaza, where hundreds of thousands of residents gather for religious or other ceremonial occasions. You will see no high-rise towers to guide you to the CBD—it may be a large market or a maze of narrow lanes. Business may be just as intense as in any major department store in Western culture, or more so. But the accommodations are quite different.

In a way the preindustrial city is the last direct link to the ancient cities of Southwest Asia, although not one preindustrial (or nonindustrial) city is completely untouched by modern times. A good example is the large Ethiopian city of Harar, east of Addis Ababa, the country's capital. Harar (or *Harer*, as it is sometimes

The Soviet Union contains several very large metropolises, with populations as high as 9 million (Moscow) and 5 million (Leningrad). But fly over and then drive into a large Soviet city, and you will be struck by an unusual feature: though huge and populous, bustling and crowded, there does not seem to be a core, a "real" CBD marked by the kind of skyline we would expect in such a large conurbation.

The reason behind this lies in Soviet urban planning. In American cities, the cost of land is a major factor in urban development, and in the commercial economy there is comparatively little control over growth. Compared, that is, to Soviet practice. In Soviet cities, urban land is controlled by government and is earmarked for specific uses; it cannot be bought and sold. In this way, Soviet urban planners can prevent the evolution of a skyscrapered central

city that requires a system of highways and railroads to move commuters in and out each working day. Rather, offices, factories, and other urban functions are assigned a location in the urban complex, and people who work there are similarly assigned nearby housing. Thus, the Soviet city is the ultimate in multiple-nuclei realization. It consists of a large number of *microdistricts* (as the planners call them) where people can live, work, go to school, shop, and play.

The central city—where we would expect a CBD—instead remains a focus for impressive public buildings, wide, uncongested avenues, and usually a great public square for ceremonial events such as military parades. During the early Communist period in China, the Soviets introduced this urban form in Beijing and other Chinese cities, and it still marks townscapes there.

---

spelled) is not a famous place, and probably many people who know the world's major cities have not heard much about Harar. The city's name is not associated with desired manufactures or raw materials. And yet, Harar has more than 100,000 inhabitants. Certainly there are signs of change: The old Moslem walled city is now surrounded by "suburbs" of more contemporary dwellings that sprawl across the plateau on which the place lies. But what influences the spatial character of the local economy? Not industry and labor and resources, but the contrasting rituals involved in the slaughtering of animals. And so, there is a Moslem marketplace and a Christian one.

Just as was the case thousands of years ago, trade is the essence of this urban place. Agricultural products from the hinterland pass through the markets; basketweaving is an important activity; some silver jewelry is made by skilled craftspeople. Life goes on without commuter streams, smokestacks, or sweatshops. People walk and ride bicycles; there is bus service, and a few residents use automobiles. But the city is little changed to accommodate these modern

invaders. The urban geography of Harar is shaped by forces quite different from those with which we are familiar.

Such circumstances still prevail in towns and cities remote from the tentacles of the Industrial Revolution. Goods still move on the backs of people and animals, on carts and wagons. Houses, hostels, workshops, and other special-function buildings lie in an apparently unplanned urban tangle. The dimensions of things are smaller than those of industrialized places: dwellings are modest, vertical development is very limited, streets are narrower, vehicular traffic is slowed by less efficient interconnections. From a vantage point overlooking such a city, it looks like a sea of roofs, all appearing much the same, with only a few more prominent buildings rising above it.

URBAN TRANSFORMATION

Some truly preindustrial cities survive, but a more common phenomenon is the evolution of a kind of twin-city arrangement representing both the old and

---

the new. Large cities such as Ibadan (Nigeria) and Lahore (Pakistan), great urban centers long before the Europeanization of the world began, were strong enough to withstand total transformation by the impact of the Industrial Age, but they were changed significantly, nevertheless. In such cities, you encounter the traditional as well as the modern: centuries-old markets and neighborhoods survive, but grafted upon the old city is a new urban form. Rising above the clusters of the traditional townscape are the multistory high-rises of banks and corporations, giving the appearance of a Western CBD. And cutting through the maze of housing and shops are wide, multilane avenues that carry heavy traffic quite reminiscent of American urban arteries. Factories employing the local labor force lie on, or near, the outskirts; or perhaps they were built there but are now engulfed by the growing urban population.

This, the in-migration of people from rural areas and smaller towns to the beckoning cities, is one of the great urban problems of the twentieth century. Where an essentially subsistence economy is being transformed, there is widespread dislocation. Agriculture, the industry that for millennia sustained region and country, suffers from governmental neglect and demographic depletion as workers seek ''better'' jobs in the cities and their industries. Millions arrive every year to find the city a harsh, uncompromising place, a place of hunger and disease, of scarce opportunities or jobs—but few return to the countryside. In Nigeria, for example, an oil boom during the 1970s brought a new economic age, or so it was believed. The cities mushroomed; Lagos came to be called the Calcutta of Africa. Then the promise of long-term progress and prosperity was broken by a sharp decline in oil revenues during the 1980s. But when the government looked toward agriculture to help overcome the economic crisis, it found trouble: organization was disrupted, lands neglected, production down. The lure of the city was a curse in the countryside.

In many ways, those skyscrapered downtowns in the large cities of developing countries are anachronisms, incongruities. They symbolize the veneer of penetration by new economic forces, and promise what—to countless millions of hopeful—cannot be delivered.

Reading about Urban Geography

Urban geography is the topic of a number of excellent introductory textbooks, Among standard works is T. A. Hartshorn, *Interpreting the City: An Urban Geography* (New York: Wiley, 1980). Some topics receive more attention in D. T. Herbert and C. J. Thomas, *Urban Geography: A First Approach* (New York: Wiley, 1982). A book that touches on many of the areas discussed in Chapter 9 is E. Jones, *Towns and Cities* (Westport, Conn.: Greenwood Press, 1982). Although this book was first published in 1966, it remains an important source for insights into preindustrial cities, urbanization processes through time, and city–region relationships. Another classic is *The City in History* by L. Mumford (New York: Harcourt, Brace & World, 1961). This book deals with urban beginnings, change and transition, form and function. Still another volume you will see cited today although it was published nearly 30 years ago is G. Sjoberg, *The Preindustrial City, Past and Present* (Glencoe, Ill.: Free Press, 1960). More recent, and worth a trip to the library, is *This Scene of Man: The Role and Structure of the City in the Geography of Western Civilization* by J. E. Vance, Jr. (New York: Harper & Row, 1977). J. Borchert's article, ''American Metropolitan Evolution,'' appeared in the *Geographical Review*, Vol. 57 (1967), pp. 301–32.

Specialized topics enjoy good coverage. For the social aspects of urban geography, see D. F. Lee, *A Social Geography of the City* (New York: Harper & Row, 1983), a book with a range of viewpoints and fine treatment of the social environments of the city and how groups relate to it. Aspects such as sense of place and perception as they relate to urban centers were explored more than a generation ago by K. Lynch in a book still worth reading, *The Image of the City* (Cambridge, Mass.: MIT Press, 1960). M. Jefferson's ''Law of the Primate City'' appeared in the *Geographical Review*, Vol. 29 (1939), pp. 226–32.

Urban regions also form the topics of individual works. Dated, but still relevant, is R. E. Murphy's *The*

371

Central Business District (Chicago: Aldine, 1972). Also see H. M. Rose, *The Black Ghetto: A Spatial Behavioral Perspective* (New York: McGraw–Hill, 1971). P. O. Muller's *Contemporary Suburban America* (Englewood Cliffs, N.J.: Prentice–Hall, 1981) discusses the suburb as "the essence of the contemporary American city." A related topic is addressed by N. Smith and P. Williams (eds.) in their volume on *Gentrification of the City* (Winchester, Mass.: Allen & Unwin, 1986). Another classic work still relevant deals with an interurban rather than an intraurban region: J. Gottmann's *Megalopolis: The Urbanized Northeastern Seaboard of the United States* (Cambridge, Mass: MIT Press, 1964).

Individual cities and conurbations likewise have been subjected to geographical analysis. See, for example, C. W. Condit, *Chicago, 1910–1929: Building, Planning, and Urban Technology* (Chicago: Univ. of Chicago Press, 1973) and the two-volume *Cities of Canada* by G. A. Nader (Toronto: Macmillan, 1975; 1976); the latter books cover the historical evolution of Canadian cities, the structure of individual cities, and urban problems in Canadian context. Still of interest is A. E. Smailes's *The Geography of Towns* (London: Hutchinson, 1965). One of the most useful sources on the American city is a set of publications sponsored by the AAG. The first is *A Comparative Atlas of America's Great Cities: Twenty Metropolitan Regions* (Minneapolis: Univ. of Minnesota Press, 1976), a volume edited by R. F. Abler, J. S. Adams, and Ki-Suk Lee. The second is a four-volume series of vignettes covering the cities detailed in the *Atlas.*

Edited by J. Adams, these volumes are (1) *Cities of the Nation's Historic Metropolitan Core*, (2) *Nineteenth Century Ports*, (3) *Nineteenth Century Inland Centers and Ports*, and (4) *Twentieth Century Cities* (Cambridge Mass.: Ballinger, 1976).

Many geographical writings concern cities in Europe, the Soviet Union, and other areas of the world beyond North America. See, for example, V. F. S. Sit (ed.), *Chinese Cities: The Growth of the Metropolis Since 1949* (London/New York: Oxford Univ. Press, 1985); M. Kosambi, *Bombay in Transition: The Growth and Social Ecology of a Colonial City, 1880–1980* (Stockholm: Almqvist & Wiksells, 1986); and, in a very different vein, *Outcast Cape Town* by J. Western (Minneapolis: Univ. of Minnesota Press, 1981), a book that is much more than an urban geography. D. J. Dwyer (ed.) published *The City in the Third World* (New York: Barnes & Noble, 1974), still a useful set of readings; a classic work on the same topic is T. G. McGee, *The Urbanization Process in the Third World* (London: Bell, 1971). A comprehensive work is S. D. Brunn and J. F. Williams, *Cities of the World: World Regional Urban Development* (New York: Harper & Row, 1983).

The importance of urban geography as a field is underscored by the existence of a specialized journal, *Urban Geography*, published in the United States along with another journal, *Environment and Planning: International Journal of Urban and Regional Research*, published in London. When you scan the journal literature in other publications, you will find urban geography a frequent topic.

Industrial
Regions
and
National
Development

The economic geography of the contemporary world is a patchwork of almost inconceivable contrasts. On the shifting farm fields in equatorial American and African forests, root crops still are grown according to ancient practices and with the most rudimentary tools, whereas modern machines plow the land, seed the grain, and harvest the wheat hundreds of acres at a time on the plains of North America, in the Soviet Union, and in Australia. Toolmakers in the villages of Papua New Guinea still fashion their implements by hand as they did many centuries ago, whereas the factories of Japan disgorge automobiles by the shipload for distribution to markets thousands of miles away. Between these extremes, the range and variety of productive activities is virtually endless.

Notwithstanding the globe-girdling impact of the Industrial Revolution, there are areas even within the industrialized countries where change is coming only slowly. Parts of the poverty-stricken rural South in the United States remain comparatively remote from the effects of U.S. economic growth. In areas of northern Japan, life has changed little during the dramatic century of Japan's modernization. In industrializing Europe, there remain regions of isolation and stagnation. Conversely, there are places in less-developed countries where industries have become established, where urban agglomeration is taking place, and where conditions differ sharply from those prevailing on most of the land. South-central, coastal Brazil is a world apart from the rural Brazilian interior. India's northwest (including western Uttar Pradesh State) presents a picture quite different from that of Bihar and West Bengal. In these countries, the kind of development that has come to prevail in the United States, Japan, and Western Europe is still insular or incipient. Again, there are countries in which, for all intents and purposes, no industrial growth has taken place at all. Apart from some extractive operations (e.g., a mine with a railroad to the coast), there is almost none of the activity associated with modern industrialization. In such countries as Mauritania, Chad, and Laos, the ripples of two centuries of world transformation have barely been felt.

• Definitions of Development •

The countries of the world, therefore, lie along a kind of development continuum that extends from the very least developed to the most industrialized, urbanized, and materially productive societies. Economic geographers and others have tried to group countries in such a way that each group shares certain characteristics, so that it might be possible to speak of the world's **developed countries (DCs), underdeveloped countries (UDCs)**, and an ill-defined group in between. This last group is given various names; sometimes the term **takeoff** is applied to their economies, suggesting that they have left the ranks of the UDCs, and are showing signs of joining the list of developed countries.

All kinds of problems arise from such a classification. In the first place, there is the question of criteria. As we noted earlier, even the DCs have pockets of underdevelopment; and there are developed regions even in the UDCs. No agreement exists even on the list of measures of development commonly used to rank countries. "Development" is an unsatisfactory concept, and it also has emotional connotations. People from UDCs sometimes argue that "developing" or "less developed" countries (LDCs) are preferable terms that have less pejorative association, and such sensitivities are understandable.

So it might be well to remember that terms such as *developed* and *underdeveloped* also mean rich and poor, haves and have-nots, and perhaps most appropriately, advantaged and disadvantaged. What the numerical measures of development fail to convey is the time factor. The DCs are the advantaged countries—but why cannot the UDCs and takeoff countries catch up? It is not, as is sometimes suggested, simply a matter of environment, resource distribution, or cultural heritage (e.g., a resistance to innovation). The sequence of events that led to the present division of our world began long before the Industrial Revolution occurred. Europe even by the middle of the eighteenth century had laid the foundations for its colonial expansion; the Industrial Revolution magnified Europe's demands for raw materials while its products

What distinguishes a developed economy from an underdeveloped one? Obviously, it is necessary to compare countries on the basis of certain measures; the question cannot be answered simply by subjective judgment. No country is totally developed; no economy is completely underdeveloped. We are comparing *degrees* of development when we identify DCs (Developed Countries) and UDCs (Underdeveloped Countries). Our division into developed and underdeveloped economies is arbitrary, and the dividing line is always a topic of debate. There is also the problem of data. Statistics for many countries are inadequate, unreliable, incompatible with those of others, or simply unavailable.

The following list of measures is normally used to gauge levels of economic development.

1 *National product per person.* This is determined by taking the sum of all incomes achieved in a year by a country's citizens and dividing it by the total population. Figures for all countries are then converted to a single currency index for purposes of comparison. In DCs, the index can exceed $10,000; in some UDCs it is as low as $100. The World Bank in its *1984 Report* used U.S. $390 as the upper limit for low-income countries (the most severely underdeveloped countries).

2 *Occupational structure of the labor force.* This is given as the percentage of workers employed in various sectors of the economy. A high percentage of laborers engaged in the production of food staples, for instance, signals a low overall level of development.

3 *Productivity per worker.* This is the sum of production over the period of a year divided by the total number of persons comprising the labor force.

4 *Consumption of energy per person.* The greater the use of electricity and other forms of power, the higher the level of national development. These data, however, must be viewed to some extent in the context of climate.

5 *Transportation and communications facilities per person.* This measure reduces railway, road, airline connections, telephone, radio, television, and so forth, to a per capita index. The higher the index, the higher the level of development.

6 *Consumption of manufactured metals per person.* A strong indicator of development levels is the quantity of iron and steel, copper, aluminum, and other metals utilized by a population during a given year.

7 *Rates.* A number of additional measures are employed, including literacy rates, caloric intakes per person, percentages of family income spent on food, and amount of savings per capita.

377

increased the efficiency of its imperial control. While Western countries gained an enormous head start, colonial dependencies remained suppliers of resources and consumers of the products of the Western industries. Thus was born a system of international exchange and capital flow that really changed little when the colonial period came to an end. Underdeveloped countries, well aware of their predicament, accused the developed world of perpetuating its advantage through neocolonialism—the entrenchment of the old system under a new guise.

There can be no doubt that the world economic system works to the disadvantage of the UDCs but sadly it is not the only obstacle the poorer countries face. Political instability, corruptible leaderships and elites, misdirected priorities, misuse of aid, and traditionalism are among the circumstances that inhibit development. External interference by interests representing powerful DCs have also had negative impact on the economic as well as the political progress of UDCs. Underdeveloped countries even get caught in the squeeze when other DCs try to assert their limited strength; when the OPEC countries, mostly under-developed themselves, raise the price of oil, energy and fertilizers slip still farther from the reach of the poorer UDCs not fortunate enough to belong to this favored group. As the DCs get stronger and wealthier, they leave the underdeveloped world ever-farther behind: the gap is widening, and the prospects for the UDCs are not bright.

DISMAL PROSPECTS FOR UDCs

The UDCs of the world far outnumber those that can be classed as DCs. In their book entitled *Population, Resources, Environment* biologists Paul and Anne Ehrlich characterize the UDCs as the "hungry" countries that are not industrialized, have inefficient (usually subsistence) agricultural systems, low gross national products and per capita incomes, high illiteracy rates, and very high rates of population growth. They describe the UDCs as "never-to-be-developed" countries in the sense of the United States today. It is not a very bright picture, but a rather accurate one. We may speak optimistically of developing, or emergent, countries when, in fact, little or no economic progress is being made in many of them. Populations continue to grow, but life expectancies do not, and all-too-many people will know hunger and malnourishment as a prominent reality of life. The cycle of poverty is not easily broken. The DCs are not immune from the problems of poverty and hunger, either. In the United States, President Lyndon Johnson proclaimed a "war on poverty" during the 1960s—a program of governmental spending that acknowledged the existence of underdevelopment even here.

The UDCs are said to have an extremely low **gross national product (GNP)**. The GNP is calculated as the total value of all goods and services produced by a country during a certain period, usually a year. This value is measured at prevailing market prices in order to make it possible to attach a worth to the many different items produced. When the GNPs of many countries of the world are compared, they are all converted to one currency (say the dollar), again at prevailing rates of exchange. Thus there are some pitfalls both in the calculation and the interpretation of GNP data. Monetary rates of conversion may not

be realistic (witness recent revaluations and fluctuations of the dollar). The purchasing power of the equivalent of, say, $100 in Nigeria may be far different from what $100 is worth in the United States. Nevertheless, GNP statistics tell much of the story of underdevelopment. Reduced to per capita figures, we quickly get an idea of the contrasts between DCs and UDCs. The World Bank, which reports annually on such matters, showed the 1983 per capita GNP in the United States to be about $13,160, in Canada $11,320, in the U.K. $9,660; but in some 50 UDCs, it averaged less than $1,000. Of course, we really should be careful with these GNP figures. Record keeping in UDCs often is not very good, and a lot of those statistics on which conclusions are based began as estimates and guesses.

This is especially true where farming is of the subsistence variety, another characteristic of UDCs. The quality of a national economy can be effectively gauged by measuring the percentage of the labor force engaged in agriculture. In a preceding chapter, we noted how that percentage has declined in the United States over the past century and how large a proportion of the workers in India and China are still engaged in growing crops to feed themselves and the small percentage of people not living directly off the land. Such differences in the kind of work done by the labor force give insights into the contrasts between DCs and UDCs.

In Chapter 6, the problem of high population growth rates in the UDCs was discussed in another context. Coupled with shorter life expectancies, this means that most of the people in the UDCs are young people. In Middle and South America in the late 1980s, *half* the population was under about 15 years old; in the Western countries of North America and Europe, the figure is under 30 percent. Most of these youngsters cannot be absorbed as workers; many remain unemployed. They are consumers but not producers, thus constituting an enormous burden and cost.

This brings us to the criterion of central interest in this chapter, the different levels of industrialization of DCs and UDCs. Of course there is some industry, however modest and local, in all countries; it is in the *kinds* of industries, their prevalence and dimensions,

The reorganization of agriculture is a pressing need in most underdeveloped countries. Land fragmentation and misuse of the soil all too often lead to scenes like this: severe erosion on a slope too steep for farming, despite an effort to terrace some of the fields. These hills are in East Africa near Meru, Kenya.

industries, metals play a major role in industry, and transport networks reflect the exchange character of an economy.

Small wonder, then, that some countries whose governments seek to accelerate their economic development have tried to go the route of massive industrialization. In the Soviet Union, planners poured great amounts of the country's resources into a determined program to industrialize. Undeveloped countries from Indonesia to Egypt want their own steel mills and national airlines as symbols of progress. But not all countries have made this decision. The governments of some UDCs realize that the criteria by which development is measured reflect not only those particular attributes, but also the degree of transformation of a whole society. A domestic steel mill or some other major industry will do little to speed that transformation, and it remains nothing more than an alien anomaly, a costly producer of misleading statistics. A good example of a different approach is the case of Tanzania: after independence, the *first* national order of business was to change the country's agriculture; to help farmers living a life of inadequacy under subsistence to move toward adequacy; to diffuse improved farming techniques, fertilizers, and tools (not necessarily tractors, but better hand hoes, for instance); and to encourage and support cooperatives. Unfortunately this program was linked to a scheme to relocate hundreds of thousands of farmers in newly planned villages, a project that met with resistance. Still, Tanzania's effort became an alternative model for countries in search of progress.

that the differences between DCs and UDCs lie. As we noted, among indicators to gauge the level of development in a country are the productivity per employee in the labor force, the consumption of power such as electricity per person, the quantity of metal used, and the transport facilities that have been developed. These indicators also reflect a country's industrialization because the output per worker goes up when industry becomes increasingly mechanized: commercial energy is consumed in huge quantities by

Industrial Intensification

It is proper to describe our modern age as one of industrial **intensification** because industrial development did not begin with the Industrial Revolution: rather, it accelerated industrialization and diffused the process from certain areas of innovation to other parts of the world. Long before the momentous events of the second half of the eighteenth century occurred, when the foundations for mass production were laid, industries already existed in many parts of the world, and trade in their products was widespread. In the towns and villages of India, workshops produced metal goods of iron, gold, silver, and brass; India's carpenters were artists as well as craftspeople, and their work was in demand wherever it could be bought. India's textiles, made on individual spinning wheels and hand looms, were acknowledged to be the best in the world. These industries were sustained not only by the patronage of local Indian aristocracies, but also by trade on international markets. So good were the Indian textiles that British textilemakers staged a major riot in 1721, demanding legislative protection against the Indian competition that had begun to dominate local British markets.

China, too, possessed a substantial industrial base long before the Industrial Revolution, and so did Japan. European industries, from the textilemakers of Flanders and Britain to the iron-smelters of Thüringen, had also developed considerably—but in terms of price and quality the European products could not match those of other parts of the world. But what European manufacturing products lacked in finesse, Europe's merchants and their representatives made up for in aggressiveness and power. Europe's commercial companies (e.g., the Dutch and British East India companies) laid the groundwork for the colonial expansion of Europe: they gained control over local industries in India, Indonesia, and elsewhere, profited from the political chaos they precipitated, and played off allies against enemies. British merchants could import about as many tons of raw fiber for the textile industries as they wanted, and all they needed to do, they knew, was to find ways to mass-produce these raw materials into finished products. Then, they would bury the remaining local industries in Asia and Africa under growing volumes and declining prices. Even China—where local manufactures long prevailed over inferior and more expensive European goods—would succumb.

During the eighteenth century, European domestic markets were growing, and there was not enough labor to keep pace with either the local trade or the overseas potentials. Machines capable of greater production were urgently needed, especially improved spinning and weaving equipment. The first steps in the Industrial Revolution were not so revolutionary, for the larger spinning and weaving machines that were built were driven by the old source of power: water running downslope. But then Watt and others who were working on a steam-driven engine succeeded (1765–88), and this new invention was adapted for various uses. About the same time, it was realized that coal could be transformed into coke and that coke was a superior substitute for charcoal in the smelting of iron.

These momentous innovations had rapid effect. The power loom revolutionized the weaving industry. Freed from their dependence on dwindling wood supplies from the remaining forests, iron smelters could now be concentrated near the British coalfields—the same fields that supplied fuel for the new textile mills. One invention led to another, each innovation made to serve some particular industry also serving other industries. Pumps could keep water out of flood-prone mines. Engines could move power looms as well as locomotives and ships. As for the capital required, plenty was available for investment. British industrialists had been getting rich from the overseas empire for many years.

Thus, the Industrial Revolution had its effects also on transportation and communications. The first railroad in England was opened in 1825. In 1830, the city of Manchester was connected to the port of Liverpool; in the next several decades, thousands of miles of rail were laid. Ocean shipping likewise entered a new age as the first steam-powered vessel crossed the Atlantic in 1819. Now England enjoyed even greater advantages than those with which it entered the period of the Industrial Revolution. Not only did England hold a monopoly of products that were in world demand, but Britain alone possessed the skills necessary to make the machines that manufactured them. Europe and America wanted railroads and locomotives; England had the know-how, the experience, the capital, and the raw materials. Soon the fruits of the Industrial Revolution were being exported, and British influence around the world reached a peak.

Meanwhile, the spatial pattern of modern industrial Europe began to take shape. In Britain, industrial regions, densely populated and heavily urbanized,

380

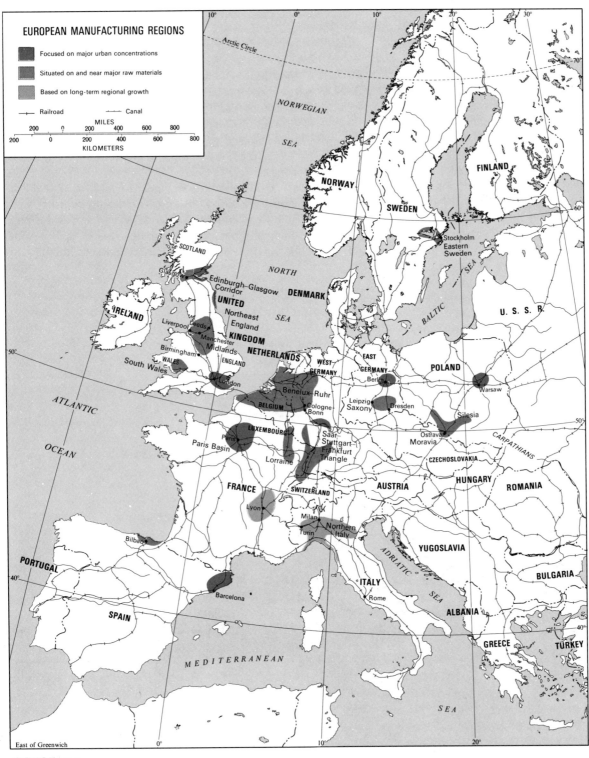

EUROPEAN MANUFACTURING REGIONS

■ Focused on major urban concentrations

■ Situated on and near major raw materials

■ Based on long-term regional growth

Railroad   Canal

MILES
200   0   200   400   600   800

KILOMETERS
200   0   200   400   600   800

Arctic Circle

NORWEGIAN SEA

NORTH SEA

BALTIC SEA

MEDITERRANEAN SEA

ATLANTIC OCEAN

ADRIATIC SEA

NORWAY
SWEDEN
FINLAND
DENMARK
IRELAND
UNITED KINGDOM
SCOTLAND
Glasgow
Edinburgh-Glasgow Corridor
Northeast England
Leeds
Liverpool
Manchester
Midlands
Birmingham
WALES
South Wales
ENGLAND
London
NETHERLANDS
WEST GERMANY
EAST GERMANY
Berlin
Benelux-Ruhr
BELGIUM
Cologne-Bonn
Leipzig
Saxony
Dresden
POLAND
Warsaw
Silesia
LUXEMBOURG
Paris
Paris Basin
Lorraine
Saar
Stuttgart
Frankfurt Triangle
Ostrava
Moravia
CZECHOSLOVAKIA
FRANCE
SWITZERLAND
AUSTRIA
HUNGARY
ROMANIA
Lyon
Milan
Turin
Northern Italy
Bilbao
PORTUGAL
Barcelona
SPAIN
ITALY
Rome
YUGOSLAVIA
BULGARIA
ALBANIA
GREECE
TURKEY
U.S.S.R.
Stockholm
Eastern Sweden
CARPATHIANS

East of Greenwich

381

FIGURE 10.1

developed in the Black Belt—near the coalfields. The largest complex was (and remains) positioned in the Midlands of north-central England. In mainland Europe, a belt of major coalfields extends from west to east, roughly along the southern margins of the North European lowland—due eastward from southern England, across northern France and southern Belgium, the Netherlands, the German Ruhr, western Bohemia in Czechoslovakia, and Silesia in Poland. Iron ore is dispersed within a broadly similar belt, and the industrial map of Europe reflects the resulting concentrations of economic activity (Fig. 10.1). Nowhere on the continent, however, were the coalfields, the iron ores, and the coastal ports located in such close proximity as they were in Britain.

Factors of Location

Industrial activity takes place in certain locations; it does not occur in others. What determines where industries develop? We observed that the Industrial Revolution itself was centered in a particular region, then diffused to other areas. Today not just one but several regions of the world are outstanding industrial zones.

Industries differ in kind. The making of steel is an industry, but so is fishing. The fishing industry is Iceland's very means of survival, more important even to that nation than our steel industry is to us. But somehow the extraction of fish from the sea is not the same

Singapore is a city-state, a developed country in underdeveloped Southeast Asia. More than perhaps any other place in this region, Singapore is the product of geography—its pivotal location at the tip of the Malay Peninsula and at the head of the Strait of Malacca an asset that has been translated into progress and prosperity. But decision-making played a crucial role: an authoritarian government has controlled the economy as well as the social life of Singapore. Population growth has been limited through penalties on families with more than two children, for example. One of the critical economic-geographic decisions involved the construction of massive refineries to handle crude from the Persian Gulf for sale in Japan, from a position about midway between supply and demand. Now Singapore is embarked on industrial diversification, its continued well-being reflected by these new skyscrapers on its crowded waterfront.

382

Chapter 10   Industrial Regions and National Development

• The Primary Industries •

The primary industries are the extractive industries, including cultivation and pastoralism as well as mining and quarrying, fishing, hunting, and forest-product industries.

Mining. Mining must, quite obviously, take place where the resources are located. An ancient industry (we refer to the Stone Age and the Iron Age of human history), mining today is strongly mechanized and employs only about 1 percent of the world's labor force.

No country in the world, no matter how large, possesses a complete range and adequate volume of all mineral resources; so, there is an active trade in minerals and mineral fuels. A constant search for new reserves goes on, and new industries establish themselves when discoveries are made. At the same time, the exhaustion of the highest-quality reserves brings lower-grade resources into production.

Fishing. Fishing is another ancient human occupation that has been made more efficient through mechanization. Improved technologies and the concentration of fish catches on continental-shelf areas now threaten the oceans' fish resources. Japan, the Soviet Union, the United States, Peru, and China harvest the bulk of what may be a declining resource through the use of fleets of fishing boats and floating canning factories.

Forestry. Forests, ancient allies of human communities, provided shelter, food, building materials, firewood, even clothing. Still today, the forests—heavily depleted through centuries of exploitation, but supplemented now by reforestation—provide building materials (lumber), paper (pulpwood), and many less important products such as cork, bamboo, resin, gums, tannin, and quinine.

kind of industry as the manufacture of a product such as steel rails. Fishers catch what exists in nature. Steelmakers mix raw materials and come up with something artificial, new. And what about the transportation industry? It produces no product at all, but moves people and goods from place to place. Still it is also an industry!

Thus three classes of industry can be established. The **primary industries** are extractive, for example, fishing. Other primary industries include forestry, mining, quarrying, and hunting. Though in a different subgroup, the various forms of agriculture discussed in Chapter 8 are also primary industries. The **secondary industries** are the manufacturing industries. Raw materials are transformed into products that range from plastic toys to railroad cars. When we speak of industrial nations, we really mean the manufacturing nations. Finally, there are the **tertiary industries**, those that involve the production not of goods, but of services. Teaching is such an industry, as is transportation. There is such a thing as the medical industry.

Dentists, bankers, bus drivers, and postal carriers all operate in the tertiary sector.

LOCATIONAL DECISION MAKING

In the primary industries, production has to take place where the resources are. The timber, the fish, and the ores and fuels leave no choice—it is necessary to come and get them where they are. But we just noted that textile industries and other industrial plants began to position themselves in clusters as Europe's Industrial Revolution gained momentum and that iron smelters, among others, were positioned near the coalfields of the Black Belt. This means that the iron ores had to be brought to the coalfields and that reality illustrates an important aspect of the secondary industries: decisions have to be made about where to locate them. Weight and bulk have a lot to do with such decisions. Is it not more sensible to refine an ore or to saw off useless waste from treetrunks practically at the site of production and then to ship the raw material for final processing into finished products than to ship them long distances first? In the case of iron ore and coal, both needed in large quantities to make steel, the alternatives are (1) to take the iron ore to the coalfields, (2) to take the coal to the iron ore deposits, or (3) to transport both to an intermediate location. Most frequently, the ore is transported to the coalfields, sometimes after the partial elimination of waste and impurities. There are exceptions: for example, coal is transported in large quantities to the French Lorraine industrial region, where coal is scarce. When an industrial complex develops near coalfields and the coal supplies become exhausted, it may be less expensive to start importing coal from elsewhere than to move the factories. Even when both coal and iron ore are shipped over large distances to some intermediate location, the iron ore usually travels the farthest.

RAW MATERIALS

Numerous considerations enter into location decision making, and such questions are of prime interest to economic geographers who want to understand the processes whereby economic activity organizes and adjusts spatially. Naturally the resources involved—the raw materials—play a major role. One example is the coastal steel industry of the United States along the Eastern Seaboard. Those industries are there in large measure because they use iron ore shipped in from Venezuela, Liberia, and other overseas mines. Instead of transferring these ores from the arriving ships into trains and railing them inland, the ores are used right where they arrive—practically at the point of unloading. So, in this case, some faraway ore deposits had much to do with the location of industry in the United States.

Thus transport costs affect the location of industry in important ways (transportation also is discussed later in a separate section because this factor involves finished products as well as raw materials). In the case of raw materials, we have already noted the spatial relationships between Europe's zone of coalfields and iron ores and the spread of manufacturing through the region. However, not all of the world's great industrial regions lie near major sources of raw materials. Japan's massive industries must import their raw mate-

rials from distant sources because Japan's domestic resource base is quite limited. This has not prevented Japan from developing into one of the world's great industrial nations, but it does present particular problems of availability and cost to Japanese manufacturers. Japan's early industrial progress was based on its own indigenous manufacturing traditions and, initially, on a set of comparatively minor domestic resources. The rapid depletion of these local raw materials was among the motives that led Japan to embark on its expansion into East Asia, where Korea and northeast China became Japanese dependencies and sources of additional industrial resources. Ultimately, Japan lost its colonial empire, but its industrial strength prevailed—aided mightily by its large, relatively cheap, highly skilled labor force. Selling its products on markets around the world, the Japanese could afford to purchase the raw materials needed to sustain its industries, from countries as far away as Swaziland in Africa and from Australia.

Raw materials, just as the term implies, are untreated commodities that often must be refined before they can be used in manufacturing. Obviously, there are some advantages to transporting a lighter load by refining (or *concentrating*) some of these raw materials at their sources. Various metallic ores are treated in this way, reducing their bulk prior to shipment. For many decades, the industrialized countries controlled the sources of raw materials that they needed because they had colonized the countries where these resources existed. Thus, these powers could organize the mining, smelting, and shipment of ores as required, and in large measure, the systems of extraction set up during that period still continue to function. However, some political factors have come into play in recent years as newly independent countries recognize the potential advantages of their control over resources vital to the industrialized nations. This is true not only for the oil-producing states (energy is treated separately later), but also for countries possessing critical strategic minerals or vital alloys.

LABOR

A second factor of industrial location is the availability of labor. Even in this age of automated assembly lines and computerized processing, the available skills of a

Europe's rapid industrialization during the nineteenth century attracted the attention of economic geographers at an early stage. What processes have channeled Europe's industrialization and produced the patterns that were developing? Much of this pioneering research was incorporated in Alfred Weber's book, *Concerning the Location of Industries* (1909). Like von Thünen before him, Weber began with a set of assumptions in order to minimize the complexities of the real Europe. However, unlike von Thünen, Weber dealt with activities that take place at particular points rather than across large areas. Manufacturing plants, mines, and markets are located at specific places, and so, Weber created a model region marked by sets of points where these activities would occur. He eliminated labor mobility and varying wage rates; this enabled him to calculate the pulls exerted on each point in his theoretical region.

In the process, Weber discerned various factors that affected industrial location, and he defined these in various ways. For example, he recognized what he called *general factors* that would affect all industries—transportation costs for raw materials and finished products, and *special factors*—perishability of foods. He also differentiated between *regional factors*—transport and labor costs—and local factors. Local factors, Weber argued, involved agglomerative (concentrative) and deglomerative forces; thus, the advantages of clustering drew manufacturing plants to large existing urban centers, thereby perpetuating their growth.

Weber singled out transportation costs as the critical determinant of regional industrial location and suggested that the point of least transportation cost is the place where it would be least expensive to bring raw materials to the point of production and from which finished products could be distributed to consumers. However, economic geographers following Weber have concluded that some of Weber's assumptions seriously weakened the usefulness of his conclusions, especially in his notions concerning markets and consumer demand. Consumption does not take place at a single location but rather over a wide (in some cases a worldwide) area. Nonetheless, practically all modern analytical studies of industrial location have a direct relationship to Weber's work, and he was a pioneer in a class with von Thünen.

Other economic geographers later extended Weber's theories. A major contribution was made by August Lösch, whose book *The Spatial Structure of the Economy* (1940), was published in English translation in 1954 as *The Economics of Location*. Lösch countered Weber's studies of least-cost location by seeking ways to determine maximum-profit locations. He inserted the spatial influence of consumer demand as well as production costs into his calculations, a major step forward in the effort to discern the forces shaping the economic landscape.

Industrial location theory is a challenging field of economic geography, a link between geography and economics.

substantial labor force remain important criteria in the location of manufacturing plants. Over decades and generations, a certain cumulative effect (a form of agglomeration) has manifested itself, whereby the people in certain areas become known for their particular talents. The Swiss make fine instruments; specialized textiles are still made in Britain; precision cameras and optical equipment are produced in Japan. The book-publishing industry in the United States remains concentrated in and near New York City, with a secondary center in San Francisco. Here the several skills needed to transform a manuscript into a finished book are available. A new company would be likely to consider availing itself of such available capacities and locate in, or near, a skill-pool center.

It may not be skills but rather the cheapness of labor that will attract some industries. In the United States, there are areas in which prevailing wages are lower than they are elsewhere; unless other factors of industrial location are of overriding importance, a manufacturing concern may locate where it can pay less for its labor needs. This has been one of the leading factors in the industrial development of Puerto Rico, to cite one recent example, where many manufacturers decided to locate plants because prevailing wage rates were low.

The effect of regional differences in prevailing pay scales may be diminishing. The continued industrial development of Western Europe is often cited in this context because labor forces there are supplemented by workers from Mediterranean countries where earnings and opportunities are lower. However, in Europe, as elsewhere, the growing strength of labor unions is serving to reduce wage differences. Unionism in the United States is having a similar effect. Furthermore, increased mechanization and automation in industry is diminishing the importance of available labor as a factor of industrial location; limited and high-priced skills, not plentiful unskilled or semiskilled labor, may soon be the chief attraction.

MARKET

Another very obvious influence on the location of industry is the market where the products are to be sold. Obvious though this influence may be, it is not so easy to understand. In the developed countries, much depends on the transport networks as well as the location of large population concentrations. A company is likely to establish itself in, or near, a major city, where a substantial local market exists and from where good communications radiate outward into the rest of the country, thus facilitating distribution. Industries that produce perishable goods (e.g., dairies) will locate as near the market as they can. Some industries make products specifically designed to be used by other industries, certain equipment for automobiles, for example. It would not be sensible for a factory making odometers for automobiles to position itself very far from Detroit. But there is still more to the issue. When a commodity is very heavy and bulky, the proximity of the market is far more significant than when it is smaller and lighter and, as is likely, more valuable per unit of weight. Even if the Swiss watchmakers sell more watches in the United States and Canada than in Europe, it is unlikely that they will

move their factories to New York for that reason alone. Other considerations such as import restrictions might impel the move, however. On the other hand, an American automobile manufacturer planning to sell a small car on the European market may decide to build an assembly plant in Europe rather than to build the cars here and export them overseas.

Whereas the influences of labor and raw material location are diminishing, the spatial character of the market is of growing importance in the location of industries; thus, decentralization (e.g., of automobile assembly plants) results. The cost of distributing finished products to potential buyers has steadily increased (note the transportation surcharge on the sale of an automobile, a rising percentage of the overall price), so that manufacturers may determine that location in close proximity to consumers is a crucial competitive advantage. If the potential market is concentrated, as for example the megalopolis of the U.S. northeastern seaboard, proximal location is, of course, desirable. If consumers are more dispersed, an approximately central location may be most efficient. This will ensure that the minimal cost is expended on distributing the product to a dispersed population of buyers.

ENERGY

Another factor in the location of industry is the availability of an energy supply. This used to be much more important than it is now. Those British textile mills, as long as they depended on water rushing down hillsides to drive the looms, had few locational alternatives. But these days power comes from different sources and can be transmitted via power lines over long distances without much loss. Industries are thus able to locate primarily on the basis of considerations other than power, except when the industry needs really exceptionally large amounts of energy— for example, certain metallurgical (aluminum and copper processing) and chemical (fertilizer production) plants. Those industries are positioned near sites where abundant energy is created, as is the case at hydroelectric plants.

The role of power supply should not be understated, however. In DCs, dense transmission networks bring power to most areas, but in UDCs, electrifica-

386

tion still is an elusive goal for whole regions. The electrification program undertaken in the Soviet Union during the present century has been one of the prideful accomplishments there. Elsewhere, hydroelectric projects and powerlines have the same aura of modernization and progress that causes UDCs to pour huge investments into the facelifting of capital cities and the maintenance of a money-losing national airline. The hope always is that the cheap electricity, once available, will attract industries. Those anticipations do not always materialize.

Industrial development in UDCs was severely affected when the price of oil rose strongly, as happened during the 1970s. Several of these countries (as we noted earlier) had invested heavily in industrial growth, thereby committing themselves to a course of economic evolution that had unexpected implications. Metallurgical and chemical industries were set back; fertilizer production declined. Representatives of some developing countries reported that their governments could not afford to buy the oil needed to run the pumps of irrigation systems, thereby endangering food supplies even further. The role of energy and the

limitations of its supply are, thus, of growing concern in UDCs as well as in DCs.

TRANSPORTATION

We have referred to the role of transport and communications several times in discussing other location factors, and it is difficult to isolate the role of transport facilities in the location of industries. There may be a huge market for a product, but if that market is not served by good transport networks, its influence is lessened. When you compare the maps of manufacturing regions in this chapter with maps of transport systems (railroads, roads, ocean routes, and air routes), you observe immediately that these highly developed industrial areas are also those served most effectively by transport facilities. Industrialization and the development of transport systems go hand in hand; in a sense the Industrial Revolution was a transport revolution—a revolution that is still going on. Every year, more freight is carried by air; in the United States, trucks haul goods formerly carried on trains.

For industry, efficient transportation systems en-

Container ships such as this increasingly move large volumes of goods between ports. The prepackaged cargoes are placed aboard, then unloaded onto flatbeds as instant trucks. Even dock space is a lesser problem when (as in this picture, taken at Suva, Fiji) the ship can "back" against the pier and lower its rear door as a bridge to its cavernous interior.

able manufacturers to purchase raw materials from distant sources and to distribute finished products to a dispersed population of consumers. Manufacturers desire maximum transport effectiveness at the lowest possible level of costs. Their location decision making will also consider the availability of alternative systems in the event of emergencies (e.g., truck routes when rail service is interrupted). Among the significant recent innovations in bulk transport is the development of container systems that facilitate the transfer of goods from one type of carrier to another (from rail to ship, ship to truck). This has lowered costs and increased flexibility, permitting many manufacturers to pay less attention to transportation in their location decisions.

It has long been known that, for most goods, it is cheapest to transport by truck over short distances. Alternatively, railroads are cheapest over medium distances, and ships are cheapest over the longest distances. However, when decisions are made relating to the location of industries, numerous aspects of transportation must be taken into account, and no single generalization can do justice to the complexities of the problem. For example, when goods are hauled, costs are incurred at the terminal where trucks, trains, and ships are unloaded. These costs vary and are much higher for ships than for trucks. Then, of course, there is the actual cost of transportation itself, which increases with distance—but at a decreasing rate, making long-distance transportation cheaper per mile. Such *long-haul economies* make it possible for a manufacturer to reach out to distant suppliers of raw materials and also to sell to faraway customers. Still, another factor has to do with the weight and the volume of the freight. Certain goods may be of light weight, but may occupy a lot of space inside railroad cars or ships' holds and, therefore, still be expensive to transport. Consider just these few aspects of transportation—and remember that transportation is only *one* factor affecting industrial location among a number of others!

ADDITIONAL FACTORS

Factors other than those we have identified may also be at work in industrial location. If several plants have already located in a certain area, others might be influenced to do the same—not only because of the advantages of that site, but also because of a certain clustering or agglomeration effect that comes into play. Had those already established plants not been there, the other industries might have been positioned elsewhere. There is also a factor of political stability and receptiveness to investment: industries are frightened away if there are signs of uncertainty in the political future of a country or when a government gives indications that it intends to nationalize industries owned by foreigners. Policies of taxation also play a role, and some countries try to attract industries by offering huge tax exemptions over long time periods. Sometimes influential industrialists can simply decide to locate a major plant in some area and can afford to ignore theories of industrial location. The directors of global corporations (so-called **multinationals**) can affect the course of regional industrial development in many countries almost at will. And some industries are located where they are because of environmental conditions: the film industry has been strongly concentrated in Southern California because of the large number of clear, cloudless days there. Industrial location is no simple matter, and there are no easy explanations to account for the maps we are about to see.

PLANNED ECONOMIES

We have discussed the factors of industrial location as they affect decision making in a capitalist, free-enterprise system. But in many noncapitalist countries, the crucial decisions are made by professional planners who may ultimately base their decisions on quite different grounds. Certainly these planners take account of the distribution and availability of resources, energy, transport facilities, and other factors, but more important may be the security of the state (dispersal of industries rather than concentration might result). Or the planners may persuade the government that a particular region needs stimulation and expansion through an infusion of economic activity. In the U.S.S.R., some industries have been built to produce goods that could have been bought more cheaply than they could be manufactured at home in order to

388

The tertiary industries do not (as the primary and secondary industries do) generate an actual, tangible product. Tertiary industries include *services* such as transportation and communications, financial services (e.g., banking), retailing, recreation, teaching, even government. Here the measure of worth is not the quantity of a commodity produced, but the quality and effectiveness with which the particular service is performed.

Although tertiary industries do not produce tangible commodities, they *do* employ large numbers of people. In the United States, for example, the service industries employ *more* workers than the primary and secondary industries combined. This should not surprise us: primary industries such as farming and mining no longer employ large numbers of people today, and the manufacturing industries of the secondary economic sector are becoming increasingly mechanized. So, the main task now is to bring the producer and the consumer together— through advertising, transportation, selling, install-

ing, servicing, lending money, and so forth. Moreover, as the economies of the most advanced countries achieve **postindustrial** status, the *information* (**quaternary**) sector becomes increasingly important—and, by one recent reckoning, is already the single largest employment sector in the United States in the late 1980s.

Tertiary industries are often quite small, so that, in number, they outrank both primary and secondary establishments. In terms of location, these are the most directly people oriented of all industries, so that a map showing population clusters (Plate 5, for example) is a good indicator of the world distribution of tertiary economic activities. Tertiary industries, quite logically, tend to be concentrated in cities and towns; unlike the primary industries, they are not tied to the location of raw materials nor are they as closely governed by locational factors as are the secondary industries. Indeed, certain tertiary industries today are often called *footloose* industries because their range of locational choices is so wide.

389

reduce Soviet dependence on external products. In the 1930s, industrial plants were located far from the major potential military threat (Germany to the west) so that they could continue to function even in the event of invasion and war. This practice of protective dispersal continues today.

• Industrial Regions of the World •

Whatever the underlying causes—and they are many and complex—a small number of major industrial regions exists today whose factories produce a large part of the entire world's industrial output. Take note that we are now considering the **secondary** industries, the industries that manufacture goods from raw mate-

rials made available mainly by the primary industries. It is hardly necessary to enumerate these industries, for when we view the major regions we will frequently refer to specific manufactures. Four major regions can be recognized:

1 Western and Central Europe.
2 The United States and Canada.
3 The core area and its eastward extension in the U.S.S.R.
4 The Japan–China industrial concentration.

EUROPEAN MANUFACTURING REGION

It is appropriate to begin with Europe for here the Industrial Revolution commenced. The industrial regions of Europe largely constitute the **European** heartland. Europe's industrial region consists of a number of districts, from the British Midlands in the west to Silesia in the east.

Nowhere is the principle of functional specialization better illustrated than in the industries of Britain. Wool garment-producing cities lie east of the Pennines, the mountain backbone of the island, centered on Leeds and Bradford. The textile industries of Manchester, on the other side of the range, produce cotton goods. Birmingham (now with 3 million inhabitants) and its neighbors concentrate on the manufacture of steel and metal products, including automobiles, motorcycles, and airplanes. The Nottingham area specializes in hosiery; Leicester adds boots and shoes to that field. In northeast England, shipbuilding and the manufacture of chemicals are the major large-scale industries along with iron and steel production and, of course, coal mining. Coal became an important British export early; with this resource's seaside locations in northeast England, Scotland, and Wales, it could be shipped directly to almost any part of the world. In Wales, coal mining is supplemented by shipbuilding and the production of iron and steel: the same complex of industries is based on the coalfields of Scotland, where Glasgow (2 million) has become one of Britain's major industrial cities. It is a measure of the impact of the Industrial Revolution that today four fifths of the population of Scotland is concentrated in the area identified on Figure 10.1 as the Edinburgh–Glasgow Corridor.

Britain, at the vanguard of the Industrial Revolution, has not always kept up with modern developments. Plants that at one time were the epitome of industrial modernization still operate today, but they are comparatively inefficient, expensive to run, slow, and wasteful. With the relative decline of the industrial cities of the Midlands and northern England, there is a tendency for industrial enterprises that can afford it and are able to do so to relocate near the old focal point of Britain, London. Here, still, is the greatest domestic market of the British Isles, and increasingly that market is important to local manufacturers as the competition for markets from elsewhere intensifies. All this reflects the decreasing importance of coal in the energy-supply picture (which is changing toward nuclear power), the desire to start afresh with up-to-date machinery, and a realization that

London, in addition to forming a huge domestic market, is also a good port through which to import raw materials. So, London, too, shows up on Figure 10.1 as an industrial district within the European region.

Paris does also. When the Industrial Revolution came to the mainland, Paris was already continental Europe's greatest city, but Paris did not (as London did not) have coal or iron deposits in its immediate vicinity. Even so, Paris was the largest local market for manufactures produced for hundreds of miles around, and when a railroad system was added to the existing road and waterway connections, the city's centrality was strengthened still further. As in the case of London, Paris itself began to attract major industries, and the city, long a center for the manufacture of luxury items (jewelry, perfumes, fashions), experienced great industrial expansion in such areas as automobile manufacturing and assembly, the metal industries, and the production of chemicals. With a ready labor force, an ideal position for the distribution of finished products, the presence of governmental agencies, a nearby ocean port (Le Havre), and France's largest domestic market, the development of Paris as a major industrial center is no accident.

Europe's coal deposits, however, lie in a belt across northern France, Belgium, north-central Germany, northwestern Czechoslovakia, and southern Poland—and it was along this zone that mainland Europe's primary concentrations of heavy industry developed. Before its fragmentation into East and West Germany, three such manufacturing districts lay in the German state: the Ruhr, based on the Westphalian coalfield, the Saxony area near the boundary with Czechoslovakia (now in East Germany), and Silesia (now in Poland). Among these, the Ruhr has become the greatest industrial complex of Europe; West Germany ranks among the world's leading coal and steel producers (Table 10.1), and is still Europe's leading industrial power.

The Ruhr, named after a small tributary of the Rhine River, provides an illustration of the advantages of a combination of high-quality resources, good accessibility, and a position near large markets of high purchasing power. The coal of the Ruhr is of excellent quality for steel-making purposes. The iron ore locally available was soon exhausted, but ores could be

TABLE 10.1 Recent Steel Production: The World and Selected Leading Countries (millions of metric tons)

	1979	1981	1983
World	746.6	706.3	663.7
U.S.S.R.	149.0	148.5	152.5
Japan	111.7	101.7	97.2
U.S.	123.7	108.8	76.8
China	34.4	35.6	40.3
West Germany	46.0	45.5	40.2
Italy	24.3	27.0	24.4
France	23.4	23.2	19.6
United Kingdom	21.6	16.2	16.2

Source: Organization for Economic Co-operation and Development, Annual Steel Market Reports.

shipped in from overseas with only a single transshipment. Long before the Ruhr emerged as Germany's major complex of heavy industries, cities had been growing nearby along the Rhine, including Cologne (Köln) and the present capital of West Germany, Bonn. The surrounding agricultural area was productive and densely populated. After the early 1870s, the Ruhr expanded rapidly and increasingly received the benefit of government support in the form of subsidies and tariff protection. The chief manufactures made here are iron and steel (and products forged from these such as railroad equipment, vehicles, and machinery), textiles, and chemicals, with an almost endless variety of associated products.

While the Ruhr region specializes in heavy industries, Saxony (now in the German Democratic Republic [East Germany]) is skill- and quality-oriented. Today this region, which includes such famous cities as Leipzig (printing and publishing), Dresden (ceramics), and Karl-Marx-Stadt (textiles), benefits from nearby coalfields, but even before the Industrial Revolution made its impact here, these places were substantial manufacturing centers.

Farther eastward, the industrial district of Silesia also was first developed by the Germans, although it now lies in Poland and extends into Czechoslovakia. A major industrial district is developing in southern and southwestern Poland, based on local, high-quality coal resources and iron ores (though less plentiful), which are, however, supplemented by imports from the Soviet Ukraine. Czechoslovakia shares the raw materials with Poland; they lie astride the gap between the Erzgebirge (Ore Mountains) and the Carpathians. In Czechoslovakia, this particular industrial district is called Moravia, and the city of Ostrava, with its metallurgical and chemical industries, lies at its center.

Our map shows other industrial districts in Europe, notably those focused on large cities such as Berlin and Warsaw, and the rapidly emerging northern Italian district. Even so, the map is but a generalization of Europe's total industrial strength. We have highlighted only the really outstanding areas. Other noteworthy concentrations of industry exist in eastern Sweden, around Stockholm (noted for paper, wood products, textiles, precision instruments); in northern and eastern Spain; and in east-central France, where the focus is on the Lyon textile complex. These industrial areas, just districts in Europe's massive industrial structure, would stand out as major concentrations in other parts of the world.

NORTH AMERICAN MANUFACTURING REGION

Europe's industrial prowess notwithstanding, the North American industrial complex has no rival in the world today. Sustained by a wide array of natural resources and spurred by networks of natural as well as man-made transport networks, remoteness from the devastation wrought by wars in other industrial regions, and positioned on the doorstep of the world's richest market. North American manufacturing developed rapidly and successfully. Ample capital, mass production, specialization, and diversification mark the growth of this region.

American Manufacturing Belt

The bulk of American manufacturing in the United States and Canada is concentrated in the rectangular-shaped region delimited on Figure 10.2, the *American Manufacturing Belt*, which extends from the northeastern seaboard to Iowa and from the St. Law-

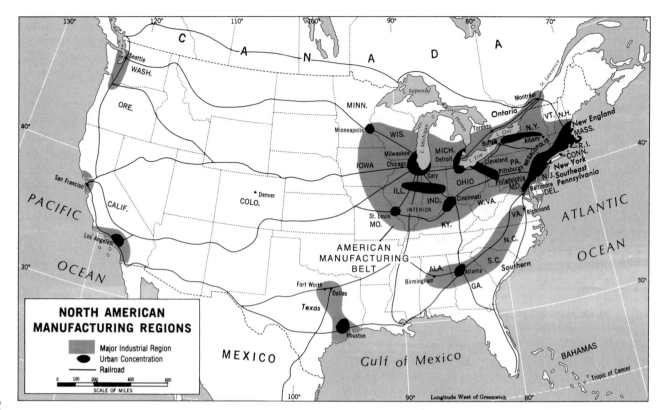

FIGURE 10.2
Leading manufacturing regions of North America.

rence Valley to the confluence of the Ohio and Mississippi rivers. Manufacturing in North America began in *New England* as early as late Colonial times, but these northeastern states are not especially rich in mineral resources. Still, this oldest manufacturing district continues to produce high-quality light manufactures. This is one of those industrial areas where skills are generated; the kind that contribute importantly to balance in manufacturing activity. However, as we know, other areas have long-since overtaken New England in terms of productivity and diversity. As the settlement frontier in North America marched westward, manufacturing spread as well. Farther west and south is where some great mineral deposits, fuels as well as ores, were discovered.

Another district of rather early importance centers on metropolitan *New York*, at the very heart of mega-

lopolis and the locus today of tens of thousands of industrial establishments. Again, New York and its immediate environs are not especially well endowed with mineral resources, but we have here a situation not altogether different from the one we encountered in the cases of London and Paris. The agglomeration process plays a major role, New York is the country's largest, most concentrated market (lying at the heart of a still larger market); the place where a huge skilled and semiskilled labor force exists; the focus of an intensive transport network; one of the world's greatest ports and break-of-bulk locations. New York is also a center for business and finance; numerous companies with nationwide and global tentacles have their head offices, their management operations, in the city and its growing suburbs. As for the character of its manufacturing, the New York district, like New

England but on a more massive scale, is an area of predominantly light industries. Clothing is made here as are books, metal goods, various kinds of food. There is also some treatment of incoming commodities from overseas—for example, the refining of petroleum and sugar.

Still farther southwest along the northeastern seaboard, the southern half of megalopolis stands out as a district of heavier manufacturing. This district centers on *Southeastern Pennsylvania*, specifically metropolitan Philadelphia, but also encompasses greater Baltimore. As we noted earlier, iron ores shipped all the way from South America and Canada are smelted right on the waterfront in tidewater steelmills; in the past, the district's raw materials came from local sources. In addition to steel mills, there are chemical industries (most notably in northern Delaware), pharmaceutical factories, and plants of lighter manufactures. A large market is clustered here; a substantial, diversified labor force; some of the major ports on the eastern seaboard; and an excellent transportation network that serves the district.

Farther west, we can discern a rather well-defined industrial district in *Upstate New York*, extending approximately from Albany on the Hudson River to Buffalo on the shore of Lake Erie. As in Britain, much local specialization occurs in this district, and a wide variety of manufactured items is produced. Industrial decline, however, has marked this district since 1970. Some familiar city names and products are known here: Rochester for its cameras and optical products, Schenectady for electrical products, and Buffalo for its steelmaking plants. The old Erie Canal, dug in the early nineteenth century to connect the east coast to the Great Lakes, engendered the growth of this district.

Canada and the Interior

On the northern and western sides of Lake Erie lies Canada's *Ontario* district, really a continuation of the Albany–Buffalo industrial zone. Toronto is the focus of this district, which is Canada's industrial heartland and extends all the way to Windsor, the city across the river from Detroit. Toronto and Hamilton manufacture steel, and various centers in this district manufacture

products ranging from heavy machinery to clothing to canned foods. Figure 10.2 also shows Montréal to lie in a zone of manufacturing. The Montréal–Ottawa area along the upper St. Lawrence River is no match for the Ontario district, but it has one big advantage: cheap hydroelectric power. This, we know, attracts such industries as aluminum refining and papermaking. Other industries have followed suit and now flour mills, sugar refineries, and textile plants have located there as well.

Returning to the United States, Figure 10.2 shows several major industrial districts to lie west of Buffalo, including those we might identify as Pittsburgh–Cleveland, Detroit–southeast Michigan, Chicago–Gary–Milwaukee and, to a lesser degree, those focused on Minneapolis, St. Louis, Cincinnati, and several smaller, but yet substantial metropolitan centers. Collectively, we might identify these as the U.S. *Interior* industrial district of the American Manufacturing Belt, but to try to identify the great variety of products fabricated here would be impractical. Many of them are very well known—who thinks of Pittsburgh without thinking at the same time of its huge steel mills, or of Detroit without the image of automobiles, or of Gary (Indiana) and Chicago without miles of steel furnaces and mills as well as Lake Michigan barges bringing loads of iron ore from Lake Superior shores? This is where U.S. industrial power really transforms the landscape, where Appalachian coal and Mesabi iron ore are transformed, and where autos, bulldozers, harvesters, armored cars, and tanks roll off the assembly lines. Refrigerators, record players, toys, cornflakes, and pills—these and thousands of other products pour from the factories of the Midwest.

South and West

Figure 10.2 reminds us, however, that this industrial heartland we have been discussing is not the only region of significant development in the United States. A region we call the *Southern* district extends from the vicinity of Birmingham, Alabama, to Richmond, Virginia. Birmingham is a name associated with iron and steel in this country—as well as in Britain—and local raw materials have sustained this now rapidly declin-

393

The final step in the preflight production line of the Boeing Aircraft Company's Renton plant near Seattle underscores the matchless technology of American industry. Like Houston's aerospace industry and California's food-processing plants, this is just an outlier beyond the boundaries of the North American industrial heartland. The dimensions of North American industry are yet unrivaled.

this industrial district as well. Denver, also outside the American Manufacturing Belt, shares substantially in the growth of energy-based industry, too. Finally, our map shows three manufacturing zones lying on the West Coast—in the Los Angeles–San Diego, San Francisco, and Seattle areas. These *West Coast* districts respond to the West's growing markets; particular physical environments here have also boosted food-processing industries (citrus products in southern California, wines in the vicinity of San Francisco). Many of the aircraft you may have traveled on are built in the Seattle area: the aerospace industry—here led by Boeing—is of world-class proportions. The San Francisco Bay Area is also home to Silicon Valley, the research and manufacturing leader in the computer industry—a leading hearth of postindustrial America and its high-tech society. Nevertheless, these large metropolises with their substantial industrial complexes are but outliers of the American Manufacturing Belt, still by far the most productive world region of its kind.

U.S.S.R. MANUFACTURING REGION

As one would expect in so vast a country, the U.S.S.R.'s manufacturing districts are widely dispersed. Two of them, the districts around Novosibirsk and Karaganda in Soviet Asia, do not appear on our map (Fig. 10.3), which focuses on the western part of the U.S.S.R. Although those distant districts show signs of developing into important centers, they are so far dwarfed by the districts shown on our map: the area around Moscow, the Ukraine industrial district, the Volga district, and the district that lies astride the Urals. These lie in what some geographers would term *European Russia*, still the most populous region of the country and demarcated to the east by the Ural Mountains.

The degree of detail on Figure 10.3, it should be noted, also obscures some smaller but nonetheless significant industrial areas. The Leningrad urban area, for example, one of Russia's oldest manufacturing centers, still retains some importance. With 4.8 million residents, Leningrad has a substantial market, but possesses none of Moscow's advantages of centrality. Nor does Leningrad lie near major sources of raw

ing U.S. industry for many years. Atlanta, rapidly rising as the South's regionwide focus, has a growing industrial base. Cotton and tobacco rank high in this district's array of products and furniture is an important product, too. On the Gulf of Mexico, the Texas district is emerging, centered on the thriving urban areas of Houston and the Dallas–Fort Worth metroplex. The oilfields generate a major petrochemical industry here, and meat-packing and flour-milling industries are also substantial. Houston's aerospace and hospital complexes employ thousands of people in

FINLAND

North Dvina

Arkhangelsk

BALTIC SEA

Tallinn

ESTONIAN S.S.R.

Leningrad

Riga

LATVIAN S.S.R.

LITHUANIAN S.S.R.

R.S.F.S.R.

POLAND

Minsk

BELORUSSIAN S.S.R.

Kirov

Rybinsk

Yaroslavl

Kineshma

Ivanovo

Kostroma

Vyshniy Volochek

Kalinin

CENTRAL

Rzhev

Moscow

Noginsk

Kovrov

Gorkiy

Cheboksary

INDUSTRIAL

Kolomna

REGION

Serpukhov

Tula

SOVIET

Dnieper

Kursk

RUSSIAN

Kazan

SOCIALIST

FEDERATED

Ul'yanovsk

Kuybyshev

Syzran

Chapayevsk

VOLGA

Saratov

Engels

REGION

Bereznikl

Serov

Krasnokamsk

Chusovov

Nizhniy Tagil

Perm

Lys'va

Votkinsk

Sverdlovsk

Izhevsk

Kyshtym

URALS

Chelyabinsk

REGION

Magnitogorsk

URAL

MOUNTAINS

Orsk

Kama

Volga

Don

KAZAKH S.S.R.

Volgograd

Gur'yev

Astrakhan

Ural

ARAL SEA

UZBEK S.S.R.

Kiev

UKRAINIAN S.S.R.

Poltava

Kharkov

Kremenchug

Voroshilovgrad

Kirovograd

Dnepropetrovsk

Makeyevka

Shakhty

Krivoy Rog

Dneprodzerzhinsk

Zaporozh'ye

Donetsk

Novocherkassk

UKRAINE REGION

Nikopol'

Taganrog

Rostov

MOLDAVIAN S.S.R.

Berdyansk

Nikolayev

Melitopol'

Odessa

Kherson

Stavropol

Krasnodar

CASPIAN SEA

ROMANIA

Sevastopol

Sochi

Groznyy

Baku

BULGARIA

BLACK SEA

Tbilisi

GEORGIAN S.S.R.

ARMENIAN S.S.R.

AZERBAYDZHAN S.S.R.

TURKMEN S.S.R.

GREECE

TURKEY

IRAN

East of Greenwich

U.S.S.R. MANUFACTURING REGIONS

Manufacturing region

Railroad

MILES

200   0   200   400   600   800

200   0   200   400   600   800

KILOMETERS

FIGURE 10.3

Manufacturing regions of the Soviet heartland.

materials; these must be brought from afar (except for bauxite, located nearby at Tikhvin). However, Leningrad was at the vanguard of the Industrial Revolution in nineteenth-century Russia, and its specialization and skills have survived. In the 1980s, the city and its environs annually contribute about 5 percent of the country's manufacturing, much of it through high-quality machine building. In addition to the expected association of industries in a large urban area (food processing, textiles, chemicals), Leningrad has major shipbuilding facilities and, of course, the activities associated with its port and the nearby naval base at Kronshlot (Kronstadt).

Central Industrial Region

Although Leningrad does constitute a notable industrial center, it is dwarfed by the dimensions of the region surrounding the capital, Moscow. This, the Soviet Union's oldest industrial district, is often called the *Central Industrial Region*. It is not especially well endowed with natural resources, but it possesses other advantages. With the present distribution of population, Moscow commands outstanding centrality; roads and railroads radiate in all directions—to the Ukraine in the south, to Minsk, Belorussia, and Eastern Europe in the west, to Leningrad and the Baltic coast in the northwest, to Gorky (Gor'kiy) and the Ural district in the east, and to the cities and waterways of the Volga in the southeast. A canal links the city to the Volga, the U.S.S.R.'s most-used river. Moscow is the focus of an area that includes some 60 million people (more than one-fifth of the country's total population), many of them concentrated in such major cities as Gorky (Gor'kiy), the automobile-producing Soviet Detroit (1.5 million); Yaroslavl' (0.7 million), the tire-producing center; Ivanovo (0.5 million), the Soviet Manchester, with its textile industries; and Tula (0.6 million), the mining and metallurgical center.

The comparative paucity of natural resources within the Central Industrial Region is counterbalanced by an excellent transport system, which facilitates the inflow of a wide range of commodities—gas from Stavropol', petroleum from the Volga and new western Siberian fields, and electricity from Volga River dams for energy; and metals, wood, wool, cotton, flax, and leather (among many other items) for processing. The major flow—other than that of food—is of these materials into the Central Industrial Region and of finished products being distributed to the consumers. Thus, the manufactures of the district are marked, as are those of the London and Paris areas, by high values in relation to bulk. Moscow has always been the leading center of skilled labor in the Soviet Union, and the available skills are those that—as the U.S.S.R. found out after Western Europe had—do not develop quickly and cannot be imparted overnight to peasant labor freed of farm work. Primacy, time, and place have favored Moscow and the Central Industrial Region.

Ukraine

In the **Ukraine** the U.S.S.R. has one of those regions where major deposits of industrial resources, fuels and ores lie fairly close together. The Ukraine began to emerge as a major region of heavy industry toward the end of the nineteenth century, and one major reason for this was the Donets Basin, one of the world's most productive coalfields. This area, known as **Donbas** for short, lies north of the city of Rostov (Fig. 10.3). In the early decades, this Donbas field produced more than 90 percent of all the coal mined in the country. Other fields have since been opened up, but even today, three quarters of a century later, the Donets Basin alone still accounts for between one quarter and one third of the total U.S.S.R. coal output. And the coal is of high grade; some of it is even anthracite, and most of it is suitable for coking purposes.

What makes the Ukraine unique in the U.S.S.R. is the location—less than 200 miles (300 km) from all this Donbas coal—of the Krivoy Rog iron ores. The quality of this ore is very high, and although the best ores have been worked out, the field still produces. Major metallurgical industries arose on both the Donets coalfield and near the Krivoy Rog iron ores. Donetsk and its satellite, Makeyevka, dominate on the Donets side of the district (this is the Soviet Pittsburgh), whereas Dnepropetrovsk is the chief center in the cluster of cities on the Krivoy Rog side. In one way or another, all the major cities located nearby have benefited from the fortuitous juxtaposition of minerals in the southern Ukraine: Rostov, Volgograd, and Kharkov (Khar'kov), and even Odessa and Kiev farther away. Like the Ruhr, the Ukraine industrial district lies in an area of dense population (hence available labor), adequate transportation, good agricultural productivity, and large local markets. And when the better ores were exhausted, somewhat lower-grade ores on the basis of which production could be maintained proved to be available. The biggest development in this context is the discovery of the so-called Kursk magnetic anomaly, positioned north of the Ukraine district and south of the Central Industrial Region (Fig. 10.3), and thus able to serve both districts. The Kursk magnetic anomaly is a large iron ore deposit whose exploitation had been delayed by technical problems. Mining began in 1959, and the ores have given the metallurgical industries from Tula to Rostov a new lease on life.

Volga and Urals Districts

Two other, elongated industrial districts appear on Figure 10.3. Their elongation relates to their uniting physiographic features: the Volga River and the Ural Mountains. The **Volga** region has witnessed major development in the decades starting during the Second World War. With the Ukraine and the Moscow area threatened by the German armies, whole industrial plants were dismantled and reassembled in Volga cities, protected from the war by distance. Kuybyshev was even the Soviet capital for a time. Then, after the war, the progress set in motion earlier continued. A series of dams was constructed on the Volga, and electric power became plentiful. Oil and natural gas proved to exist in quantities larger than anywhere else in the U.S.S.R. Canals linked the Volga to Moscow and the Don River, making the importing of raw materials easy. The cities along the river Volga, spaced at remarkably regular intervals, developed specializations: Kazan is known for its leathers and furs, Volgograd for its metallurgy, Saratov for chemicals and Kuybyshev for its huge oil refineries. This district's contribution to the Soviet economy is still increasing.

The last major industrial district on our map is the **Urals** region, the easternmost district of major importance (still farther east, the Kuznetsk–Novosibirsk area is gaining ground). This area, too, developed rapidly during the Second World War. But there has been nothing artificial in its growth. The Urals yield an enormous variety of metallic ores, including iron, copper, nickel, chromite, bauxite, and many more. The only serious problem is coal. There is not enough of it, and what there is does not have the required quality. So, coal is railed in all the way from the Kuznetsk Basin, 1000 miles (1600 km) to the east. In the cities of the Urals district, metals, metal goods, and machinery are produced in great quantities. With the Siberian centers, the Urals district now produces as much as half of all the iron and steel made in the entire U.S.S.R. It is a sign of the eastward march of the country's center of gravity and a part of the national planners' grand design.

The Eastern Interior

Three industrial regions are developing in the Soviet east, islands of industrial and mining activity located within the ribbonlike eastward extension of the Soviet heartland. About 600 miles (1000 km) east-southeast of the Urals region lies the Karaganda–Tselinograd area, where iron and steel plants, a chemical industry, and a variety of other manufacturing plants have emerged. More important is the region another 600 miles eastward, the Kuznetsk Basin or *Kuzbas*. In the 1930s, this area was developed as a supplier of raw materials, especially coal for the Urals, but this function has steadily diminished in importance as local industrial growth has accelerated. The original plan was to move coal from the Kuzbas to the Urals and to let returning trains carry iron ore to the coalfields, but, subsequently, good iron ores were discovered in the area of the Kuznetsk Basin itself. As the resource-based Kuzbas industries grew, so did its urban centers: Novosibirsk (with 1.6 million inhabitants) stands on the Trans-Siberian Railroad at the crossing of the Ob River as the symbol of Soviet enterprise in the vast Siberian interior. To the northeast lies Tomsk, one of the oldest Russian Siberian towns in the whole eastern region, founded three centuries before the Bolshevik takeover and now caught up in the modern development of the Kuznetsk area. Southeast of Novosibirsk lies Novokuznetsk, a city of nearly 700,000 people specializing in the manufacture of such heavy engineering products as rolling stock for the railroads. Aluminum products, using Ural bauxite, are also fabricated here.

Finally, there is the region called the Soviet Far East, long focused on the Pacific port of Vladivostok but developing now in, and around, such newer cities as Komsomolsk (the first steel producer in the region) and Khabarovsk, now the leading center in the Soviet Far East (with metal and chemical industries). The remote Far East is finally beginning to yield its rich raw materials, and there is enormous potential for development beyond the low-grade coal and iron ore found to date. Already the zinc and tin deposits mined here constitute the country's largest sources of these metals. Development of this region, which is on China's doorstep, is a top strategic as well as economic priority.

JAPAN–CHINA MANUFACTURING REGION

Japan now is one of the world's three leading industrial powers; by some criteria, it already ranks in first or second place. This is all the more remarkable when it is realized that Japan has a very limited domestic natural resource base. Much of what Japan manufactures is made out of raw materials bought all over the world and shipped to that small archipelago off the East Asian coast. Japan's national territory is just 1/25 the size of the United States and 1/60 that of the Soviet Union. Its population of 121 million is barely more than half the U.S. total. Its transformation into a world-class industrial giant has rightfully been described as an economic miracle.

Japan rose from comparative obscurity, from UDC to a modern industrial power, in just one century. But we should not forget that Japan, prior to its modernization era, did have a domestic manufacturing industry. During the first half of the nineteenth century, manufacturing of the handcraft variety was widespread in Japan. In cottage industries and community workshops, the Japanese produced silk- and cotton-textile manufactures, porcelain, wood products, and metal goods. During those days, the small domestic re-

sources were enough (Japan does have some ores) to supply the local industries. Power came from human arms and legs and from wheels driven (as in England) by water. The chief source of fuel was charcoal. Thus there was an industry, there was a manufacturing-oriented labor force, and there were known manufacturing skills.

Japan's push toward industrial modernization began with the integration and enlargement of these community and home workshops, the export to world markets of light manufactures, and the adoption of the techniques generated by the Industrial Revolution. There was some coal, and Japan's topography produced numerous sites for hydroelectric power development. As industry grew and the Kyushu coalfields were worked out, it was realized that raw materials would have to come from overseas. The age of the factory had arrived. The coastal cities were in the best position to receive the raw materials and to convert them to finished products. Now, the regional industrial pattern shown in Figure 10.4 took shape.

Kanto Plain

Figure 10.4 reveals that the largest district of manufacturing in Japan is the **Kanto Plain**, which contains not only Tokyo (the capital and largest city) and Yokohama, its major port, but also several smaller but substantial industrial towns such as Kawasaki and Chiba. The Kanto Plain and its gaint conurbation produces between one fourth and one fifth of the country's industrial output. Tokyo is the leading steel producer, with ores coming from as far as the Philippines, Malaya, Australia, India, and even Africa. Although hydroelectric power is brought in from interior mountain sites, coal is shipped to Tokyo from a new field in Hokkaido (the northernmost major island of Japan) and from Australia and America. Petroleum comes from Southwest Asia and Indonesia. The Japanese are able to pay for all this by selling enormous quantities of their manufactures on the world's markets.

Kinki District

The second-ranking manufacturing district in Japan is the **Kobe–Osaka** area, with Kyoto as an outlier (Kyoto is the ancient Japanese capital, still a significant light-manufacturing center). This district, long an area of

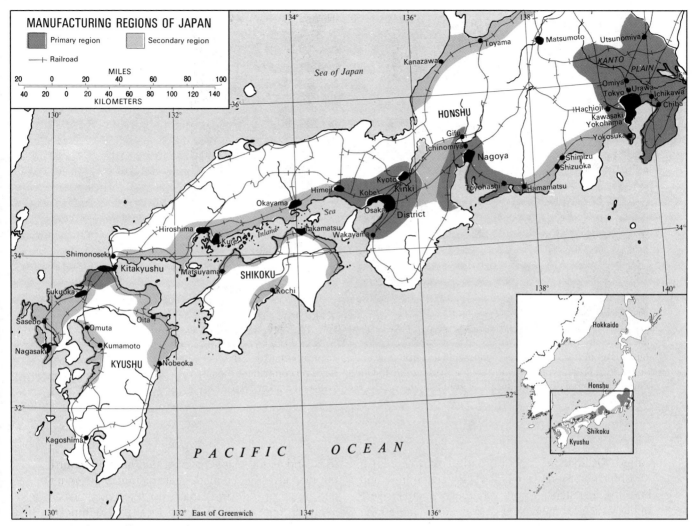

FIGURE 10.4

Manufacturing regions of Japan.

diversified manufacturing, has been developing heavier industries in recent decades and now produces not only textiles and chemicals but also iron and steel, machinery, ships, and airplanes. The Kobe–Osaka district, as Figure 10.4 shows, lies at the eastern end of the Inland Sea. The Inland Sea in many ways is the heart of Japan. Overlooking its waters are dozens of large and small cities, and crisscross traffic is as busy as on any of the world's major rivers.

Other Districts

Between the Kanto Plain and the Kinki district (the general name for the Kobe–Osaka–Kyoto triangle) lies the **Nagoya** district, where a wide variety of textiles are made and heavy industries are also developing. Many of the Japanese automobiles you see on the world's roads are built at Nagoya.

At the opposite end of the Inland Sea from Kobe–Osaka lies a fourth major industrial district,

Japan is an industrial giant, its products ranging from tiny toys to giant supertankers. This shipyard at Nagasaki manufactures oil tankers for sale to United States oil companies. Metals, energy, and other needs are imported from afar; the products go to distant markets. Japanese skills are the key.

400

called **Kitakyushu**, a conurbation of five North Kyushu cities. Here were Japan's first coal mines, and here the first steel mills of modernizing Japan were built—which for many years remained the largest the country had. Look for this district to develop even faster than it has should the trade normalization with China continue. No place in Japan is located in a better position to do business with mainland Asia. Presently, heavy industries dominate, with shipbuilding and iron- and steelmaking in the lead, supplemented by a major chemical industry and numerous lighter manufacturing plants.

Only one manufacturing district in Japan, in the area depicted by our map, lies outside the belt extending from Tokyo in the east to Kitakyushu in the west. This is the district centered on **Toyama**, on the Sea of Japan. The advantage here is cheap electricity from nearby hydroelectric stations, and the cluster of industries reflects it. Paper manufacturing, chemical indus-

tries, and textile plants have located here. Of course, our map gives but an inadequate picture of the variety and range of industries that exist throughout Japan, many of them oriented to the local (but significant) market. Thousands of manufacturing plants operate in the cities and towns other than those on the map, even in the cold, northern island of Hokkaido. As we did in the case of Europe, North America, and the U.S.S.R., we have focused on the really outstanding manufacturing regions of Japan.

China

The People's Republic of China has eight times the number of people Japan has, but its industries cannot yet match those of its island neighbor. Still, China, too, is on the move, and we now examine those areas where its industrial progress is most evident. In the past, China's industries were established and maintained by foreigners, first the Europeans, who built factories in the coastal cities to process goods to be sold at home, and then the Japanese, who invaded Korea and established a sphere of influence in Manchuria, China's northern province (Fig. 10.5). As in the case of the Soviet Union but on a smaller scale, some dispersal of industries into the interior was occasioned by the Second World War. Chongqing (Chungking) and Chengdu (Chengtu), for example, grew substantially during this period. But the major industrial development of China has occurred during the modern, Communist period. When the Communist planners took over in Beijing (Peking), one of their priorities was to develop as rapidly as possible China's own resources and industries.

China is a very large country, and it is likely that some of its natural resources have yet to be discovered. Even so, China has already proved to possess a substantial domestic resource base. In terms of coal, there is hardly a limit on industralization in China: the quality is good, the quantity enormous, and many of the deposits are near the surface and easily extracted. China's iron ores are not so productive and are generally of rather low grade—but new finds are frequently made and the picture has improved steadily. This is also true of China's oil potential. In recent years, China, with the aid of Western companies, has intensified its search for oil reserves, and in two areas—the

FIGURE 10.5

Leading manufacturing regions of China.

western interior and the continental shelf—the effort has achieved some results. Nonetheless, in the late 1980s, China's reserves were still estimated to constitute less than 3 percent of the world total.

China's industrial heartland extends from the Northeast (the region formerly called Manchuria [now the Northeast]) to the Lower Chang Jiang (Yangtze River), from the vicinity of Shenyang (Mukden) to the environs of Shanghai. As Fig. 10.5 shows, this industrial region is not continuous, but consists of three major and several minor districts. The *Northeast* has become China's industrial core, and this district's largest city, Shenyang, with a population of 3.6 million, has emerged as the Chinese Pittsburgh. This is based on the region's coal and iron deposits, which lie in the basin of the Liao River. There is coal within 100 miles (less than 150 km) to the west of Shenyang and about 30 miles (50 km) to the east, at Fushun, as well. To the southwest, just 60 miles (100 km) away, lie the iron ores of Anshan. Since this iron ore is of rather low grade, the coal is hauled from Fushun to the iron deposits. Under the circumstances, this is the cheapest way to make iron and steel here. Anshan has become China's leading iron and steel producing center, but Shenyang lies amidst the largest and most diverse overall industrial complex, with metallurgy, machine fabrication, and other large engineering works. The city of Harbin is just becoming part of an expanding continuous Northeast industrial region. Already Harbin (2 million), China's northernmost large city, has plants making farm machinery, textile factories, food-processing industries, and light manufacturing plants of many kinds.

The *North China* district, including the capital of Beijing and the port city of Tianjin (Tientsin), also benefits from nearby coalfields. As one might expect, here in the productive basin of the Huang He, food processing and textile making rank next to the heavy industries. Still farther to the south lies the third district of China's industrial heartland, the *Chang (Yangtze)* district. On our map, the Chang district is divided into two areas, one centered on Shanghai, China's largest industrial city, and the other focused on Wuhan, astride the Chang Jiang several hundred miles upstream. In fact, still farther upstream along the Chang Jiang lies Chongqing (Chungking), the city that grew so strongly during the Second World War.

Whether we view this Chang district as one region or as three, it is a pacesetter for Chinese industrial growth, if not in terms of iron and steel production alone, then in terms of its diversification of production and its local specializations. Railroad cars, ships, books, foods, chemicals—an endless variety of products comes from the burgeoning Chang district.

Industrial growth by agglomeration and in response to the expansion of urban markets has also generated manufacturing centers in and near the Guangzhou (Canton)–Hong Kong area in the south, and in China's larger interior cities. The fortuitous wide dispersal of the country's raw materials places many cities near natural resources, stimulating industrial development and broadening China's industrial base as the modernization program launched after Mao's death in 1976 brings increasing national progress and prosperity in the 1980s.

INDUSTRIAL REGIONS ELSEWHERE

Compared to the regions discussed so far, industrial growth in other areas of the world is of smaller proportions (Figure 10.6). Previously in this chapter, industrialization in LDCs was described as insular—resembling islands of modernization in otherwise slow-to-change economies. That situation prevails, to a greater or lesser degree, in countries of South and Southeast Asia, Africa, and Middle and South America.

India

When India became independent in 1947, it inherited the mere beginnings of an industrial framework. After 90 years of British control over the economy, only 2 percent of India's workers were involved in industry, and manufacturing and mining combined produced only about 6 percent of the national income. Textile and food-processing industries dominated; what little heavy industry there was had been hastily launched during the Second World War. Manufacturing was concentrated in the major cities; Calcutta led, Bombay was next, and Madras ranked third.

The pattern of manufacturing today still reflects these beginnings, and industrialization and urbaniza-

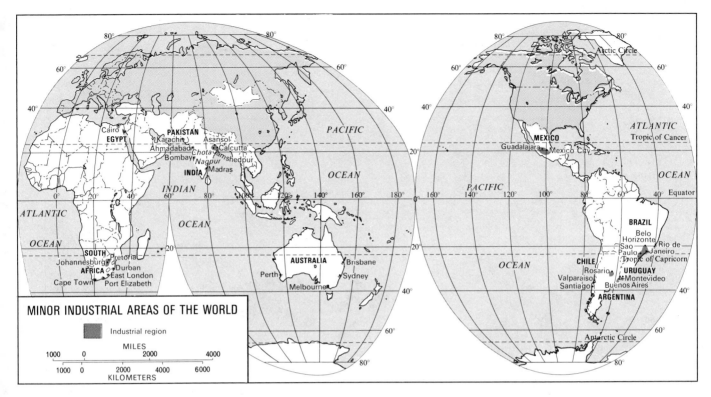

FIGURE 10.6
Smaller industrial areas of the world.

tion in India have proceeded slowly even after independence (Fig. 10.7 and Plate 9). Calcutta now forms the center of India's eastern industrial region, in the states of Bihar and West Bengal, where the jute industry dominates but engineering, chemical, and cotton industries also exist. In the adjacent interior Chota–Nagpur district, coal mining and iron and steel manufacturing have developed, and Jamshedpur is the nucleus of heavy industry. On the opposite side of the subcontinent, two industrial areas combine to form the western industrial district: one centered on Bombay and the other focused on Ahmadabad. This district, lying in the states of Maharashtra and Gujarat, specializes in cotton and chemicals, with some engineering and food processing. The cotton industry has long been an industrial mainstay in India, and it was one of the few industries to derive some benefit from

the new economic order brought by the British after 1857. With the local cotton harvest, the availability of cheap yarn, abundant and cheap labor, and the power supply from the Western Hills' hydroelectric stations, the industry thrived, and today outranks Britain itself in the volume of its exports. Finally, the southern industrial region centers on Madras and emphasizes textiles and light engineering.

When India achieved independence, the new government immediately set out to develop its own industries to lessen Indian dependence on imported manufactures and to cease being an exporter of raw materials to the developed world. In the process, emboldened by the early results of the Green Revolution (which raised crop yields), Indian planners actually overspent on their industrialization programs—but the problem of rapid population growth bedeviled indus-

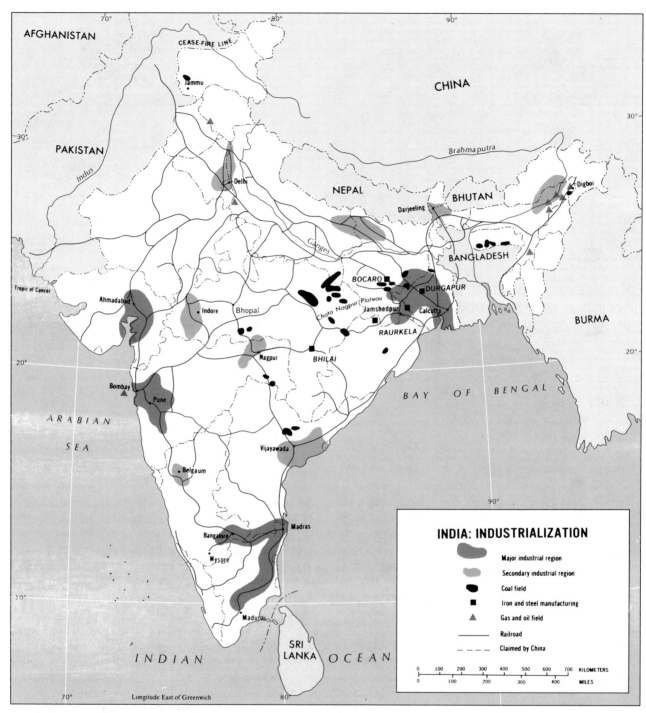

FIGURE 10.7

try as it did agriculture. Unemployment was to be reduced as industrialization progressed; instead, unemployment rose. Per capita incomes would rise as high-volume products came from new assembly lines; but incomes rose very little and today they still average under $260 (U.S.) per year per person. India has limited coking coal deposits as well as large, but lower-grade, coalfields in the Chota–Nagpur; but in the absence of known major petroleum reserves (some oil comes from Assam, Gujarat, offshore from Bombay, and the Punjab), it must spend heavily on energy every year. Major investments have been made in hydroelectric plants, especially multipurpose dams that provide electricity, make irrigation possible, and facilitate flood control. India's iron ores in Bihar and Karnataka may be among the largest deposits in the world, but despite the emergence of Jamshedpur and environs as India's Pittsburgh, the country still exports iron ore as raw material to DCs. Entrenched patterns of this kind are difficult to break.

Such obstacles notwithstanding, India's industrial progress is noteworthy. In addition to its iron and steel industries at Jamshedpur, the cotton industry centered at Bombay and Ahmadabad has elevated India to rank among the world's leading textile exporters (only Japan outranked India during the 1970s). Calcutta and environs have developed into a major center of diversified manufacturing, including railway assembly plants, food-processing factories, jute mills, chemical industries, printing plants, and a host of other industries mainly oriented toward the domestic market.

Africa and South America

In Africa, a growing industrial complex is located in South Africa's interior, in the southern Transvaal and the northern Orange Free State. Johannesburg lies at the center of this industrial district of some 4 million people, which includes the iron and steel works near Pretoria, and steel mills in the Orange Free State. An important element of this industrial complex is the facility at Vanderbylpark that converts coal to oil in this petroleum-poor country. This complex had its origins in the Witwatersrand goldfields and subsequently, benefited from the presence of other industrial resources nearby. Lesser industrial districts have developed around Cape Town, Durban, and Port Eliz-

abeth, all on the South African coast. South Africa is blessed with a wide range of mineral resources, and its industrial strength is considerable—far ahead of any other African country. In second place might be Egypt, which boasts a steel mill built about 20 miles (30 km) south of Cairo, where iron ore from the country's southern deposits is smelted with coking coal imported via the Nile. Egypt has also developed a rather substantial textile industry, based on its large cotton crop.

In South America, substantial industrial development is quite limited, and the pattern of light manufactures for local markets, clustered around the major cities, prevails. The exceptions are concentrated in just four South American countries. Brazil, which can count coking coal and iron ore among its resources, has a developing industrial region in the Rio de Janeiro–São Paulo district that includes steelmaking and other heavy industries. Light manufacturing dominates in the core area of Argentina around Buenos Aires and neighboring Uruguay. Central Chile's industries also serve principally the local market, although some Chilean plants process mineral ores prior to export.

Local Markets

The role of local markets as opposed to international trade is significant in any assessment of world industrialization. Even in DCs such as Australia and New Zealand as well as developing countries such as Nigeria and Brazil, manufacturing is encouraged primarily *not* because it is hoped that the products of the factories will immediately sell well on the world market. Rather, the hope is to reduce the country's dependence on foreign manufactures, which are increasingly expensive to buy. Light manufacturing, then, dominates in those areas and tends to concentrate in, and around, the largest local markets: the urban centers. In Australia, manufacturing centers on Sydney, Melbourne, Brisbane, and Perth; production for the domestic market includes the making of vehicles, clothing, shoes, foods and beverages, and similar products. Australia does have a substantial export trade in certain foodstuffs (dairy products, meats) and wool, and some of the country's manufacturing involves the processing and packing of these goods.

The Development Dilemma

We now return to the theme that opened this chapter: the global problem of uneven development. In the industrial as in the agricultural world, the advantages lie with the DCs. But how has this situation become so extreme, and what perpetuates it?

Regional contrasts in development have been attributed to many factors. Climate has been blamed for it (as a conditioner of human capacity) as has cultural heritage (e.g., resistance to innovation) and, of course, colonial exploitation—more recently, neocolonialism. Obviously, the distribution of accessible natural resources and the territorial location of countries are part of it. Opportunities for interaction and exchange always have been greater in certain areas than in others. Isolation from the mainstreams of change has been a disadvantage. These factors, to a greater or lesser degree—depending on location—certainly have played their roles. Countries once well watered have fallen victim to climatic change. Colonial oppression thwarted development for long periods in many countries, and the end of colonial rule did not automatically bring an economic upsurge. The causes for development disparities can be found in physical as well as human geography.

CORE AND PERIPHERY

In the first five chapters, we viewed the physical world and the distributions of its climates, soils, and raw materials for industry. In this second part, the focus has been on the human world—the emergence of cultures and clusters, networks and systems. From the very beginning of this discussion of human geography, we have observed the evolution of **core areas** of human activity: the ancient cities, the culture hearths, the heartlands of empires, the sources of the Industrial Revolution, the growth of manufacturing regions. Consider this: such cores would not have developed without contributions from the surrounding areas. One of the first developments in the ancient cities was the creation of organized armed forces to help rulers secure taxes and tribute from the people in the coun-tryside. So, core and periphery (margin) became functionally tied together. The core had requirements, and the periphery gave.

Have modern times ended this core–periphery relationship? The answer is not only negative but that the situation has actually worsened. You can see core (have) and periphery (have-not) relationships in every country in the world. In many African countries (Kenya, Senegal, and Zimbabwe to name just three examples in different regions) the former colonial capital now lies at the focus of a core area. The tall buildings and governmental centers symbolize a concentration of privilege, power, and control. Here are the trappings of development: the paved roads and street lights, hospitals and schools, corporate headquarters and businesses, industries and markets. Surrounding the city (you will be reminded of von Thünen) lie farms producing high-priced goods, especially perishables, for the urbanites who can afford them. But travel into the countryside and life changes. In the periphery, the core is distant, even ominous, always a concern. What will be paid for crops on the markets in the core? What will the cost be of goods made there and needed here, in the countryside? How high will taxes be set?

This core–periphery arrangement is a system that keeps regional development contrasts high. The core may give the impression of a thriving metropolis, and may remind one of a city in a developed country. But the periphery is the land where underdevelopment reigns.

Looking at the world as a whole from this perspective, we see that core-periphery systems now encompass not just individual countries, but the entire globe. Today the whole of Western Europe functions as a core area, North America also forms such a core area, and Japan is a third core area in the Western, capitalist world. And the western Soviet Union constitutes a fourth global core area, the dominant core in the Communist world.

What was said about that powerful city in an otherwise UDC applies, in larger terms, to these global core areas as well. Whether in Western Europe, North America, Japan, or the western Soviet Union, here is where the power, money, influence, and decision-making organizations are concentrated. Imagine

406

that you, as a geographer, are asked to locate a factory (say a textile plant) to sell products on core-area markets. Salaries and wages in the core area are very high, so you look for an alternative—in the periphery, where wages are much lower. Raw materials also can be bought there. So, you advise the corporation to establish its factory in a certain country in the periphery, thus initiating a relationship between core-based business and periphery-based producers. Now, however, the corporate management expresses concern over political stability in that periphery country. Soon you are talking to representatives of the country's government about guarantees, and the economic connection also becomes a political one.

Is all this necessarily disadvantageous to the countries in the periphery, the mainly UDCs, the countries that we have called the Third World? After all, the core-area corporations and businesses make investments, stimulate economies, and put people to work in the countries of the periphery. The problem is that a dependency relationship develops that tends to diminish any such advantage. The profits made return mostly to the core and benefit the periphery much less. When you see that high-rise, modern hotel towering above the palm trees of a Caribbean island and watch its staff at work, you observe both sides of the issue. True, an investment was made here and jobs were created, but the hotel's profits go back to the corporate bank account—in New York, London, or Tokyo.

Another problem has to do with the sociopolitical effects. That textile factory mentioned previously needs a director, manager, and other administrative personnel. These people tend to be (or become) part of the elite, the upper class, in that periphery country, a kind of corps of representatives of the core area in the Third World. Because they represent core-area interests, these people sometimes have divided loyalties. What is best for core-area enterprises is not always best for their country.

Such examples of core–periphery interactions exist in almost countless number, and the entire world is interconnected to a far greater degree than many of us realize. We discussed the durability of tradition in the context of cultural geography and the landscapes these traditions create, even in this modern age. But superimposed on those regions is a global network of interaction, reaching even into the smallest village store in the most remote part of a UDC. When we recognize geographic realms and regions (and they certainly exist), we should also remember that they are enmeshed in a worldwide economic-geographic system—a system oriented to the great core areas.

And this includes the Soviet Union. Economic geographers sometimes blame the core–periphery dilemma on capitalist economics; in fact, the relationships between the Soviet Union and its Second World neighbors is not totally different. Like the three great capitalist core areas, Soviet economic enterprise (state directed, of course) aims to lower costs, strengthen the core, and influence events in the periphery.

PERSISTENT INEQUALITY

Thus, the countries of the periphery find themselves locked into a global economic system over which they have no control. More than 160 years ago, von Thünen noted the power of the central market over the outlying areas of production in his Isolated State; that principle still applies. Even countries that find themselves possessing commodities in high demand in the core (such as the OPEC states) have difficulty converting their temporary advantage into permanent or even long-term parity with core-area powers. Countries that depend heavily for their income on the export of raw materials such as strategic minerals (or single crops such as bananas or sugar) are at the mercy of those who do the buying.

But there are other reasons for the condition of many periphery countries. Like the city (Chapter 9), the core beckons constantly, siphoning off the skilled and professional people from the periphery countries. How many doctors, engineers, and teachers from India are working in the United Kingdom and the United States? And how India needs such trained people! Every loss is magnified—but oppressive political circumstances in periphery countries contribute as push factors.

Another side to the continued underdevelopment of periphery countries (by the measures given earlier) relates to cultural, traditional attitudes and values. The introduction of core-area tentacles in regions where traditional culture is strong may lead to a reaction, perhaps a resurgence of fundamentalism. This, in turn, inhibits further economic change. The invasion of modern ways can be unsettling, and not just in UDCs. Modern, secular values have also stimulated a fundamentalist religious reaction in the United States in recent years. Do not forget that the long traditions of societies in the periphery have generated large bureaucracies, whose interests are threatened by the kind of change development brings with it. Occasionally the tangible presence of the external (core) installation, such as an office building, airline agency, or retail establishment, is the target of violent attack. Such incidents symbolize ideological opposition to the intrusion they represent.

The countries of the periphery, therefore, confront problems of many kinds. They are pawns in a global economic game whose rules they cannot change. Their internal problems (e.g., tribalism) are intensified by the involvement of core-area interests. Their resource utilization (e.g., allocation of soils to food crops or export crops) is strongly affected by foreign interference. They suffer—more than core-area countries do—from environmental degradation, overpopulation, and mismanagement. They have inherited disadvantages that have grown, not lessened, over time. The resulting gaps between enriching cores and impoverished peripheries are a threat to the future of the world.

Reading about Economic Geography

Economic geography, as Chapter 10 proves, incorporates the study of many forms of productive activity: extractive, manufacturing, and service industries; fishing, farming, building; transport and trade; and much more. Economic geography naturally overlaps with urban geography (cities are hubs of economic activity) and other fields of the discipline. Much useful theory in geography has come from economic geography. The problems of national development are much debated in the geographic literature.

A good place to begin reading on economic geography is in some general texts on the topic, for example, J. O. Wheeler and P. O. Muller, *Economic Geography* (New York: Wiley, 1986). This book displays the wide range of the field and includes discussions of agriculture, industry, and service activities as well as energy, transportation, and urbanization. Another valuable work is also called *Economic Geography*, by J. W. Alexander and L. J. Gibson (Englewood Cliffs, N.J.: Prentice-Hall, 1987). An older book that still has much value was written by A. R. de Souza and J. B. Foust, *World Space Economy* (Columbus, Ohio: Merrill, 1979). This book contains some comments by geographers who are opposed to the capitalist system, and this makes the discussions of development problems especially interesting. A different book is T. J. Wilbanks's *Location and Well-Being: An Introduction to Economic Geography* (San Francisco: Harper & Row, 1980). This book considers some topics not stressed in our chapter, including governmental and public policy in economic-geographic context as well as planning issues.

More specialized aspects of economic geography also form the subjects of entire books. The problem of regional disparities is the focus of S. R. Jumper, T. L. Bell, and B. A. Ralston, *Economic Growth and Disparities: A World View* (Englewood Cliffs, N.J.: Prentice-Hall, 1980). A book by G. J. D. Hewings concentrates on industrialization and development: *Regional Industrial Analysis and Development* (New York: St. Martin's Press, 1977). This book uses models and theories to discuss regional industrial analysis. Another book that focuses on industrial aspects is P. J. McDermott and M. Taylor, *Industrial Organization and Location* (London/New York: Cambridge Univ. Press, 1982), although this book uses British industries as examples. Also important is a book by D. M. Smith, *Industrial Location: An Economic Geographical Analysis* (New York: Wiley, 1981). Here you can get acquainted with the theorists and theories of this field. On the topic of transportation, see J. C. Lowe and S. Moryadas, *The Geography of Movement* (Prospect Heights, Ill.: Waveland Press, 1984). In the area of markets and marketing, an older book by J. E. Vance, Jr., still is worth reading: *The Merchant's*

World: A Geography of Wholesaling (Englewood Cliffs, N.J.: Prentice-Hall, 1970).

Development is a topic of concern in many disciplines, and what has been written by scholars in other fields is relevant to students of economic geography. A classic is E. Boserup's *The Conditions of Agricultural Growth: The Economics of Agrarian Change under Population Pressure* (Hawthorne, Ill.: Aldine, 1965). This book has relevance to our earlier discussions of population (Chapter 6) and Agriculture (Chapter 8) as well as the present chapter. Also see M. Chisholm, *Modern World Development: A Geographical Perspective* (Totowa, N.J.: Barnes & Noble, 1982), a book that discusses the advantaged and disadvantaged states in the core–periphery context we have discussed in our chapter. Notice that Chisholm feels that raw materials and cultural traditions are more important in explaining the economic state of the world than exploitation of countries in the periphery by core-area states. A volume by an African geographer, A. L. Mabogunje, studies *The Development Process: A Spatial Perspective* (New York: Holmes & Meier,

1981). The author discusses development in rural, urban, and national contexts. A very readable book is D. M. Smith's *Where the Grass Is Greener: Living in an Unequal World* (Baltimore: Johns Hopkins Press, 1982), a book that focuses on inequalities of different kinds, in different settings, and at different scales. An interesting article entitled ''Core–Periphery Systems as a Framework for Cross-Cultural Perspectives in World Regional Geography'' by geographer J. P. Allen appears in a book edited by L. Baskauskas, *Unmasking Culture: Cross-Cultural Perspectives in the Social and Behavioral Sciences* (Novato, Calif.: Chandler & Sharp, 1986).

Economic geography is the subject of many articles in various geographic journals. One of these is actually called *Economic Geography* and is published at Clark University (Worcester, Mass.). If you are working on a paper, the index to that journal's contents will be very useful. Articles in English also are published in a Dutch journal, TESG *(Ti jdschrift voor Economische en Sociale Geografie)*, published in Amsterdam.

409

CHAPTER ELEVEN

Wo das Volk
die Macht ausübt,
ist der Frieden
in sicheren Händen

The State
and the
World:
Geography
and
Politics

In previous chapters we have observed the imprints made on the earth's surface by farming, urbanization, industrial development, and cultural traditions. From the carefully terraced hillslope paddies of Asia to the giant circular pivot-irrigation fields of America, from Europe's great industrial complexes to Japan's giant conurbations, human activity is etched on the ground.

Political behavior, too, leaves imprints on the earth. Political geography is the study of such behavior in spatial context, including the marks it makes on the land. We are all familiar with some of these marks of political activity: the Berlin Wall, a county line, a boundary checkpoint. Other spatial-political features may not be as obvious in the landscape, but they affect our lives nevertheless: electoral districts and their boundaries, for example.

To many of us, the most constant reminder of our political world is the map of states and nations. Just looking at that map reminds us of the divisive nature of politics, and the way our earth is fragmented. A careful count will yield more than *200* political units . . . on an earth that is only about 30 percent land. More than 160 of these political entities are independent countries or, to use the technical term, *states*. Nearly 50 are *dependencies*, territories that are not independent and are governed by states. Most of these dependencies are quite small, including numerous islands in the Pacific Ocean.

A closer look at the world political map reveals that some states have huge territories, whereas others are very small (Fig. 11.1). The largest territorial state in the world is the Union of Socialist Soviet Republics (U.S.S.R.), which has an area of 8.6 million sq miles (22.3 million sq km). Among the world's smallest states are those with a territory of less than 1000 sq miles (2500 sq km)—for example Luxembourg in Europe, Bahrain in the Middle East, Barbados in Middle America, and Kiribati in the Pacific realm. The world's largest state, therefore, is more than ten thousand times as large as the smallest state!

In Chapter 6 we noted the enormous range in national populations. Here it is not the U.S.S.R., but China that leads the world, with 1.1 billion inhabitants. In these terms, China also is about ten thousand times as large as the smallest of states.

In a world of limited resources, such enormous disparities create problems in the economic and political relationships among states. Size alone, of course, is no guarantee of a diversified resource base. But countries with areas as large as China, the United States, Australia, and Brazil are much more likely to possess a substantial share of raw materials than, say, Malta, Israel, Djibouti, or Fiji. People in countries of the underdeveloped world often refer to the unfair nature of the international economic order to which they are tied. Neither, unfortunately, is there anything fair about the division of earth resources among countries.

If the framework of the world map is so unsatisfactory, why does it persist? The answer lies in the political behavior of state (national) governments. Many would like to see change, but only in their favor; this means that other countries would have to accept further disadvantage or even annexation. Historical geography has shown that once established and functioning, states will resist any such encroachment. The boundaries of many states on the world map are a legacy of European colonialism. A large number of those boundaries could be logically relocated to reunite peoples with a common cultural heritage and to separate those with different traditions. But almost invariably, the governments of newly emergent, postcolonial states have resisted boundary change. And so the map of Africa, more than a generation after the start of decolonization, is little changed today.

The second half of the twentieth century has been a period of momentous politicogeographic change. After the end of the Second World War, European colonial empires began to break down. Even before midcentury, British India achieved sovereignty. The Dutch could not regain control over their Southeast Asian colonies, and the state of Indonesia was born. In Africa, the European colonial powers yielded to nationalism from Algeria in the north to Zimbabwe in the south. Belgium abandoned its Congo, and Zaïre appeared on the map. Portugal fought a bitter war against the forces of independence, but it, too, lost control in Moçambique, Angola, and Guinea-Bissau. Even on small islands in the Caribbean and the Pacific, the political winds of change transformed colo-

412

nies into independent countries represented in the United Nations. The number of states grew; the number of dependencies dwindled. A new map of the world took shape. At midcentury, hundreds of millions of people still lived under colonial control. By 1988, that total had shrunk below 2 million.

Today, therefore, the overwhelming majority of the world's more than 5 billion people are citizens of one of the 160 states. On the map, they identify their homeland as a part of the earth's surface marked off by boundaries on land and at sea. These boundaries appear as lines on our world map, but, as we will see later, they represent much more than that. What we, in ordinary language, call a *country* (and what political geographers define as a state) is more than a piece of territory inhabited by a population. A *state* is a complex system of many interacting parts. That system functions to serve the people, to bind them together as a *nation*.

The Nation-State

The state is more than a share of the earth's available territory and a number of people to inhabit it: it is a region of cities, towns, and hinterlands; communications networks, administrative subdivisions, schools, hospitals. It is a maze of circulation, movement, exchange: of people, raw materials, finished products, foodstuffs, newspapers. It must exist for the people, requires their taxes, demands their adherence, often compels their service in military forces. To succeed, it must foster a sense of national unity and pride. Does the incorporation of an aggregate of diverse people within a single boundary automatically create a nation? Not exactly. The term **nation** implies much more than a people simply existing within certain boundaries. The point can be proved rather easily through a check of some major dictionaries of the English language. The dictionary definitions seek to convey a sense of unity, commonality, homogeneity in the context of the idea of nation. Many states, however, include within their boundaries people of different racial sources and people who speak different

languages and adhere to different religions. These are *plural* societies; some have developed far toward nationhood, others have not. There must be something those definitions fail to transmit.

The variable the definitions say little or nothing about is an intangible but vitally important one: peoples' attitude, their emotional posture toward the state and what it stands for. Thus we might profitably differentiate between **legal** nationhood—proved by birth certificate, military record, voter's registration, and other attributes of citizenship—and **emotional** nationhood—the level of commitment of the people, individually and collectively, to the state as they perceive it. No matter how homogeneous a population, serious, divisive forces will cause the national cohesion to degenerate. But a people divided by race, religion, and history can, nevertheless, be welded into a nation of strong fiber.

The *nation-state* is a term used in political geography to designate a state in which close coincidence exists between the legal perimeters of the nation and the people's demonstrated commitment to the state. No absolute measure of this is possible, of course, and many definitions of the nation-state have been written. Here we define the nation-state as a political unit comprising a clearly defined territory and inhabited by a body of people, both of sufficient size and quality and sufficiently well organized to possess a certain measure of power; the people shall consider themselves to be members of a nation with certain emotional and other ties that are expressed most tangibly in law and government or ideology.

This is a very broad definition; even so, several of the world's prominent large states may not qualify as *nation*-states. And note that there is no attempt to be quantitatively precise in this model. We say "of sufficient size," and "a certain measure of power," and "the people shall consider themselves to be members of a nation." Obviously, since there is not a state in the world where *all* the people consider themselves committed and emotionally involved in their nationality, we are saying that the vast majority of the people must feel this way. Whether it is 90 percent or 80 percent, the definition does not say. But it does give us a sense of what the ingredients of the nation-state must be.

413

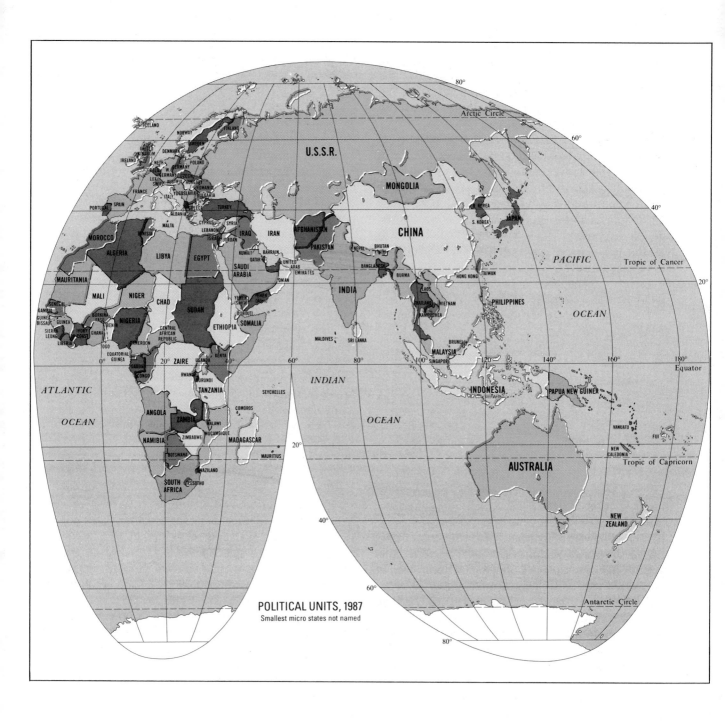

POLITICAL UNITS, 1987
Smallest micro states not named

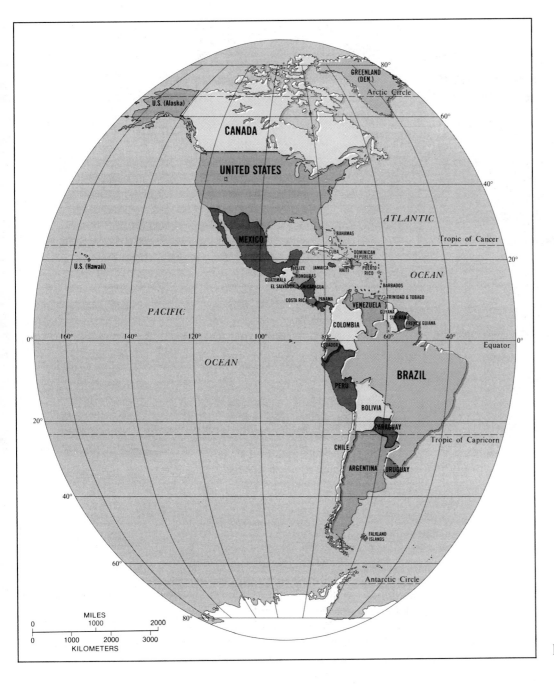

FIGURE 11.1

416

The anniversary of the October Revolution in the Soviet Union is celebrated with great fervor and zeal. Large paintings of patriotic scenes adorn the frontages of entire city blocks. Armed forces and military hardware (including rockets and missiles) are paraded before the assembled crowds in a massive demonstration of power and national strength. This is part of the plan to forge a cohesive Soviet nation, to underpin the nation-state.

It is possible, on the world political map, to discern states whose boundaries do enclose true nations. France is often cited as the prime example of a nation-state; others so identified include Poland, Japan, Egypt, and Uruguay. But the jigsaw of boundaries on the map is no guide to the distribution of nations by whatever definition. Often such boundaries divide rather than unify national groups. Europe's colonial powers constructed Africa's political boundaries with only occasional regard for the distribution of homogeneous culture groups. As a result, numerous African peoples were divided into two or even more segments separated from each other by closed borders.

Elsewhere (e.g., in Nigeria) the new boundaries threw together peoples of different heritage, traditions, and outlook. To the empire builders, this mattered little; these were boundaries of administrative convenience. At the time that they were established, no one foresaw that some day these would have to function as the limits of independent states. Thus were sown the seeds of territorial conflict that will bedevil Africa for decades to come. Nor is Africa alone in this respect. Disputes over boundaries and territories (hence, the allegiance of peoples) also afflict South and Southeast Asia and Middle and South America. And, of course, there are the age-old boundary quarrels of Europe.

Within the past decade, Yugoslavia was again calling for the incorporation of parts of Austria (notably an area known as Burgenland) because Slovenes and Croats form a majority of the people there. The world's political boundaries are no guide to its nations.

It should also be recognized that there are peoples who consider themselves a nation but have long been without a national or state territory. Before Israel was established under U.N. auspices in 1948, the Jewish "nation" was in this situation. Today the 3.5 million Palestinians living in seven Middle Eastern countries are in a similar position, having lost their homeland during the decolonization and partition of Palestine following the Second World War. The Kurds, a homogeneous if dispersed nation numbering more than 7 million, constitute minorities in Turkey, Iraq, Iran, Syria, and Soviet Armenia. They have from time to time rebelled against the governments of the countries they partially occupy, but they have never achieved a national state.

Comparatively few states, then, are nation-states on the European model: Europe itself exhibits prominent exceptions (Belgium, Yugoslavia). Nevertheless, the decolonization period witnessed the creation of newly independent states along European lines that possessed all four of the European model's essential ingredients: a clearly defined territory, a substantial population, certain organizational properties, and a measure of strength. We now examine these four pillars of the state.

• Centrifugal and Centripetal Forces •

Political geographers use the terms *centrifugal* and *centripetal forces* to identify forces that, within the state, tend—respectively—to pull the system apart and to bind it together. *Centrifugal* forces are divisive forces. They cause deteriorating internal relationships. Religious conflict, racial strife, linguistic division, and contrasting outlooks are among major centrifugal forces. During the 1960s, the Indochina War became a major centrifugal force in the United States, the aftermath of which is still felt almost three decades later. Newly independent countries find tribalism a leading centrifugal force, sometimes strong enough to threaten the very survival of the whole state system, as the Biafra conflict did in Nigeria. *Centripetal* forces are those that tend to bind the state together, to unify and strengthen it. A real or perceived external threat can be a powerful centripetal force, but more important and lasting is a sense of commitment to the system, a recognition that it constitutes the best option. This commitment is sometimes focused on the strong charismatic qualities of one individual, a leader who personifies the state, who captures the popular imagination. (The origin of the word *charisma* lies in the Greek word that means divine gift.) At times, such charismatic qualities can submerge nearly all else; Perón's lasting popularity in Argentina is a case in point. Kenyatta in Kenya, Tito in Yugoslavia, de Gaulle in France, Mao Zedong in China, Pandit Nehru in India all possessed similar charisma, and played disproportional roles (as individuals) in binding their nations and states.

The degree of strength and cohesion of the state depends on the excess of centripetal forces over the divisive centrifugal forces. It is difficult to measure such intangible items, but some attempts have been made in this direction—for example, by determining attitudes among minorities and by evaluating the strength of regionalism as expressed in political campaigns and voter preferences. When the centrifugal forces become excessively strong and cannot be checked (even by external imposition), the state breaks up—as Pakistan did in 1971 and as Malaysia did during the mid-1960s with the secession of Singapore. Or, it will undergo revolutionary internal change that makes it, in effect, a new entity, as Nicaragua became after the ouster of Somoza and as Cuba became after the victory of Castro's forces.

417

Territorial Morphology

No state can exist without territory. Although states exchange resources and raw materials, the opportunities provided by their domestic bases have a fundamental impact on their well-being. Britain's industrial surge came after its local resources already had been put to effective use. Japan possessed a substantial network of home industries based on local coal and metals before it became a world economic power. The exploitation, organization, and fullest development of its own territory's soils and other resources is a state's top priority.

We do not need a very detailed politicogeographic map of the world to prove that states are unequally endowed with territorial and environmental assets. Two aspects of the map stand out: the enormous range in the **size** of states—their total area—and the variety of **shapes (morphology)** they possess.

SIZE

It is tempting to assume that the more territory a state owns, the better off it is in all respects. But that is not necessarily the case. True, a very large state has a greater chance that a wide range of environments and resources exist within its borders than a small state has. But much depends also on location; in Chapter 5 we viewed the distribution of known mineral resources and found them clustered in some areas, virtually absent in others. No matter how large a state, if it is underlain by desert sand or by a crystalline shield, the mineral-resource picture may not be all that favorable. Shield areas do not contain fuels; deep sand can obstruct exploration and mining. A much smaller state that lies astride the Eurasian mineral backbone may be much better endowed. Then there is the question of population distribution and size as it relates to areal dimension. A very large territory may be a liability rather than an asset if the population is too small to organize and exploit it. A government must be able to exercise effective national control over the state's total area; areas that lie beyond the reach of its administration and influence can produce problems. In the first place, they are not productive; in the second, such areas may come to harbor elements hostile to the national government. Neighboring states may even seek to lay claim to such isolated reaches of territory. In recent decades, the government of Brazil, for example, has been trying to open up the Amazonian interior and other lands west of that country's coastal heartland. Brazil is larger than the coterminous United States, but its population is just about half that of the United States. And that population is far more strongly concentrated in the east than that of the United States. The Brazilian government even went so far as to create a whole new capital—400 miles (650 km) into the interior—to draw national attention westward. Now attempts are being made to build a trans-Amazonian highway system. Brazil wants to integrate its whole territory into its national framework. For all intents and purposes, Brazil for many years has consisted of the occupied eastern periphery of its huge land area.

It was noted earlier that the range of sizes among the world's more than 160 states is so great as to defy generalization. The continent-size U.S.S.R. possesses about one sixth of the entire world's land area; at the other extreme, there are states so tiny that they are referred to as **microstates**. But not all these microstates are inconsequential or merely picturesque anomalies such as Liechtenstein and San Marino. Quite another kind of microstate is Singapore, where a small island of 232 sq miles (602 sq km) at the southern end of the Malay Peninsula is inhabited by nearly 2.7 million people who have converted their minuscule territory into one of the great trade hubs of the world. Singapore's great resource lies not in its tiny area, but in its geography—its relative location. On the southeastern corner of mainland Asia, at the narrow end of the Strait of Malacca, nearly midway between oil-consuming Japan and the oil-producing Persian Gulf states, Singapore is a world crossroads, historically an entrepôt, but also a growing industrial complex.

Size alone, thus, is no criterion of a state's potential for economic growth and development. There are strong states and weak, rich and poor, well fed and hungry in all size categories, even in the larger ones. Several countries possess as many as 3 million sq miles (8 million sq km): Australia, Brazil, Canada,

China, and the United States cluster around this figure. Another group has an average of 1 million sq miles (2.5 million sq km): Argentina, India, Sudan, Zaïre. These are not lists of uniformly strong and developed countries.

In recent years, the size and territorial configuration of states has taken on a new significance that relates to the rising competition for control over the adjacent ocean waters and the submerged continental shelf. Larger states tend to have longer coastlines; thus, the world's large states have benefited most from international agreements that gave coastal states rights not only to their adjacent seafloor, but also to the ocean's waters as far as 200 nautical miles from shore. In 1982, the concept of an Exclusive Economic Zone (EEZ) gained international approval. The countries that stood to gain most from this development were the United States, Australia, Indonesia, New Zealand, Canada, the Soviet Union, Japan, Brazil, Mexico, and Chile. Five of these 10 countries have exceptionally large territories, and 2 more (Indonesia and Mexico) are considerably larger than average.

SHAPE

A second quality of states' territories evident from the world political map is their shape, or spatial form. Some states, such as the Philippines, lie on a group of islands, their territories broken by extensive waters. Other states—for example Chile—seem to consist of a long, narrow strip of land. When at first we look at the map, there appear to be so many different kinds of territorial forms that classification seems impossible. But when the states of the world are examined carefully, a typology does emerge. The Philippines and Chile suggest two types, and there are others.

Compact States

Many states—Belgium, Uruguay, Kenya, and Kampuchea, among others—have territories that are somewhere between round and rectangular. There are no islands, peninsulas, or major indentations such as bays or estuaries. These states' territories are **compact**, which means that the distance from the geometric center of the area to any point on the boundary does not vary greatly. The case of Hungary (Fig. 11.2)

FIGURE 11.2
Hungary: compact state.

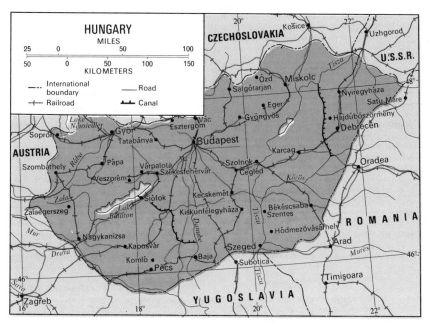

illustrates this situation. The compact state encloses a maximum of territory within a minimum length of boundary, an obvious asset.

Fragmented States

Quite the opposite situation prevails in the case we mentioned, the Philippines, and in Indonesia, Japan, and Malaysia. In those states, the territory is **fragmented** into numerous pieces, and it is possible to go from one part to the other only by water or by air. We can recognize three different kinds of fragmented states: those whose national territory lies entirely on islands (Japan, the Philippines); those whose territory lies partly on a continental landmass and partly on islands (Malaysia, Italy); and those whose major territorial units lie on the mainland, separated by the territory of another state (the United States, whose state of Alaska cannot be reached overland from the coterminous states except through Canada).

Fragmented states have problems of internal circulation and contact, and often suffer from the friction of distance (Fig. 11.3), so that national integration is made difficult. Far-flung Malaysia, for example, was unable to accommodate the pressures placed upon it by mushrooming, Chinese-dominated Singapore. In 1965, Singapore seceded from Malaysia to become the city-state it is today. After years of rising tension and a destructive civil war, East Pakistan broke away from West Pakistan to become the separate state of Bangladesh. Indonesia has faced rebellions on a number of its widely scattered islands, including West Irian (on New Guinea) and Timor (where the Portuguese handed control to Djakarta). The Philippines, another state that is entirely insular and strongly fragmented, is facing opposition in the south, where Moslem influence is strong and regional autonomy is a frequently expressed demand. The islands of the Sulu Archipelago and parts of Southern Mindanao are among Manila's major problems from this point of view, and as the map (Fig. 11.3) shows, they also lie farthest from the capital and the country's leading island, Luzon.

Elongated States

Still another spatial form is represented by Chile, Norway, Malawi, Panama, and Vietnam. The territory of each of these states is **elongated**. Such elongation (or **attenuation**, as it is sometimes called) also presents certain difficulties. If a state has a large territory or if it lies astride a cultural transition zone, its elongated shape may endow it with contrasting lifestyles. An example is Togo in Africa, not a very large state but extending from the west coast across the forest and savanna to the Moslem-influenced interior. In Italy, which is an elongated as well as a fragmented state, there are north-south contrasts that are related to the different exposures of those two regions to European mainstreams of change. In Norway, the distant, frigid north, inhabited by Laplanders, is another world from that of the south, where the capital city and the country's core area are positioned. And Chile, of course, provides the classic example of environmental, if not cultural, contrasts (Fig. 11.4). Northern Chile is desert country, the barren Atacama prevailing there (Fig. 11.4). Central Chile has a large area of Mediterranean climatic conditions, and here the majority of the people are clustered. Southern Chile is dominated by marine west coast conditions and by rugged topography. The effects of distance and isolation are very evident in the far desert north and in the remote south. In many ways Chile *is* that central area between the two, the rest of the country still awaiting effective integration in the state.

Prorupt States

A different situation is exhibited by states for which Thailand could serve as a model (Fig. 11.5). Thailand would be a compact state, except for one peculiarity. From its main body of territory there extends a long peninsula, several hundred miles southward to the boundary with Malaysia. In fact, Thailand's neighbor Burma has the same feature, except that its southward extension is a bit shorter. This spatial form is called **prorupt**, and any time you see such a situation on the map some investigation is warranted. Proruptions often have noteworthy histories. They also create special problems. In the case of Thailand, those border areas near Malaysia are 600 miles (1000 km) away from the capital, Bangkok, farther than any other area in the country is removed from the government's headquarters. But at least there are good surface communications, even a railroad along the length of the

420

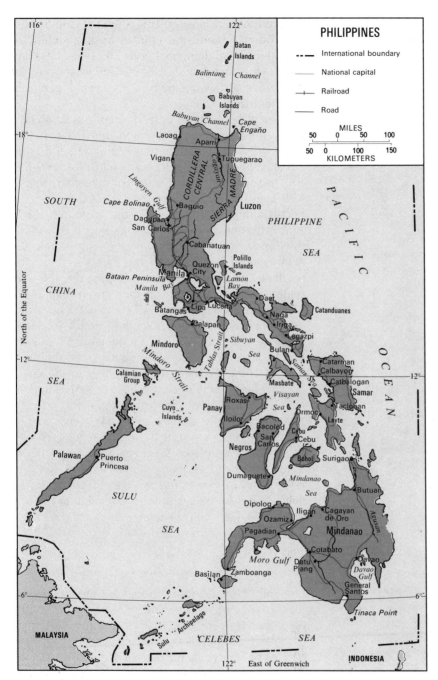

421

FIGURE 11.3
Philippines: fragmented state.

FIGURE 11.4

Thai territorial extension. This is not the case in adjacent Burma. There, the railroad leading southward from the capital (Rangoon) ends 300 miles (500 km) short of the boundary, and for the land extension's last 150 miles (240 km), there is not even a good all-weather road!

Not all proruptions or land extensions lie on peninsulas. Afghanistan's proruption, the Wakhan Corridor, lies squeezed between the U.S.S.R. and Pakistan, the product of Russian and British imperialism. In South America, a small but potentially significant land extension called the Leticia Corridor connects Colombia to the banks of the Amazon River. In Africa, the most prominent example is undoubtedly the Caprivi Strip, the narrow proruption that links the main area of Namibia and the Zambezi River, the result of a series of negotiations during the 1884–85 Berlin Conference. Also, on the African map, Zaïre can be observed as having two comparatively narrow extensions of land, one into the interior in Shaba Province (formerly Katanga, the mineral-rich interior that has been the country's mainstay), and the other leading along the Zaïre (Congo) River to the coast, creating the country's ocean outlet. These proruptions have their origins in political negotiation and even territorial tradeoffs, as in the case of Western Zaïre.

The significance of such corridors tends to change over time. The role of Southern Asia's inner tier of states as buffers between colonial empires has vanished (although other pressures buffet them now). The Leticia Corridor means little to Colombia at the moment, but that could change with Brazil's development of its Amazonian interior.

Perforated States

One other spatial form should be mentioned, although its significance hardly matches that of the other four. In rare instances the territory of a state completely surrounds that of another state. That state, then, is perforated by the surrounded country. A look at the map of South Africa will make this clear (Fig. 11.5). The country of Lesotho (Basutoland on older maps) is entirely surrounded by South Africa. Lesotho is about the size of Belgium, so it takes a substantial piece of territory out of the heart of South Africa.

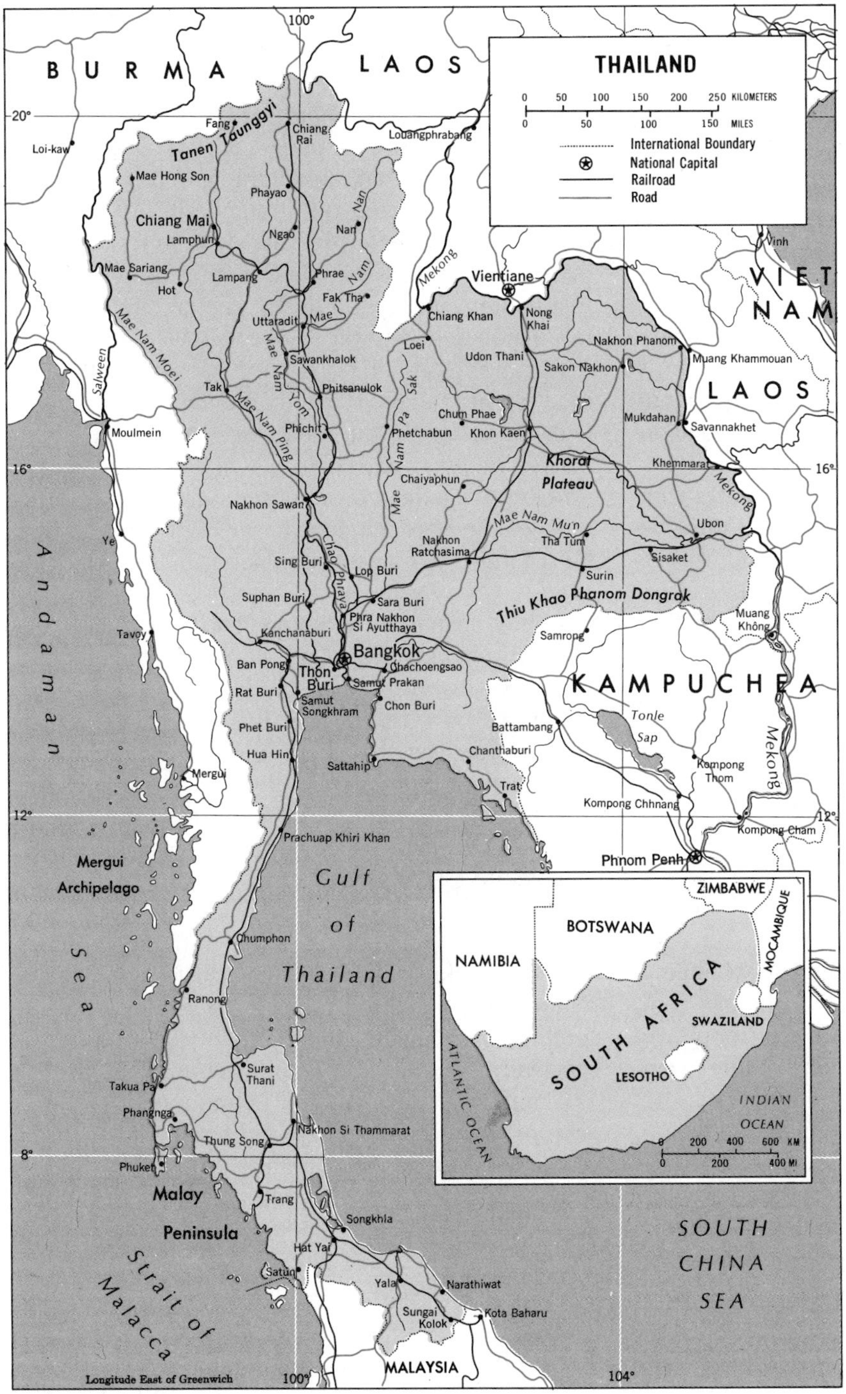
THAILAND
0 50 100 150 200 250 KILOMETERS
0 50 100 150 MILES
International Boundary
National Capital
Railroad
Road
BURMA
LAOS
VIETNAM
LAOS
KAMPUCHEA
MALAYSIA
Tanen Taunggyi
Chiang Mai
Khorat Plateau
Thiu Khao Phanom Dongrak
Malay Peninsula
Mergui Archipelago
Andaman Sea
Gulf of Thailand
Strait of Malacca
SOUTH CHINA SEA
Fang
Chiang Rai
Loi-kaw
Louangphrabang
Mae Hong Son
Phayao
Lamphun
Ngao
Nan
Mae Sariang
Hot
Lampang
Phrae
Fak Tha
Uttaradit
Vientiane
Nong Khai
Chiang Khan
Loei
Udon Thani
Nakhon Phanom
Sakon Nakhon
Muang Khammouan
Vinh
Sawankhalok
Tak
Phitsanulok
Phichit
Chum Phae
Khon Kaen
Phetchabun
Mukdahan
Savannakhet
Moulmein
Khemmarat
Chaiyaphun
Nakhon Sawan
Mae Nam Mun
Tha Tum
Ubon
Nakhon Ratchasima
Sisaket
Surin
Ye
Sing Buri
Lop Buri
Suphan Buri
Sara Buri
Phra Nakhon Si Ayutthaya
Tavoy
Kanchanaburi
Samrong
Muang Không
Bangkok
Ban Pong
Thon Buri
Chachoengsao
Samut Prakan
Rat Buri
Samut Songkhram
Chon Buri
Phet Buri
Battambang
Tonle Sap
Hua Hin
Chanthaburi
Sattahip
Mergui
Kompong Thom
Trat
Kompong Chhnang
Kompong Cham
Phnom Penh
Prachuap Khiri Khan
Chumphon
Ranong
Surat Thani
Takua Pa
Phangnga
Nakhon Si Thammarat
Thung Song
Phuket
Trang
Songkhla
Hat Yai
Satun
Yala
Narathiwat
Sungai Kolok
Kota Baharu
Salween
Mae Nam Moei
Mae Nam Ping
Mae Nam Yom
Mae Nam Pa Sak
Chao Phraya
Mekong
Longitude East of Greenwich
100°
104°
20°
16°
12°
8°
ZIMBABWE
BOTSWANA
NAMIBIA
MOCAMBIQUE
SOUTH AFRICA
SWAZILAND
LESOTHO
ATLANTIC OCEAN
INDIAN OCEAN
0 200 400 600 KM
0 200 400 MI

FIGURE 11.5

423

Elsewhere the perforators are smaller, as in the case of San Marino, which perforates Italy. But only the South African case really has importance in political geography for Lesotho is an independent black state and South Africa remains white-controlled. In other cases, perforation is a curiosity rather than a vital aspect of territorial morphology

POPULATION

The size and shape of national territories take on more meaning when we view them in association with their respective populations. As we noted earlier, there is great variety in the size and character of states' populations. The largest state in the world, by population numbers, is China—now a state of more than 1 billion. At the other end of the spectrum are states with populations of less than a half million. It would be impossible to say just what the minimum number of people is that can constitute a nation; conditions vary too greatly. For example, among Iceland's 230,000 residents there may be more persons with specialized education and particular skills than in an underdeveloped country with 10 times that number of inhabitants. The capacities of the nation, as well as its size, are relevant.

In Chapter 6, we discussed some of the problems related to the interpretation of data on population density. Averages for individual states are not very meaningful unless we know something about the agglomeration, the clustering of the people. Consider again the example of Brazil: with the overwhelming majority of that state's 150 million inhabitants concentrated in the eastern margins of the territory, we might easily misinterpret the statement that the average density of people per square mile there is 46. Thus, we want to know not only the total population of a nation-state, but the way this population is distributed over the national territory. When we have this input, then our knowledge of states' spatial forms takes on more meaning. For example, we know that Burma and Zaïre are both prorupt states. In the case of Burma, a state of 39 million people, the core area lies in the main body of territory, and the extension of area that forms the proruption is one of the least effectively integrated parts of the country. Zaïre is a much larger

state territorially, and contains a population of 33 million. When, however, we consider the distribution of population within Zaïre, we realize that the two proruptions of that state are no mere extensions of area, but are vitally important, productive regions. The southeastern proruption constitutes the former Katanga (now Shaba) region, the famous copper-producing area centered on Lubumbashi. The western proruption connects the vast Zaïre interior to the ocean—it is the state's sole outlet to the sea. Within that western proruption lie the capital city, Kinshasa; the major port, Matadi; a developing transport network; a substantial population cluster; and much more. In Burma, there may in the future be problems with the isolation, the remoteness of the proruption that lies on the Malay Peninsula. In Zaïre, the problem is not so much the proruptions, it is the vastness of the intervening Zaïre (Congo) Basin itself. The distribution of these two states' populations tells much of this story.

Thus spatial morphology and population distribution should always be viewed in conjunction. When we discern that a state possesses a compact shape, we may assume that it has certain organizational advantages, but these may be negated by the peripheral position of its dominant population cluster. Multiple clusters may also have the effect of neutralizing the advantages of minimum internal distances. Nigeria, for example, is a compact state—but it is one of those products of European colonialism that threw a boundary around diverse and different peoples. Nigeria has three major cultural clusters and several minor ones; the three major clusters dominate the southwest, southeast, and north of the country, respectively. Nigeria is a populous country, it has over 110 million inhabitants, and in contemporary Africa each of those three primary clusters could form the core of an independent country. Moreover, the two southern clusters, near Nigeria's coast, were strongly influenced by European colonialism. The third one, in the Moslem-influenced interior, has always been remote and different. On independence in 1960, Nigeria became a federal state to provide a measure of autonomy to its diverse regions. The system did not work well enough to prevent one of them, the southeast (called Biafra), from trying to break away, an attempt that led to a destructive civil war during the late 1960s. Nigeria's compactness was not enough to overcome its internal cultural divisiveness, although the Biafran secession movement was defeated. Always compare Figs. 11.1 and Plate 5: even at this level of generalization, it is possible to identify relationships of spatial morphology and population density and distribution. Nigeria's pattern is clearly outlined, as are those of other major states.

ORGANIZATION

The state is a system with many subsystems. From transport networks to educational systems, national health plans to tariff barriers, tax-collection methods to the armed forces, the state has a multitude of organizational arms. The taking of a census, the calling of an election, the distribution of funds to the elderly— all these evince the effectiveness of organization in the state.

On the map, the most clearly evident manifestation of state organization is the state's administrative system, the hierarchy of its government-based regions. Other organizational attributes are revealed by maps of economic regions and transport networks. State organization of land use, mineral exploitation, educational facilities, health-care systems, military installations, and other functions appears on thematic maps that reveal much about the level of its development. In the LDCs, organization is poor and often rudimentary; a true surface communications network, for example, is not present in such states.

In political geography, then, the most obvious manifestation of organization in the state is administrative. Except for some of the smallest microstates, all states are divided into subunits such as provinces, states, departments, or other designations. In the U.S.S.R. and China, complex hierarchies of such subdivisions were created following the appearance of the new orders. In the U.S.S.R., each of 15 Soviet socialist republics broadly corresponds to one of the major nationality groups in the huge state. Most Soviet republics include smaller clusters of people with some cultural identity; for them, 20 autonomous republics were created—republics *within* republics. Several additional levels of subdivision exist, includ-

ing 8 autonomous regions and 10 national districts. It is obvious that the complicated Soviet political map was created: (1) to give Russia preeminence; (2) to permit culturally distinctive people to retain their identity while providing them with avenues for representation in Moscow; and (3) to organize population and territory in such a manner that state control over the resource base and the means of production would be effective.

The Chinese system was modeled after that of the U.S.S.R., also with the aim of giving the minority peoples certain measures of identity and autonomy. Three areal levels of administration exist. Interestingly, the three municipalities (including Beijing), the 22 provinces, and 5 autonomous regions (such as Xizang [Tibet], Xinjiang [Sinkiang], and Nei Monggol [Inner Mongolia]) are all at the top level. The second and third levels exist for minority peoples.

Large and populous countries are naturally more likely to have complex administrative systems, and those of the U.S.S.R. and China are more intricate than most. Among other large countries, the United States' politicogeographic map includes 50 states, many with lengthy geometric (straight-line) boundaries. In the United States, the organizational hierarchy places the county below the state (all states are divided into counties except Louisiana, which has parishes), and municipalities come next. Federal, state, county, city, and lesser municipal governments exercise a variety of sometimes overlapping jurisdictions, but the United States does not have a spatial commitment to minorities comparable to the U.S.S.R. or China (the Indian reservations come closest, but one of the largest, the Navajo Reservation in northeastern Arizona, has fewer than 140,000 residents).

Canada's administrative map divides the country into 10 provinces and 2 territories, the provinces governed by a premier and an elected legislature; the territories administered by a (federal) government commissioner and a council of appointed or elected members. Brazil's politicogeographic map reveals 22 states, 4 territories, and the federal district of the capital city, Brasilia. In the United States, Canada, and Brazil, the historic sequence of European settlement is reflected by the geographic pattern of administrative units. Eastern provinces and states tend to be smaller;

425

western units are larger and are more often bounded by geometric borders, revealing the frontier character of the interior during the evolution of modern settlement.

Unitary and Federal States

Another common organizational feature of Canada, the United States, and Brazil is that they are all **federal** states. The Romans were the first to establish a kind of federal system in their large and varied empire; for centuries, political philosophers have studied that early example of regional autonomy and partial self-determination within the greater national frame-work. Actually, the federal idea never took hold to any great extent in Europe, where countries are comparatively small and a tradition of centralization of authority has long prevailed. Only three modern European states are federations: Switzerland (a durable case) and Yugoslavia and West Germany (much more recent ones). In Switzerland, the common interest of the old Alpine cantons led to a voluntary federal association to protect their shared advantage. In Yugoslavia, federation was a compromise to overcome historical, political, and cultural diversity within the state. West Germany became a federal state after the Second World War. The remainder of Europe's nation-states are **unitary states**, implying that their governments are centralized and regional autonomy is minimal.

Federalism proved to be a most useful form of government in the New World, however, and not just in the Americas. In an article entitled "Sixty Years of Federation in Australia" (1961), K. W. Robinson described the federation as the most geographically expressive of all political systems, based as it is on the existence and accommodation of regional differences. "Federation," the author states, "does not create unity out of diversity; rather, it enables the two to coexist." Indeed, the federal concept was employed in the politicoterritorial organization of several states with strong internal contrasts. Such countries as Nigeria and India attained independence as federations, and India's federal framework (Plate 9) has often been credited as being the critical element in India's survival as a unified state. As we noted in Chapter 7,

India is a multilingual country in which language plays a vital role as a cultural anchor. The imprint of India's language map can be seen on the politicogeographic map (Plate 9), but India's 22 states and 8 union territories remained within the national state, each retaining enough autonomy to make participation acceptable. Any attempt to make India a unitary state and to force a single language on the whole population, would undoubtedly have led to its destruction. Federalism proved to be the spatial solution.

The modern world's 160 states, thus, present a patchwork of governmental organizations. All the states in the largest category (the U.S.S.R., Canada, the United States, Brazil, and Australia) are technically federal states, except one—China. Many populous states have a form of federal organization, including not only India, but also Nigeria and Mexico. In other states, however, the federal idea was untenable. Traditional African forms of political organization tend to be highly centralized, and most African states became unitary states on independence, including (notably) Ghana and Kenya. In Middle and South America, unitary (and dictatorial) traditions from Iberia prevailed in a majority of countries; in others, federal systems were eroded by increasing centralization of power, for example, in Argentina. Any map of a state's administrative-political subdivisions, therefore, should be viewed as a significant organizational element and always in a functional context, that is, how it is presently being made to work.

Economic Organization

The economic organization of the state is a vital link in the total system. All states have resources; some have very few, others have a wide range. Even location can be a resource. For example, centuries ago, the Swiss found themselves positioned near the Alpine passes across which Venice and the north of Italy had to trade with Flanders and neighboring areas. The Swiss controlled the transit of goods and bounded a state in the process—a state spawned, in large measure, by the relative location of the Alps.

How has the state managed to mobilize its resources—natural, spatial, and human? Probably the most dramatic example of economic reorganization as a state-building process comes from Japan. The archipelago that is Japan does not have a very large or varied resource base. There were mineral resources to sustain its handicraft and home-workshop industry, but the modernization envisaged by the leaders of Japan's Meiji Restoration required the acquisition of far greater stores. The first step had to be at home: the reorganization of the handicraft industries. Then Japanese goods would have to be sent to world markets, to undersell competing products. With the funds so gained, investments were made in the industrial sector—and in another area of state organization, the military. Soon the Japanese military was capable of driving into neighboring territories (Korea, Manchuria [now China's Northeast]), where resources needed in Japan were located and where goods made in Japan could be sold. The same Japan that lay isolated and remote some decades earlier was now an aggressive colonial power. For Japan, the European model did its work.

No student of society's capacity for organization can escape the conclusion that organizational ability and preference are cultural attributes. The diffusion of the products and techniques of the modern agricultural and industrial revolutions and the dissemination of political practices require, for their adoption, a certain aptitude for organization, one that is not the same among all cultures. In Africa, this was one major factor that contributed to the strengthening of some societies and the weakening of others. On the slopes of Mount Kilimanjaro, there were farmers who saw the advantages of commercial agriculture and, even in the face of European competition, organized themselves into highly successful cooperatives. Elsewhere, similar opportunities went unused, mainly because traditional organizational systems could not be modified. Organization is the key to the maximization of opportunity, an essential ingredient of national strength.

POWER

National power is an ingredient of the state system. Power is a product of organization in the state, and it is expressed not only in military terms, but also in the overall strength of the state politically, economically, and otherwise. A state's capacity to police and protect national interests is a measure of its power.

Power cannot be measured exactly, so the subject of state power is a subject of much speculation. Obviously, small states such as Luxembourg or Iceland do not have the strength that larger states possess; but when it comes to closer matches, the differences are much less clear. Is present-day France stronger than the United Kingdom? Does Japan's economic strength and proven capacity for military organization outweigh China's much greater territorial size and population? These are the sorts of questions that students of power and politics try to answer. This is done by allotting a quantitative measure to every feature of the state that is deemed to have an effect on its strength and power: size of the armed forces, quantity of military equipment, industrial production, length of roads and railroads, availability of energy, and much more. However, even these calculations must take into account some intangibles, such as the technical efficiency of the population, its morale, and the potential food supply in times of crisis. Recent calculations place the United States and the Soviet Union in a tie for first place, but with the Soviet Union gaining more rapidly. China, Japan, the United Kingdom, West Germany, and France follow. Still, speculation ocntinues to play a major role, for information of this sort is often of uncertain validity.

Power, whatever its measure, comes from organization. In his book, *The Might of Nations* (1961), J. G. Stoessinger defines power as "the capacity of a nation to use its tangible and intangible resources in such a way as to afect the behavior of other nations." Again, this does not mean solely or even primarily by military presence or threat; rather, it involves possessing and producing ineconomic spheres. A state can win concessions or reciprocal agreements with other states through its economic strength. It can outbid other states in the competition for exploitation rights in the territories of less developed countries by being able to offer superior conditions. Growth and development require organization; organization generates power.

• Boundaries: Limits of the State •

The four pillars of the state, as just defined, are territory, people, organization, and power. We now turn to the spatial limits of the first of these. The politicogeographic map of the world shows a veritable maze of boundaries, evidence of the way our earth's limited living space has been divided and redivided. That irregular grid has become our frame of reference, our **mental map**. Notice that when the news reports an event somewhere in the world, it is this mental map that is our first spatial impression of its location. "In West Africa today," the report may say, "tens of thousands of Ghanaians, who have been working in Nigeria, are being sent back to their homeland." As geographers, we know why: Nigeria's oil-dependent economy has suffered from declining oil prices during the 1980s. We also know that Nigeria and Ghana are not immediate neighbors. How will the countries between Nigeria and Ghana react to this exodus?

In this broad context, our mental map is defined by the region's boundaries. We can draw the map of the United States: its coasts, the straight-line boundary with Canada between Pacific and Great Lakes, and the Rio Grande that separates Texas from Mexico. We have seen it many times. Having a good mental map of the world is a strong asset for any geographer.

However, boundaries are more than meets the eye. In the first place, a boundary may appear as a line on a map, but in the real world, a boundary is a plane—a vertical plane that cuts through airspace and underlying geology, with all this implies. The line on the map is only the place where the vertical plane intersects the ground. For every mile of amicably negotiated boundary, there is a mile of contested border. As we know too well, these contests are not over yet. In Middle and South America (Guyana–Suriname), in Asia (Kashmir), in the Middle East (Israel–Syria), in Africa (Ethiopia–Somalia), the disputes continue. Boundaries are under pressure in many parts of the world, and even in Europe, with its old and apparently stable states, numerous changes have taken place in the last 50 years alone. The boundary that now divides Germany into an East and a West appeared on the map in 1945. The city of Berlin was also divided after the Second World War, but the division was strengthened by the building of the Wall between its East and West sectors in 1961. Poland's boundaries changed after the war also, and there were adjustments along the Dutch–German and French–German borders as well.

The boundary framework with which we are familiar in the 1980s, then, is the result of innumerable modifications made over time, and although the broad outlines of this framework are likely to remain intact, every decade brings changes of detail. Still, what we see on the map today is a far cry from the situation of just a century ago. In 1875, Bolivia still possessed the northern Atacama (and thus an outlet to the sea); there was no Pakistan; the whole African jigsaw puzzle was just beginning to develop. The ill-defined sphere of influence was a different reality in the nineteenth century from what it is today.

FRONTIERS

Sometimes the terms **boundary** and **frontier** are used interchangeably, as synonyms. But there is a difference, and an important one. The frontier throughout history served as a zone of separation, an area between states, a territorial cushion to keep rivals apart. Such a frontier might have been a dense, vast, almost impenetrable forest, or a swampland, or a disease-infested river basin, or a range of mountains. Sometimes a frontier was defined by distance as much as by inaccessibility. The world map of the mid-nineteenth century was full of frontiers, in the Americas, in Africa, in Asia, in Australia, even in Europe. But then the effects of Europe's technological and political revolutions, the colonial scramble, explosive population growth, and the rush for resources and empires combined to produce competition for even the most remote frontier areas. Before long, the frontiers disappeared, divided and parceled out among states and colonial acquisitions.

Thus the frontier has *area*; it is a spatial phenomenon. The boundary, on the other hand, should be thought of as a vertical plane separating two states; it is linear on the ground. It appears on our maps as a line, but it also separates airspace above the ground and the rocks and resources below. Today, though,

428

only one frontierlike land area, Antarctica, remains. There, several states claim zones of preeminence, some of them overlapping. But Antarctica is mainly a region of research and investigation, and the claims have not yet led to serious conflict. A major discovery of some critical resource could change that picture considerably, however.

Boundaries, then, are the limits of states. They mark the extent of the state's territory, the limit of jurisdiction. They separate state from state, but they are also the place where states make territorial contact. This dual function, of separation and association, has led geographers to investigate the ways boundaries have come about, how they function (and how those functions have changed in modern times), and how they might be usefully classified.

BOUNDARY EVOLUTION

In ideal circumstances, boundaries evolve through three stages. Imagine a frontier area about to be divided between two states. First, agreement is reached on the rough positioning of the border. Then the exact location is established through the process of **definition**, whereby a treaty-like, legal-sounding document is drawn up in which actual points in the landscape are described (or, where a straight-line boundary is involved, points of latitude and longitude). Next, cartographers, using large-scale maps and referring to the boundary line as defined, put the boundary on the map in a process called **delimitation**. And, if either or both of the states so desire, the boundary is actually marked on the ground by steel posts, concrete pillars, fences, or some other visible means, sometimes even a wall. That final stage is the **demarcation** of the boundary. All boundaries on the world map are by no means demarcated. There are thousands of miles where you could cross from one state into the other without ever knowing it. Demarcating a lengthy boundary in any way is expensive, and it is hardly worth it in inhospitable mountains, vast deserts, frigid polar lands, or other places where there is virtually no permanent human population.

Peoples have sought to demarcate their territories for thousands of years. The Romans built walls in northern Britain to mark the limits of their domain; the

Chinese Wall is still one of the wonders of the world; the pre-Inca Peruvian Indians used stone lines to confirm their territorial claims. More often, linear physiographic features such as rivers and mountain chains became trespass lines in practice, then boundaries in fact. Until as recently as the Second World War, boundaries were viewed as viable lines of defense. The French in the late 1930s fortified their northeastern boundary with Germany, called it the Maginot Line, and hoped that it would withstand the German onslaught. Modern warfare has more or less eliminated this function of the boundary, except in places where guerrilla warfare occurs. There, a river or a mountain range may still present some strategic advantages. But the concept of the boundary as the state's ultimate and practical line of defense no longer applies.

In the Old World especially, but also in Middle and South America and in Africa, boundaries seem to zigzag endlessly and, it would seem, pointlessly. Why are all boundaries not like those between the United States and its neighbors, along a major river, through the middle of lakes, along a line of latitude? The answers are many. From Roman times on, Europe's cultural and political patterns have changed almost continuously. In places (such as along the Pyrenees between Spain and France), a fairly stable and rather permanent boundary evolved; elsewhere states and nations fought over territory time and time again and boundaries were defined, altered, defined once more, wiped out, reestablished. Even boundaries in mountainous areas and along rivers were subject to pressure: in the mountains, there would be argument over the sources of river water and control over watersheds; in the rivers, there would be conflict over who would own the banks, the navigable channel, the river mouth. There is a piece of historical geography attached to almost every one of those bends and turns you see boundaries make on maps. Yet today, after all those centuries of adjustment and readjustment, there are still Italian-speaking people in France, French-speakers in Switzerland, Dutch-speakers in Belgium, Hungarians in Romania, Germans in Poland.

If that is the situation in crowded and culturally complex Europe, what about the Americas, where the population numbers are so much less, or what about

Africa, where external powers could lay down boundaries with impunity? In the Americas, as the daily press will confirm, the appetite of states for their neighbors' territory is far from satisfied. Numerous Middle and South American states consider sections of their boundaries temporary and subject to future redefinition. Venezuela has a potential dispute with Guyana, Guyana has a claim in Suriname, and Guatemala has thinly veiled hopes for the absorption of a whole country, Belize. Bolivia considers the question of its exits through northern Chile unresolved. And in Africa, the colonial powers negotiated with each other, contesting their empires' boundaries in only a few places—but they created many borders that proved untenable after decolonization. African states have begun to dispute some of these boundaries already, as in the case of Somalia and Ethiopia, Morocco and Mauritania (over former Spanish Sahara), and Malawi and Moçambique (involving islands and waters in Lake Malawi).

So we can observe the very adjustments that result in the tortuous maze we see on the map. Israel acquires a small piece of Syria called the Golan Heights. Guatemala proposes a land exchange to defuse its designs on Belize. Panama and the United States negotiate the transfer of the Canal Zone. China redefines its boundary with Burma and starts a dispute with India over the Simla Conference, where many decades ago the China–India boundary was defined. Cold, desolate country can provoke some hot, intense confrontations.

BOUNDARY FUNCTIONS

If boundaries today (with a few exceptions) no longer serve in a defensive context, how do they serve the modern state system? Boundaries may not be trespass lines any longer, but they do serve as the symbol of state inviolability. They are a factor in the development of that intangible emotional bond of national consciousness, especially if contrasts between societies and cultures on opposite sides are strong. On the front pages of newspapers, in schoolbooks, on stamps, even on flags, states display maps of their national territories. A real or even an imagined threat to the boundary can be used to arouse national feeling. But there are many states whose relationships with their neighbors across the boundary are such that the boundary hardly functions as a separator. People move across with only a minimum of control, and some even commute across daily. Social and economic conditions on the two sides are so similar that the boundary seems an unnecessary impediment. The U.S.–Canadian boundary comes to mind, and the boundaries between the Netherlands and Belgium, Norway and Sweden, Austria and Switzerland.

Even these boundaries (note that they are *within*, not *between* geographic realms) still have important functions. One of these is obvious enough. The whole state system, with all its parts—education, taxation, law, conscription, and so forth—is built within this framework. The boundary marks the limit of state jurisdiction in all these and countless more spheres. The boundary signifies the limit of sovereignty; weakening its role implies yielding of some of that sovereignty. As we shall see, a number of states in the postwar period have proved themselves willing to do just that: to reduce the divisive functions of the boundary in order to join with other states in economic, political, and strategic blocs.

Although boundaries no longer serve as fortified protectors of states, they are in places demarcated to shield the state against other threats. The Berlin Wall gave substance to the concept of an Iron Curtain; it was built, primarily, to stop a flood of emigration by East Germans under Communist rule to West Germany. Along the southern boundary of the United States, fences have been erected against the flow of illegal immigrants from Mexico. In 1986, President Reagan announced a war against drugs, and again part of this war involved the strengthening of U.S. boundaries—at sea as well as on land, this time to inhibit the flow of smuggled contraband. In these contexts, the boundary remains the state's ultimate line of defense.

The most consequential function of the national boundary today, however, relates to its role in commerce and economics. We could argue that this is not strictly a boundary function; in fact, the state's political boundary is the limit of the domestic market—a market that can be closed against foreign goods. A government's economic policymakers can erect a tar-

A superimposed boundary: the dividing line (marked, between Soviet and non-Soviet sectors, by a wall) between East and West Berlin.

iff wall against products made by foreign countries, thus protecting local industries. Such a tariff wall makes foreign goods more expensive and helps domestic manufacturers sell theirs at an advantage. But such a policy has its repercussions. Other countries, too, can establish protective regulations. That, in turn, hurts producers who export their products. In recent years, the United States and Japan have engaged in this kind of competition. American automobile manufacturers want Japan to limit its vehicle exports to the United States; Japan's market (one of the world's richest) has not been as open to U.S. products as this country would like. It is a dispute between governments, but the line of economic defense is the national boundary.

Certain states have discovered that there is much to be gained in the other direction, that is, by lowering economic barriers rather than by raising them. After the end of the Second World War, Western Europe (aided by the Marshall Plan) witnessed the emergence of such economic unions. At first, six states (France, Italy, West Germany, Belgium, the Netherlands, and Luxembourg) formed the European Economic Community, also known as the Common Market. Goods flowed across Common Market boundaries with unprecedented ease. The mutual advantage of core-area membership was so plain that other states later joined the EC (Economic Community, as it is now known) as well. States that have economic interests in common also have shared political objectives, and soon a movement toward European unification gained momentum. This process, known as **supranationalism**, is a very significant twentieth-century politicogeographic phenomenon because states involved in it must agree to give up a measure of their sovereignty, their independence of action, for a common objective. This, of course, lowers the barrier effect of boundaries, and the ability of a state to use its boundary as an economic device.

Thus we see boundaries functioning in many different ways. Some states are still strengthening and confirming their borders (notably those states that have recently become independent); some still seek to move their boundaries outward to acquire part of their

431

neighbors' territory; and other states can afford the luxury of reducing the divisiveness and absoluteness of their borders in order to participate in and open ways to a wider, international economic and political sphere.

BOUNDARY CLASSIFICATION

Earlier, we examined the shape (morphology) of state territories and found that certain states share certain spatial properties. From the resulting grouping of states, we derived some useful generalizations about centripetal and centrifugal forces. In the same way, the world boundary framework can be subjected to geographic classification. The global maze of boundary lines consists of segments that exhibit recurrent properties.

Geometric Boundaries

One of the most obvious aspects of the world boundary framework lies in its alternation between straight-line boundary segments and twisting, winding ones. The United States has straight-line boundaries with both of its major neighbors, Canada and Mexico. These boundaries are called **geometric boundaries**. They follow lines of latitude (as in the U.S.–Canada boundary west of the Great Lakes) or longitude (see the boundary between Egypt and Libya) or they coincide with a great circle in other directions. Such geometric boundaries (1) were often established (defined and delimited) before major settlement developed in an area and (2) resulted from imperial or colonial rule. Clearly, they are not adjusted to physical (landscape) features or cultural clustering.

Geometric boundaries have been used especially for internal purposes (as in U.S. state and Canadian provincial borders). Colonial powers used them, too, to divide their empires. Then those internal boundaries (e.g., between Mauritania and Mali, once parts of France's West African Empire) became national boundaries when decolonization took place. Note the prevalence of geometric boundaries in various regions of Africa.

Geometric boundaries on land are mostly a legacy of another century, but a new surge of geometric boundary making is now in progress. States are extending their control over the oceans, and the ocean surface does not provide landmarks for boundary negotiators to use. Hence, all maritime boundaries are, in one way or another, geometric: either straight lines or arcs constructed in accordance with international agreement. Again, the old sequence of events (definition, delimitation, in some cases demarcation) is in force.

Physiographic-political Boundaries

Long segments of international boundaries coincide with physiographic features in the regional landscape: the crests of mountain ranges, the courses of rivers, the middle of lakes. Such boundaries are called **physiographic-political**, and many of them were established centuries ago when linear landscape features such as rivers and mountain ranges formed natural trespass lines. At first glance, it would seem that such natural boundaries would make good political borders, too. But carelessly written definitions (treaties) have caused many problems. Rivers change course over time, as we noted in Chapter 2. Mountain ranges often consist of several parallel ranges; using the highest crests can lead to a zigzag boundary course. And rivers that have their sources in one area of mountains can become vital for irrigation on the mountains' other side. If a river's source is controlled by one country and the other state needs its waters, there is potential for trouble. Physiographic-political boundaries (e.g., that along the Pyrenees between France and Spain) are not without problems.

Anthropogeographic Boundaries

Political boundaries also coincide with changes in the cultural landscape. Compare the map of European languages to the political map; over long segments, linguistic and political borders coincide, for example between Romania and Bulgaria, between Czechoslovakia and Germany, between Hungary and the Soviet Union. The map of religions, too, provides such examples. Where culture, tradition, and political transition conform, the boundary is referred to as **anthropogeographic**. These are the kinds of boundaries the South African government tried to establish with its program of separate development, which was to

432

The Zambezi River's gorge forms the physiographic-political boundary between Zambia and Zimbabwe. The boundary thus crosses the mid-point of the bridge at left, near Victoria Falls. Recently, when a political conflict required solution but neither side would meet in the other's territory, a railroad car was placed exactly on the center of the bridge, with a conference table inside. Representatives of the two countries (plus South African mediators) met around this table—each on his "own" side of the table and thus without leaving their countries.

provide all major cultural sectors in the country with their own homelands.

Anthropogeographic boundaries occasionally emerge as separatist objectives in states with plural societies. Cyprus is presently (1987) divided between Turkish and Greek-Cypriot communities in this way. The state of Sri Lanka has witnessed the rising demand of a separate state in that country's north and east for its Tamil (Hindu) minority. The proposed boundaries of that state would have been anthropogeographic in nature.

An Alternative System

As has been noted in the preceding discussion, some boundaries are very old, others are just emerging or (as in the case of maritime boundaries) in the definition stage. Another way to understand the functioning of boundaries, then, is to view them in temporal perspective.

Old, long-existing boundaries may have been defined and delimited long before the major elements of the present-day cultural landscape in the boundary area evolved. That was the case with the boundary between the United States and Canada west of the Great Lakes. Such boundaries are called **antecedent**, meaning prior or preceding.

Other boundary segments were negotiated as the cultural landscape took shape, or even later. Such boundaries are referred to as **subsequent**, or following. That may not be the most fortunate term because the boundary evolved during, not after, the development of the major elements of the cultural landscape. The boundary between the United States and Canada east of the Great Lakes and the border between Belgium and the Netherlands in Europe are examples of this type.

The map reveals still another form of boundary,

one that results from war and its aftermath. Such boundaries are **superimposed** on societies, sometimes dividing unified nations. The boundary that separates North and South Korea is one of today's most divisive superimposed boundaries; we already know the Berlin Wall in this context. Superimposed boundaries are created by external forces.

That was the case in Vietnam, the elongated Southeast Asian coastal state that once was a part of the French colonial empire and then became the scene of an ideological conflict. Vietnam became divided into Communist North Vietnam (capital Hanoi) and South Vietnam (capital Saigon). The boundary between North and South was on the map from 1954 to 1976, when the North claimed victory in the conflict. The two parts of the country were reunited, and the superimposed boundary disappeared and became a **relic boundary**. Relic boundaries, however, are not erased quickly or easily from the cultural landscape. In Africa, where relic boundaries once separated two parts of the states of Somalia, Ethiopia, and Cameroon, the evidence of their existence still exists on the map and on the land. Toponyms, languages, even local traditions reveal how much impact the temporary boundary had, and continues to have.

BOUNDARIES ON SEAS AND OCEANS

International competition for space on our limited earth is not confined to the land areas. Although boundary disputes and territorial conflicts still continue on the continents, the real focus of rivalry among states has shifted to the waters offshore and to the ocean bed beneath them. Even Antarctica—the remote, frozen wasteland—and the waters that surround it are now in the process of being drawn into this contest for the planet's last remaining frontiers.

Although national claims over waters adjacent to states' coasts are, in several instances, centuries old, nothing has ever matched the intensity of present-day competition for control over the earth's maritime regions. Since these oceanic zones comprise about 70 percent of the globe's surface, there is much at stake here—but not just in the form of surface area. Not only the waters themselves, but also the rock strata beneath the ocean floor contain crucial resources, ranging from food to fuel. Improved technologies of exploration and exploitation have revealed the existence of large reserves of petroleum and natural gas and made their recovery feasible. Metallic minerals are brought to the surface from the ocean floor. In a protein-deficient world, nations compete for control of exclusive fishing zones. And so, in their national interest, the governments of states push their maritime claims ever farther into the high seas, partly on the principle that others will annex what they do not. In large part, the current scramble for the oceans has resulted from our growing awareness of the resource potential of the marine realm.

But there are other reasons. One is the explosive growth of the earth's human population and its mushrooming demands for food, energy, and other commodities. We are straining the planet's productive capacity as never before, and our numbers continue to increase the pressure. Thus the search for supplies extends even to the most remote of the planet's frontiers. Another reason lies in the changed politicogeographic situation discussed previously in this chapter. For centuries, the seas and oceans were controlled, essentially, by a rather small number of powerful nation-states with maritime concerns and capacities. The era of decolonization that began during the 1950s raised the number of claimants from a few dozen to more than 160. A third reason involves the much wider range of national objectives in the modern-day maritime scramble. The great powers of yesteryear needed a strip of water adjacent to their coasts for purposes of a territorial sea where their undisputed sovereignty would prevail. But even more important was the principle of the free and open high seas, and none would demand so wide a territorial sea as to interfere with maritime commerce. Today the world's states with maritime claims include even the landlocked countries that have no coastline at all and others that have no commercial or fishing fleets and certainly no navy. The landlocked states demand the partition of the oceanic realm on a fair-share basis. Many underdeveloped countries have claimed huge territorial seas even though they cannot exploit these areas themselves; they sell exploration rights to the companies of the more developed countries and lease parcels for exploitation if resources are found. Thus the majority of the present-day claimant states have little concern for the principle of open and free high seas. Their objectives differ greatly from those of the older maritime powers.

Early Claims

The evidence from Asia, Africa, and South America indicates that the Europeans were not the first to put the concept of a demarcated linear land boundary into practice. It does appear, however, that the idea of a national maritime boundary first gained adoption in northwestern Europe. As early as the thirteenth century, the rulers of Norway decreed that no foreign vessel would be permitted to sail north of the latitude of Bergen without a Norwegian royal license. In the 1500s, the state of Denmark, then a regional power, instituted the novel rule that a continuous belt of water, 8 nautical miles wide, would henceforth constitute part of the Danish national territory. This was the first formally delimited territorial sea, established nearly four centuries ago. Later, the Danes decided to annex a wider zone, and eventually they claimed a maximum of as much as 24 miles of territorial waters, including most of the good fishing grounds off Scandinavia and Iceland, then under Danish hegemony.

While the Danes were busy closing off their extensive North Sea waters, the English were establishing **baselines** along the coast of Britain, connecting capes and promontories and thus closing off bays and estuaries. They then demanded that Dutch fishing boats obtain permits to fish within those lines, some of them 50 nautical miles in length. Since the Dutch did not have indented, embayed coastlines similar to those of Britain, their position was to argue that the British (and Danish) closures were violations of customary international law. And so, in the seventeenth century, a legal issue came under debate that has yet to be resolved. A Dutch jurist, Grotius, published a treatise entitled *The Free (Open) Sea*. The British legal expert John Selden soon had an answer prepared, and his book, predictably, came out under the title *The Closed Sea*. One interesting concept emerged during this exchange: the idea that the width of any zone of territorial waters should be determined by the reach of weapons positioned on the shoreline.

In other words, if a cannon standing on a beach dune could shoot a cannonball 3 nautical miles, then that state would have jurisdiction over 3 miles of territorial sea.

Time was on the side of the Dutch. As the English fleet grew in size and capacity, the British realized that a general application of the closed-sea concept would adversely affect their interests. In due course, they shifted their position toward that of the Dutch and agreed that a narrow belt of maritime sovereignty would be better than massive closures. Now another set of issues arose. Should such a belt be continuous, as the Danes had it, or should the belt in fact consist of a set of noncontiguous semicircles and take on reality only where shore-based cannon could enforce it? The Dutch jurist Cornelius van Bynkershoek suggested just such a solution, arguing in 1702 that no state should be permitted to claim adjacent waters over which it had no control by force of arms. Then he introduced another principle still being debated today, that of "intended and continuous occupation." Van Bynkershoek felt that any state should be allowed to claim permanently any part of the sea or ocean where it could establish permanent control. In his day, that would have meant the assignment of a fleet, but in modern times, intended and continuous occupation can be achieved by the construction of permanent steel structures such as offshore drilling rigs and platforms.

One result of the discussion was the limitation of claims. The Danes, who had consistently increased their territorial sea width, agreed to observe a continuous zone of just 1 league, equal to 4 nautical miles. This was in 1745; the 4-mile limit became the general rule in Scandinavia for the next two centuries. The Dutch and British observed a 3-mile limit, and countries of the Mediterranean generally declared a 6-mile limit to their territorial waters. With Europe's colonial expansion, these rather modest claims were applied also in Europe's overseas dependencies. During the nineteenth century, the United States and Canada adhered to a 3-mile limit, as did Australia. As late as 1951, the self-interest of maritime powers in such modest claims could be read from the statistics: in that year, 80 percent of the merchant-shipping tonnage of the world was registered in countries that subscribed

435

to the 3-mile limit, and another 10 percent belonged to states that adhered to the 4-mile zone. As we will see, the situation today is very different.

Beginnings of the Scramble

The first occasion nation-states with maritime interests convened was in Monaco in 1930 under the auspices of the League of Nations. This Conference for the Codification of International Law was attended by 47 countries; the United States, a nonmember of the League, sent a delegation. The chief objectives of this first conference involved the determination of a uniform, worldwide limit for the territorial sea (3 miles was the preference of most participants) and the creation of a zone immediately outside the 3-mile limit, the contiguous zone, where states would control sanitation, customs, and overall security. Discussions also focused on states' exclusive rights over certain fishing grounds—the record of the conference foreshadows the events of the present. Certain countries (those with large commercial fleets and global fishing operations) wanted narrow territorial seas and open fishing everywhere. Others insisted on vast, exclusive fishing grounds—protected, if necessary, by wide territorial seas—whether this would interfere with maritime commerce or not. The conference, which continued in The Hague (Netherlands), did not produce a global treaty. But it did define major problems, issues, and viewpoints.

The international crises of the 1930s and the outbreak of the Second World War focused international attention elsewhere. The League of Nations collapsed and what progress had been made in the codification of international law, as far as the oceans were concerned, was moot. And before another international conference could be convened, the unilateral action of one government—that of the United States—gave the scramble for the oceans its first major boost. This occurred in September 1945 when President Harry S Truman issued twin proclamations concerning U.S. intentions beyond its shores. The first proclamation announced that the United States would take regulatory action with regard to high-seas fishing in zones adjacent to its territorial sea. The second and more consequential proclamation announced U.S. jurisdiction over the resources on, and in, the continental shelf adjacent to the national territory, to a depth of 200 ms (somewhat over 100 fathoms). The Truman Proclamation of 1945 thus asserted U.S. control over a marine region of about 1 million sq miles (2.5 million sq km).

This action had immediate effect. It riveted the attention of other countries on the marine realm—not only as a source of fish, but also as a reserve of potential riches yet undiscovered. Within a matter of weeks, Mexico issued a proclamation claiming, in turn, all that the United States had claimed, off Mexican shores. In 1946, Argentina exceeded even the

Truman terms by claiming not only the continental shelf, but also the waters above it. In the following year Chile and Peru, countries on the western coast of South America where the continental shelf is very narrow, simply claimed total jurisdiction (including exclusive fishing rights) over a 200-mile (370-km) territorial sea. Other countries followed suit. The scramble was on in full force.

U.N. Conferences

In the meantime, the United Nations had been established and efforts were begun to convene another conference on the law of the sea. The first of what proved to be a series of such conferences met in 1958 in Geneva and was attended by 86 states, nearly double the 1930 participation. This first U.N. Conference on the Law of the Sea (UNCLOS I) addressed the same problems taken up in 1930 at The Hague and scored some successes, notably in the approval of four conventions. These four conventions codified such concepts as the territorial sea, the continental shelf, high seas, and fishing and conservation of living resources in the seas. But in critical areas, UNCLOS I failed. There was no agreement on the width of the territorial sea, nor was the issue of exclusive fishing grounds resolved—just as had been the case at The Hague in 1930. Still, now there was greater urgency in view of the Truman Proclamation and its aftermath. So the United Nations called another conference to try once more to achieve international agreement. UNCLOS II (convened in Geneva in 1960) was attended by 87 states. The principal hope was that agreement could be reached on a 6-mile territorial sea, a compromise among states still adhering to the 3-mile limit and others demanding as much as 12 miles. Even with the provision of an adjacent 6-mile exclusive fishing zone as further encouragement, enough states voted against the proposal to defeat it. UNCLOS II was a failure, and as the 1960s wore on, states everywhere on earth pushed their claims outward. In equatorial Africa, for example, Gambia's claim went from 3 miles to 6 in 1968, to 12 in 1969, and to 50 in 1971. In South America, Brazil's modest claim of 3 miles went to 6 in 1966, to 12 in 1969, and to 200 miles in 1970.

The expansion of national claims was fueled by resource discoveries on, and beneath, the continental shelf (the value of manganese nodules on the ocean floor that contained manganese, nickel, copper, and cobalt among other components, was being realized) by growing national needs, by intensified international ideological competition, and by the rapidly growing number of competitors in the decolonizing world. When the United Nations organized UNCLOS III, which opened in Caracas, Venezuela, in 1974, there were as many as 150 delegations representing national interests, nearly double the number of UNCLOS II just 14 years earlier.

UNCLOS III continued from 1974 until 1982, when the conference closed successfully in Jamaica. The resulting treaty had major implications. Most important of all, it legalized the creation of a 200 nautical-mile Exclusive Economic Zone (EEZ) for all coastal states, a zone within which the coastal state would have control over the exploitation of resources in water or subsoil. The treaty also established legal rules for the closing of bays and of waters between islands in an archipelago, the recognition of islands, historic waters, and other special cases. The United States and other industrialized countries did not, however, join the 117 states that signed the treaty. The reason had to do with deep-sea mining in the marine regions that remained high seas, beyond the EEZ (Fig. 11.6). UNCLOS III wanted to limit this activity and established a U.N. authority to govern it. Several industrialized countries could not accept this restriction.

The EEZ concept created a new maritime map; it also required many states to define (or redefine) their common borders at sea. The conference recognized territorial-sea claims up to 12 nautical miles, but maritime boundaries now had to be defined and delimited to the 200-mile line as well. The United States and Canada, for example, had a dispute over the boundary in the Gulf of Maine, over the Georges Bank, and on to the EEZ limit. That issue had to be settled by the World Court in The Hague. Thus Figure 11.6 is only an approximation; the whole process of definition and delimitation must take place for hundreds of boundary segments. The United States, for one, has neighbors in the Atlantic, Caribbean (Fig. 11.6), Arctic, coastal Pacific, and interior Pacific. The

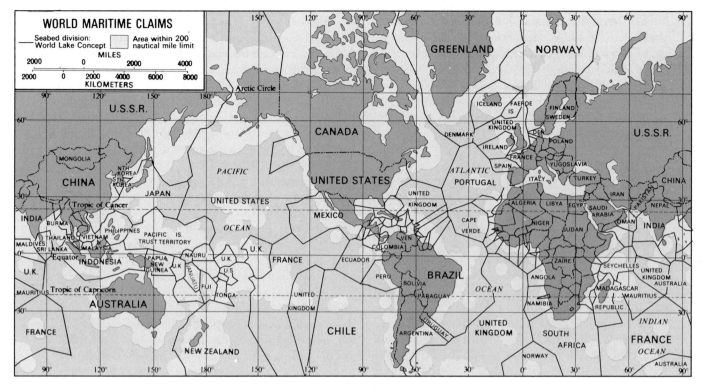

FIGURE 11.6

Effect of the 200-mile limit on the high seas.

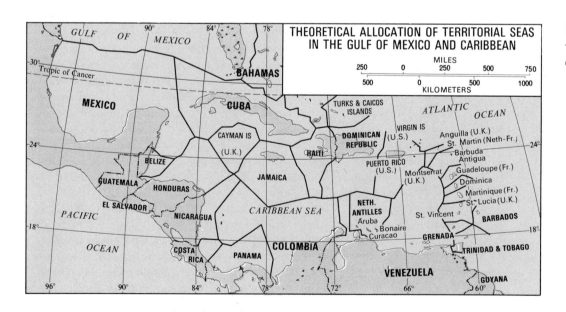

FIGURE 11.7

The Gulf and Caribbean Allocated.

EEZ regulations have opened a new era of boundary negotiation in which lawyers, political and historical geographers, and politicians are already participating. The global boundary-making process is far from over.

In the background looms the prospect that the EEZ will be only a stage in the ultimate absorption of all the oceans by coastal states. The world-lake concept would award some control over all that remains of the high seas—as has already happened in marine regions where distances between countries and islands are below 400 nautical miles, for example in the Caribbean (Fig. 11.7), in Southeast Asia, and in the North Sea. UNCLOS III may not have solved the complete list of potential problems after all.

As usual in our world of maldistributed advantage, the major benefits from the marine scramble have accrued to the developed states. These are the countries whose fishing fleets already ply the distant oceans and seas, whose technology permits the recovery of deep-ocean manganese nodules, whose oil companies can negotiate concessions for the exploitation of reserves in the continental shelves of the less-favored countries. And so the marine common heritage of humanity diminishes—rich countries be-

---

• The Southern Realm •

Even as EEZs reduce the high seas, still another geographic realm faces the threat of rising international competition: Antarctica and the Southern Oceans.

Antarctica is the world's most remote continent, and its environments are the harshest on earth. The region has been described as the ''white desert'' and as the ''home of the blizzard.'' The Southern Oceans' waters are frigid, filled with icebergs, and whipped by raging storms.

Yet the region has attracted international attention, first by whalers and sealers in the Southern Oceans, then by explorers planting national flags on the icy mainland. Reaching the South Pole became an obsession that motivated scientific societies and individual adventurers to make overland journeys into the interior. A Norwegian, Roald Amundsen, reached the pole first in 1911; the British explorer Robert Falcon Scott found the Norwegian flag when he reached the same point just a month later.

Such explorations and heroics led to national claims. Coastal claims on the Antarctic perimeter were extended to the pole as pie-shaped sectors of Antarctica. Australia's was the largest, followed by the claims of Norway and New Zealand. Inevitably, some claims overlapped. Where Antarctica extends (in the Palmer Peninsula) toward South America, the claims of three states (Argentina, Chile, and the United Kingdom) clash. France has a narrow sector,

and a large area named Marie Byrd Land remains unclaimed (Fig. 11–8).

After the Second World War, scientific research in the southern region expanded. States recognized the need for international cooperation, and in 1959, a group of 13 states signed and ratified an Antarctic Treaty, including all 7 claimant states. The treaty took effect in 1961 and was to remain in effect for 30 years. Under its chief provisions, all states agreed to reserve Antarctica for peaceful, scientific purposes; territorial claims were held in abeyance, and no new claims were recognized.

Significantly, the Antarctic Treaty did not address resource exploitation. Fishing for krill and other marine life in the Southern Oceans intensified. During the global energy crisis of the 1970s, interest in the mineral and fuel potential of the frozen continent and its deeply submerged continental shelves also grew. As technology brings more such raw materials within reach, the southern realm will attract more attention.

Rules for maritime-boundary delimitation established by UNCLOS III do not apply here, and the Antarctic Treaty will expire in 1991. In the late 1980s, meetings were in progress to negotiate an extension. But attitudes toward the issue vary among member states. What was deemed desirable in the 1950s may not be attainable again in the 1990s.

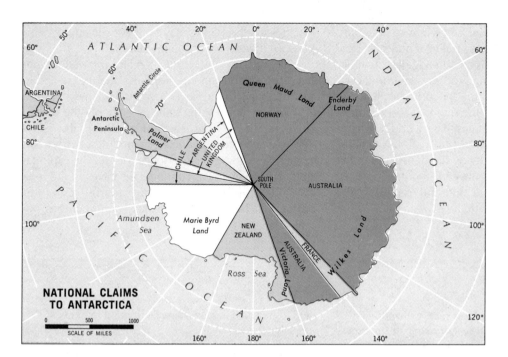

FIGURE 11.8

440

come still richer, economic gaps widen. The Geographically Disadvantaged States (a formal appellation that includes the approximately 30 landlocked countries), though a large majority at the UNCLOS conferences, have been unable to stem the tide of expansion and have, in fact, contributed to it. International maritime law still is in a formative stage, not rigorous enough to contain or channel the demands of powerful developed states. The freedom of the high seas and the shared heritage of the oceans are fading ideals in a hungry, overpopulated world where national self-interest still comes first.

Core Areas of States

In Chapter 10 we studied the core–periphery concept in economic-geographic context. It soon became clear that core areas are more than economic regions.

They have political relevance, too; their political influence protects the economic advantages, in fact. Every functioning state has such a regional focus, a heartland. When we think of France, what comes to mind is the great city of Paris and its environs, the Paris Basin, for centuries the heart of the French Republic. In the United States, the core area still lies in the East, including the great cities of the megalopolis, and extending westward to the vicinity of Chicago. An American geographer, D. Whittlesey, wrote in his book *The Earth and the State* (1939) that many states typically "crystallized about a nuclear core that fostered integration . . . in nearly all states the nuclear core is also the most populous part—its ecumene. [The core area is] the portion of the state that supports the densest and most extended population and has the closest mesh of transportation lines . . . it is more richly endowed by nature than the rest of the state."

It is interesting that Whittlesey chose dense population and an intensive transport network to characterize the core area. Actually, it is one thing to form a general idea of the core area, but quite another to determine exactly what constitutes it. One problem is that of **scale**. Obviously, we cannot expect a state such as Tanzania or Thailand to possess a core area as highly developed as France or the United States. On the other hand, there are territorially quite large states that simply have not developed strong core areas. Look at an atlas map of Mauritania, Central African Republic, Saudi Arabia, Mongolia. You will be hard put to find the population agglomeration and denser transport networks that would indicate the existence of a core area. The core area not only reflects the substantial population our nation-state definition required, but also manifests organization.

Core areas, then, have the following qualities. They contain the largest cluster of population in the state, but the size of that cluster varies from state to state. In a country of 50 million people, we might expect to find 14 or even 20 million people in the core area. In a state that has only 4 million inhabitants, the core area might contain fewer than 1 million people. And, as maps of surface communications show, well-developed core areas are characterized by webs of railroads and roads that radiate outward to connect the heart of the state to more outlying areas. The interconnections are many within the core area itself, then become sparser with distance away from it. But again, we should keep the scale of things in mind. In a country where dirt roads and paths connect most of the towns and villages, a few miles of busy hard-surface, all-weather highway may be as important to the state as a whole network of roads and railroads is to a more developed state.

Core areas, too, are the most urbanized parts of states. Not only does the core area contain a large agglomeration of people, it also constitutes a substantial part of the urbanized population. In the United States, cities within the core area include not only the megalopolis between Boston and Washington, D.C., but also such places as Pittsburgh, Cleveland, Detroit, and Chicago (Fig. 11.9). In France, Paris, the largest city in that state by far, is the focus of the core area. In the Soviet Union, the core area centers on Moscow but also includes Leningrad, Gorky (Gor'kiy), and other major industrial centers. China's core area lies in the northeast and incorporates Beijing, Shanghai, and so many other urban foci that the core area is far more highly urbanized than China as a whole.

This brings us to the capital city and its relationship to the core area. Obviously, the normal situation is for the capital city to lie within the core area, if indeed it is not the very focus of that region. In numerous states—France, the United Kingdom, Chile, Mexico, Egypt, and Thailand, for example—the capital city is also the largest city and, therefore, the obvious center of gravity of the core area as well as of the state as a whole. But there are cases where the capital city is not by any means the largest, as in the United States, Australia, South Africa, and Nigeria. Note that these states have something in common: they are federally organized, and their capital cities were either created or selected specifically to function as the states' governmental headquarters. In these cases, the capital cities lie within the core area, but they are not the dominant foci. In a very few special cases, the capital actually lies *outside* the core area. The outstanding example is Brazil, whose new (also federal) capital of Brasilia was created and located specifically to draw the attention of Brazilians away from their eastern, coastal core area and into the vast interior.

The core areas of states also are (or long were) the most productive regions of their states. Some core areas rose to prominence because they contained a combination of mineral resources that could sustain the forward march of the Industrial Revolution. Towns grew into cities, labor streamed in, and communication systems intensified. In other instances, core areas emerged on the basis of exceptionally favorable conditions for agriculture. Argentina's Pampa is such a case. Combinations between the two circumstances also occurred, as in South Africa, where the major industrial heartland lies very near the most extensive and productive farming areas. And then, of course, there is the factor of self-perpetuation. Once in the lead, a core area can survive as the state's leading region even when other areas in the state prove to have equal or even greater advantages. London found itself rather far away from the focus of change when the Industrial Revolution transformed Britain, but the city was, and remained, the state's largest population agglomeration and market, the center of governmental offices and agencies, and a major transport hub. In the United States, people for many decades have been moving west, and California has become the largest

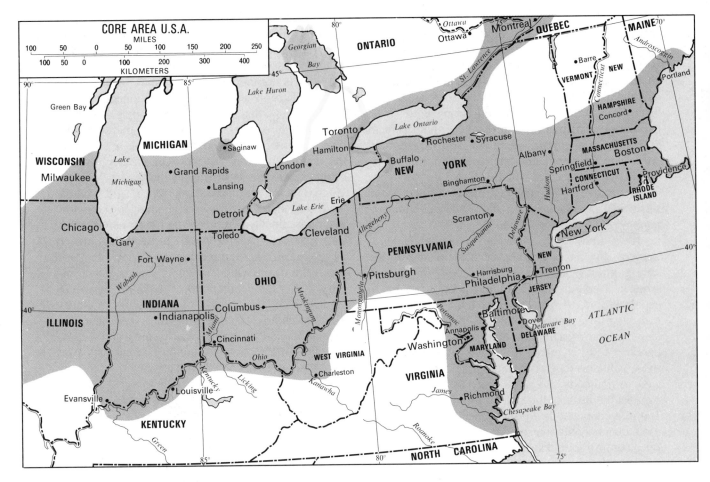

FIGURE 11.9

The contiguous core areas of the United States and Canada.

state in terms of population, but the core area still is where it always was, in the east.

Our scrutiny of the world political map in connection with the core areas of states reveals another significant reality. Some states have not one but two or even three distinct core areas. Normally the situation is that one core area will clearly be dominant and the other(s) might be viewed as secondary or subsidiary (or emerging) core areas, as California obviously is. But there are instances where the core areas within one state may actually be nearly equal in content and importance. In Nigeria, for example, there are core areas in the southwest, centered on Ibadan and containing also the federal capital of Lagos; in the north,

focused on Kano and constituting the Moslem stronghold of the state; and in the southeast, where Biafra tried to secede from the federation. In Ecuador, too, there is a core area on the coast and one in the Andean interior. The implications of such a situation are naturally serious. Competition between the people who form the majority in the state's multiple core areas can endanger the cohesion of the state system, as happened in Nigeria.

CAPITAL CITIES

After the state's external boundaries, the capital city is perhaps its most prominent politicogeographic element. The capital city is the focus of the nation-state, its national headquarters, the seat of government and power, the center of its national life. Often the name of the capital city is used to identify an entire country, when it is said that "London's position has changed" or "Moscow's conquest of Afghanistan proceeds."

The world distribution of approximately 160 capital cities (a majority shown on Fig. 11.10) is further evidence of the diffusion of the European state idea to all parts of the globe. European nation-states had capital cities, the foci on their power during the colonial-imperial era. Those capital cities, many of them primate cities (note again Mark Jefferson's "The Law of the Primate City" in this context), personified the colonial powers in numerous ways. When these powers organized the administration of their colonial possessions, the centralization of governmental functions in a particular city or town occurred. Sometimes the city chosen was already a major traditional urban center (e.g., Mombasa in Kenya, capital of British East Africa before Nairobi was established) or a conveniently located smaller place (as in the case of Calcutta, then the village of Kalikata, chosen by the British as their capital for India before they created New Delhi). Thus, the concept of a capital city became entrenched, and on independence many emerging states spent huge sums of money to embellish their national headquarters, to make them representative of national modernization and aspirations for the future. Some of the emergent states even decided to move their capital-city functions to a new location, to signify the beginning of a new era or for spatial reasons. Tanzania, Malawi, and Botswana are among the African states that have established new capital cities; Nigeria in the mid-1980s was in the process of doing so. In Asia, Pakistan moved its government from coastal Karachi to the northern interior at Rawalpindi, relocating it again to the new nearby city of Islamabad.

Capital cities are of interest to cultural as well as

South Africa is among the few countries that have *divided* capitals. Pretoria and Cape Town share the administrative and legislative functions of government: Pretoria, the old Boer headquarters, and Cape Town, the country's oldest city. This photograph shows the modern central business district and part of the port of Cape Town, as seen from Table Mountain.

443

CAPITAL CITIES OF THE WORLD

POPULATION

MEXICO CITY ▲	Over 2,000,000
Lisbon △	500,000–2,000,000
New Delhi •	100,000–500,000
Luxembourg ○	Under 100,000

MILES

1000   0   1000   2000

1000   0   1000   2000   3000

KILOMETERS

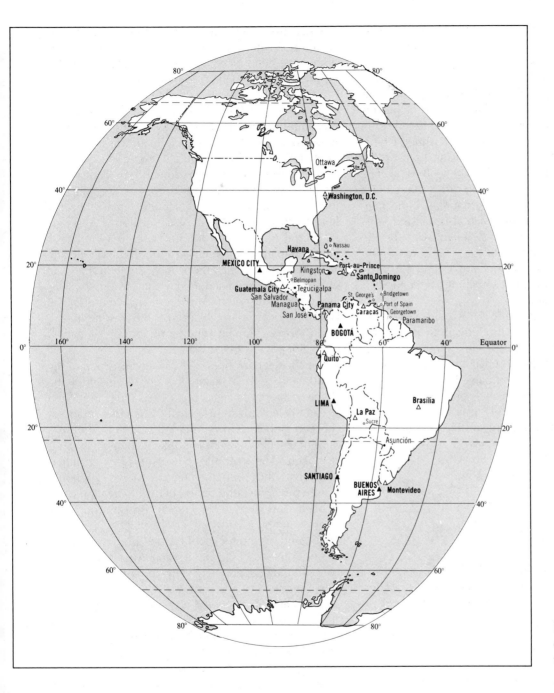

445

FIGURE 11.10
Capital cities of the
world.

political geographers because they are occasionally used to focus a society's attention on a national objective. In Pakistan's case, the two-step transfer of the capital's functions to Islamabad (indeed an appropriate name for the capital of an Islamic republic) was part of a plan to orient the nation toward its historic focus in the interior, and toward the contested north where the country narrows between Kashmir, China, and Afghanistan. In Brazil, the decision to move the capital from historic Rio de Janeiro to Brasilia (Fig. 11.10) was made, in large part, to direct the nation's attention toward the huge underpopulated and ineffectively integrated interior. Japan also transferred its capital functions, from Kyoto to Tokyo, with specific national objectives. The capital city, therefore, can be used as a device to achieve national aims, to spearhead change. Geographers sometimes refer to such cities as **forward capitals**.

In general, the capital city is the pride of the state, and its layout, prominent architectural landmarks, public art, historic buildings and monuments, and often its religious structures reflect the society's values and priorities. It may be employed as a centripetal force, a binding element; it can assert a state's posture internally as well as externally. It is the focus of the state as a political functional region.

From Colonialism
to Supranationalism

Friedrich Ratzel, who suggested that states must absorb territory and population to remain vital (Chapter 1) and who indirectly gave impetus to the field of geopolitics, did his geographic research at a time when colonialism was the norm for strong European powers. Few moral constraints confronted those who furthered the aims of nation-states in other areas of the world. Indeed, when the colonial forces convened in Berlin in 1884 to delimit the map of Africa, this action was viewed more as a civilizing mission than as aggressive expansion. When the British found themselves embroiled in a war against outnumbered Boers in South Africa, world condemnation followed—but

not because Britain was subjugating black African societies. Protest rallies in the United States and elsewhere denounced London for fighting another European people in the colonial world.

Colonialism was the ultimate expression of state expansionism. Europe's core-area states, enriched by mercantilism and empowered by the Industrial Revolution, occupied the periphery. Spain was in the forefront, acquiring a huge empire in the Americas. Portugal took Brazil and large domains in Africa. The Netherlands (Holland) gained control over the East Indies, the Indonesia of the future. France secured large empires in Southeast Asia and North, West, and Equatorial Africa. The "sun never set" on the British Empire. Japan conquered Korea, northeast China, and Taiwan, following the European example in this as in other ways. Belgium acquired its famous Congo. Italy, latecomer in the colonial wars, established itself in the Horn of Africa and in Libya across the Mediterranean. And Ratzel's Germany, finding opportunities, carved out major possessions in Africa.

Such excercise of power changed the map of the world, and with it began the development of the global system of linkages that perists today. Although the age of colonization actually was quite short, its effect on global networks will last far into the future—and not just because of the politicogeographic map it produced. The colonial powers further enriched themselves, thus strengthening cores at the expense of peripheries. When the period of decolonization commenced, and the Dutch, French, British, and other colonists withdrew from their colonial capitals, they left behind economic connections from which the former colonies could not escape. That relationship (of continuing dependency of the former colony on its former master) became known as *neocolonialism*.

In Chapter 8 we noted one of the features of this situation: the continued use of good cropland for luxury, export crops rather than food production. The fact is that states need outside income just as families do (and cities and provinces, too). They must sell goods to produce that income. One kind of product that brings substantial income is a luxury crop: tea, cocoa, coffee. The former colonial power established the plantations and bought the harvests; now the colonial power may have withdrawn, but the state still needs to sell its products. The most likely buyer is the former colonial ruler, a market the newly independent state cannot afford to lose. And so, although there may well be malnutrition in the country, soils still carry crops for export.

COLONIALISM IN THE CULTURAL LANDSCAPE

Before African, Asian, and American colonies achieved independence, they were endowed with lasting contributions by their ruling powers. Some of these colonial legacies were institutional (a civil service, a judiciary, a unitary or federal governmental structure). Others were economic (e.g., an economy organized to facilitate the exporting of raw materials to overseas consumers.) Many colonial dependencies experienced three stages of occupation: (1) a period of intervention and the establishment of machinery for governmental control; (2) a period of colonial administration accompanied by exploitation, sometimes modified in response to local demand or pressure from the "home" government; and (3) a preindependence period of preparation for sovereignty, either with the cooperation of the departing colonial power or attended by a campaign of armed resistance and insurgency.

All three stages left their imprint on the geography of the states that emerged from the colonial period. The majority of the tangible contributions (roads and railroads, medical facilities, urban development, agricultural organization) were made during the second stage, but the third stage frequently witnessed the hasty preparation of a national capital, the demarcation of boundaries, the improvement of internal communications systems, and the provision of other needed, often long-delayed facilities. Thus, the cultural landscape of the colonial power remains etched in distant locales, sometimes strongly provoking an atmosphere of Paris or London or Lisbon.

These vestiges reflect culture and tradition in the colonizing state. France, until decolonization became an inevitable reality, viewed its overseas domains as *La France d'Outre-mer* (Overseas France). Major efforts were made to transfer French cultural traditions and images to the colonies. French-colonial cities such as Dakar, Algiers, Abidjan, and Saigon were adorned with French architectural styles. Religion, schools, and other institutions were modeled after the French version. The French language was strongly promulgated. To this day, the former French dependencies of Africa are referred to as **Francophone** Africa, and French remains a *lingua franca* there.

The French also brought their concept of a highly centralized unitary state to their colonies. They poured resources into the modest core areas and capital cities of their colonies, and in virtually everything—from road networks to electoral systems—organized matters on a single-focus model.

In this and in other respects, the British colonial experience (and legacy) was different. There was no overt effort to create a second England overseas. British institutions were exported, certainly, but so were notions of flexibility and adaptability. The British tried to install their version of an efficient and incorruptible civil service in the colonies. Well acquainted with regional problems at home, the British accommodated regional pressures effectively in their colonies. While France's former colonies, almost without exception, became unitary states, various federal systems were left behind by the British in such states as multicore Nigeria, plural India, and fragmented Malaysia. These systems did not always work well, but in the decolonization period their negotiated birth was a major factor in the generally peaceful transition.

A visit to a formerly British-developed city in Middle America (say Kingston), West Africa (Lagos), East Africa (Nairobi), or southern Asia (Bombay) will bring far fewer reminders of Britain than the French colonial cities do of France. The British cultural imprint has been less overt, less pervasive than that of the French, and if there are remnants of a French cultural landscape from Vietnam to Senegal, you will discover no British equivalent. However, this is more than compensated by the British contributions to government and administration, education, sports (the West Indies have, for decades, fielded a powerful cricket team, as have India and Pakistan), and other institutions.

Portugal acquired small dependencies in Asia (Macao in China, Timur in Indonesia, Goa in India), but its major empire, after the early loss of Brazil, lay in Africa. Angola and Moçambique were the cornerstones of Portugal's colonial empire, and in West Africa Portugal ruled over what is today Guinea-Bissau and the Cape Verde Islands.

Dakar, capital of Senegal, was one of France's leading African cities. As in Abidjan (Ivory Coast) and Brazzaville (Congo), and in the cities of Algeria, the French imprint on the townscape was (and remains) strong.

Luanda and Lourenco Marques (now Maputo), the capitals of Angola and Moçambique, respectively, were given a strong Portuguese flavor, complete with inlaid sidewalks, meticulously manicured gardens, and plentiful monuments to notables of Portuguese history and culture. The Roman Catholic faith was encouraged as a matter of colonial policy, and local people could achieve "civilized" status by attaining certain prescribed standards of competence in various areas, including language and history. If education in the French colonies was oriented to France, Portuguese education was even more exclusively focused on Portugal and things Portuguese.

Portugal in Europe had had a long tradition of authoritarian government; that tradition, too, was imposed on the colonies. Portugal's colonial practices gained worldwide notoriety even before colonialism became subject to worldwide disapproval. Forced la-

bor, compulsory farming, and other excesses produced profit for European Portugal, but they also sowed the wind that would harvest the storm of the 1970s. During the period of decolonization in other parts of Africa, armed resistance to Portugal's rule in both Moçambique and Angola broke out and a costly African war followed.

These events go far to explain why Portugal's two major dependencies chose Marxist courses after independence. Wartime circumstances destroyed old connections, and new economic and political ties were needed. Neither Moçambique nor Angola fared well under the circumstances; the Communist world is just as core dominated as the capitalist world is.

Knowledge of the colonial experience is essential in any assessment of state behavior even now, a generation after the decolonization of many countries. The oppressive Dutch "culture system" that made the East

Indies a giant plantation, and the Belgians' paternalism in ''their'' Congo, which made later national integration very difficult, still hang heavily over the independent states of Indonesia and Zaïre. The colonial legacy is deep and durable.

Persistent State Expansionism

The end of the colonial age has not terminated all expansionist behavior by state governments. Whether for purposes of ''security,'' water rights, maritime access, or to redress grievances over old treaties, states engage in expansionist behavior. Spatially, the pattern of this behavior has changed. Today it is confined (with the exception of potential competition in the southern realm, Antarctica and its waters) mainly to *contiguous* territories—areas immediately adjacent to the aggressive state. The days are over when a state could with impunity begin a campaign of territorial acquisition in a distant continent.

But the problem of expansionist behavior remains a global threat to international order. States in the world's core areas have the resources and organization to impose their will on neighboring states. The sole remaining colonial empire of global proportions, the Soviet Union, is a contiguous entity, and its major campaign for the extension of its direct influence also is happening in an adjacent state, Afghanistan. The United States, for all its might, failed in its ideological and military support of more distant South Vietnam—a sign of the times.

In the late 1980s, the inventory of state conflict, ranging from boundary disputes to full-scale war, remains depressingly long. In addition to the strife in Afghanistan, a war between Iran and Iraq raged with undiminished ferocity, and Lebanon's future as a state was threatened by internal division and external interventions. Ethiopia was at war in Eritrea (technically its own domain); Morocco's expansion into Spanish Sahara gave rise to armed opposition. Boundaries between India and Pakistan in Kashmir, the Soviet Union and China, the two Koreas, Venezuela and Guyana, and many other neighbors continued to constitute potential sources of conflict. Add to this the prospects for disputes over Antarctica and over the allocation of ocean space, and it is clear that fertile ground for state expansionism still exists on our crowded globe.

INTERSTATE COOPERATION

Although states face internal problems of regionalism, they also engage in quite a different activity: external cooperation and even mergers. The phenomenon of interstate cooperation is quite old (the city-states of ancient Greece formed *leagues* to promote common advantage), but the degree to which this idea has now taken root is unprecedented. The twentieth century has witnessed the establishment of numerous international groups, blocs, associations, and alliances in political, economic, and cultural as well as military spheres. The term **supranationalism** is used to define such multistate efforts in which formal association among three or more states results from a desire to secure common goals.

The modern resurgence of internationalism came with the conferences that followed the end of the First World War. The concept of an international organization that would include all the nations of the world became a flawed reality in 1919 with the League of Nations (the United States was among countries that declined membership). In all, 63 states participated in the League, although the total membership at any single time never reached that number. Costa Rica and Brazil left the League even before 1930; Germany departed in 1933, shortly before the Soviet Union joined (in 1934). The League was born of a worldwide desire to repudiate any future aggressor, but the failure of the United States to join dealt the organization a severe blow. Then the League had its big opportunity to stand on principle when Ethiopia's Haile Selassie made a dramatic appeal for help in the face of an invasion by Italy, a member state until 1937. The League failed to take action, and amid the chaos of the beginning of the Second World War, it collapsed.

Nonetheless, the interwar period witnessed significant progress in the domain of interstate cooperation. The League of Nations spawned other international organizations, and a prominent one was the Permanent Court of International Justice, created to adjudicate legal issues between states (e.g., disputes over fishing rights). After the end of the Second World War, the international community once again formed an organization designed to foster international security and cooperation: the United Nations. Just as the United Nations in many ways was a renewal of the

449

ICELAND 1946
NORWAY
FINLAND 1955
DENMARK
UNITED KINGDOM
SWEDEN 1946
IRELAND 1955
NETH.
WEST GERMANY 1973
EAST GERMANY 1973
BELG.
POLAND
LUX. 1973
GERMANY
CZECHOSLOVAKIA
SWITZERLAND AUSTRIA HUNGARY
FRANCE
1955 1955
YUGOSLAVIA
ROMANIA 1955
BULGARIA 1955
PORTUGAL 1955
SPAIN 1955
ITALY 1955
ALBANIA 1955
GREECE
TURKEY
CAPE VERDE IS 1975
MALTA 1964
CYPRUS 1960
SYRIA
LEBANON
ISRAEL 1949
JORDAN 1955
IRAQ
AFGHANISTAN 1946
MOROCCO 1956
TUNISIA 1956
IRAN
KUWAIT 1963
QATAR 1971
BAHRAIN 1971
PAKISTAN 1947
NEPAL 1955
BHUTAN 1971
ALGERIA 1962
LIBYA 1955
EGYPT
UNITED ARAB EMIRATES 1971
BANGLADESH 1974
BURMA 1948
Tropic of Cancer
SAUDI ARABIA
OMAN 1971
INDIA
MAURITANIA 1961
MALI 1960
NIGER 1960
CHAD 1960
SUDAN 1956
YEMEN 1947
YEMEN (P D R) 1967
LAOS 1955
VIETNAM 1977
SENEGAL 1960
GAMBIA 1965
BURKINA FASO 1960
DJIBOUTI 1977
THAILAND 1946
GUINEA BISSAU 1974
GUINEA 1958
NIGERIA 1960
KAMPUCHEA 1955
BENIN 1960
SRI LANKA 1955
GHANA 1957
CENTRAL AFRICAN REPUBLIC 1960
ETHIOPIA
PHILIPPINES
SIERRA LEONE 1961
IVORY COAST 1960
MALDIVES 1965
LIBERIA
TOGO 1960
CAMEROON 1960
MALAYSIA 1957
BRUNEI 1984
Equator
EQUATORIAL GUINEA 1968
UGANDA 1962
SOMALI REPUBLIC 1960
SINGAPORE 1965
GABON 1960
SAO TOME & PRINCIPE 1975
CONGO 1960
ZAIRE 1960
RWANDA 1962
BURUNDI 1962
KENYA 1963
SEYCHELLES 1976
INDONESIA 1950
PAPUA NEW GUINEA 1975
SOLOMON IS 1978
TANZANIA 1961
VANUATU 1981
ANGOLA 1976
MALAWI 1964
COMOROS ISLANDS 1975
FIJI 1970
ZAMBIA 1964
MOZAMBIQUE 1975
NEW CALEDONIA
NAMIBIA
ZIMBABWE 1980
MADAGASCAR 1960
Tropic of Capricorn
BOTSWANA 1966
SWAZILAND 1968
SOUTH AFRICA
LESOTHO 1966
MAURITIUS 1968
AUSTRALIA
NEW ZEALAND

U.S.S.R.
(including UKRAINIAN S.S.R. and BYELORUSSIAN S.S.R.) 1971
MONGOLIA 1961
NORTH KOREA
CHINA 1972
JAPAN 1956
SOUTH KOREA
TAIWAN 1945-71

ATLANTIC OCEAN
INDIAN OCEAN

THE UNITED NATIONS

Charter Members 1945

Members after 1945 with dates of entry

Non-Members

MILES
1000   0   1000   2000

KILOMETERS
1000   0   1000   2000   3000

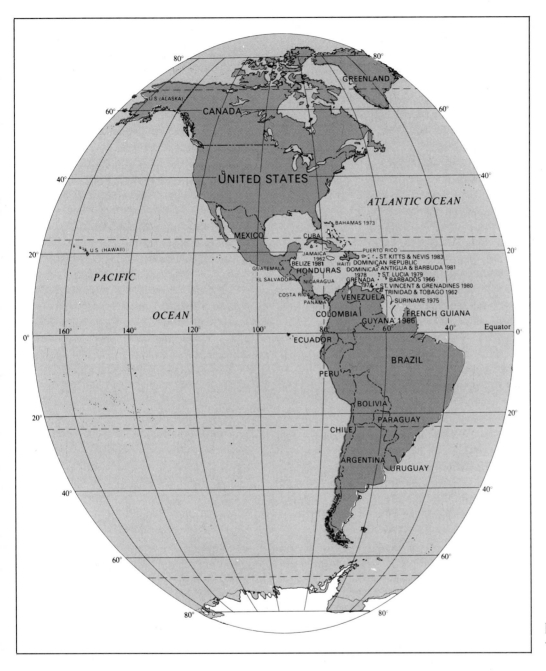

451

FIGURE 11.11
The United Nations.

League of Nations, so its International Court of Justice succeeded the Permanent Court of the interwar period.

The representation of countries in the United Nations has been more universal than that of the League (Fig. 11.11). A handful of states, for various reasons, did not belong to the United Nations in the middle 1980s, but the admission of the People's Republic of China in 1971 gave the organization unprecedented constituency; in mid-1987, the United Nations had 160 member states. The U.N.'s General Assembly and Security Council, the newsmaking units, have also overshadowed the cooperative efforts of numerous less-visible but enormously productive subsidiaries such as the FAO (Food and Agricultural Organization), UNESCO (United Nations Educational, Social, and Cultural Organization), and WHO (World Health Organization). Membership in these organizations is less complete than in the parent body, but their work has been of benefit to all humankind.

Regional Supranationalism

The League of Nations and the United Nations constitute manifestations of a twentieth-century phenomenon that is expressed even more strongly at the regional level. This century has been a period of cross-national ideologies, growing cultural-regional awareness and assertiveness, and intensifying economic competition. States have begun to join together to further their shared political ideologies, their economic objectives, and their strategic goals. In 1987, more than 50 such organizations existed, many of them with subsidiaries designed to focus on particular issues or areas. Today interstate cooperation is so widespread around the world that a new era has clearly arrived.

The first major experiments in interstate cooperation were undertaken in Europe. Three small countries—the Netherlands, Belgium, and Luxembourg—led the way. It had long been thought that all three might benefit from mutual agreements that would reduce the divisiveness of their political boundaries. And certainly they have much in common. Northwest Belgium speaks Flemish, a language very close to Dutch; Luxembourg speaks French, like southeast Belgium. But even more important, the countries have considerable economic complementarity. Dutch farm products sell heavily on Belgian markets. Belgian industrial goods go to the Netherlands and Luxembourg. Would it not be reasonable to create common tariffs, to eliminate import licenses and quotas? Representatives of Benelux (as the organization came to be called) thought so, and even before the end of the Second World War, they met in London to sign an agreement of cooperation. Other European countries watched the experiment with great interest, and soon there was talk of larger, more comprehensive economic unions.

This movement proved to be crucial to the reconstruction of early postwar Europe, and it was given an enormous boost in 1947 when U.S. Secretary of State George Marshall proposed that the United States finance a European recovery program. A committee representing 16 Western European states plus West Germany presented the U.S. Congress with a joint program for economic rehabilitation, and Congress approved it. From 1948 to 1952 the United States gave Europe about $12 billion under what became known as the Marshall Plan. This investment not only revived European national economies, it also constituted a crucial push toward international cooperation among European states.

Out of that original committee of 16 was born the Organization for European Economic Cooperation (OEEC), and this body, in turn, gave rise to other cooperative organizations. Soon after the OEEC was established, France proposed the creation of the European Coal and Steel Community (ECSC), with the principal objective of lifting the restrictions and obstacles that impeded the flow of coal, iron ore, and steel among the mainland's six primary producers: France, West Germany, Italy, and the three Benelux countries. This proposal was also implemented, but the six participants did not stop there. Gradually, through negotiations and agreement, they enlarged their sphere of cooperation to include reductions and even eliminations of certain tariffs and a freer flow of labor, capital, and nonsteel commodities. Ultimately, in 1958, they joined in the European Economic Community (the EEC), also known as the **Common Market**.

Since 1958, the European Community (as the EEC

is now known) has developed further. Not only has its economic role within Europe and the free world expanded, but other European countries have joined (Fig. 11.12). The United Kingdom (which had initially declined to participate), Denmark, and Ireland joined in 1973, Greece in 1981, and Spain and Portugal were admitted at the beginning of 1986.

The existence of the Iron Curtain, and the regional ideological division of Europe that it represents, has deterred still wider European integration. In 1949, the Soviet Union took the lead in establishing a counterweight to the Western European unification movement by establishing COMECON, the Council for Mutual Economic Assistance (Fig. 11.12). This union was designed to promote industrial specialization in its Eastern European satellite countries, to finance investment projects planned jointly by two communist countries or more, and to encourage economic integration generally.

Supranationalism is not unique to Europe, and this phenomenon now has wide global expression. One of the most interesting interstate unions is ECOWAS, the Economic Community of West African States. This organization aims at economic cooperation and integration, joint development efforts, and a reduction in barriers among West African countries generally. A less successful venture was the union of three East African states (Kenya, Tanzania, and Uganda); disputes among these three states eroded this union after some considerable progress had been made. Other less successful interstate unions have come and gone in Southwest Asia and North Africa. For a time, Egypt and Syria were linked together in a political union called the United Arab Republic

FIGURE 11.12

European unions.

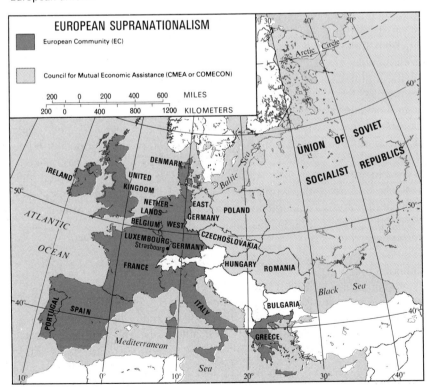

453

(UAR). Again, political change led to the dissolution of this state. Today, Egypt and Syria have retaken their old names, and the UAR has disappeared from the map.

The phenomenon of supranationalism takes several forms, which comprise three broad categories: economic, military-strategic, and multipurpose or cultural.

The EC is the prime example of a successful economic union, but ECOWAS, too, is such a union. In the Americas, CARICOM is a community of Caribbean states; CACM stands for the Central American Common Market; and LAFTA is the Latin American Free Trade Association.

Among military-strategic unions, NATO (the North Atlantic Treaty Organization) is the most consequential. Formed in 1949, NATO ties the United States, Canada, and European countries west and south of the Iron Curtain into a formidable alliance. In turn, the Soviet Union and its Eastern European neighbors formed the Warsaw Pact (see map again) to coordinate their military and strategic operations. Again, not all military-strategic organizations have been successful. The Central Treaty Organization (CENTO) once linked NATO to countries extending from Turkey to Iran, but political and cultural developments in Southwest Asia since the 1960s overtook this union.

Multipurpose organizations have numerous, less clearly defined objectives. The Organization of American States (OAS) discusses Western Hemisphere matters ranging from political and cultural to economic. The Organization of African Unity (OAU) and the League of Arab States (LAS) include the great majority of African and Islamic countries, respectively, but internal squabbles periodically lead to the temporary ousters of some members. Egypt, for example, was ousted from LAS in the late 1970s when it agreed to negotiate with Israel.

The obvious question is: will this phenomenon of supranationalism lead to lasting unions among states and will the state system prove, after all, to have been a temporary phase in world politicogeographic organization? It is too early to make such predictions. To date, there has been only one truly political organization that has met with some success: the Council of Europe and its elected European Assembly. This highly significant union of European states, with more than 20 members, meets in a "European capital," Strasbourg, France. Its Assembly is elected by member countries, and its budget is supported by these members. The European Assembly does not yet have the power to legislate (i.e., its decisions cannot supersede those of the governments of member states). But in a number of international matters affecting many or most European states west of the Iron Curtain, the Assembly's decisions have been accepted. Thus, the first steps in European unification have been taken, and in this respect, European supranationalism has no peer.

The supranational phenomenon may offer a ray of hope in a world full of political conflict. Strong, unyielding nationalism lies at the root of many international disputes. Collective action—and collective restraint—may just enable states to overcome the centrifugal forces confronting them today.

•   Is the State System Failing?   •

In the preceding pages we have viewed the nation-state as a flexible system, capable of modification to adjust to changing national needs. Underlying it all, however, is an impression that in several respects, the state system as it has emerged on the European model cannot accommodate the rising pressures placed upon it. Although it may be too early to predict that the world's patchwork of national states will prove inadequate to preserve world order, we might well ponder the following danger signs.

1 INTENSIFYING REGIONALISMS

Nation-states from Europe to Africa, from the Americas to Asia are being torn by conflicts that are rooted in the desire of clustered minorities for greater regional identity and autonomy. This problem affects developed as well as underdeveloped states. The U.K., which has served as a model for numerous newly independent states during the transfer of sov-

454

ereignty, is buffeted now by rising nationalism in Wales (Cymru) and Scotland (Caledonia), and by a violent crisis in Northern Ireland. In London, the response has been to prepare for a gradual adjustment in the country's governmental structure to accommodate these pressures for change, and the process is described as **devolution**—signifying a reversal of the sequence of events that brought the "United" Kingdom under one flag in the first place.

The British experience is not unique, even in Europe. In Spain, a minority of Basque separatists have demanded autonomy for their small northern domain, and they have perpetrated violent actions to promote that goal. France has faced intermittent problems with a secessionist movement on its Mediterranean island of Corsica. Cyprus fractured into two sectors, one populated by Greek-Cypriots, the other by the Turkish minority. In Sri Lanka, a violent campaign by Tamils for an independent state has destroyed the

island-state's reputation for peaceful coexistence. Many other examples could be cited to underscore a growing problem: to many people in plural societies, the state system no longer is acceptable. Growing political and ideological awareness, ideological conflicts, real and perceived discrimination, underrepresentation in government, regional economic disparities, and other circumstances encourage peoples to take up the cause against their own state. It is a global problem, magnified by intercore rivalries, and it has major portents for the future.

2 FAILURE OF THE EUROPEAN MODEL

When the European colonial powers conferred independence upon their overseas territories, their final actions generally involved tutelage in the structuring of government and administration. A prominent example was Ghana, where the British worked hard to

Basque separatism has strengthened in recent years as Basques demand autonomy and independence from Spain. This movement, which threatens to change the map of Spain, is only one of many separatist campaigns being waged against states of the world.

assist in the creation of a government on the Westminster model, but British involvement also led to Nigeria's initial federal constitution, to Kenya's first two-party government, and to other variants in India, Pakistan, Malaysia, and elsewhere. France exported the concept of the unitary state to Africa as well as Asia, and other European powers in various ways provided the model for the newly independent states.

But the European model frequently proved incompatible with the traditions and needs of those former colonies. The Biafran crisis in Nigeria resulted in large measure from a British failure to heed Nigerian warnings that the federal states should not be placed in a position of strength that might motivate them to secede. In Ghana, Kenya, and elsewhere, the two-party system soon succumbed to the one-party state; African traditions of a strong central authority had not been changed by the colonial interlude. Pakistan's federal system collapsed, unable to accommodate the East; Singapore seceded from Malaysia's complicated federal state. The colonial powers had thrown boundaries around diverse peoples, but they failed to achieve workable state systems when these plural societies approached self-government.

3 MULTIPLYING SUPRANATIONALISM

So numerous are the associations, organizations, and blocs that tie states together strategically, economically, and in some cases politically that the sovereignty of many is impaired. The mushrooming of supranationalism constitutes evidence that a need exists for other frameworks to serve the interests of nation-states. In part, this phenomenon of the second half of the twentieth century relates to the newfound independence of many countries that fall far short of nation-states in almost every aspect and whose security and other interests are dependent on the fortunes of other, stronger states. It is a measure of the corruption of the nation-state idea that small islands in the Pacific and Indian Oceans (and in the Caribbean Sea) are abandoned to an independence that exists in name only.

Supranational unions consisting of states that are themselves poorly organized and weak cannot be effective in furthering constituent states' aims and objectives. Such supranational political unions as the short-lived United Arab Republic (Egypt and Syria) and an incipient merger between Morocco and Libya only serve to emphasize the failure of the state idea in those countries.

4 CORE-CORPORATE INFLUENCE

In a growing number of nation-states (even in some of the less impoverished ones) the influence of multinational corporations is so great that governments' independence of action is threatened. Today there are corporations—industrial concerns, oil companies, and others—so powerful that they can influence the behavior of nation-states and their governments: in effect, taking over part of the government's role. In the boardrooms of those corporations in New York or Tokyo or London sit small groups of directors who control far more money and exercise much more power than the governments of many entire countries. They have representatives in cities all over the world who control, on behalf of the corporate headquarters, the industries of foreign countries. These corporations are called multinationals; in fact, they mostly are less than that. Generally there is neither a genuine partnership among countries in which the corporation functions nor are the citizens of those countries much involved in the decision making and actual directing of corporate affairs. Of the 10 largest multinational corporations in 1988, 8 had their headquarters in the United States. Every one of these corporations had a total production that was larger than the entire national product of half the world's nation-states.

It has been estimated that perhaps as much as two thirds of the investment by multinational corporations takes place in developed countries, but the remaining one third (spent in the UDCs) has an enormous impact on weaker economies and governments. In a country where there is limited industrial development, the ownership of factories by multinational corporations can impair the government's ability to control domestic industry; the corporation might even begin to dictate to the government. At least one multinational corporation has been accused in recent years of actively sabotaging a legitimate national government when other attempts at control failed.

It is true that multinational corporations can speed the diffusion of technological advances across otherwise inhibitive political boundaries, that they can provide capital and do employ labor where none of this might otherwise have occurred. But the corporations also throttle local competition, develop worldwide monopolies at the consumer's expense, and sometimes use improper methods to achieve their aims. It is not altogether inconceivable that such global corporations (as they are better called) might use their capital strength not merely to buy off politicians, but also to develop their own strike forces to protect their interests. Some states have been able to confront and contain the global corporate machine, as Brazil did in the 1970s. But many states cannot do so, another symptom of the weakening of the nation-state system around the world.

5 PERSISTENT BOUNDARY PROBLEMS

In virtually all parts of the world, national boundaries are under stress and nation-states seek redefinition of their boundaries at the expense of neighbors. This is true in South America, Africa, and Asia; but significantly, it also continues in Europe (especially in Eastern Europe). It is evident that no revision of the existing boundary framework could ever satisfy all nation-states because conflicts prevail even among countries within the same culture realm and between states with otherwise similar objectives (e.g., in changing the international economic order). The manner in which boundaries have been defined and delimited and the voluntary and enforced movement of peoples across existing boundary lines render many boundaries liabilities rather than assets in the world order. If, after such long maturation of the concept of the territorially defined and bounded nation-state the system still does not function adequately, it is clear that another weakness lies here—in the boundary framework.

These factors suggest an affirmative answer to the question raised earlier: is the state system failing? Disaffection with state policies, internal as well as external, is rising, and this disaffection can take the form of violent struggle. Governments coping with regional rebellions have an impaired ability to pro-

mote the well-being of the state as a whole. The time is long past when the nation-state, its flag, and its government symbolized national pride and enjoyed overwhelming adherence; the fall of Europe's monarchies represents this change. Respect for the symbols of state dwindles in many countries (witness public behavior during the playing of the national anthem). Access to weapons and explosives has greatly increased the potential impact of even splinter-group opposition. On the other hand, the consensus nation-state is a receding prospect. A government that accommodates one segment of society may antagonize another, and the cycle of dissent begins anew. Many nation-states, some old and durable, are under politicogeographic stress, a signal that the system itself is in jeopardy. Change may lie ahead—but in a form we cannot foresee.

• Reading about •
Political Geography

457

As we noted in this chapter, political geography is a wide-ranging field of geography. Research has focused on broad topics such as the nation-state; other research has concentrated on electoral behavior in small voting districts. You may want to read for yourself what Friedrich Ratzel had to say in his first statement of the organic theory in the journal *Petermanns Mitteilungen* in 1896. R. L. Bolin translated Ratzel's complicated German, and his interpretation appears in a book that still is worth a trip to the library: R. E. Kasperson and J. V. Minghi (eds.), *The Structure of Political Geography* (Chicago: Aldine, 1969). That volume contains a number of other classical papers and gives a good idea of the status of political geography about two decades ago.

Recently published books in this field evince a continued interest in the state. A book of importance (difficult to read, but worth the effort) is *State Apparatus: Structures and Language of Legitimacy* by G. L. Clark and M. J. Dear (Boston: Allen & Unwin, 1984).

Also see a book by R. J. Johnston, *Geography and the State: An Essay in Political Geography* (New York: St. Martin's Press, 1982), which insists that better theoretical bases are needed in politicogeographic research on the state. A book with a tempting title is R. Paddison's *The Fragmented State: The Political Geography of Power* (New York: St. Martin's Press, 1983), a volume that touches on several of the topics covered in our chapter. A general textbook on political geography is M. I. Glassner and H. J. de Blij, *Systematic Political Geography* (New York: Wiley, 1980). Older, but still of value, is the volume by W. G. East and J. R. V. Prescott, *Our Fragmented World: an Introduction to Political Geography* (London: Macmillan, 1975). For a good historical overview of political geography as a field see the article by J. C. Archer and J. Shelley in the book edited by M. Pacione, *Progress in Political Geography* (Dover, Del.: Croom & Helm, 1985).

Other books reveal the many sides of political geography. Power is discussed by J. D. Stoessinger in *The Might of Nations* (New York: Random House, 1963). The core–periphery concept as it manifests itself in this context is discussed by a series of authors in a volume edited by J. Gottmann, *Centre and Periphery: Spatial Variation in Politics* (Beverly Hills, Calif.: Sage, 1980). The differences between the approaches of geographers and nongeographers writing in this volume are abundantly clear; as a result, the book is a good plug for the field. In a book published by AAG, R. L. Morrill and C. G. Knight edit a set of articles on *Political Redistricting and Geographical Theory* (Washington, D.C.: AAG, 1981). A very different presentation of politicogeographic issues can be found in the volume edited by D. G. Bennett, *Tension Areas of the World: A Problem-Oriented World Regional Geography* (Champaign, Ill.: Park Press, 1982). Still worth going through is S. D. Brunn's *Geography and Politics in America* (New York: Harper & Row, 1974). And no bibliography of political geography would be complete without reference to the excellent works of J. House, for example, *Frontier on the Rio Grande: A Political Geography of Development and Social Deprivation* (London/New York: Oxford Univ. Press, 1982).

Many of the topics discussed in this chapter have been covered in one or more of the many works authored by J. R. V. Prescott of the University of Melbourne, Australia. See especially his *Maritime Political Boundaries of the World* (New York: Methuen, 1985) and the new edition of his earlier book on limits of the state, *Political Frontiers and Boundaries* (London: Allen & Unwin, 1987). If you have a term paper to write in political geography, check first to see what Dr. Prescott has written about it, and consult his citations. He has written on topics as varied as Antarctica and the *Map of Mainland Asia by Treaty* (Melbourne, Austral.: Melbourne Univ. Press, 1975).

Another place to look for contributions in political geography, of course, is the journal literature. Articles on political geography, some of them seminal to the field, have appeared in the *Annals*, Association of American Geographers, in the *Geographical Review*, and elsewhere. There also is a relatively new journal called *Political Geography Quarterly*; it began publication in 1982 and is now well established.

Opportunities
in Geography

Chapter 1 described the history and development of geography as a discipline. The chapters that followed provided an idea of the wide range of topics geographers pursue. But there are specializations within each of those topics and, in an introductory book such as this, there was not enough space to discuss all of these. This concluding section is designed to help you should you decide to major or minor in geography or to consider it as a career option.

Areas of Specialization

As in all disciplines, areas of concentration or specialization change over time. As we noted, there was a period when most geographers were physical geographers and the physical landscape was the main objective of geographic analysis. Then the pendulum swung toward human (cultural) geography, and students everywhere focused on the imprints made by human activity on the surface of the earth. Still later, the spatial manifestations of human behavior became a major area of interest. In the meantime, geography's attraction for some students lay in technical areas: in cartography, in computer-assisted analysis of mass data, in remote sensing.

All this meant that geography posed—and continues to pose—a challenge to its professionals. New developments require that we must keep up to date; but also, we should continue to build on established foundations.

REGIONAL GEOGRAPHY

One of these established foundations is regional geography. Regional geography encompasses a large group of specializations and these are not highlighted in this book, for obvious reasons: this is a *topical* geography. Regional geography and its areas of specialization focus on one of geography's key concepts. This is done in several ways. Some geographers specialize in

the theory of regions: how they should be defined, how they are structured, how their internal systems work. This led in the direction of regional science and some geographers preferred to call themselves regional scientists. But make no mistake: regional science is regional geography.

Another, older approach to regional geography involves specialization in an area of the world, ranging in size from a geographic realm (see Chapter 7) to a single region (say Western Europe) or even a state or part of a state. There was a time when regional geographers, because of the interdisciplinary nature of their knowledge, were sought after by government agencies and other offices. Courses in regional geography abounded in universities' geography departments. Regional geographers played key roles in international study and research programs. But then the drive to make geography a more rigorous science and to search for universal (rather than regional) truths contributed to a decline in regional geography. The results were not long in coming, and you have undoubtedly seen the issue of 'geographic illiteracy' discussed in newspapers and magazines. Now the pendulum that has affected geography throughout its existence is swinging back again and regional geography and regional specialization are reviving. It is a good time to consider regional geography as a professional field.

YOUR PERSONAL INTERESTS

Geography, we have learned, is united by several bonds, of which regional geography is only one. Regional geography exemplifies the spatial view that all geographers hold. The spatial approach to study and research binds physical and human geographers, regionalists, and topical specialists. Another unifying theme is our interest in the relationships between human societies and natural environments. We have referred to this topic throughout this book; as an area of specialization it has gone through difficult times. Perhaps more than anything, geography remains a field of synthesis, of understanding interrelationships.

Geography also is a field science, using "field" in another context. In the past, almost all major geography departments required a student's participation in a "field camp" as part of a Master's degree program, and many undergraduate programs included field experience. It was one of those bonding practices where students and faculty with diverse interests met, worked together, and learned from each other. Today, fewer field camps of this sort are offered but that does not change what geography is all about. If you see an opportunity for field experience with professional geographers—even just a one-day reconnaissance—take it. But realize this: a few days in the field with geographic instruction may hook you for life.

Some geographers, in fact, are far better field-data gatherers than analysts or writers. In this respect they are not alone: this also happens in archeology, in geology, and in biology (among other field disciplines). This does not mean that these field workers do not contribute significantly to knowledge. Often, in a research team, some of the members are better in the field and others excel in subsequent analysis. From all points of view, field work is important.

Geography, then, is practiced in the field and in the office, in physical and human contexts, in generality and detail. Small wonder that so many areas of specialization have developed. If you check the bulletin of your university or college, you will see some of these specializations listed as semester topics. But no geography department, no matter how large, could offer them all.

How does an area of specialization develop, and how can one become a part of it? The way in which geographic specializations have developed tells us much about the entire discipline. Some major areas, now old and established, began as research and theory-building by one scholar and his or her students, as we saw in Chapter 1. The students dispersed to the faculties of other universities and began teaching what they had learned. So W. M. Davis's physical geography and C. O. Sauer's cultural geography spread from Harvard and Berkeley.

But it is one of the joys of geography that the basics and methods, once learned, are applicable to so many features of the human and physical world. Geographers have specialized in areas as disparate as shopping-center location and glacier movement, tour-

ism and coastal erosion, real estate and wildlife, suburbs and sports. Many of these specializations started with the interests and energies of one scholar; some thrived and grew into major geographic pursuits. Others remained one-person shows, but with potential. When you discuss your own interest with an adviser, you may refer to a university where you would like to do graduate work. "Oh yes," the answer may be, "Professor X is in their geography department, working on just that." Or perhaps your adviser will suggest another university, where a member of the faculty is known to be working on the topic in which you are interested. Then is the time to write a letter of inquiry. What is the professor working on now? Are graduate students involved? Are there research funds available? What are the career prospects after graduation?

The Association of American Geographers (1710 Sixteenth Street, N.W., Washington, D.C. 20009) recognizes several dozen so-called Specialty Groups. The 1987 list of these groups, and their objectives, are printed on pages 464 and 465. The list is by no means a comprehensive summary of geography's areas of specialization, but it does provide insight into the leading ones; to be recognized, a Specialty Group must have 50 members. You may well find one of your own interests listed here. To establish contact with the current Chairperson of any Specialty Group, write to the AAG and your letter will be forwarded.

All this may seem far in the future. Still, the time to start planning for graduate school is now. Applications for admission and financial assistance must be made shortly after the *beginning* of your senior year! That makes your junior year a year of decision.

• An Undergraduate Program •

The most important concern for any geography major or minor, however, is basic education and training in the field. An undergraduate curriculum contains all or several of the following courses (titles may vary):

• Specialty Groups of the Association of American Geographers, 1987 •

AFRICA   To further geographic research on Africa. To promote geographic education on Africa.

AGING   To support research, teaching, and service pertaining to the geography of aging and the aged.

APPLIED   To increase the visibility of applied geography in the profession and the general population.

ASIA   To promote geographic research and to facilitate teaching geography of Asia.

BIBLE   To encourage geographical study and teaching in biblical geography.

BIOGEOGRAPHY   To promote the interests and development of biogeographers.

CANADA   To stimulate a more visible series of activities and increased research on Canadian topics.

CARTOGRAPHY   To encourage cartographic research, promote education in cartography and map use, and facilitate the exchange of ideas and information about cartography.

CHINA   To promote the study of the geography of China, including Taiwan.

CLIMATE   To support research and teaching in climatology.

COASTAL AND MARINE   To encourage the intellectual exchange of knowledge related to coastal and marine environments and resources.

CONTEMPORARY AGRICULTURE AND RURAL LAND USE   To promote the common interests of geographers working on agriculture and rural land use problems in developed countries.

LATIN AMERICA   To promote education, research, and other scholarly activity relating to Latin American geography.

MATHEMATICAL MODELS AND QUANTITATIVE METHODS   To aid and enhance research on mathematical models, statistical and other techniques, and to promote scientific methodology in all fields of geography.

MEDICAL   To provide a forum for those with interests in geographical epidemiology, spatial aspects of medical care delivery, and ethnomedicine.

MICROCOMPUTERS   To investigate ways in which microcomputers can be used as a tool in geographical research and teaching.

NATIVE AMERICANS   To encourage and disseminate research that contributes to an understanding of the problems, concerns, and issues of Native Americans.

POLITICAL   To provide a focus and organization for political geographers through which they can achieve scholarly growth and to improve the status of the field.

POPULATION   To promote research, teaching, and service in the general field of population geography.

RECREATION, TOURISM, AND SPORT   To foster the teaching of and research on the subject matter areas of recreation, tourism, sport, and leisure.

REGIONAL DEVELOPMENT AND PLANNING   To publish and distribute newsletters featuring activities and items of interest in this field.

CULTURAL ECOLOGY   To promote and conduct scholarly activities on cultural ecology topics.

ENERGY   To promote communication and interaction among energy geographers (and) to increase the contribution of geographers in energy research.

ENVIRONMENTAL PERCEPTION   To advance the academic and applied interests of environmental perception within the profession of geography.

GEOGRAPHIC INFORMATION SYSTEMS   To promote research, development, and practice in computer-based hardware, software, and graphic capabilities.

GEOGRAPHIC PERSPECTIVES ON WOMEN   To promote geographic research and education on topics relating to women and gender.

GEOGRAPHY IN HIGHER EDUCATION   To promote research, development, and practice in the learning and teaching of geography and (to) strengthen the role of geography in education.

GEOMORPHOLOGY   To foster better communication among those working in the geomorphic sciences, especially in geography.

HISTORICAL   To promote the common interests of persons (working in) historical geography.

INDUSTRIAL   To facilitate the exchange of information and to stimulate research, teaching, and applications in industrial geography.

REMOTE SENSING   To foster an understanding of the nature and application of remote sensing science.

RURAL DEVELOPMENT   To promote sharing of ideas and information among geographers interested in facets of rural development.

SOCIALIST GEOGRAPHY   To develop socialist perspectives, theories, and analyses of geographical questions, and to investigate the issue of radical change toward a more collective society.

SOVIET UNION AND EASTERN EUROPE   To promote the professional competence of, and to enhance communication among, geographers working in these regions.

TRANSPORTATION   To develop transportation geography as a field of the discipline.

URBAN   To facilitate communication of information and ideas among urban geographers and other urban specialists.

WATER RESOURCES   To provide a forum for discussion, reinforcement, and exchange of information for geographers interested in the use, development of, and physical characteristics of water resources.

1 Introduction to Physical Geography I (Landscapes, landforms, soils, elementary biogeography)

2 Introduction to Physical Geography II (Climatology, elementary oceanography)

3 Introduction to Human Geography (Theory and topics)

4 Introduction to Regional Geography (Major world realms)

These beginning courses are followed by more specialized courses, including both substantive and methodological ones:

5 Introduction to Quantitative Methods of Analysis

6 Beginning Cartography

7 Analysis of Remotely Sensed Data

8 Cultural Geography

9 Political Geography

10 Urban Geography

11 Economic Geography

12 Historical Geography

13 Geomorphology

14 Geography of Europe (and/or other realm)

You can see how Physical Geography I would be followed by Geomorphology, and how Human Geography now divides into such areas as cultural and economic geography. As you progress, the focus becomes even more specialized. Thus Economic Geography is followed for example by:

15 Industrial Geography

16 Transportation Geography

17 Agricultural Geography

At the same time, regional concentrations may come into sharper focus:

18 Geography of Western Europe (or other region)

In these more advanced courses, you will use the technical know-how from courses numbered 5, 6, and 7 (and perhaps others). Now you can avail yourself of the opportunity to develop these skills further. Many departments offer such courses as:

19 Advanced Quantitative Methods

20 Advanced Cartography

21 Advanced Modeling

From this list (which represents only a part of a comprehensive curriculum), it is evident that you cannot, even in four years of undergraduate study, register for all courses. The geography major in many schools requires a minimum of only 30 (semester) credits—just 10 courses, fewer than half those listed. This is one reason to begin thinking about specialization at an early stage.

Because of the number and variety of possible geography courses, most departments require of their majors completion of a core program that includes courses in substantive areas as well as theory and methods. That core program is important, and you should not be tempted to put these courses off until your last semesters. What you learn in the core program will make what follows (or *should* follow) much more meaningful.

You should also be aware of the flexibility of many undergraduate programs, something that can be especially important to us geographers. Imagine that you are majoring in geography and develop an interest in Southeast Asia. But the regional specialization of the geographers in your department may be somewhere else—say Europe. However, courses on Southeast Asia are indeed offered by other departments such as anthropology or political science. If you are going to be a regional specialist, then those courses will be very useful and should be part of your curriculum. But you are not able to receive geography credit for them. After taking these courses successfully, however, you may be able to register for an independent study or reading course in the geography of Southeast Asia if a faculty member is willing to guide you. Always discuss such matters with the chairperson of your geography department.

466

By now, as a geography major, you will be thinking of the future—either in terms of graduate school or a salaried job. In this connection, if there is one important lesson to keep in mind, it is to *plan ahead* (a redundancy for emphasis!). The choice of a graduate school is one of the most important you will make in your life. The undergraduate preparation you acquire as an undergraduate will affect your competitiveness on the job market for years to come.

CHOOSING A GRADUATE SCHOOL

Your choice of a graduate school hinges on several factors, and the geography program it offers is one of these. Possibly you are constrained by residency factors and your choice involves the schools of only one state. Your undergraduate record and grade-point average affect the options. As a geographer you may have strong feelings in favor of (or against) particular parts of the country. And although some schools may offer you financial support, others may not.

Certainly the program and specializations of the prospective graduate department are extremely important. If you have settled on your own area of interest, it is wise to find a department that offers opportunities in that direction. If you have yet to decide, it is best to select a large department with several options. Some students are so impressed by the work and writings of a particular geographer that they go to his or her university solely to learn from, and work with, that scholar.

In every case, information and preparation are crucial. Many a prospective graduate student has arrived on campus eager to begin work with a favorite professor, only to find that the professor would be away for a year of sabbatical leave!

Fortunately, information can be acquired with little difficulty. One of the most useful publications of the AAG is the *Guide to Departments of Geography*, published annually at a cost (to nonmembers) of $15. A copy of this book should be available in the office of your geography department, but if you plan to enter graduate school, a personal copy would be an asset. Not only does this *Guide* describe the programs, requirements, financial aid, and other aspects of geography departments in the United States and Canada, it also lists all faculty members and their areas of specialization. You will find the *Guide* indispensable in your decision making.

You may discover one particular department that stands out as the most interesting, most appropriate to your plans. But do not limit yourself to one school. After careful investigation, it is best to rank a half-dozen schools (or more), write for admission application forms to the registrar of all of them, and apply to several. Multiple applications are costly, but the investment is worth it.

Assistantships and Scholarships

One reason to apply at several universities has to do with financial (and other) support for which you may be eligible. If you have a reasonably well-rounded undergraduate program behind you and a good record of achievement, you are eligible for a position known as TA (teaching assistant) in a graduate department. Such assistantships usually offer full or partial tuition plus a monthly stipend (during the nine-month academic year). Conditions vary, but they may make it possible to attend a university that would otherwise be out of reach. A tuition waiver alone can be worth more than $10,000 annually. Application for an assistantship is made directly to the department. Write to the department chairperson, who will either respond directly to you or forward your inquiry to the departmental committee that evaluates applications.

What does a TA do? The responsibilities vary, but often teaching assistants are expected to lead discussion sections (of a larger class taught by the professor), laboratories, or other classes. They prepare and grade examinations and help undergraduates tackle problems arising from their courses. It is an excellent way to determine your own ability and interest in teaching.

In some instances, especially in larger geography departments, there are RAs (research assistantship) available. When a member of the faculty is awarded a large grant for some research project (e.g., by the

467

National Science Foundation), that grant may make possible the appointment of one or more research assistants. These RAs perform tasks generated by the project and are rewarded by a modest salary. Normally, RAs do not receive tuition waivers, but sometimes the department and the graduate school can arrange a partial waiver to make the research assistantship more attractive to better students. Usually, RAs are chosen from among graduate students already on campus, students who have proven their interest and ability. But sometimes an incoming student is appointed. Always ask about opportunities.

Geography students are among those eligible for many scholarships and fellowships offered by universities and colleges. When you write your introductory letter, be sure to ask about other forms of financial aid.

Jobs for Geographers

You may decide to take a job on completing the bachelor's degree rather than going on to graduate school. Again this is a decision best made early in the junior year for two main reasons: (1) to tailor your curriculum toward a vocational objective and (2) to start looking for a job well before graduation.

INTERNSHIPS

A very good way to enter the job market—to become familiar with the working environment—is by taking an internship in an agency, office, or firm. Many organizations find it useful to take interns. It helps train beginning professionals; it gives the company an opportunity to observe the performance of trainees. Many an intern has ultimately been employed by his or her company. Some employers have even suggested what courses the intern should take in the next academic year to improve future performance. For example, an urban- or regional-planning office that employs an intern might suggest that the intern add an advanced cartography course or an urban-planning course to his or her course program.

Some offices will appoint interns on a continuing basis, say two afternoons a week around the year; others make full-time, summer-only appointments available. Occasionally, the internship can be connected to your departmental curriculum, yielding academic credit as well as vocational experience. Your department Chair will be the best source of assistance.

One of the most interesting internship programs is offered by the National Geographic Society in Washington, D.C. Every year, the society invites three groups of about eight interns each to work with the permanent staff of various departments. Application forms are available in every geography department in the United States, and competition is strong. The application itself is a useful exercise since it tells you what the society (and other offices) look for in your qualifications.

PROFESSIONAL OPPORTUNITIES

A term you will sometimes see in connection with jobs is *Applied Geography*. This, one supposes, is to distinguish geography teaching from practical geography. But, in fact, all professional geography in education, business, government, and elsewhere, is applied. In the past, a large majority of geography graduates became teachers—in elementary and high schools, colleges and universities. More recently, geographers have entered other areas in increasing numbers. In part, this is related to the decline in geographic education in the schools, but it also reflects the growing recognition of geographic skills by employers in business and government.

Nevertheless, what you as a geographer can contribute to a corporation is not yet as clear to the managers of many corporations as it should be. The anybody-can-do-geography attitude is a form of ignorance you will undoubtedly confront. (This is so even in precollegiate education, where geography was submerged in social studies—and taught by teachers many of whom *never* took a course in the discipline!)

So, once employed, you may have to prove not only yourself, but also the usefulness of the skills and capacities you bring to your job. There is a positive side to this. Many employers once dubious about

468

hiring a geographer learn how geography can contribute—and become enthusiastic users of geographic talent. Where one geographer is employed, whether in a travel firm, publishing house, or planning office, you will soon find more.

Geographers are employed in every area: in business, government, and education. Planning also is a profession that employs many geographers. Employment in business has grown in recent years. Government at national, state, and local levels always has been a major employer of geographers. And education, where a decline was long suffered, now is reviving, and the demand for geography teachers will grow again.

To secure more detailed information about employment, write for the booklet entitled *Careers in Geography* from the AAG (address given earlier) at $1 per copy.

Business

With their education in global and international affairs, their knowledge of specialized areas of interest to business, and their training in cartography, methods of quantitative analysis, and good writing, geographers with a bachelor's degree are attractive business-employment prospects. Some undergraduate students already have chosen the business they will enter, and for them there are departments that offer concentrations in their area. Several geography departments, for example, offer a curriculum that concentrates on tourism and travel—a business in which geographic skills are especially useful.

These days, many companies want graduates who have a strong knowledge of international affairs and fluency in at least one foreign language in addition to their skills in other areas. The business world is quite different from the academic, and the transition is not always easy. Your employer will want to use your abilities to enhance income and profit. You may, at first, be placed in a job where your geographic skills are not immediately applicable, and it will be up to you to look for opportunities to do so. One of my students was in such a situation several years ago. She was one of more than a dozen new employees doing what was, essentially, clerical work. (Some companies will use this procedure to determine a new employee's punctuality, work habits, adaptability, and productivity). One day she heard news that the company was considering establishing a branch in East Africa. She had a regional interest in East Africa in undergraduate school and had even taken a year of Swahili language training. On her own time, she wrote a carefully documented memorandum to the company's president and vice presidents, describing factors that should be taken into consideration in the projected expansion. That report evinced her regional skills, locational insights, knowledge of the local market and transport problems, the probable cultural reaction to the company's product, the country's political circumstances, and her capacity to present such issues effectively. She supported her report with good maps and several illustrations. Soon she received a special assignment to participate in the planning process, and her rise in company ranks had begun. She had seized the opportunity and proved the utility of her geographic skills.

Geography graduates have established themselves in businesses of all kinds: banking, international trade, manufacturing, retail, and more. Should you join a large firm you will be pleasantly surprised by the number of other geographers who hold jobs there—not under the title of geographer but under countless other titles, ranging from analyst and cartographer to market researcher and program manager. These are positions for which the appointees have competed with other graduates, including business graduates. As we noted earlier, once an employer sees the assets a geographer can bring, the role of geographers in the company is assured.

Government

Government long has been a major employer of geographers at the national and state as well as the local level. Dr. S. J. Natoli, in *Careers in Geography* (p. 9), estimated that in 1983 more than 2500 geographers were working for the government, about half of them for the federal government. In the U.S. State Department, for example, there is an Office of the Geographer staffed by professional geographers. Other agencies where geographers are employed in-

469

clude the Defense Mapping Agency, the Bureau of the Census, the U.S. Geological Survey, the Central Intelligence Agency, and the U.S. Corps of Engineers. Other employers are the Library of Congress, the National Science Foundation, and the Smithsonian Institution. Many other branches of government also have positions for which geographers are eligible.

Opportunities also exist at state and local levels. Several states have their own Office of the State Geographer; all states have agencies engaged in planning, resource analysis and protection, environmental matters, and transportation questions. All these agencies need geographers who have skills in cartography, remote sensing, computer use, and the quantitative analysis of mass data.

Securing a position in government requires early action. If you want a job with the federal government, start at the beginning of your senior year! Every state capital and many other large cities have a Federal Job Information Center (FJIC). The Washington office is at 1900 E Street, N.W., Washington, D.C. 20415. You may request information about a particular agency and its job opportunities; it also is appropriate to write directly to the Personnel Office of the agency or agencies in which you are interested. Also write to the Office of Personnel Management (OPM), Washington, D.C. 20415. OPM offices exist in most large cities.

Planning

Planning has become one of geography's allied professions. The planning process is a complex one in which people trained in many fields participate. Geographers, with their cartographic, locational, regional, and analytical skills, are sought after by planning agencies. Many an undergraduate student gets that first professional opportunity as an intern in a planning office.

Planning is done by many agencies and offices at levels ranging from the federal to the county. Cities have planning offices, as do regional authorities. Working in such an office can be a rewarding experience because it involves the solving of social and economic problems for the future, the conservation and protection of the environment, the weighing of diverse and often conflicting arguments and viewpoints, and much interaction with workers trained in other fields. Planning is a learning experience.

A career in planning can be much enhanced by a background in geography, but you will have to adjust your undergraduate curriculum to include courses in such areas as public administration, public financing, and other related fields. Thus a career in planning requires early planning on your part. At many universities, the geography department is closely associated with the planning department, and your department Chair can inform you about course requirements. But if you have your eye on a particular office or agency, you should enquire directly from its director about desired and required skills.

Planning is by no means a monopoly of government. Government-related organizations such as the Agency for International Development (AID), the World Bank, and the International Monetary Fund (IMF) have planning offices, as do nongovernmental organizations such as banks, airline companies, industrial firms, and multinational corporations. In the private sector the opportunities for planners have been expanding and you may wish to explore these possibilities. An important address is the planners' equivalent of the geographers' AAG, the American Planning Association, at 1776 Massachusetts Avenue, N.W., Washington, D.C. 20036. The AAG's *Careers in Geography* provides additional information on this expanding field and where you might go to study for it.

Teaching

If you presently are a freshman or sophomore, your graduation may coincide with the end of a long decline in the geography teaching profession. Just 25 years ago, teaching geography in elementary or high school was the goal of thousands of undergraduate geography majors. But then began the submergence of geography into the hybrid field called social studies, and prospective teachers no longer needed to have any training in geography in many states. In Florida, for example, teachers were required to take courses in regional geography and conservation (as taught in geography departments), but in the early 1970s those requirements were dropped. Education

470

planners fell victim to the myth that the teaching of geography does not require any training. What was left of geography often was taught by teachers whose own fields were history, civics, or baseball coaching.

If you have read the papers, you have seen reports of the predictable results. "Geographic illiteracy" has become a common complaint (often by the same education planners who pushed geography into the social studies program and eliminated teacher-education requirements). Now the pendulum is swinging the other way again. States are returning to the education requirement so that teachers will learn some geography. And geography is returning to elementary and high school curricula. The need for geography teachers will soon be on the upswing again.

So this may be a good time to consider teaching geography as a career for the 1990s. But you should do some research because states vary in their progressiveness in this area. Opportunities will become more widespread as the decade proceeds, but it will take time. You should visit the School or College of Education in your college or university and ask questions about this. Also write not only the AAG, but also the National Council for Geographic Education (NCGE) at Western Illinois University, Macomb, Illinois 61455. Your state geographic society also may be helpful; ask your department Chair for details.

The opportunities in geography are many, but often they are not as obvious as those in other fields. You will find that good and timely preparation produces results and, frequently, unexpected and satisfying rewards.

471

Consult the index for location of the fuller discussion in the text of these terms and concepts.

Abrasion Erosive action by running water, glacial ice, waves, and wind. The process involves the grinding action of rock fragments on rock surfaces (and on other rock fragments), wearing these down and creating even finer sediments.

Abrasion platform Gently sloping surface sculpted from bedrock during abrasion caused by wave action, also called wave-cut platform. Such nearly flat surfaces can be seen at the base of a coastal cliff, extending seaward under the breakers; sometimes movement in the rocks along the coast elevates the abrasion platform so that it appears as a terrace fringing the exposed cliff.

Absolute base level A much-debated concept of the lowest level to which stream erosion can continue. It is actually a few meters below sea level, since a sediment-laden stream can erode a valley whose bottom lies well below the ocean's surface. Absolute base level varies from stream to stream, depending on the speed with which the running water reaches the sea, the sedimentary load, the rock materials at the river's mouth, and other factors.

Absolute location Absolute location is determined by means of the earth grid. The north–south lines are

called lines of *longitude* and the east–west lines of *latitude*. In addition, one must state whether the location is north or south of the Equator and, since an east–west reference is also needed, whether it lies east or west of the line that runs through Greenwich, England—the place where the north–south line has a value of zero (*see* Prime Meridian). Values are given in degrees and minutes, and sometimes seconds for great accuracy. Thus Chicago, Illinois, lies at 41° and 49 min *north* of the Equator and 87° and 37 min *west* of Greenwich. If we wanted to pinpoint a city block, we would have to calculate the seconds as well.

Abyssal plains Extensive, nearly flat ocean-floor surfaces located at depths of as much as 15,000–20,000 ft (4500–6000 m) below sea level.

Accessibility The degree of ease with which it is possible to reach a certain location from other locations. Some places can be reached by road, rail, and air; others are served by a single road that may become impassable during the wet season. Accessibility varies from location to location and can be measured.

Acculturation When peoples with different cultures meet, they begin to adopt certain of the practices and even values of their new contacts. The term *acculturation* is applied to this process of cultural modification as a result of intercultural borrowing. In anthropology and cultural geography, it is especially used for the change that oc-

curs in indigenous peoples when contact is made with a technologically more advanced society.

The local area within which a person moves in the course of daily activity. The territorial range of this daily effective environment depends on the mode of transportation to which an individual has access.

Adiabatic lapse rate *See* Wet and Dry adiabatic lapse rate.

Aesthetic Relating to things beautiful.

Agglomerated (nucleated) settlement A compact, closely packed settlement (usually a hamlet or larger village) that is sharply demarcated from adjoining farmlands.

Agglomeration The clustering or grouping together of people and services to their joint benefit. The availability of specialized equipment, a technologically sophisticated labor force, and a large-scale, comparatively wealthy market combine to attract many more people and additional services to existing urban places. As a result, the attractiveness of these urban centers increases, producing still further agglomeration.

Aggradation The building up (rather than wearing down, or *degradation*) of a stream bed or other surface through the deposition of sediment that is accomplished by streams, glaciers, waves, and wind.

Agrarian Relating to the alloca-

472

tion and use of land, to rural communities, and to agricultural societies.

Agricultural density The number of inhabitants per unit of agricultural land. As used in population geography, agricultural density excludes the urban residents so that it reflects the pressure of population in the rural areas. *Physiologic density* measures the total population, urban and rural, against the agricultural land.

Agriculture The purposeful tending of crops and livestock.

Air mass An extensive body of air in the lower troposphere that has acquired relatively uniform properties of temperature and moisture in terms of horizontal distribution as well as vertical gradients.

Albedo The percentage of incoming radiation reflected by a surface. An extensive snow-covered surface can reflect as much as 90 percent of incoming solar radiation; a forest zone may reflect as little as 15 percent.

Alfisols A soil order, formerly called pedalfers, consisting of soils found in humid as well as subhumid zones. These soils can evolve under a wide range of environmental conditions and are marked by a lower horizon of clay accumulation.

Alluvium Sediment laid down in a stream valley or channel. Deposition occurs when the stream's velocity decreases and the valley fills with a veneer of unconsolidated material. Streams are subject to seasonal flooding, and during the flood stage, a wide stretch of the valley is inundated by sediment-laden water. When the floodwaters subside, another layer of alluvium has been added to the valley floor and floodplain. Since soil particles washed from slopes in the drainage basin form a large part of the deposits, alluvial soils often are fertile and productive.

Alpine glacier A long, narrow glacier that occupies a valley in mountainous topography.

Altiplano High-elevation plateau, basin, or valley between even higher mountain ranges. In the Andes Mountains of South America, altiplanos are at 10,000 ft (3000m) and even higher.

Animism The belief that inanimate objects such as hills, rocks, rivers, and other elements of the natural landscape (including trees) possess souls and can help as well as hinder human efforts on earth.

Antecedent *Antecedent*, like *subsequent* and *superimposed*, is a term used in human as well as physical geography. Antecedent is something that goes before; from Latin *ante* (before) and *cedere* (to go).

Antecedent boundary A political boundary that existed before the major elements of the modern cultural landscape evolved.

Antecedent stream A stream channel that existed before the rock deformations in an area developed. It is conceptualized as a stream that maintained its course and remained in place while a structural impediment (e.g., a dome or anticline) emerged. Its erosive power enabled the stream to cut down as fast as, or faster than, the obstacle that rose in its path.

Anthracite coal Highest carbon-content coal (therefore, of the highest quality) that is formed under conditions of high pressure and temperature to eliminate most impurities. Anthracite burns almost without smoke and produces high heat.

Anthropogeographic boundaries Political boundaries that coincide substantially with cultural discontinuities in the human landscape such as religious or linguistic transitions.

Anticline A fold in layered rocks in which the axis of the structure lies at the highest elevation and the two limbs face away from each other and slope downward. An anticline thus has the appearance, in cross section, of an upfold.

Anticyclone A roughly circular or oval pressure cell in the atmosphere in which the highest pressures exist near the center and pressure decreases outward. An anticyclone, therefore, is a "high."

Antipodal Every place on the Earth's surface has a point that is exactly opposite on the "other" side: the North Pole is antipodal to the South Pole. This term is used for *areas* as well as points on the globe. On the globe opposite Africa, the Pacific Ocean is antipodal to Africa; southeast England lies about opposite (is antipodal to) New Zealand.

Apartheid Literally, apartness; the Afrikaans term given to South Africa's policies of racial separation. The term no longer has official sanction and has been replaced by "separate development."

Aquaculture The use of a pond or other artificial body of water for the growing of food products, including fish, shellfish, and even seaweed. Aquaculture also may be carried out in controlled natural water bodies such as estuaries or river segments. Japan is among the world's leaders in aquaculture. When the activity is confined to the raising and harvesting of fish, it is usually referred to as "fish farming."

Aquifer An underground reservoir of water contained within a porous, water-bearing rock layer. This rock stratum is often overlain by impervious layers to prevent upward seepage of groundwater by capillary action.

Arable land Land fit for cultivation by some method. Not all the ara-

473

ble land in the world is presently under cultivation, despite the enormous food shortages and nutritional deficiencies that exist in many places.

Archeology (also spelled Archaeology) The systematic study of ancient humanity through fossil remains and through the analysis of artifacts left behind.

Archipelago A cluster of islands. The Greek *pelagos* actually means sea, so, it is appropriate not only to define archipelago as a group of islands, but also to include the intervening waters.

Area A term that refers to a part of the earth's surface with less specificity than *region*. *Urban area* alludes very generally to a place where urban development has taken place, whereas urban region requires certain specific criteria upon which a delimitation is based.

Areal interdependence A term related to *functional specialization*. When one area produces certain goods or has certain raw materials or resources and another area has a different set of resources and produces different goods, their needs may be complementary and by exchanging raw materials and products, they can satisfy each other's requirements. The concepts of areal interdependence and *complementarity* are related; both have to do with exchange opportunities among regions.

Arête In an area of mountain glaciation, a sharp, jagged ridge that formed from the intersection of two or more retreating cirques.

Aridisols A soil order consisting of soils formed under arid conditions, found (in the United States) over much of the dry Southwest. The world's most widespread soils, they are marked by calcification and salinization.

Arithmetic density The number of persons per unit area, irrespective of the habitability of a country's various environmental zones. An unreliable measure and an unsuitable index for comparisons between countries.

Artesian Artesian water rises to the surface through a well under pressure. It is contained in the pore spaces of a porous rock (an aquifer) that is confined between impermeable layers. The water supply is replenished by precipitation on an exposed section of this aquifer, which then dips below the ground. Artesian conditions thus depend on this particular structural situation. The name derives from the type area near the town of Artois, France.

Aryan From the Sanskrit *Arya* (noble), a name applied to people who speak an Indo-European language and who moved into northern India. Although properly a language-related term, Aryan has various additional meanings, especially racial ones.

Asthenosphere The comparatively soft, plastic layer of the Earth's upper mantle on which the plates of the rigid lithosphere rest.

Atmosphere The Earth's envelope of gases that rests on the ocean and land, penetrating the open spaces within soils. This layer of gases is densest at the Earth's surface and thins with altitude. It is held against the planet by the force of gravity.

Atmospheric pressure The weight of a column of air at a given location, determined by the force of gravity and the composition and properties of the atmosphere at that location.

Attenuation The quality of elongation.

Axis A line of motion, or activity. The axis of rotation of the Earth is an imaginary line that connects the North Pole and the South Pole and on which the earth rotates one full turn every 24 hours. In physical geography and geol-ogy, the term is also used to identify a line or zone of particular behavior of the crust. An axis of uplift, for example, is a line across a certain region (even a continent) marked by high elevations that show evidence of upward movement compared to the adjacent zones.

Balkanization Fragmentation of a region into smaller, often hostile, political units.

Band A territorially based group of people (usually numbering no more than several dozen) living together. This was the smallest community of early human society, smaller and much lass complex than what has come to be known as a tribe.

Bantustan In South Africa, term formerly used to denote one of its African territories designated for ''independence'' under the terms of separate development (see *apartheid*). The official designation today is Bantu Homeland.

Bar (sand bar) A sandy ridge that forms across the seaward opening of an estuary or river mouth, blocking it. Such a deposit is laid down when a current, carrying sediment, is slowed down and, hence, drops some of the load of mud it is carrying. As time goes on, the offshore bar, or sand bar, grows bigger until its top is about even with mean sea level. Then, the only way through it is cut when water from the river that flows into the estuary breaks through the bar. Sometimes a breach is made when tidewater, which has flowed over the bar, retreats as the tide goes out and cuts into the top of the sand bar. Many of the world's rivers have bars or partial bars called spits across their mouths.

Barchan A wind-fashioned dune whose shape (seen from above) resembles a quarter moon with the points of the crescent lying downwind.

474

Barometer Scientific equipment designed to measure and record atmospheric pressure.

Barrio Slum development in a Middle American city; increasingly applied to inner-city concentrations of Hispanics in such southwestern U.S. cities as Los Angeles.

Base level *See* Absolute and Local base level.

Baseline A straight line extending from headland to headland, thus "straightening" coastlines and cutting bays and indentations from the high seas; a subject of debate in coastal political geography. Baselines serve as devices to extend the territorial sea; waters on the landward side are designated as internal waters.

Basic activities Economic activities whose products are exported beyond a region's limits. Nonbasic, or service, activities involve production and consumption within the region.

Batholith An intrusive igneous rock mass—formed well beneath the earth's surface and subsequently exposed by erosion—with a surface area exceeding 40 sq miles (100 sq km).

Bauxite An ore of aluminum; an earthy, reddish-colored material that usually contains some iron as well. Soil-forming processes such as leaching and redeposition of aluminum and iron compounds contribute to the building up of these ores, and many of them are found at shallow depths in areas of considerable precipitation.

Bay A large and wide body of water that penetrates the land for a considerable distance. Examples are Hudson Bay and Chesapeake Bay.

Biome Largest unit in the subdivision of the earth's total ecosystem, consisting of a group of plant and animal ecosystems and their interactions with land and water. A biome is characterized by its overall appearance, especially its prevailing vegetation, and by the climatic conditions that sustain it.

Birthrate (crude) The *crude birthrate* is expressed as the number of live births per year per 1000 persons in the population.

Bituminous coal A coal of higher grade than lignite (it has a higher fixed carbon content) but not as high in quality as anthracite. Usually found in relatively undisturbed, horizontal layers between rock beds, it is also called soft coal. Bituminous coal can be heated and converted to coke, which is used in the great blast furnaces.

Bicultural Of two cultures.

Blocky structure Soil structure in which irregularly shaped aggregates of earth tend to have straight edges.

Bolsheviks The name taken by the radical left wing of the Russian Social Democratic Labor Party (Lenin's adherents) when they gained a plurality on that party's executive committee in 1903—from the Russian *bol'she*, meaning large or majority. The minority (the Mensheviks) were later to regain a majority and actually represented a larger faction of the party, but the Bolsheviks had a politically advantageous rallying point in their new name; thus the 1917 revolution became known as the Bolshevik Revolution. In 1918, the party changed its name to the Communist (Bolshevik) Party, and henceforth the Bolsheviks held center stage in Russian and Soviet affairs.

Boundary *See* Political boundaries.

Brahman A Hindu of the highest caste, most often a priest, a person believed to possess sacred knowledge and to be of the greatest purity. Brahmans (or Brahmins) alone are believed capable of carrying out particular religious rituals and tasks. In India, Brahmans have for many centuries been religious, intellectual, and even political leaders.

Braiding The process in which a stream carrying a high sediment load divides into several channels that reunite some distance downstream. As the channels shift position, stretches of deposited alluvium become divided between them, giving the "braided" appearance.

Break-of-bulk A point or location where a cargo must be transferred from one carrier to another, typically at ports where goods are transferred from water to land transport.

Buffer zone A higher-order frontier, a *shatter belt*; an area consisting of a country or countries (possibly parts of countries) that serves as a separating zone between actual or potential enemies or adversaries. Historically, Eastern Europe has been such a zone; more recently, a buffer zone across south-central Africa has separated black Africa from South Africa for more than 20 years.

Butte Small, steep-sided, caprock-protected hill, usually in dry or semiarid areas, an erosional remnant of a plateau.

Calcification The accumulation of calcium carbonate in soils (such as aridisols), usually in the B horizon but, under very arid conditions, in the lower A horizon as well. Where moister conditions prevail the calcium carbonate can be found in a layer in the C horizon.

Caldera Steep-sided, roughly circular, troughlike depression formed during the collapse of a large composite volcano. Such a collapse may accompany the explosion of the volcano, but it may have other causes;

475

certain calderas may be related to the evolution of rift valleys.

Caliche The layer of white, stony calcium carbonate that accumulates during calcification in certain soils.

Caprock Erosion-resistant layer of hard rock that, in an area of layered rocks, impedes erosion and protects softer rocks below. The result is a landscape of *mesas* and *buttes*, steep-sided, table-shaped hills whose flat surfaces are remnants of a hard caprock.

Capture (River capture) Process in which the headward-eroding waters of one stream intercept and divert the waters of another stream. Imagine a river that is flowing from east to west, from the interior of a certain area to the coast. In addition, assume that another river begins nearby and flows to the south. This second river is slowly eating away toward the first river. Eventually, its valley will reach and intersect that of the first river, and the waters of the upper portion of that first river's course may be diverted to the valley of the second river. Thus part of the waters of the first river have been "captured" by the second. This process has occurred in hundreds of places all over the world and has affected major rivers as well as small streams. If you think you see evidence of this on a wall map, say a suspicious-looking "elbow" in a river's course, you might follow it up by looking at some topographic sheets: you may find that your conclusion was correct and that capture did indeed take place.

Cartel An international syndicate formed to promote common interests in some economic sphere through the formulation of joint pricing policies and the limitation of market options for consumers. The Organization of Petroleum Exporting Countries (OPEC) is an example.

Caste system The strict social segregation of people (specifically in Hindu society) on the basis of ancestry and occupation.

Central business district (CBD) The "downtown" region of a city, the core of the urban area, the zone with the highest accessibility, the tallest buildings, most of the city's offices, the largest retail establishments, the greatest concentration of automobile traffic and pedestrian flow.

Central place theory Geographic theory in search of explanations for the spatial distribution of urban places.

Centrality A condition of high accessibility, the quality of position at the center of transportation systems.

Centrifugal forces Political geographers use the terms *centrifugal* and its opposite, *centripetal*, to identify forces that tend, respectively, to pull states and nations apart and bind them together. Centrifugal forces are divisive forces. Language barriers, religious differences, and even distance between population centers are among centrifugal forces.

Centripetal forces Forces that unite and bind a country together—such as a strong national culture, shared ideological objectives, and a common faith (Islam constitutes a centripetal force in Arab countries).

Cephalic index Ratio of the length of the human skull to its width, a measure once popular with anthropologists seeking to unravel human origins and dispersals.

Chaparral Regional name for a vegetation type that consists of hard-leaf scrub and low-growing woody plants, usually evergreen with thick waxy leaves, adapted to the long dry summers of the Mediterranean climate. Such sclerophyll scrub and dwarf forest is called *chaparral* in southern California, *maquis* in Mediterranean areas of Europe, *mattoral* in Chile, *fynbos* in South Africa.

Charisma In Greek this word means *divine gift*—a special talent for healing, prophecy. It has come into use to describe the qualities some people have to capture the popular imagination, to secure the allegiance and even devotion of the masses. *Charismatic* leadership at its best combines such appeal with administrative competence.

Chemical weathering Chemical alteration of minerals in rocks through exposure to air and moisture, mainly through oxidation, hydrolysis, and carbonation.

Chinampa cultivation A system of cultivation developed by Indian peoples in indigenous Middle America that involved the creation of "floating gardens," a form of land reclamation in shallow lakes and marshly areas. Strips of aquatic vegetation, cut from the lake margins, were sunk in layers until the top one just reached the water surface. On top of the uppermost layer, the Indian farmers spread fertile mud from the lake bottom to create a planting surface. This resulted in long artificial farm fields 6 to 9 ft (2 to 3 m) wide and more than 33 yd (30 m) long, separated from each other by narrow canals.

Chinook A strong local air flow (wind) on lee (eastern) slopes of the Rocky Mountains, with dry, rapidly warming air that can quickly raise winter temperatures and melt and evaporate snow.

Cirque Bowl-shaped depression that forms (or formed) the source area of an alpine (or mountain) glacier.

City-state An independent political entity consisting of a single city with (and sometimes without) an immediate *hinterland*. The ancient city-states of Greece have their modern equivalent in Singapore.

Clastic sediments Sediments made up of particles of other rocks.

476

Cliff A steep, vertical (or nearly vertical) rock wall carved from bedrock by weathering and erosion. A marine cliff is the product of wave action.

Climate A term used to convey a generalization of all the recorded weather observations at a certain place or in a given area. It represents an average of all the weather that has occurred there. In general, a tropical location such as the Amazon Basin has a much less variable climate than areas located, say, midway between the equator and the pole. In low-lying tropical areas, the *weather* changes little, thus the climate is rather like the weather on any given day. But in the middle latitudes, there may be summer days to rival those in the tropics and winter days so cold that they resemble polar conditions. Here, when we refer to climate, we must combine the warm summers and the cold winters in our generalization.

Climax vegetation The final, stable vegetative community that has developed at the end of a succession under a particular set of environmental conditions. This vegetation is in dynamic equilibrium with its environment, including other elements of the ecosystem such as its animal life.

Cloud A heavy concentration of water (or ice particles of very small size) in suspension in the atmosphere.

Collectivization The reorganization of a country's agriculture that involves the expropriation of private holdings and their incorporation into relatively large-scale units, which are farmed and administered cooperatively by those who live there. This system has transformed Soviet agriculture, has been resisted in Communist Poland, and has gone beyond the Soviet model in China's program of communization.

Colloid Substance within the soil in a state of fine subdivision, the individual particles having a diameter of less than 0.001 mm. Both mineral and organic colloids exist.

Common Market Name given to a group of European countries that formed an association to promote their economic interests. Official name is European Community (EC).

Compact A politicogeographic term to describe a state that possesses a roughly circular, oval, or rectangular territory in which the distance from the geometric center to any point on the boundary does not vary very much. Poland, Belgium, and the state of Colorado are examples of this shape category.

Complementarity Regional complementarity exists when two regions, through an exchange of raw materials and finished products, can specifically satisfy each other's demands.

Composite volcano *See* Volcanoes.

Concentric Having a common center.

Concentric zone theory A geographic model of the city that suggests the existence of five concentric regions

Conceptualization A thought process whereby we arrive at a general idea or generalization by observing and recording specific instances related to each other. Thus we see that nearly all cities have ''downtown'' areas, where tall buildings, businesses, busy sidewalks, traffic jams, and so on, are common. Instinctively we think we can identify just where such an area is in the city or cities we know best, and we expect to see the same in cities we have not even visited. Thus we conceptualize that a central business district (CBD) forms the core of any American city.

Condensation Process by which a substance changes from the gaseous state (such as water vapor) to the liquid state (water) or the solid state (ice).

Condominium A term applied to buildings owned jointly by the occupants of the apartments that constitute them, but also a term that has political connotation. It is not difficult to see what it means: con (together) *dominium* (domain): a realm governed jointly.

Confluence The joining of a river (or a glacier) with its tributary; the place where one river flows into another or where two streams of approximately equal size come together to become one. By looking at the angle of convergence on the map, it is possible under normal circumstances to determine the direction of flow when neither stream's source is on the map. This angle is usually considerably less than 90° unless the underlying rock structures cause an abnormal convergence.

Coniferous forest A forest of cone-bearing trees, consisting of needleleaf, evergreen trees that have short branches and are of very regular shape. There are two great belts of these forests: one in Eurasia, the other in North America.

Connectivity The degree of direct linkage between one particular location and other locations in a transport network.

Contagious diffusion The distance-controlled spreading of an idea, innovation, or some other item through a local population by contact from person to person—analogous to the communication of a contagious illness.

Contiguous A term of some importance to geographers, for whom it has spatial implications. It means, literally, to be in contact with, adjoining, or adjacent. Sometimes we hear the continental United States without Alaska referred to as contiguous. Alaska is not contiguous to these states, for Canada lies between; neither is Hawaii, separated by thousands of miles of ocean. The ancient Hellenic civilization was not contiguous either, for its settlements were separated by sea, peninsulas, and rugged mountains.

Continental glaciation Glaciation of large areas (e.g., the Canadian Shield) by enormous sheets of ice that spread outward from centers where the ice is thousands of feet thick. Such glaciers are called sheet glaciers to distinguish them from alpine (*valley-confined mountain*) glaciers.

Continental rise The slightly inclined zone at the foot of the continental slope where the ocean floor declines gently toward the extensive ocean plains known as the abyssal plains.

Continental shelf Beyond the coastlines of the continents, the surface beneath the water in many offshore areas declines very gently until a depth of about 100 fathoms (somewhat under 600 ft/200 m). Thus, the edges of the landmasses today are inundated, for only at the 100-fathom line does the surface drop off sharply to the deeper ocean bottom along the *continental slope*. The submerged continental margin is called the continental shelf: it extends from the shoreline to the continental slope.

Continental slope The scarplike margin of the continental shelf, the actual edge of the continental landmass where the ocean floor drops off to the greater depths of the continental rise and the abyssal plains.

Contour plowing Plowing along the contour line (along horizontal rows) rather than up and down the slope.

Conurbation Term used to identify an urban area marked by some degree of coalescence of two or more urban centers.

Convection In the atmosphere, the upward movement of air in strong drafts.

Convection cell In the atmosphere, a cluster of columns of air

driven upward by convection, usually attended by cloud formation and often by rainfall.

Copra Meat of the coconut; fruit of the coconut palm.

Cordillera A mountain chain that consists of a series of more-or-less parallel ranges.

Core area In geography, a term with several connotations. *Core* refers to the center, heart, focus; in physical geography, the core area of a continent identifies the ancient shield zone. In human geography, the core area of a nation-state is constituted by the national heartland, the largest population cluster, most productive region, the area with greatest centrality and accessibility, probably containing the capital city as well.

Coriolis force The force that, owing to the rotation of the earth, tends to deflect all objects moving on the surface of the earth away from their original path. For example, if a pressure situation arises so that a wind would blow due south (a northerly wind), this wind will be diverted to the *right* of its north–south path in the Northern Hemisphere. It is thus likely to become a *northeast* wind rather than a north wind. In the Southern Hemisphere, it would be the reverse; there, moving objects (including air) are deflected to the left of their original path. Everything is affected: you when you are walking, your moving car, the flowing water in a river.

Corridor A land extension that connects an otherwise landlocked state to the ocean. History has seen several such corridors come and go. Poland once had a corridor (it now has a lengthy coastline); Bolivia lost a corridor to the Pacific Ocean between Peru and Chile; Finland possessed a corridor to the Arctic Ocean. Secondary corridors may lead to another means of egress, for example, a navigable river:

Colombia, although it has coasts on the Caribbean Sea and the Pacific Ocean, also has a secondary corridor to the Amazon River.

Corrosion Erosion by solution: in a stream valley, the river water acts to dissolve certain minerals in the bedrock. Chemical solutions in the water corrode other minerals as well, acting as mild acids.

Coterminous A word of some geographic importance that means "having the same boundaries."

Counterradiation Longwave radiation emitted by the atmosphere and by clouds and directed toward the earth's surface.

Crop rotation The cultivation of a succession of different crops on the same piece of land. A certain crop, if grown year after year on the same piece of soil, may demand too much from that soil and not replenish the nutrients it removes. As a result, the soil may become poor, yields will drop, and deterioration sets in. However, farmers know that although some crops remove certain ingredients from the soil, others seek different nutrients and replenish those that were taken away. By finding out which crops should be grown in succession, the farmer is able to keep up the soil and yet get good harvests each year. In some instances, the farmer may leave the soil fallow for a season, to rest. The whole cycle is one of crop rotation.

Crust The Earth's crust is the planet's outermost solid shell, usually between 6- and 25-miles (10- and 40-km) thick, composed of the earth's lightest rocks.

Crustal spreading The divergent contact between tectonic plates, usually (but unsatisfactorily) called *seafloor spreading*.

Cul-de-sac Literally, the bottom of the bag or pouch; applied to situations where there is no possibility of further movement—a dead-end street is sometimes called a cul de sac. From the point of view of southward migration, southern African was obviously a cul de sac.

Cultural diffusion The spreading and dissemination of a cultural element from its place of origin over a wider area.

Cultural ecology The interactions between cultural development and prevailing conditions of the natural environment.

Cultural landscape The totality of the human imprint on the natural landscape, especially as manifested by a particular society.

Culture area A distinct, culturally discrete spatial unit; a region within which certain cultural norms prevail. *Culture region* might be more appropriate.

Culture complex A related set of culture traits such as prevailing dress codes, cooking and eating utensils.

Culture   hearth Heartland, source, core area of a culture. An area of cultural growth, of innovations and ideas.

Culture realm A cluster of regions in which related culture systems prevail. In North America, the United States and Canada form a culture realm, but Mexico belongs to a different culture realm.

Culture trait A single element of normal practice in a culture such as the wearing of a turban.

Cycle of erosion Conceptualization of denudational processes as having an orderly sequence leading from the original landscape thrown up by geologic forces to a nearly flat surface called a *peneplain* or *pediplane*.

Cyclic movement Movement (e.g., migratory) that has a closed route repeated annually or seasonally.

Cyclone A roughly circular or oval pressure cell in the atmosphere in which the lowest pressures exist near the center and pressure increases outward. A cyclone, therefore, is a *"low."*

Death rate (crude) The crude death rate is expressed as the number of deaths per 1000 individuals in the population during a given year.

Deciduous A deciduous tree is one that loses its leaves during a certain season of the year. The word is derived from the Latin *deciduus* (to fall off). In the middle latitudes, such trees tend to lose their leaves during the autumn as protection against the coming winter. But some deciduous trees also occur in tropical areas where they shed their leaves to void excessive moisture losses through evapotranspiration. Most of the trees in the low latitudes, however, are not deciduous; nor are trees of the very high latitudes, which are appropriately called evergreen trees. In our part of the world, familiar deciduous trees are the oak, elm, beech, and maple.

Definition In political geography, the description (in legal terms) of a boundary between two countries or territories.

Deflation Removal by wind of clay and silt-sized particles.

Deflation hollow Shallow depression created by persistent deflation in a location.

Degradation Wearing down of the landscape by stream erosion.

De jure **and** *de facto* *De jure* means something that is sanctioned and, indeed, supported by the law. *De facto* means something that is actually happening, whatever the law. For example, *de jure* segregation would mean segregation according to legal prescription. *De facto* segregation is segregation that may not be according to the law; it may even be illegal, but it is happening nevertheless.

Delimitation In political geography, the translation of the written terms of a boundary treaty (the *definition*) into cartographic representation.

Delta Alluvial lowland at the mouth of a river, formed when the river deposits its alluvial load on reaching the sea. Often triangular in shape—hence, the use of the Greek letter whose symbol is Δ.

Demarcation In political geography, the marking of an international political boundary on the ground by means of barriers, fences, posts, or other markers.

Demographic variables Births (fertility), deaths (mortality), and migration are the three basic demographic variables.

Demography The quantitative study of human population: size, distribution, structure, growth or decline, movement, and related aspects.

Desert An arid area supporting very sparse vegetation, usually exhibiting extremes of heat and cold because the moderating influence of moisture is absent.

Desertification The encroachment of desert conditions on moister zones along the desert margins, where plant cover and soils are threatened by desiccation: through overuse, in part by humans and their domestic animals and, possibly in part because of inexorable shifts in the earth's environmental zones.

Determinism See Environmental determinism.

Devolution In political geography, the disintegration of the nation-state as a result of emerging or reviving regionalism.

Dew point The dew point temperature is the temperature at which a parcel of air is completely saturated by the amount of moisture it contains.

Dhow From the Arab word *dāwa* (boat). These wooden sailing boats plied (and still sail) the waters of the Indian Ocean between the Persian Gulf, South Asia, and East Africa. They carried slaves, contraband, and legitimate commerce; to this day they play a role in the illegal trade in rhinoceros horns and ivory from East Africa.

Diffusion The spatial spreading or dissemination of a culture element (e.g., a technological innovation) or some other phenomenon (e.g., a disease outbreak). See also contagious, expansion, hierarchical, and relocation diffusion.

Dike A thin layer of intrusive igneous rock that penetrates older rock strata at a vertical or near-vertical angle or dip, often inserting itself along joint planes.

Dispersed settlement In contrast to agglomerated or nucleated settlement, dispersed settlement is characterized by the wide spacing of individual homesteads. This lower-density pattern is characteristic of rural North America.

Dissected tableland A landscape that has been attacked by a river or river system. Much of the original surface has disappeared, and valleys crisscross the area, exposing rock layers below like the muscles and bones of a body. A plateau that has been attacked and is in the process of being worn down by rivers appears "dissected."

Distance decay The various degenerative effects of distance on spatial processes. The degree of spatial interaction diminishes as distance increases, but there is no simple way of saying exactly how this happens. In migration, for example, the information flow from the new homelands of the emigrants proved less accurate the farther they had to travel.

Diurnal Daily.

Divided capital In political geography, a country whose administrative functions are carried on in more than one city is said to have divided capitals.

Doldrums Equatorial belt of variable winds and frequent extended calms. The name is associated with the problems of sailing vessels that were often becalmed, sometimes with disastrous results, in this zone of undependable winds.

Domestication The transformation of a wild animal or wild plant into a domesticated animal or cultivated crop. The process involves changes in the physiology and tameness of the animal and modifications in the growing habits of plants.

Double cropping Taking two harvests per year from the same piece of land.

Doubling time The time required for a population to double in size.

Drift, glacial All forms of rock debris deposited by ice sheets.

Dry adiabatic lapse rate Cooling rate of expanding (rising) air when no condensation takes place, 5.5°F/1000 ft (1.0°C/100 m).

Dune Mound or ridge of loose sand particles possessing a characteristic shape and usually subject to wind-induced motion.

Earthquake The shaking or trembling of the earth's surface as a result of the passage through the crust of seismic waves that originate at the focus or source of the earthquake.

Ecology Strictly speaking, ecology refers to the study of the many interrelationships between all forms of life and the natural environments under which they have evolved and continue to develop; the study of ecosystems.

The term is also used in more specific ways, as in cultural ecology and urban ecology, involving the study of the relationships between human cultural manifestations and the natural environment in which they are occurring.

Economies of scale The savings that accrue from large-scale production whereby the unit cost of manufacturing decreases as the level of operation enlarges. Supermarkets operate on this principle and are able to charge lower prices than small groceries.

Edaphic factors Factors that influence the development of soils and the condition of the soil as it affects plan growth.

Elevation Altitude of a point or area above sea level.

Elite A small but influential, upper-echelon social class whose power and privilege give it control over a country's political, economic, and cultural life.

Eluviation A soil-forming process that involves the removal from the A horizon and downward transportation of soluble minerals and microscopic, colloidal-size particles of organic matter, clay, and oxides of aluminum and iron.

Emigrant A person migrating away from a country or area; an out-migrant.

Empirical Relating to the real world as opposed to theoretical abstraction.

Enclave A piece of territory that is surrounded by another political unit of which it is not a part.

Entisols A soil order marked by recent soils without the vertical horizonation expected of more mature soils.

Entrepôt A place, usually a port city, where goods are imported, stored, and transhipped. Thus, an entrepôt is a *break-of-bulk point.*

Environmental determinism The view that the natural environment has a controlling influence over various aspects of human life, including cultural development and even physical attributes. Also referred to as ''environmentalism.''

Environmental lapse rate The rate of temperature decrease with altitude in the troposphere's stationary air, 3.5°F/1000 ft or 6.4°C/1000 m). Also known as the normal lapse rate.

Environmental perception An area of study in cultural geography that focuses on the idea that peoples in different societies perceive their natural environment in different ways, so that dissimilar cultures will make different decisions about the use and exploitation of their natural environments.

Epeirogeny The process of mountain building through regional uplift without significant deformation or tectonic activity.

Epicenter The place at the surface of the earth's crust directly above the focus of an earthquake.

Equatorial This term refers to phenomena astride or near the earth's equator and denotes a latitude lower than *tropical*. Thus the equatorial currents in several oceans move the waters in latitudes 0 to about 12 north and south; equatorial rain forests cover regions astride the equator in South America, Africa, and Southeast Asia. The equatorial trough is a belt of low atmospheric pressure that normally lies on or near the equator.

Erosion The removal of weathered rock material and the associated processes of earth denudation.

Erosion cycle An orderly sequence in the complex of erosional processes that involves a staged reduction of landscape from mountainous to near-plain topography.

Erosion surface The landscape resulting from a prolonged period of erosion.

Escarpment The sheer or very steep and wall-like edge of a plateau or mountainous highland.

Esker Ribbonlike ridge of glacial material extending across the countryside, sometimes for several miles, formed by the clogging of a stream that formed below the glacier's ice and the deposition of the material it contained.

Estuary The drowned (submerged) mouth of a river. On the map, an estuary appears to be an especially wide river mouth. Sometimes such a river mouth is funnel-shaped, and the sea penetrates up the river valley for some distance. Thus tidal fluctuation is recorded in estuaries. Here the freshwater of the river mixes with the saltwater of the sea. Estuaries are usually the result of the drowning, or submergence, of the coastal area where the river mouth is located. This often happens along coastlines: the coastal land sinks slowly downward and the ocean ''invades'' the shore. Thus a river mouth will become a branch of the sea as well—an estuary.

Evaporation Process in which a liquid changes from the liquid (water) or solid (ice) state into the vapor state (water vapor, humidity).

Evapotranspiration The loss of moisture to the atmosphere through the combined processes of evaporation from the soil and plants as well as from transpiration from plants.

Evergreen vegetation Trees and other plants that never lose all their leaves at the same time or during the same season; they are never bare and always green.

Exclave A bounded piece of territory that is part of a particular state but lies separated from it by the territory of another state. Note: An island does not qualify as an exclave under this definition.

Exfoliation Process of rock disintegration made possible by a concentric shell-like joint pattern, in which the rock surface breaks away in thick shells, thinner sheets, or scales, often leaving rounded domes, hills, or boulders to be subjected to the process in the future.

Expansion diffusion The spread of an innovation or an idea through a population in an area in such a way that the number of those influenced grows continuously larger, resulting in an expanding area of dissemination.

Extrusive igneous rock Rock formed by solidification from the molten state at the earth's surface, such as solidified lava.

Fault A fracture in rocks of the earth's crust along which displacement, often accompanied by earthquakes or tremors, occurs.

Favela Shantytown on the outskirts or within an urban area in Brazil.

Fazenda Coffee plantation in Brazil. In Middle America, the word used is *finca*.

Federation The association and cooperation of two or more nation-states or territories to promote common interests and objectives.

Ferroalloy A metallic mineral smelted with iron to produce steel of a particular quality. Manganese, for example, provides steel with tensile strength and the ability to withstand abrasives. Other ferroalloys are nickel, chromium, cobalt, molybdenum, and tungsten.

Ferrous metals Metals consisting of, or containing, iron.

Fertile crescent Semicircular zone of productive lands extending from near the southeast Mediterranean coast through Lebanon and Syria to the alluvial lowlands of Mesopotamia. Once more fertile than today, this is one of the world's great source areas of agricultural innovations.

Feudalism   A socioeconomic and sociopolitical system in which the landowning classes controlled the lives of the masses of the people living on the land. In return for subsistence tenancy, the landless peasants were forced to provide goods and services to the landowners in the form of crops and harvests as well as domestic and military service.

Fjord   Steep-sided, often deep glacial valleys that extend below sea level in coastal areas of glaciated parts of the world.

Flora and fauna   Flora encompasses all the plant life—trees, shrubs, flowers—of an area. *Fauna* is used to describe all the animal life, from the largest to the smallest.

Fluvial   Relating to streams and running water. Fluvial *landforms* are landscape elements sculptured by running water.

Floodplain   Nearly flat surface created by the lateral erosional movement of a stream whose floodwaters periodically (often annually) inundate the entire area, thus contributing additional alluvium to the plain.

Focus, earthquake   The location within the earth's crust where the movement of rocks initiates the seismic waves of an earthquake.

Fog   Cloud layer at, or near, the earth's surface.

Food web (food chain)   All plants and animals are in some way connected to this organic sequence in which each life form consumes the form below and, at the same time, is consumed by the form above. Plant life constitutes the foundation of the food web; the meat-eating animals (carnivores) stand at the top. The food web also prevails in the oceans, where certain fish eat smaller fish but are, in turn, the prey of still larger fish. Interruption of the cycle—for example, by overfishing a certain species or by exterminative hunting—can have far-reaching consequences and destroy the entire web over a large region.

Forced migration   Human migration flows in which the migrants have no choice whether or not to move to a new abode.

Foredeeps   Another name for the deep-ocean trenches often positioned adjacent to an island arc, marking a subduction zone where an oceanic lithospheric plate pushes beneath a continental lithospheric plate. Depth of the ocean in these trenches may exceed 30,000 ft (9000m).

Formal region   A type of region marked by a certain degree of homogeneity in one or more phenomena; also called uniform region or homogeneous region.

Formation classes   A subdivision of a biome based on the character of the plants (size, shape, structure, height, etc.) that gives the assemblage its dominant aspect.

Forward capital   Capital city positioned in actually or potentially contested territory, constituting an advance bastion that confirms the state's determination to maintain its presence in the region in contention.

Fragmented state   A state whose territory consists of several separated parts, not a contiguous whole. The individual parts may be separated from each other by the land area of other states or by international waters.

Francophone   A term used to identify countries or areas where French is the lingua franca or the official language, or perhaps the language of the elite, the people who make the important decisions. For some time the term *Francophone Africa*, for example, has been used to identify those African states formerly under French colonial control whose educated people communicate in French (they also know their African languages, of course).

Front   Surface or plane of contact between air masses possessing different properties such as a warm and a cold air mass.

Frontier   An area of uncertain disposition, of contention between two or more states or power cores. The world's frontier areas are diminishing in number and size, but historically they have been marked by great distances, sparse populations, uncertain authority, and often by difficult natural environments. In the past, frontier areas were unmarked by boundaries or similar lines of jurisdiction, but even in today's compartmentalized world some remaining frontier areas can be recognized, for example, in Baluchistan (Iran), in Pushtunistan (Pakistan), and in Antarctica.

Functional region   A region marked less by any internal consistence or sameness than by its integrity as a functioning system; because of its focus on a central node, also called nodal region.

Functional specialization   The production of particular goods or services as a dominant activity in a particular location. Certain cities specialize in producing automobiles, computers, or steel; others serve tourists.

Fynbos   *See* Chaparral.

Generic   Relating to, or characteristic of, an entire group or class of phenomena.

Geometric boundaries   Political boundaries defined and delimited as straight lines or arcs.

482

Geomorphology The study of the evolution of landscape, including the development of erosion surfaces.

Geopolitics (Geopolitik) A school of political geography that involved the use of quasi-academic research to encourage a national policy of expansionism and imperialism. Its origins are attributed to Rudolf Kjellén, but its most famous practitioner was a German named Karl Haushofer. The misuse of *Geopolitik* does not necessarily invalidate its position as a field worthy of academic pursuit.

Ghetto An urban region marked by particular ethnic, racial, religious, and economic properties, usually (but not always) a low-income area.

Glacial advance A period during which a continental or mountain glacier expands, covering a hitherto unglaciated area with ice.

Glaciation The overall process of ice sheet (and mountain glacier) advance and erosional activity by the agent ice. More specifically, a period of earth history that witnessed the development and spread of ice sheets and mountain glaciers (such as the Pleistocene glaciation).

Glacier An accumulation of ice on a land area, either as an ice sheet or as a confined mass in a valley, capable of movement.

Graded profile The smooth downward slope or profile characteristic of a graded stream.

Graded stream A stream in which slope and channel characteristics are adjusted over a long period of time to provide just the velocity required for the transportation of the load supplied by the drainage basin.

Granular structure In soils, a high degree of separation of individual soil grains.

Gravity model A mathematical formulation to describe the reality of reduced contact and interaction that comes with increasing distance.

Great circle The circle formed when a sphere is cut in half by a plane. The shortest distance between two points on the surface of a sphere lies along a great circle.

Great Dyke A belt of mineralization in Zimbabwe along a ridge more than 200 miles (320 km) long and running from near Harare south-southwest across the country to near Bulawayo, which has become a zone of economic activity, with mines, towns, and farms (there are good soils here as well). Although the exact geologic origin of this belt, which stands as a fairly prominent ridge above the plateau countryside, is in doubt, it has long been thought to be a *dike*, a vertical intrusion of igneous rocks through a giant crack in the crust. Thus, the Great *Dyke*, to use the British spelling.

Green Revolution Name given to the development of high-yielding strains of rice, wheat, and certain other grain crops that permit significant increases in agricultural production in certain areas (but requiring considerable investment in fertilizers).

Greenhouse effect The blanket effect created by the atmosphere resulting from its absorption of outgoing longwave radiation and consequent accumulation of heat.

Gross national product (GNP) The sum of all goods and services produced by a population during a given year.

Growing season The number of days between the last frost in the spring and the first frost in the fall. Some localities on the earth have no frost and a full-year growing season; at the other extreme are polar areas and high-mountain zones where there are no frost-free days.

Growth pole An urban center with certain attributes that, if augmented by a measure of investment support, will stimulate regional economic development in its *hinterland*.

Gyres Large, circular ocean currents resembling atmospheric anticyclones in terms of rotation and behavior. The system usually is large enough to fill an entire hemispheric segment of an ocean (such as the North Atlantic and South Atlantic gyres).

Hacienda Literally, a large estate in a Spanish-speaking country. Sometimes equated with *plantation*, but there are important differences between these two types of agricultural enterprise.

Hail A form of precipitation that involves ball-shaped fragments of ice with concentric, layered structures and ranging in size from drop-sized pellets to those with the diameter of tennis balls.

Hamlet A small clustered settlement, sometimes consisting only of several farmsteads at an intersection, but in other instances somewhat larger and with a few services; smallest in the hierarchy of settlements, less than a village.

Hanging valley Valley contact formed by the intersection of a tributary glacier with a larger glacier, which comes about when valley floors are not adjusted to each other.

Hardpan A rock-hard layer of concentrated iron and aluminum oxide that develops in certain tropical soils and interferes with agriculture.

Headland An exposed, prominent, often cliff-edged promontory that juts out into the ocean and is sustained by especially resistant rocks.

Headward erosion The lengthening of a river valley in which the river erodes in its uppermost reaches by un-

483

termining its banks and causing mass movement.

Heartland The core area of a nation-state. The term came into use with an article by Halford Mackinder that identified such a region in interior Eurasia.

Heat island Prevailing and persistent concentration of higher temperatures over an urban area where longwave radiation from streets and rooftops is intense.

Hegemony The domination of peoples or nations by stronger nations or states. For example, much of Europe was under German hegemony during the Second World War.

Hematite Iron-rich rock that forms the ore body of many iron mines around the world.

Herb Nonwoody, small, and often fragile annual or perennial plant resembling a soft grass.

Herbivore An animal that lives on plants; often used to describe the grazing animals.

Hierarchical diffusion A form of diffusion in which an idea or innovation spreads by trickling down from larger to smaller adoption units. Urban hierarchies are usually involved, encouraging the leapfrogging of innovations over wide areas, with geographic distance a less important influence.

Hierarchy A term used to identify an order of gradation of phenomena, each grade being subordinate to the one above it and superior to the one below it, like the steps of a ladder. It can be established on the basis of many criteria and can be applied to many phenomena. Originally a religious *term (hierarch* [religious leader]) denoting the organization of a ruling body of clergy into ranks.

Highveld A term used in southern Africa to identify the grass-covered, high-elevation plateau that dominates much of the region. *Veld*, sometimes misspelled *veldt* means grassland in Dutch and in Afrikaans. The lowest areas in South Africa are called lowveld; areas that lie at intermediate elevations are the middleveld.

Hinterland Actually "country behind"—applied to areas that lie around cities, where that city dominates the commercial life and is the chief service center and major urban influence.

Histosols A soil order consisting of soils characterized by a thick upper layer of organic matter, occurring in relatively small areas in marshes and bogs.

Holistic A term that pertains to a total, overall, all-encompassing view of a subject.

Horizonation The distinct layering of a soil; its division into several horizons possessing discrete properties.

Horn A glacial landform that occurs in mountainous topography and results from the intersection of three or more cirques, producing an often sharp-edged, sheer-sided peak that towers above the countryside.

Horse latitudes Subtropical high-pressure belt between the tropical easterlies and the midlatitude westerlies in the Northern and Southern hemispheres.

Horticulture Strictly, the growing of tree crops; used also to describe the cultivation of vegetables in small gardens as well as the cultivation of fruits and flowers.

Hours of insolation The number of hours during a given year that a particular location receives direct sunlight.

Humus Decomposed and partially decomposed organic matter in the upper layer (and sometimes on top) of the soil, contributing importantly to the soil's fertility and giving it a characteristic dark brown-to-black color.

Hurricane Tropical cyclone that forms in the southern part of the North Atlantic Ocean and in the Caribbean Sea, with winds attaining at least 74 miles (115 km) an hour.

Hydraulic action The force of moving water in a stream sufficient to dislodge and move material from the stream bed and drag it along the valley floor and sides, thus contributing to erosion.

Hydrologic cycle The complex system of exchange involving water in the vapor, liquid, and solid state as it circulates among atmosphere, lithosphere, and hydrosphere.

Hydrolysis The chemical combination of water molecules with minerals in rocks, producing compounds and causing expansion.

Hydrosphere The earth's sphere of water in various forms—the saltwater of the oceans, freshwater of lakes and rivers, ground water, and water held as vapor in the atmosphere.

Hygrophytes Plants adapted to such areas where an excess of water prevails as swamps, marshes, and bogs.

Ice sheet A large and thick layer of ice that flows outward in all directions from a central area where continuous accumulation of snow and thickening of ice occurs. Also called continental ice sheet or continental glacier.

Ice shelf Sheet of floating ice that protrudes from a land-supported ice sheet and is fed by flow from the land area. The largest ice shelves occur around the margins of Antartica.

Iconography The identity of a region; its particular cultural landscape and atmosphere.

484

Igneous rock Crustal rock formed from the cooling and solidification of molten *magma*.

Illuviation A soil-forming process in which downward-percolating water carries soluble minerals and colloidal-size particles of organic matter and minerals into the lower (B) horizon, where these materials are deposited in pore spaces and against the surfaces of soil grains.

Immigrant A person migrating into a particular country or area; an in-migrant.

Imperialism The drive toward the creation and expansion of a colonial empire and, once established, its perpetuation.

Inceptisols A soil order consisting of soils with poorly developed horizonation; "new" soils such as those forming on recently deposited alluvium.

Indentured workers Contract laborers who sell their services for a stated period of time. Tens of thousands of South Asians came to East Africa as indentured workers at British behest; working conditions often did not match contract stipulations, but indentured workers did not have the option to cancel their agreements.

Independent invention When an innovation appears in different parts of the world, it is tempting to surmise that it diffused rapidly from one source area to another place. In some instances, however, an invention was made almost simultaneously in comparatively isolated locations, without borrowing—in which cases, independent invention has occurred.

Industrial Revolution A series of related innovations and inventions that had the effect of intensifying industrial production in England during the eighteenth century through the introduction of machines and inanimate (nonhuman, nonanimal) power.

Infrastructure The foundations of a society: urban centers, communications, farms, factories, mines, and such facilities as schools, hospitals, postal services, and police and armed forces.

Initial landform Landform created by earth forces and not by erosional activity, for example, a volcanic mountain or a scarp produced by faulting. Erosional agents immediately begin to modify initial landforms.

Inner core The innermost, extremely heavy "ball" that forms the core of the planet Earth; extremely heavy and dense, with a diameter of about 1550 miles (2500 km).

Innovation The introduction of something new—a new idea, new tool, or new way of doing something. Geographers are especially interested in the way in which innovations are spread through populations, or how the process of *diffusion* takes place.

Insolation Arrival of sunlight (shortwave radiation) on an exposed surface.

Insularity An island position; from the Latin *insula* (island). It also implies an isolation or detachment from the outside world and, in the case of Britain, a protection from the outside world.

Insurgent state Territorial embodiment of guerrilla activity, the antigovernment insurgents establishing a territorial base in which they exercise permanent control; thus, a state within a state.

Interface The boundary surface or plane between two or more systems involving different conditions, where these systems make contact and exchanges of energy and matter take place. The Earth's surface is such an interface, and most human activity occurs here.

Interglacial A stage or period of temporary warming and reduction of glaciers between glacial advances, a stage in which the Earth may be today.

Intermontane An area (e.g., a basin) that is surrounded by, or lies between, mountains. From *inter* (between) *montane* (the mountains).

Internal migration Migration flow within a nation-state such as westward and northward movements in the United States and the eastward movement in the Soviet Union.

International migration Migration flow involving movement across international boundaries.

Intertropical convergence (ITC or ITCZ) Belt of surface convergence of air masses in the northeast and southeast trades along the axis of the equatorial trough.

Intervening opportunity The presence of a nearer and perceived-to-be better opportunity that diminishes the attractiveness of sites even slightly farther away.

Introduced capital city A capital city chosen to further certain national objectives such as seaward orientation, territorial redirection, or centralization; a disputed term.

Intrusive igneous rock Crystalline rock solidified from molten *magma* within the earth's crust while surrounded by preexisting rock; the cooling rate is normally slower than for extrusive igneous rock.

Inversion (temperature inversion) Condition in which temperature increases with altitude rather than decreases—a positive lapse rate.

Ion An atom or cluster of electrically charged atoms formed by the breakdown of molecules.

Irredentism A policy of cultural

485

extension and potential political expansion aimed at a national group living in a neighboring country.

Irrigation The artificial watering of croplands. Basin irrigation is an ancient method that involves the use of floodwaters that are trapped in basins on the floodplain and released in stages to augment rainfall. Perennial irrigation requires the construction of dams and irrigation canals for year-round water supply.

Isobar A line connecting points of equal pressure. *Iso* means the same; *bar* refers to the barometer's reading of atmospheric pressure. When a weather map depicts a high or a low pressure system, the nearly circular lines that surround the center are isobars.

Isogloss Line marking the spatial limit of a particular linguistic usage such as an individual word or term or prevailing use.

Isohyet A term used in the context of the transition zone between farmer and nomad in Mongolia. An isohyet is a line that connects points or locations of *equal* recorded precipitation.

Isolation A condition of being cut off or set apart from a main stream or streams of thought and action, a lack of receptivity to outside influences and stimuli—and the reverse as well. Nothing much emanates from a people or place that is in isolation. Isolation can have many causes such as distance or inaccessibility, and it can affect areas and peoples large and small.

Isoline Line connecting points of equal value. An *isotherm* connects points of equal temperature; an *isobar* connects points of equal atmospheric pressure; and an *isohyet* connects points of equal rainfall totals.

Isostasy Principle of balance between sial and sima in the earth's crust, in which "light" sialic mountain ranges

are thought to possess "roots" that extend deeply into the sima below.

Isotherm Line connecting points of equal temperature.

Isthmus A land bridge, a comparatively narrow link between larger bodies of land. Central America forms such a link between North and South America.

Jet stream Fast-flowing, high-altitude current of air in the Northern and Southern hemispheres. Important jet streams are positioned above the subtropical high-pressure zone, within the upper-air westerlies, and above the polar front in each hemisphere.

Jointing Pattern of natural fracturing existing within bedrock, normally occurring in parallel and intersecting sets of planes but sometimes in particular patterns (e.g., as six-sided columns in basalt); weathering processes proceed most effectively with the aid of these intrinsic planes of weakness.

Jungle Term often misused to describe the dense, almost impenetrable vegetation that occurs in certain equatorial and tropical areas. This is not the normal or natural state of equatorial forests but results from regeneration after destruction of the original forest by humans.

Juxtaposition Two places or areas located side by side, or nearly so, are said to be juxtaposed or in close juxtaposition.

Kame Comparatively small ridge or mound of glacial debris or stratified drift that developed at the edge of a glacier.

Karma In Hinduism and Buddhism, the force generated by a person's actions that affects transmigrations into a future existence, determining condi-

tions and position in the next earthly stage of life.

Karst terrain The topography of surface features (including sinkholes and caves) associated with underground chemical erosion of bedrock limestone.

Katabatic wind (cold air drainage) Downward flow of near-surface air that has cooled because of radiation loss and has become colder and heavier than stationary air.

Kilometer A distance measure in the metric system, approximately 0.62 statute mile; 1 mile equals 1.609 km.

Krill Small, shrimplike marine animal that exists in enormous quantities in Antarctic waters; principal food of whales. Improved technology has recently made harvesting krill for human consumption possible.

Lagoon A normally shallow body of water that lies between a barrier reef or barrier island and the original shoreline.

Land alienation A term often used when land is expropriated and acquired by one culture's representatives from another—the land going to alien, foreign invaders from its original owners or users.

Land breeze A wind that blows from land to water, usually at night during radiation loss.

Landform The individual product of the Earth's internal geologic forces (producing an *initial landform*) or the result of weathering and erosion (e.g., as a monadnock or a horn), or the product of aggradation (a dune or esker). A land*form* is a single feature in the land*scape*.

Landlocked Term for a country or state surrounded everywhere by land and therefore without coasts. Examples: Switzerland, Austria, Czechoslovakia, Hungary, Bolivia, Zimbabwe, Nepal.

Landmass   A major segment of the earth's continental crust that lies above sea level; often used as a synonym for continent.

Land reform   The spatial reorganization of agriculture through the allocation of farmland (often expropriated from landlords) to peasants and tenants who never owned land; also, the consolidation of excessively fragmented farmland into more productive, perhaps cooperatively run farm units.

Landscape, cultural   *See* Cultural landscape.

Landscape, natural (physical)   The total scenery of a part of the Earth's surface, including all the various landforms that create this overall impression.

Landslide   Rapid movement (sliding) of a mass of rock and soil on steep mountain slopes.

Lapse rate   Rate of temperature decrease with altitude. *See* specific categories such as *Environmental lapse rate*.

Lateral moraine   Moraine in a valley glacier positioned between the *medial moraine* and the glacier's valley sides, leaving a ridgelike strip of debris when the glacier melts.

Laterization   A soil-forming process that involves the decay and removal of the organic layer and the leaching of minerals, leaving oxides of iron and aluminum to color the soil a characteristic red. The process is associated with warm and moist climates of low latitudes and produces soils such as oxisols and ultisols.

Latitude   Latitude lines are aligned east–west across the globe. Thus the equator is 0° latitude, and the North Pole is 90° north latitude. Areas of low latitude lie near the equator and in the tropics, whereas areas of high

latitude lie near the pole and Arctic. The closer to the pole, the higher the angle of latitude and thus, by common usage, the latitude itself. The United States and Europe lie in the middle latitudes.

Latosolic soil   Latosolic (and the name *latosol*) have to do with the end product of the process whereby the important plant nutrients are carried away from the upper layer of the soil in certain excessively wet, tropical regions. (The term *laterite* used to be applied to such reddish soils.)

Lava   Magma extruded onto the earth's surface through a volcanic vent or fissure.

Leaching   The process in which percolating water dissolves and removes many of the mineral substances from the upper layer of a soil.

Leaching, soils   The removal of ingredients from the soil by percolating water. Rainfall makes the ingredients in the soil usable to plants, and the water in the soil promotes the development and maturing of that soil. But, as in much of Southeast Asia, the soil can get too much of a good thing—an excess of rainfall and too much soil moisture. When there is such an excess, the percolating waters tend to carry away vital bases, including calcium, sodium, magnesium, potassium; the process is called eluviation. Iron and aluminum may be left behind, coloring the soil a characteristic red, a sign of low fertility and reflecting that the soil is heavily leached.

League   An association of units united by a common purpose or interest. In European history, there were many attempts by city-states, port cities, and other political entities to join in alliances (known as leagues) with a common purpose—defense, protection, regulation of profit making, and so on.

Lebensraum   Literally, "space for

living"; a German word that became part of the language of *Geopolitik*.

Levee   River-girdling ridge of alluvial material. When a silt-laden river such as the Hwang Ho in China reaches the lower part of its course, it is slowed down, and the sediment it carries begins to settle on the side of the channel through which the river is flowing. Thus, a sort of natural dike forms, which contains the river and whose edges slope gently outward into the flood plain. This the so-called *natural levee*; as it is built up year after year, it becomes the highest part of the entire floodplain. In times of severe flood, the river may break through its own levees and inundate the surrounding area. Engineers often take a hand in the process and try to strengthen a flood-prone river's natural levee—and even build *artificial* levees to keep rivers within their channels.

Lignite (brown coal)   A woody variety of coal, just a little better than peat but not nearly as good as the next higher grade, bituminous coal. Lignite is not good enough to be usable in the big steel furnaces—it is not a coking coal. Some lignite is less woody than other varieties and of better quality. As a fuel, lignite is by no means unimportant.

Lingua franca   A "common language" that can be spoken and understood by many peoples although they speak another language at home. The term derives from "Frankish language" and applies to a trade tongue spoken in Mediterranean ports and consisting of a mixture of Italian, French, Greek, Spanish, and even some Arabic. In Africa, where the African peoples speak hundreds of African languages, after colonization English became the lingua franca over large areas as did French elsewhere on the continent.

Lithosphere   The Earth's solid crust; more specifically, the uppermost,

487

solid layer of the Earth, above the asthenosphere.

Littoral   Coastal; along the shore.

Llanos   The savanna-like grasslands of the Orinoco River's wide basin in parts of Venezuela, Guyana, and Colombia, elsewhere called *Chaco*, as in northern Argentina. In South America, the savanna has several different names.

Loam   Soil that contains sand, silt, and clay.

Local base level   The lower level below which a stream cannot under existing conditions erode its valley, for example the level of a lake it enters. Such a situation is temporary.

Location   The position or place of a certain phenomenon on the surface of the Earth. Since the Earth is round (or nearly so), it is necessary to create a grid of lines, identified by number, over all of its surface. Thus we can "locate" a place by stating that it is at the intersection of vertical (north–south) line x and horizontal (east–west) line y.

Location theory   Comprehensive term for the theories and models developed by geographers to explain the relative location of various human establishments, including towns and cities, factories, and shopping centers.

Loess   A deposit of very fine silt or dust laid down after having been blown some distance (perhaps hundreds of miles) by the wind. Such deposits may be related to the Pleistocene glaciation, after which material that was ground into very fine particles by the glaciers was later picked up by the wind. Loess is characterized by its vertical strength—its ability to stand in steep vertical walls when a river erodes it. It is found in North America as well as across much of inner Asia. Loess is very porous, but fertile when irrigated.

Longitude   Angular distance (0°

to 180°) east or west as measured from the prime *meridian* (0°) that passes through the Greenwich Observatory in London, England. For much of its length, the 180th meridian functions as the International Date Line.

Longitudinal dune   A type of sand dune shaped like a narrow; elongated ridge whose axis lies parallel to the prevailing wind.

Longwave radiation   Radiation emitted by the Earth, effective in warming the atmosphere and therefore sometimes called heat radiation (as opposed to the sun's light radiation).

Magma   Mass of rock at an extremely high temperature and in the molten state within the earth; cooling and solidification produce igneous rock from magma.

Malthusian   Designates the early nineteenth-century viewpoint of Thomas Malthus who argued that population growth was outrunning the Earth's capacity to produce sufficient food. Neo-Malthusian refers to those who subscribe to such positions in modern contexts.

Mantle   One of the Earth's major structural components, a shell about 1800 miles (2900 km) thick that lies above the outer core and below the crust.

Maquis   *See* Chaparral.

Marchland   An area through which, over the centuries of history, diverse powers and forces have moved, taking control and yielding it again to the next competitor. Literally, then, an area through which armies have "marched"; it may be a contested frontier or an area that has been especially open to competing powers. Between Italy and Yugoslavia lies an area called the Julian March, the whole region simply described as a marchland in its very name, so much has it been contested.

Mesopotamia, too, has been something of a marchland: it was far less protected by natural barriers than was Egypt's desert- and sea-protected Nile Valley.

Maritime influences   The effect of influences from the sea upon adjacent land areas, produced by air masses that become warmed and/or moisture-laden over the oceans and are then carried over continental areas. A nearby warm ocean may moderate the climate of a coastal region that might otherwise be extremely cold.

Market   A location where goods and services are demanded and exchanged.

Mass movement (mass wasting)   The movement of lithospheric surface material under the influence of gravity.

Material diffusion   A diffusion process that involves the dissemination of an actual object or tool.

Matrillineal   Referring to a society in which descent is traced through the female line as are inheritance, succession, and other functions.

Mattoral   *See* Chaparral.

Meander   The smooth, rounded curves and bends of a graded stream as it flows across its alluvium-filled floodplain.

Medial moraine   In a mountain or alpine glacier, the morainal material positioned approximately in the middle of the ice-filled valley, leaving a ridge of debris when the glacier melts.

Megalithic   Referring to the erection of structures made of huge stone blocks by ancient societies.

Megalopolis   A term used to identify those large, coalescing supercities that are forming in several parts of the world. A prominent example is that of the U.S. northeastern seaboard from Boston to Baltimore–Washington.

Megatherm   A plant adapted to conditions of high heat.

Mental map The image of a map of the world or of a city or neighborhood carried in the mind of an individual. Such an image inevitably contains distortions.

Mercantilism Protectionist policy of European states during the sixteenth to the eighteenth centuries that promoted a state's economic position in the contest with other countries. The acquisition of gold and silver and the maintenance of a favorable trade balance (more exports than imports) were central to the policy.

Meridian; parallel The lines of the grid whereby places on the earth's surface are identified by their location; called lines of longitude (north–south, passing through the poles) and lines of latitude (east–west, paralleling the equator). The terms *meridian* and *parallel* denote these two sets of lines. Meridians are the north–south lines, and each is a so-called great circle. Parallels are the east–west lines; thus the equator is itself a parallel.

Mesa Flat-topped, steep-sided mountain capped by resistant rock layer.

Mesoamerica *Meso* means middle or in-between; Mesoamerica is another way of saying Middle America. Since it would be confusing to identify the old culture hearth as well as the geographic region as Middle America, Mesoamerica is used to denote a historically significant part of Middle America.

Mesopause Boundary surface between the mesosphere and the thermosphere.

Mesophyte A plant adapted to conditions of intermediate water availability and continuous, all-year moisture availability.

Mesosphere The atmospheric layer above the stratosphere (thus above the stratopause), marked by decreasing temperatures with height above the earth.

Mesotherm A plant adapted to conditions of intermediate temperature.

Mestizo A person of mixed ancestry; from the Latin for *mixed*. Specifically used to denote a person whose ancestry is both European and American Indian.

Metamorphic rocks Rocks formed from preexisting rocks that have been altered in structure and chemical composition by increases in temperature and pressure and by the infusion of hot chemical solutions.

Mexica Empire The name the ancient Aztecs gave to the domain over which they held hegemony in Middle America.

Microstates Category of smallest nation-states, smaller than the ministates.

Microtherm A plant adapted to very low temperature conditions.

Midocean ridge Submarine, oceanfloor mountain ranges that mark oceanic plate contact zones and possess characteristic longitudinal rift valleys. The Mid-Atlantic Ridge is a prominent example.

Migration *See* Internal, International, Forced, and Voluntary migration.

Migratory movement Human movement from source to destination without a return journey, as opposed to *cyclic movement*.

Millenium A period of 1000 years.

Millibar (mb) A unit of atmospheric pressure equal to 1000 dyn/sq. cm.

Milpa agriculture Middle and South American subsistence agriculture in which forest patches are cleared for temporary cultivation of corn and other crops.

Mineral A naturally occurring chemical element or compound with a characteristic crystalline structure.

Model An abstract construction developed by geographers to simulate real-world conditions, used to isolate and study causes and factors underlying the development of the spatial arrangement of phenomena on the Earth's surface.

Moho discontinuity The surface of contact between the earth's crust and the mantle below.

Mollisols A soil order consisting of soils marked by a dark, humus-rich upper layer that supports extensive grasslands and grain crops.

Monadnock A remnant hill of exceptionally resistant rock (or one protected by remoteness from stream erosion) that rises above a peneplain.

Monotheism The belief in and worship of a single god.

Monsoon A seasonal wind pattern in which winds blow into a continental area during the summer and outward during the winter, often accompanied by heavy rain during the summer months.

Monsoon rain forest A particular assemblage in the forest biome consisting of evergreen as well as deciduous trees adapted to the wet–dry seasonality of monsoon climatic conditions.

Moraine A ridge or mound of glacial debris deposited during the melting phase of a mountain or continental glacier. *See* Lateral, Medial, Recessional, and Terminal moraine.

Morphology The study of forms or features for their own sake, without necessary reference to the processes whereby these forms came about. We see this word in geo*morphology*; it is often placed opposite genetics, implying something static against something dynamic. The *morph* part comes from the Greek (having a certain form). In boundary morphology, we look at the boundaries as they are—straight, or bent, or irregular—as a prelude to later study of genetics.

Mountain glaciation Ice accumulation and glacier formation in mountainous areas, the glaciers transforming the stream-carved topography by erosion as well as deposition.

Mulatto A person of mixed white (European) and black (African) ancestry.

Multinationals The multinational (or global) corporations that strongly influence the economic and political affairs of many nation-states.

Multiple-nuclei model A model that shows a city to consist of several nuclear growth points.

Nation Legally a term encompassing all the citizens of a state. It has also taken on other connotations, and dictionary definitions tend to refer to bonds of language, ethnicity, religion, and other shared cultural attributes. Such homogeneity prevails within very few states.

Nation-state A state whose population has the properties of a nation.

Natural gas A combination of hydrocarbon compounds (chiefly methane) in the gaseous state, which are contained in layers of porous rock that form a reservoir enclosed by impermeable strata.

Natural increase Population growth measured as the excess of live births over deaths per 1000 individuals per year. Natural increase of a population does not reflect either immigration or emigration.

Natural resource As used in this book, any valued element of the environment, including minerals, water, vegetation, and soil.

Nautical mile A measure used at sea; equals 6080 ft or 1852 ms, amounting to 1.15 statute miles.

Needleleaf forest Plant formation class within the forest biome that consists of evergreen trees with leaves shaped like needles.

Negative landform A landform that results from low resistance to erosion such as a valley between ridges, a basin, or a mountain gap; a "down" in the landscape.

Network, transport The entire physical system of connections and nodes through which movement can occur.

Nomadism *Cyclic movement* among a definite set of places. Nomadic people are mostly pastoralists.

Nonclastic sediments Sedimentary rocks formed from chemical deposits or organic activity.

Nonferrous metals Metals that do not contain iron.

Nonrenewable resources Natural resources that do not replenish themselves during exploitation and are, therefore, exhaustible. The world's oil reserves are nonrenewable. Additional reserves may be discovered for some time to come, but the quantity of oil is finite and oil is nonrenewable.

Nonvariant gases Gases in the atmosphere that are present in the same proportion whether at sea level or dozens of miles above the Earth's surface.

Norden A regional concept that includes the three Scandinavian countries, Finland, and Iceland.

Northeast trades Surface winds with a predominantly northeast direction that blow in latitudes between the doldrums and the subtropical high pressure belt in the Northern Hemisphere. The name derives from these winds' importance during the era of sailing vessels.

Nuclear fusion The formation of an atomic nucleus by the union of two other nuclei having a lighter mass, a process that yields a huge amount of energy which, if harnessed, may solve the world's power problems during the twenty-first century. A virtually inexhaustible source of fuel (hydrogen) exists, and the process does not produce dangerous radio-activity as nuclear fission does.

Nucleated settlement A clustered pattern of population distribution. Driving through or flying across the American Midwest, you may see a countryside with scattered farmsteads. Such settlement would be described as dispersed, that is, farms and homes are scattered or distributed with fairly even frequency over each square mile of the area. On the other hand, there are regions where people live in farm villages, their dwellings close together; each working day, they walk or ride to the land they are farming. When settlement is thus clustered into villages, it is said to be nucleated: the landscape has nuclei of settlement. Such nucleation can be induced by the physical landscape itself. In an area of high relief, only the valleys may be habitable and productive, and human settlement may cluster there. Or it may be a pattern that arises from an ancient tradition such as common defense.

Oasis An area, small or large, where the supply of water transforms the surrounding desert into a green cropland; the most important focus of human activity for miles around. It may encompass a large, densely populated zone along a major river where irrigation projects stabilize the water sup-

490

ply—as along the Nile River in Egypt, which is really an elongated set of oases.

Ocean currents and drifts Linear movements of water in the oceans. Like the air of the atmosphere, the water of the oceans is constantly in motion. Coriolis helps generate this movement, and atmospheric pressure also has to do with it. The waters of the North Atlantic Ocean circulate clockwise; those of the South Atlantic Ocean counterclockwise. This pattern is repeated in the other oceans—north of the equator, clockwise; south of it, the reverse. But this is only a generalization; there are offshoots and countercurrents.

Oceanic plate A tectonic plate consisting of a segment of crust beneath the ocean.

Oligarchy Political system involving rule by a small minority, an often corrupt elite.

Oral history Historical information transmitted by word of mouth. History is seen by some historians as the study of humanity's relatively recent past, as revealed by the written record of it; there are still those who argue that there is no history where there was no writing. Many modern historians, however, have turned to orally transmitted records (i.e., by mouth, through legends, myths, folktales, plays, etc.) to seek to unravel the past of peoples who did not write. Some astonishing results have already been obtained through this kind of oral history research.

Organic theory A determinist view that states resemble biological organisms with life cycles that include stages of youth, maturity, and old age; now largely discredited.

Oriental, Occidental The root of the word *oriental* is the Latin for *rise*; thus it has to do with the direction in which one sees the sun "rise"—the east; *oriental* therefore means eastern.

Occidental originates from the Latin for *fall*, or the "setting" of the sun in the west; *occidental*, then, means western.

Orogenic belt Zone of a present or former mountain range.

Orogeny Times of crushing, pressure, and folding of thick rock layers of the earth's crust. The mountain ranges that are built during such times are called orogenic belts. Europe's Alps constitute an orogenic belt, and geologists recognize an Alpine orogeny as part of the geologic history of Eurasia.

Orographic precipitation Precipitation caused by the cooling and condensing of moisture-laden air forced to rise over a mountain range or against a plateau escarpment.

Outer core A major component in the structure of the Earth, believed to consist of extremely heavy metallic rock material in a liquid state. The outer core lies between the solid inner core and the mantle and is nearly 1400-miles (about 2200-km) thick.

Oxic soil horizon Soil horizon in tropical soils from which weathering and leaching have removed or altered a large part of the silica previously combined with iron and aluminum, leaving mainly oxides of these two metals in this layer.

Oxisols A soil order consisting of old, highly weathered soils in low latitudes that have undergone a long period of laterization and that possess an oxic horizon.

Ozonosphere A layer in the earth's stratosphere between about 15 and 50 km (10 and 30 m) above the surface, containing a concentration of ozone. Also known as *ozone layer.*

Pangaea A vast landmass consisting of most of the areas of the present continents, including the Americas, Eu-

rasia, Africa, Australia, and Antarctica, that existed until near the end of the Mesozoic era when plate divergence and continental drift broke it apart. The "northern" segment of Pangaea is called *Laurasia*, the "southern" part *Gondwana.*

Parallel An east–west line of latitude that is intersected at right angles by a meridian, a line of longitude.

Paramós Highest zone in the Latin American altitudinal zonation, above the tree line in the Andes; sometimes called *tierra helada* (frozen land).

Parent material The rocks and their minerals from which the overlying soil is formed.

Particulate pollutants Very small particles in solid and liquid form, capable of being suspended in the air for long periods of time.

Pastoralism A form of economic pursuit: the practice of raising livestock, although many peoples described as pastoralists actually have a mixed economy—they may fish, or hunt, or even grow a few crops. Few peoples are "pure" herders.

Peasants In a stratified society, peasants are the lowest class of people who depend on agriculture for a living, but they often own no land at all and must survive as tenants or day workers.

Peat Compacted mass of plant matter, only partially decomposed, located in a bog environment.

Pecuniary Having to do with money.

Ped A lump of naturally occurring soil particles, often with a characteristic structure.

Pedestal rock Erosional landform in which wind action plays a major role through the impact of wind-driven particles against the lower part of the feature, creating a mushroom-shaped landform whose "foot" is smaller and narrower than the overlying "cap."

Pediment Gently sloping, often rock-floored surface resulting from erosional retreat of a mountain range or plateau escarpment.

Pediplane Erosion surface formed from coalescing pediments following erosional reduction of mountains in an area.

Pedology Soil science; the study of soils.

Pedon A total soil column extending from the surface to the bedrock or parent material.

Peneplain A nearly flat surface formed by long-term erosion.

Peninsula A comparatively narrow, fingerlike stretch of land extending from the main body into the sea. Italy constitutes a peninsula.

Peon Term used in Latin America to identify people who often live in serfdom to a wealthy landowner; landless peasants in continuous indebtedness.

Per capita *Capita* in this context means individual; income, production, or some other measure is often given per individual.

Percolation To penetrate, to diffuse through, to pass through a permeable substance.

Periodic market Village market that opens every third or fourth day or at some other regular interval. Part of a regional network of similar markets in a preindustrial, rural setting where goods are brought to market on foot (or perhaps by bicycle) and barter remains a major mode of exchange.

Periodic movement A form of migration that involves intermittent but recurrent movement such as temporary relocation for college attendance or service in the armed forces.

Permafrost Permanently frozen water in the soil, regolith, and bedrock, as much as 1000 ft (300 m) in depth, producing the effect of completely frozen ground. Permafrost may begin a few inches below seasonally warmed and "soft" surface soil or deeper.

Permanent capital Capital city that has historically been the headquarters of the state, for example, Rome or Athens.

Physiographic-political boundaries International political boundaries that coincide with prominent physiographic breaks in the natural landscape such as rivers or the crests of mountain ranges.

Physiographic region (province) A region within which there prevails a substantial degree of landscape homogeneity.

Physiography The total physical landscape—topography, relief, drainage, soils, vegetation, prevailing climate.

Physiologic density The number of persons per unit of land under cultivation.

Pidgin A language that consists of words borrowed and adapted from other languages, originally developed from commerce among peoples speaking different languages.

Pilgrimage A journey to a place of great religious significance by an individual or by a group of people. When Islam entered West Africa, pilgrimages to Mecca attracted tens of thousands of Moselms annually. Many took up permanent residence along the way and the human map of savanna Africa was transformed.

Pipe Volcanic vent leading from the magma chamber to the crater.

Pithecanthropus Ancient apeman. In 1890, the remains of a very ancient fossil man were discovered in central Djawa (Java). The Dutch doctor who found these remains named this creature *Pithecanthropus erectus*, or the erect apeman. This ancient forerunner of modern humans is now called *Homo erectus*, and he seems to have inhabited Djawa during the early Pleistocene. Not far from Beijing in north China, another variety of *Homo erectus* was discovered, and the early humanity seems to have been quite widely dispersed over Eurasia by the mid-Pleistocene.

Plane of the ecliptic The plane in space defined by the earth's orbit around the sun.

Plantation A large estate owned by an individual, family, or corporation and organized to produce a cash crop. Almost all plantations were established within the tropics; since 1950 many have been divided into smallholdings or reorganized as cooperatives.

Plateau An upland surface sustained by resistant rocks and bounded by an escarpment, scarp, or cliff leading down to lower topography in surrounding areas.

Plates (tectonic) See Tectonic plates.

Platy structure In soils, peds that have a flakelike appearance with the soil particles arranged in overlapping horizontal planes.

Pleistocene The name given to that epoch of the Earth's existence covering about the last 3 million years. The Pleistocene includes the last of the great ice ages, and the evolution of humanity on the earth. For archeologists, the Pleistocene epoch is the critical geologic stage in the earth's life.

Podzolized Refers to a soil that is undergoing one of the soil-forming processes; the nature of these processes depends on the climate, drainage, bedrock, and vegetative association involved. In the case of podzolization, there are likely to be coniferous trees; the climate is likely to be continental

492

(cold winters, moderate precipitation well distributed throughout the year). There is a lot of leaf matter and humus on the ground, and the acids from this material leach the upper soil of iron and aluminum oxides. Result: an ash-gray upper layer or A horizon in the soil, and a hard, iron-rich lower layer or B horizon.

Polar front   A front lying between the cold air masses of polar areas and the warmer tropical air masses of lower latitudes.

Polar front jet stream   Jet stream that forms in the middle latitudes in association with the polar front.

Polder   A Dutch word for a low-lying piece of land that has been reclaimed from the sea and is therefore likely to lie below sea level and be ringed by dikes. (Technically, reclamation from a lake also creates a polder.)

Pollution dome   Dome-shaped mass of polluted atmosphere that forms over an urban area during times of still air.

Polytheism   The worship of more than one (sometimes numerous) gods.

Population density   The number of individuals per unit area without regard for habitability or productive capacity.

Population explosion   The rapid growth of the world's human population during the past century, attended by ever-shorter doubling times and increasing rates of increase.

Population structure   Representation of a population by sex and age, usually in the form of a population pyramid.

Positive landform   An element in the landscape that stands out, normally by virtue of the higher resistance of its rocks, above lower, less resistant and more easily eroded areas nearby.

Possibilism   Geographic viewpoint—a response to determinism—that

holds that human decision making is the crucial factor in cultural development, not environmental limitation or restricted options. Cultural heritage and learning are the key attributes in society's use of the natural environment.

Postindustrial economy   Emerging economy in the United States and a handful of other highly advanced countries as industry gives way to a high-technology productive complex dominated by services, information-related, and managerial activities.

Prairie   Plant formation class in the grassland biome, consisting mainly of tall grasses.

Precipitation   The collective name for the forms in which water is returned from the atmosphere to the earth's surface, including rain, snow, hail, and sleet.

Predator   A species that preys on other species for its survival.

Pressure   *See* Atmospheric pressure.

Pressure cell   A center of high or low atmospheric pressure, marked by specific patterns of wind circulation.

Pressure gradient force   The force that tends to move air from areas of higher atmospheric pressure toward areas of lower pressure.

Prevailing winds   Winds that blow from a particular direction most of the time (but not always), such as the westerlies in the Northern Hemisphere that blow mainly from the southwest quadrant.

Primacy   First place, first in rank, at the top in a hierarchy.

Primary industries   Industries engaged in the direct exploitation of natural resources such as mining, fishing, lumbering, and farming.

Primary waves   Directly propagated earthquake waves that radiate outward in all directions from the focus

and are first to arrive at distant seismic recorders.

Primate city   A country's largest city, most expressive of the national culture, its focus, usually (but not always) also the capital city.

Prime meridian   The meridian that passes through the Royal Observatory in Greenwich, England; designated 0° longitude.

Prismatic structure   In soils, a ped structure in which the soil particles are arranged in columns.

Probablism   A geographic theory in answer to environmental determinism; it holds that human choice among options is the crucial element in the relationship between natural environment and cultural development.

Process   Casual force that shapes spatial pattern or structure as it unfolds over time.

Proletariat   Lower-income working class in a community or society; people who own no capital or means of production and who live by selling their labor.

Promontory   A prominent headland, usually associated with coastal topography.

Prorupt   A type of state territorial morphology that exhibits a narrow, elongated land extension.

Protectorate   In Britain's system of colonial administration, the protectorate was a designation that involved the assurance of certain rights and guarantees to peoples who had been placed under the control of the Crown. One of these guarantees related to the restriction of European (white) settlement and land alienation. In some protectorate areas, whites were not allowed to own land at all. It did not always work out that way, but such were the intentions.

Public architecture   The architecture of buildings such as governmental headquarters, state offices, and other official structures.

Push–pull concept Forces that impel human migration. Cultural geographers have long been interested in the process of human migration. Migration is so essentially geographic a phenomenon that it has occupied the attention of many. One concept that has been used a good deal relates to the forces that make people migrate. Sometimes people leave their abode because something pushes them away. It may be an oppressive government, unacceptable social conditions, discrimination against a minority by a majority, and so forth. At the same time, there may be attractions in the distance: the appeal of the glamorous, skyscrapered, job-generating city, or some colonial territory where land may be had cheaply and labor to work it even more so. The problem, of course, is not only to recognize these push–pull factors for what they are, but also to measure their strength in some way. That, as the literature will prove to you, has been a tall order.

Pyroclastics Volcanic debris formed by the explosive extrusion of viscous magmatic material containing pent-up gases.

Quartz A common mineral, a silicon dioxide. Contributes importantly to the resistance to erosion of crystalline rocks such as granite because it has a high hardness.

Quaternary economic activity Activities engaged in the collection, processing, and manipulation of information.

Race A much-debated concept that divides humanity into groups on the basis of physical appearance or characteristics.

Rain forest Evergreen forest in equatorial and tropical areas where high temperature and humidity prevail.

Rain shadow effect The relative dryness in areas downwind of or beyond mountain ranges caused by orographic precipitation, whereby moist air masses are forced to deposit most of their water content in the highlands.

Recessional moraine When glaciers melt during a warming phase, debris carried during the most recent advance is deposited *in situ*. Recessional moraines are ridges and piles of such glacial debris that mark the still-stand and the waning of the glacial ice.

Regime, climatic A term used in physical geography to describe the variations in behavior of changeable elements of the environment such as climate or a river. Regime implies that there is some regularity in the pattern, but recognizes change within it. The regime of a river, thus, would be its flood stages and low stages, slow and rapid flow. A climatic regime involves the four seasons and their associated conditions.

Region A geographic concept of prime importance: a section of the Earth's surface marked by an overriding sameness or homogeneity. Areas outside such regions are substantially different—they may themselves be *other* regions. Regions are artificial constructs or concepts: you cannot find a regional boundary on the ground except under unusual circumstances. It is always best to qualify the kind or type of region being referred to. For example, any country or state is, in effect, a political region. The laws of that state, the taxes collected, the citizenship—all these things apply only within the political boundary of that state. Thus the political boundary (which may, incidentally, be marked on the ground for some distance) is also a regional boundary of a sort.

Regolith Weathered, broken, loose material overlying bedrock, usually derived from the rock below but sometimes transported to the area. First

stage in the conversion of bedrock to soil.

Relative humidity The ratio (expressed as a percentage) of humidity actually in the air to the amount that would completely saturate the air at that particular temperature.

Relative location The position or situation of certain towns or other places *relative* to the position of other towns or places.

Relic boundary An international political boundary that has ceased to function but the imprint of which on the cultural landscape can still be detected.

Relief The nature of the unevenness in the landscape of an area and the degree thereof. An area of low relief is nearly flat or gently undulating; an area of high or great relief is one with towering mountains and deep valleys. Relief is not *elevation*: an area of low relief could lie at a high elevation—the top of a nearly flat plateau, for example.

Relocation diffusion Sequential diffusion process in which the things being diffused evacuate the old areas as they move (relocate) to new areas. A disease can move from population cluster to population cluster in this way, running its course in one area before fully invading the next. A population movement can carry the process, for example, the movement of the black population from the South to northern cities in the United States.

Renaissance A word that means "rebirth," applied to the period in Europe from the fourteenth into the seventeenth centuries, when the arts, literature, and science had a true intellectual revival.

Renewable resources Resources capable of regeneration, such as fish, fauna, a forest, the soil.

Rift valley The trough or trench that forms when a strip of crust sinks between two parallel faults.

Ring of fire   Zone of crustal instability surrounding the Pacific Ocean.

Rocks   *See* Igneous, Metamorphic, Sedimentary rocks.

Rural density   A measure that indicates the number of persons per unit area actually living in rural parts of the country, excluding the urban concentrations; provides an index of rural population pressure.

Sahel   Semiarid zone across most of Africa between the southern margins of the arid Sahara and the moister forest and savanna zone to the south. Chronic drought, desertification, and overgrazing have contributed to severe famines in this area since 1970.

Salinization   Deposition of soluble salts in the soil.

Saltation   The rolling or bouncing motion of small rock particles, driven by stream water or by wind action.

Satellite state   A state whose sovereignty has been compromised by the dominating influence of a larger power. Such a situation can arise when the ruling political party in such a country represents the ideology of the larger power or when the economic influence of the larger power is so great that it is in virtual control of production in the satellite state.

Saturation   The amount of humidity an air parcel can contain at a certain temperature, producing 100 percent humidity.

Savanna   A term that denotes both a vegetative and a climatic type. In terms of vegetation, the savanna is a tree-studded grassland. In places the trees are so well spaced and substantial and the grasses so full that the term parkland savanna is used. The savanna (AW) climatic type lies immediately adjacent to tropical rain forest regions and is marked by warmth and by a pro-

nounced dry season (in certain areas two moist and two dry seasons annually), and generally by lower total rainfall figures than in rain forest areas.

Scale   The representation of real-world phenomena on any map must involve reduction of the size of those phenomena. The maps in the book include the entire world on one sheet of paper. But we can also represent one country on a sheet of about the same size, or a city, or the campus of a university. When we show a campus or a city, we can show much more detail than if we show an entire country or a continent. This means that the reduction is less, and the *scale* is larger. The scale is usually represented by a fraction (say 1:1,000,000), which means that 1 in. on such a map equals 1 million in. of the real world. That's a rather small scale, but when we make a map of a city, it can be much larger, say, 1:10,000 (remember that one-one millionth is a *smaller* fraction than one-ten thousandth). Thus the scale gets larger as the detail the map provides becomes greater.

Scandinavia   Denmark, Sweden, and Norway.

Scarp   A nontechnical term commonly used to identify a small escarpment, not necessarily bounding a plateau.

Sclerophylls   Evergreen trees and shrubs with hard, leathery leaves capable of surviving long periods of high heat and low humidity.

Sea breeze   A wind that originates over water and blows onto the land, usually during warm, sunny days.

Seasonal hunger   The shortage of food that comes during the nonproductive dry season or winter.

Secondary industries   Industries that process raw materials and transform them into finished products; the manufacturing industries.

Sector model   A structural model

of the American central city that suggests land-use areas conform to a pie-shaped pattern focused on the downtown core.

Secular   Worldly, nonreligious, nonspiritual. Secularism holds that ethical and moral standards should be formulated and adhered to for life on earth, and not to accommodate the prescriptions of a deity and promises of a comfortable afterlife. The secular state is the opposite of a theocracy.

Sedentary   Permanently attached to a particular area, a population fixed in its location; the opposite of nomadic.

Sedimentary rock   Rock formed from the deposition and accumulation of matter from other rocks, chemicals, or organic sources through compaction and cementation.

Selva   The name for the rain forest of the Amazon Basin, especially along the great river itself.

Sequent occupance   The notion that successive societies leave their cultural imprints on a place, each contributing to the cumulative cultural landscape.

Shaman   In traditional societies, a shaman is deemed to possess religious and mystical powers, acquired directly from supernatural sources. At times an especially strong shaman might attract a regional following; many shamans, however, remain local figures whose influence is comparable to that of a witch doctor in a village community.

Shantytown   Unplanned slum development on the margins of Third World cities, dominated by crude dwellings and shelters mostly made of scrap wood, iron, and even pieces of cardboard. Brazil's favelas, Mexico's barrios, North Africa's *bidonvilles* and India's *bustees* are examples.

Sharecropping   Relationship between a large landowner and farmers on the land in which the farmers pay rent for the land they farm by giving the landlord a share of the annual harvest.

Shatter belt *See* Buffer zone.

Shield Ancient, crystalline core area of a continental landmass.

Shield volcano A broad, gently sloping volcanic dome formed by the accumulation of liquid lavas emanating from the crater and from flank fissures and capable of flowing over long distances and large areas.

Shifting cultivation Farming on patches of land, usually cleared from a forest, until the soil is exhausted and the clearing must be abandoned in favor of another.

Sial Lightest material of the Earth's crust, made of rocks rich in silica and aluminum compounds.

Sierra A range of mountains, especially a long and narrow one with a rough, uneven crest.

Sill Concordant intrusive igneous rock that has been wedged, in the form of a sheet, along the bedding plane between layered strata or along a joint plane. Sills, more often than discordant dikes, tend to lie horizontally or at slight dip angles; sometimes they form the caprock of a mesa or butte.

Sima Crustal rocks that are heavier than sialic rocks, containing mainly silicates of magnesium; on the landmasses the sima lies below the sial.

Single/multiple feature regions Regions defined on the basis of one or more phenomena. These terms mean exactly what they suggest: in the definition of the region, we can use only one category of phenomena or we can use several. For example, a railroad system or a river system, both single features, can mark a functional region. But functional regions (as we have seen) are usually much more complex than that and involve not only merely a railroad system, but also the roads, airlines, radio communications, telephone connections, newspapers circulated, com-

mercial relationships, and much more. The point is, of course, to try to distinguish those indexes upon which the functional region is based. As for the formal region, it may well be a single-feature region, though the example we used above involves more than one category of phenomena.

Sinicization Acculturation of a people to Chinese ways and customs. This has happened to several of the minority peoples in China to a high degree; it happened to the old Manchus of Manchuria, who were just about submerged by China's culture.

Site The physical setting of a place, its local relief, and other physiographic attributes.

Situation The relative location of a place; its position with reference to other places and to resources and areas of productive capacity.

Slash-and-burn agriculture The system of patch agriculture that involves the cutting down and burning of the vegetation in a part of the rain forest and the temporary cultivation of crops on the clearing. Also see *Milpa* agriculture and Shifting cultivation.

Smog The term derives from *smoke* and *fog*; a combination of atmospheric pollutants (particulate and chemical) in the lower atmosphere.

Soil A mixture of fragmented and weathered grains of minerals and rocks with variable proportions of air and water that is capable of supporting plant life. The mixture usually has more-or-less distinct layering or horizonation.

Soil consistency A field measure of a soil's stickiness, plasticity, hardness, and degree of cementation.

Soil creep Slow, imperceptible downslope movement of soil and regolith in response to gravity and through lubrication by moisture, temperature changes, and alternate freezing and thawing.

Soil horizon A distinct layer of the soil, part of a soil profile, identified by its particular physical properties and chemical composition, structure, organic content, color, and other characteristics.

Soil profile A vertical cross section of a soil from top to regolith or bedrock that displays its horizons.

Soil solution The water (slightly modified by mixture with the prevailing chemicals and organic material) held in a soil and available to plants.

Soil structure The characteristics of the lumps or aggregates of particles in a soil. *See* Blocky, Granular, Platy, Prismatic structure.

Soil texture The proportion of sand, silt, and clay in each soil horizon.

Solum The A and B horizons of a soil; the critical layers for plant growth and interaction between soil and plant roots.

Source area For air masses, an area where a temporarily stationary air mass attains certain qualities of temperature and humidity.

Southeast trades Surface winds with a predominantly southeast direction that blow in latitudes between the doldrums and the subtropical high pressure belt in the Southern Hemisphere.

Spatial Pertaining to space on the Earth's surface; synonym for geographic(al).

Spatial model *See* Model.

Splash erosion Soil erosion caused by the impact of raindrops; on slopes, the result is a downslope movement of soil particles.

Spodosols A soil order consisting of soils affected by podzolization processes; they are not extensive in the United States but do occur in northern areas where they support needleleaf forests.

496

Stack A steep-sided landform found along high-relief shorelines subject to rapid erosional retreat, where a section of resistant bedrock has become separated from the headland and rises from the water as a column.

State *See* Nation-state.

Steppe A term that applies to more than one kind of area. A climatic zone that surrounds deserts but is not quite as dry. Annual rainfall may be as low as 10 in. (25 cm) and as high as 30 in. (75 cm); the letter identification is BS, with an *h* added for warmer areas and a *k* for colder zones. The vast midlatitude grasslands of Eurasia, from southern Siberia into the lower Danube area, called *prairies* in North America and *pampas* in South America; much of Southern Africa's *veld* is steppe country. Steppes are either treeless grasslands or studded with a few sparse trees, plainlands in aspect, mostly under a clear sky, as rain is not plentiful in these areas.

Strait A narrow zone of sea that connects larger water bodies on both sides and lies between land areas that approach each other closely. Between North Africa and southwestern Europe, for example, there is a very narrow strait connecting the Mediterranean Sea to the east and the Atlantic Ocean to the west, the *Strait* of Gibraltar.

Stratified drift Material transported by glaciers and later sorted and deposited by the action of running water either within the glacier or during the melting state.

Stratified society A society marked by very distinct classes and/or castes: privileged people, professional classes, artisans, the poor. Such stratification may be based on wealth, or according to race or skin color, or there may be other criteria. In any case, the term implies a certain rigidity in the structure, and it may be nearly imposs-ible for a person to improve his or her status.

Stratopause Upper limit of the stratosphere.

Stratosphere Atmospheric layer lying directly above the troposphere.

Striation Scratch or gouge caused by the abrasive movement of a rock fragment dragged across solid bedrock by a moving glacier.

Structure, population *See* Population structure.

Structure, soil *See* Soil structure.

Subarctic A term that means below the Artic and includes areas lying immediately south of the Arctic Circle, usually including the zone approximately between 55° and 60° latitude, a transitional zone between the middle latitudes and the Arctic. The subarctic climatic type is marked by long winters, short summers (during which the days of sunshine are long), and extreme temperature ranges. The vegetation is generally a needleleaf forest, often called boreal forest.

Subsequent A word that has specific meaning in physical geography—just the opposite of antecedent. The river formed *after* the geologic forces had created the main regional structure.

Subsequent boundary An international political boundary that developed contemporaneously with the evolution of the major spatial elements of the cultural landscape in an area.

Subsistence Existing on the minimum necessities to sustain life.

Subsoil In marine geography, the sediments and rocks beneath the sea floor.

Substitution The introduction of one material when another has become scarce, for example, the use of aluminum as a metal where steel had formerly been employed. Today the emergence of nuclear power for power generated by oil and coal is a form of substitution.

Subtropic, subtropical A term that is not consistently applied. Some use it as though it means below (actually *within*) the tropics, that is, lying between the Tropics of Cancer and Capricorn. A better definition probably is that subtropical areas are intermediate between tropical areas and temperate or midlatitude zones. This would include much of the C climatic class.

Superimposition The impress of one element of the physical or cultural landscape upon another. Another physical geography term borrowed by the political geographers. This one is not so simple. Imagine a mountain range with a series of parallel ridges, say the Appalachians. Imagine, next, that the whole area is inundated by the sea and that the valleys between the ridges are filled by sandy sediments. Eventually the whole range is buried! Now the sea begins to withdraw, perhaps as the area is uplifted again, and rivers begin to flow toward the coast. Slowly these rivers form deeper valleys and they cut downward through the sedimentary materials laid down by the sea. Finally, they begin to reach the crests of the buried mountain ranges, but they just continue to erode downward, slicing through the ridges as they are being exhumed. Such rivers are being *superimposed* on those mountains. The political geographers use the term to identify a boundary that has been laid down across a cultural landscape by an outside, superior power, perhaps cutting a unified people in half. Think of the Berlin Wall—a superimposed boundary, first defined by the Allies and then demarcated by the Soviets and East Germans.

Supranationalism The movement, shared by many of the world's states, to seek avenues toward unifica-

497

tion or at least greater formal cooperation. Like the old Hanseatic cities, certain states obviously have common interests and goals; they form economic, cultural, political, and military alliances to attain them.

Suspension In stream and wind transportation, the removal of particles supported by eddies strong enough to keep them above the stream bed or bedrock surface for prolonged periods.

Syncline A fold in stratified rocks in which the axis of the structure lies at the lowest elevation and the two limbs face toward each other and slope upward. Thus a syncline has the appearance, in cross section, of a downfold.

System Any group of objects or institutions and their mutual interactions; geography treats systems that are expressed spatially—such as *regions*.

Taconite A rock that has a flint-like appearance, high enough in iron content to constitute a valuable iron ore, but not as rich as other iron ores.

Taiga The vegetation that marks subarctic regions. It consists of the boreal forest and the open woodland adjacent to it in the north. In the U.S.S.R., the taiga lies between the tundra to the north and the dry, cold steppes of the Eurasian interior to the south. One characteristic of the taiga is its swampiness: during the spring, the northward-flowing rivers melt in their southern courses while their northern courses are still frozen. The meltwater is blocked by the ice up north and flows across the surrounding countryside.

Takeoff Economic concept to identify a stage in a country's development when conditions are set for a domestic industrial revolution, as happened in the United Kingdom in the late eighteenth century and in Japan after the Meiji Restoration; it is happening in China today.

Tectonic plates (lithospheric plates) Segments of the earth's lithosphere that move relative to each other, diverging along midocean ridges, converging at subduction zones and sliding laterally along transcurrent (transform) fault boundaries.

Tell Coastal slopes fronting the Mediterranean Sea in the Maghreb. The Atlas Ranges in North Africa, especially in Algeria, consist of several parallel chains that extend Alplike, from west-southwest to east-northeast across the Maghreb. The ranges on the south side are aptly named the Saharan Atlas, and the ranges on the Mediterranean side are called the Tell Atlas. The *Tell* itself is the coastal plain and lower slopes that lead from the Mediterranean shores to the Tell Atlas—the lower northern flank, then, of the Atlas of Algeria.

Terminal moraine In sheet (continental) as well as alpine (mountain) glaciers, the glacial debris being carried in the leading edge of the glacier and marking its farthest line of advance before melting began.

Terracing The transformation of a hillside or mountain slope into a steplike sequence of horizontal fields for cultivation.

Territoriality A community's sense of property and attachment toward its territory as expressed by its determination to keep it inviolable and strongly defended.

Territorial sea Zone of water adjacent to a country's coast, held to be part of the national territory and treated as a segment of the sovereign state.

Tertiary industries Industries and activities that engage in services such as transportation, banking, and education.

Texture, soil *See* Soil texture.

Theocratic, theocracy A theocratic state is one whose government is deemed guided by divine action. The rulers of such a state are priests and their actions are final and beyond question since they represent divine will. A modern government that bases its authority on religious grounds is Iran's Islamic state.

Thermosphere A layer of the earth's atmosphere that lies above the mesopause, marked by rising temperatures.

Threshold A place or line of new beginning, of entry, or transition. Everyone knows what a threshold in a house is: on one side of it you're in the corridor, on the other, you're in the room. But the world threshold has geographic applications, too. In establishing hierarchies for central place settlements, for example, the concept of threshold is used. For example, a place with, say x number of gas stations, drugstores, and doctors may be classified as a village, but a place with $x + 1$ of these services may be classed as a town. Thus $x + 1$ is the threshold—where population and its demands for services are such that the village becomes a town. You will see comments about threshold values in numerous contexts in geography.

Tier (of states) The states that lie north of India—Afghanistan, Nepal, formerly independent Sikkim, and Bhutan—are often called a *tier* of states; in this context, the word tier means row. The word has probably come to apply to these states because tier also means layer, and a look at the map of Eurasia shows that this row of states lies almost in a layer between the U.S.S.R. and China in the north and India in the south. In this position, these states have functioned as buffers.

Tierra caliente The lowest of three vertical zones into which the topography of Middle and South America is divided according to elevation. The *tierra caliente* is the hot, humid coastal

498

plain and adjacent slopes up to between 2500 and 3000 ft (750 and 900 m) above sea level. The natural vegetation is the dense and luxuriant tropical forest; the crops are bananas, sugar, cacao, and rice in the lower areas and coffee, tobacco, and corn along the somewhat higher slopes.

Tierra fria The highest zone below the snow line in Middle and South America, over about 7000 ft (2100 m) and up to around 10,000 ft (3000 m). Coniferous trees stand here, and upward they change into scrub and grassland; there are important pastures on the *tierra fria*. Wheat can also be cultivated. Several major population clusters in the region lie at these altitudes. Above the *tierra fria* lies the *tierra paramós*, the highest and coldest zone.

Tierra templada The intermediate altitude zone in Middle and South America between about 2500 to 3000 ft (750 to 900 m) and 6000 to 7000 ft (1800 to 2100 m). This is the ''temperate'' zone, with moderate temperatures compared to the *tierra caliente*. Crops include tobacco, coffee, corn, and some wheat.

Till Unsorted mixture of glacial debris—ranging in particle size from clay to boulders—laid down beneath moving ice or deposited when the glacier's advance stopped and melting began.

Time-distance decay The declining degree of acceptance of an idea or innovation with increasing time and distance from its point of origin or source.

Topography The configuration of the surface of an area. A topographic map shows this configuration by means of contour lines. The physiography of an area, however, involves much more —it is the total physical landscape, including the relief, climate, soils, and vegetation.

Toponym A place name.

Tordesillas, Treaty of Treaty signed in 1494 at Tordesillas, Spain, under which Spain and Portugal divided the non-Christian world into two zones of influence. The demarcation line, along a circle 370 leagues west of the Cape Verde Islands, gave Portugal claim to Brazil.

Totalitarian regime Absolute control by a government over its subjects, usually implying a despotic, dictatorial rule. Such a regime was that of the Ottomans in Eastern Europe.

Traction Process undergone by material being dragged along a river bed.

Traditional Term used in various contexts (e.g., traditional religion) to indicate originality within a culture or long-term part of an indigenous society. It is the opposite of modernized, superimposed, changed; it denotes continuity and historic association.

Transculturation Cultural borrowing that occurs when different cultures of approximately equal complexity and technological level come into close contact. In *acculturation*, by contrast, an indigenous society's culture is modified by contact with a technologically more developed society.

Transhumance A form of periodic migration; the practice of moving livestock herds according to seasonal climate changes. In Alpine Europe, herds are moved to upland summer pastures and to lowland winter pastures; in Africa, herds follow the cyclic rain pattern in plains and against mountain slopes.

Transition zone *Transition* means passage from one state or stage to another; *zone* has spatial significance and is therefore of interest to geographers. Many spatially expressed phenomena in the world change in intensity from one place to another. When there is a German-speaking country adjacent to a French-speaking country, it may well be that there is a zone between them where both French and German are in use. It is highly unlikely that a sharp line could ever divide exactly all the German speakers from all the French speakers. Thus, there is a *transition* from one to the other. This happens in countless instances. A dense forest does not normally change to grassland along a line: it is much more common that the forest thins out and then disappears. There will again be a belt or zone of change or *transition*. We find this phenomenon time and again in the world's ethnic groups, between cultures, and in political systems.

Transportation In physical geography, the movement of eroded particles of all sizes in a stream valley by all the operative processes of removal.

Transverse dunes Ridgelike sand dunes whose axial crests lie at approximately right angles to the prevailing wind direction in an area.

Tropical Within the tropics. On any globe or atlas map of the world there are two lines, drawn at 23 1/2° south latitude, the same distance north and south of the equator. These imaginary lines represent the northernmost and southernmost limits of the sun's vertical ray as it heats the equatorial regions of the earth. They are known as the Tropic of Cancer (north) and Tropic of Capricorn (south). The word tropic itself has probably been derived from the Greek, *tropikos* (turn)—for here is where the sun turns back. Technically, then, *tropical* refers to that part of the world lying between the tropics. In fact, common usage prefers the word tropical to apply to areas where the warm, humid environmental conditions of low-lying, equatorial zones prevail. The Amazon and Congo basins are examples of this. Although the great Andes Mountains in Peru and the high Plateau of Ethiopia in Africa technically lie within the tropics, one hardly ever hears them called tropical.

Tropical rain forest *See* Rain forest.

Tropopause The boundary surface between the troposphere below and the stratosphere above.

Troposphere Lowest layer of the layered earth's atmosphere, where temperature decreases with increasing altitude.

Truncated spur A spur or protruding rock mass around which a stream formerly curved its way, but was cut off by a valley-straightening mountain glacier.

Trypanosomes The single-celled agents of sleeping sickness, carried in the blood of infected animals.

Tundra A treeless plain, lying astride the Arctic Circle along the northern margins of the coniferous forest, closer to the pole than the subarctic zone. This plain extends to about the 75° latitude, and it is covered with grasses and lichens along with some shrubs of willow. Toward the south, birch trees appear, and then the needleleaf forest takes over. The line of contact between the tundra and the evergreen trees and forests of the subarctic is often quite sharp, and the geographer who formulated the climatic system represented on Plate 3 used this contact line to differentiate between the D and E climatic classes. The tundra has also given its name to a climatic type and is lettered ET since it is marked by a warmest month whose average is below 50°F.

Typhoon Tropical cyclone that develops high wind velocities in the western Pacific Ocean and the waters off Southeast and South Asia.

Ultisols A soil order consisting of soils that are very old; located in warm subtropical environments with a pronounced wet season, subject to seasonal laterization, and not very fertile.

Underdeveloped countries (UDCs) Countries that, by various measures, suffer seriously from negative economic and social conditions, including low per capita incomes, poor nutrition, inadequate health, and related circumstances.

Unitary state A nation-state that has a centralized government and administration.

Universal religions Religions that have adherents not just in a localized area, but in several regions of the world; the major faiths.

Unstable air Air that contains a large quantity of water vapor, usually at a warm temperature, capable of sustained convection leading to the formation of large vertical cloud masses, heavy rains, and thunderstorms.

Upper-level inversion A reversal in temperature decrease with height at a certain level above the ground where temperature begins a temporary increase.

Urban geography A subfield of geography that focuses especially on urban places, their characteristics, processes of genesis and growth, their systems, relative location, and interrelationships. Like economic geographers and political geographers, urban geographers share a body of theory and deal with a particular form of spatial expression of human behavior—the city, town, or other clustered settlement.

Urban morphology The layout, structure, appearance of cities and towns.

Urbanization A term with several connotations. The proportion of a country's population living in urban areas is its level of urbanization. The process of urbanization involves the movement and clustering of people in towns and cities (it continues to affect most countries today). Another kind of urbanization occurs when a sprawling city absorbs rural countryside and transforms it into suburbs, developments, and squatters' settlements.

Urbanized area The entire built-up, nonrural area and its population, including the most recently constructed suburban developments. Gives a better picture of the dimensions and population of such an area than the delimited municipality or municipalities that form its heart. The population of a "city proper" is always less than the entire urban area within which it lies.

Valley deepening Erosion of the bottom of a stream bed, mainly through abrasion by rock fragments dragged along by the stream.

Valley lengthening Process in which a stream extends its course by headward erosion at its source, increased meandering in its major valley, and delta extension near its mouth.

Valley widening Process that has the effect of enlarging a stream valley's cross section through lateral (sideways) erosion, especially prevalent in its middle course where meanders develop.

Variant gases Gases that exist in the atmosphere in varying amounts in various places, at various levels, and at various times. Two principal variant gases are water vapor and ozone.

Vertisols A soil order consisting of soils with a high clay content that develop characteristic wide cracks during the dry season.

Volcano Cone-shaped structure built by the repeated eruption of lava flows and associated explosive emissions of gas and projectiles. A variety of landforms are created, including *com-*

500

posite volcanoes (represented by some of the world's great volcanic mountains including Vesuvius and Fuji) and *shield volcanoes* with their broader, more gently sloping morphology.

Völkerwanderung Literally, a wandering of peoples. The word is German, and has come into general use to identify the mass migrations in Europe following the breakdown of the Roman Empire.

Voluntary migration Population movement in which people relocate as a result of perceived opportunity and its attractions rather than compulsion.

Water balance An element in the Thornthwaite system of climate classification. It is derived from data on precipitation, evapotranspiration, and heat flow.

Water table Surface below ground below which saturation prevails.

Wave-cut platform *See* Abrasion platform.

Weather The prevailing conditions of temperature, humidity, wind direction and speed, and related aspects of the environment at a given moment at a particular location. *See* Climate.

Weathering The disintegration and decomposition of rocks, mainly by mechanical and chemical (but also by biological) processes, without movement by an erosional agent other than gravity.

Wet adiabatic lapse rate Adiabatic lapse rate that obtains when condensation is taking place in rising air. Value varies between 2° and 3°F/1000 ft (3° and 6°C/1000 m).

Witwatersrand Literally, whitewaters-ridge. The Witwatersrand (an Afrikaans word) is the exposed, tilted wing of a former basin of deposition. Like the Great Dyke, it rises above the surrounding Highveld (but unlike its Rhodesian version, its structure is nonintrusive). The layered rocks of the Witwatersrand contain the gold that has

been so important in the development of the South African economy.

Xerophytes Plants adapted to dry environmental conditions.

Zoogeography The analysis of the spatial factors involved in processes of migration and dispersal of species or faunal associations. In this work, geographers bring to bear their knowledge of climatology (e.g., air mass theory has much to do with the distribution and spread of disease-transmitting insects), oceanography, and other relevant fields. A related subdiscipline is phytogeography, in which the analysis of plant distributions is carried on. Zoogeography and phytogeography together are sometimes referred to as biogeography; what this means, of course, is the application of research techniques in geography to phenomena that are central to the disciplines of biology. There is a fairly substantial literature in this field.

501

507

509

511

515

516